AUTOMATIC TRANSMISSIONS

HOWARD F. TUCKER

AUTOMATIC TRANSMISSIONS

HOWARD F. TUCKER

LIBRARY OF CONGRESS CATALOG CARD NUMBER: 78-62623
ISBN: 0-8273-1648-8

10 9 8 7 6 5 4

Printed in the United States of America
Published simultaneously in Canada
by Nelson Canada,
A Division of International Thomson Limited

DELMAR PUBLISHERS INC. • ALBANY, NEW YORK

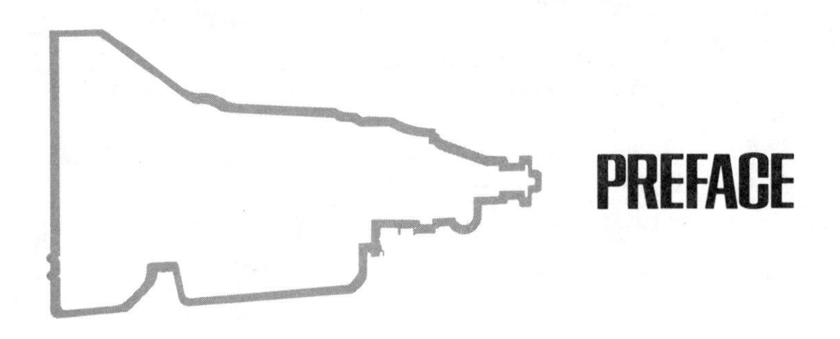

PREFACE

Automatic transmissions are used in the majority of American passenger cars and are now increasingly utilized in trucks, as well. As the implementation of automatic transmissions increases, so does the demand for personnel who are trained in the technology of these complex systems. AUTOMATIC TRANSMISSIONS is written for students who require technical competency in automatic transmission service, diagnosis and repair, as exemplified by the National Institute for Automotive Service Excellence Certification Program.

The text first explains the basic concepts of force, work, and power, and then relates these principles to speed ratio in planetary gear sets, fluid couplings, hydraulic control and pressure regulation. Applications are further developed through operational descriptions of the most commonly used automatic transmissions. Throughout the text, the content reinforces the need for efficiency in the shop, featuring systematic diagnostic and troubleshooting procedures. The presentation includes detailed descriptions of transmission service, diagnosis, valve body overhaul, and complete transmission overhaul and modification. A unit covering transmission adaptations for front-wheel drive vehicles completes the text.

The text provides unit objectives which outline the concepts and skills to be learned in the subsequent material, plus extensive illustrations which explain in step-by-step fashion the procedures best suited for efficient transmission service and repair. Review questions are presented in the National Institute for Automotive Service Excellence certification test format whenever appropriate to the material under discussion. Additional features include shop safety procedures, the inclusion of metric equivalents for all measurements, and extended study projects.

A fully developed Instructor's Guide accompanies the text. In addition to the answers to all review questions, this manual includes comments on the extended study projects contained in the text. An audiovisual package dealing with hydraulic principles and hydraulic circuit diagrams complements the text and supplements the Instructor's Guide.

Howard Tucker, the author of AUTOMATIC TRANSMISSIONS, has taught this subject to two-year auto mechanics students for the past nine years at State University Agricultural and Technical College at Delhi, New York. In addition to this vital classroom experience, he brings to his field a wealth of practical shop experience as a technician and owner-operator of an independent service facility.

AUTOMATIC TRANSMISSIONS is the most recent addition to the Delmar series of texts developed for automotive instruction. Other texts in the series are:

AUTOMOTIVE AIR CONDITIONING
AUTOMOTIVE OSCILLOSCOPE
AUTOMOTIVE SERVICE AND REPAIR TOOLS
AUTOMOTIVE SCIENCE
INTERPRETING AUTOMOTIVE SYSTEMS
PRACTICAL PROBLEMS IN MATHEMATICS FOR AUTOMOTIVE TECHNICIANS
SMALL GASOLINE ENGINES

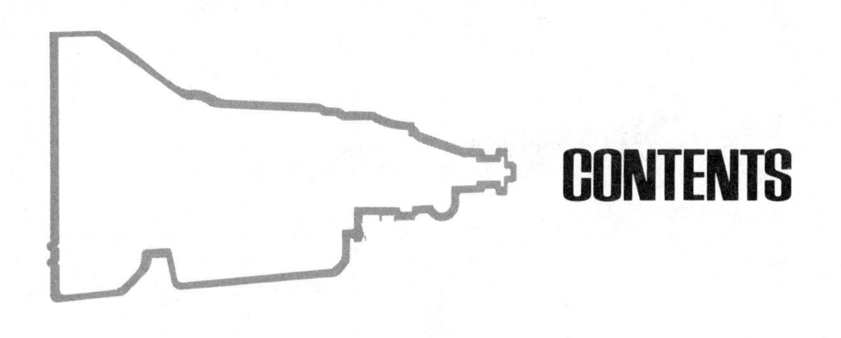

CONTENTS

A current catalog including prices of all Delmar educational publications is available upon request. Please write to:

Catalog Department
Delmar Publishers Inc.
2 Computer Drive – W
Box 15-015
Albany, New York 12212

ACKNOWLEDGMENTS

Sponsoring Editor: William W. Sprague

Editor: Mary L. Wright

Technical Reviewers: George D. Moore, Aims Community College
Harry G. Hill, Milwaukee Area Technical College
Marc Ruggeri, Albany Dodge, Inc.

Consulting Editor, Automotive Series: George D. Moore

Contributions of Content and Illustration:

Chrysler Corporation
Ford Motor Company
Chevrolet Motors Division, General Motors Corporation

Special appreciation is expressed to:

Jim Kendall, Products Assurance, Publications, Chevrolet Motors Division, General Motors Corporation
Patricia K. Hawkins, Manager of Public Relations, Hydra-Matic Division, General Motors Corporation
Ron E. Madej, Art Director, Technical Services Dept., Service and Parts Sales Div., Chrysler Corporation

* * * * * *

The author also extends special thanks to his entire family – Mary, Mary Ellen, Jim, Stephen, Dan, Richard, Allen, Edward, and Kathleen – all of whom helped in so many ways.

Portions of this text have been classroom tested at the State University Agricultural and Technical College at Delhi, N.Y.

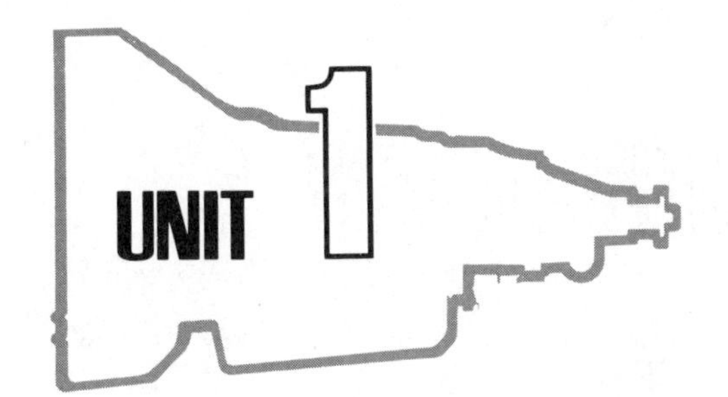

UNIT 1 TRANSMISSION OF POWER

OBJECTIVES

After studying this unit, the student will be able to:

- Define the terms work, power, and force.
- Calculate the amount of work and power.
- Explain and calculate mechanical advantage.
- Describe the three basic types of levers.
- Describe how the wheel and axle, pulleys, and gears operate.
- Determine the speed ratio of paired pulleys.

Automatic transmissions are an essential part of many of the motor vehicles in use today. Automotive mechanics must know how automatic transmissions work, and how to repair them. To do this, they must first understand the relationship between work, power, force, and distance. The operation of automatic transmissions depends on the scientific principles that link these four factors.

FORCE

A *force* is a push or a pull. When a force is applied to an object, it may change the shape of the object or change its motion. For example, when a person pushes (applies force to) a door, it swings open. One force may keep other forces from acting on the object. If a second person is holding the door from the other side, the force applied by the first person may not open the door.

WORK

Work is done when motion is caused by the application of force. An automobile engine performs work when it moves a car over the road. A mechanic performs work when he pushes on the lever of a jack to raise a car off the floor. The jack also performs work by lifting the car. If, however, a mechanic tries to loosen a rusted bolt and cannot move it, work is not performed. Even though the mechanic applies force that could cause work, no work is done. To do work, force must cause motion, and the bolt does not move.

Measurement of Work

To find the amount of work done, the force applied to an object is multiplied by the distance through which the force acts. Work can be calculated using either the English or metric system of measurement. (An explanation of metric measurement can be found at the end of this unit.) The common unit for measuring work in the English system is the *foot-pound* (ft-lb). Raising one pound a distance of one foot is one foot-pound of work. A person carrying 50 pounds of bricks up a 10-foot ladder does 500 foot-pounds of work.

$$\text{Work} = \text{Force} \times \text{Distance}$$
$$\text{Foot-pounds} = \text{Pound} \times \text{Feet}$$

As a simple example, calculate the work involved in lifting a 20-pound weight 5 feet.

Work = Force × Distance
Work = 20 pounds × 5 feet = 100 foot-pounds

POWER

Power is defined as the rate of doing work. In other words, it is the amount of work that can be done in a specific amount of time.

Measurement of Power

Power is measured in *foot-pounds per second* or in *foot-pounds per minute.* Refer again to the 20-pound weight that was lifted 5 feet. How much power is required to lift the weight in 5 seconds?

Work = Force × Distance
Work = 20 pounds × 5 feet = 100 foot-pounds

$$\text{Power} = \frac{\text{Work}}{\text{Time}}$$

$$\text{Power} = \frac{\text{100 foot-pounds}}{\text{5 seconds}} = \text{20 foot-pounds per second}$$

As another example: if an elevator lifts 3,500 pounds a distance of 40 feet and it takes 25 seconds to do it, what is the rate of doing work?

Work = Force × Distance
Work = 3,500 pounds × 40 feet
= 140,000 foot-pounds

$$\text{Power} = \frac{\text{Work}}{\text{Time}}$$

$$\text{Power} = \frac{\text{140,000 foot-pounds}}{\text{25 seconds}}$$

Power = 5,600 foot-pounds per second

Power can also be expressed in foot-pounds per minute. If a pump needs 10 minutes to lift 5,500 pounds of water 60 feet, it is doing 300,000 foot-pounds of work in 10 minutes. This is a rate of 30,000 foot-pounds per minute.

MECHANICAL ADVANTAGE

There are several ways in which force can be multiplied so that a smaller force can be used to move a larger force. For example, by exerting a force of 20 pounds on the handle of a jack, a mechanic can raise a 2,000-pound car one inch off the ground. (Here, the force exerted by the person is working against the force exerted on the car by gravity.) However, to do this, the mechanic must move the handle of the jack up and down through a total distance of 100 inches. The extra distance moved by the jack handle creates a mechanical advantage. Whenever a loss of motion results in a gain of force, it is called a *mechanical advantage*.

SIMPLE MACHINES

Mechanical advantage is provided by many simple machines such as the lever and pulley. Both of these machines, in addition to the gear, which is a combination of the two, are used in the automatic transmission to provide mechanical advantage.

The Lever

Very early in history, people learned to use poles or logs for moving heavy objects. A pole used in this way is called a *lever.* The mechanical advantage that it provides is called *leverage.*

The pivot point on which a lever rests is called the *fulcrum,* figure 1-1. The distance from the fulcrum to the force, or *effort,* is called the *effort arm* of the lever. The distance from the fulcrum to the weight being moved is called the *resistance arm.* These parts can be arranged in three different ways, as shown in figure 1-2. Study the placement of the parts and the direction of the effort and resistance in each type.

A good example of a first class lever is the clutch release fork on a standard transmission-

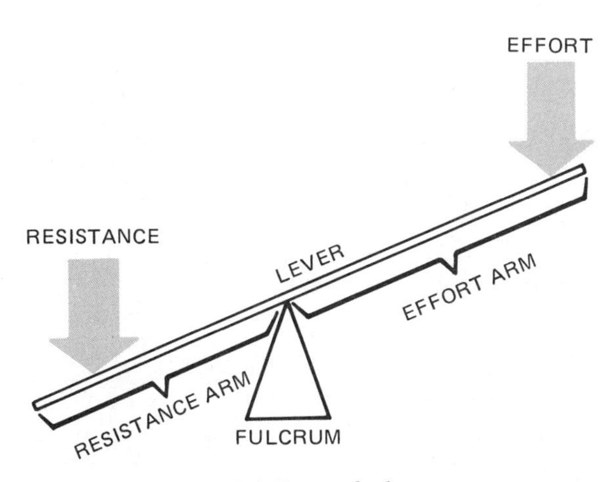

Fig. 1-1 Parts of a lever.

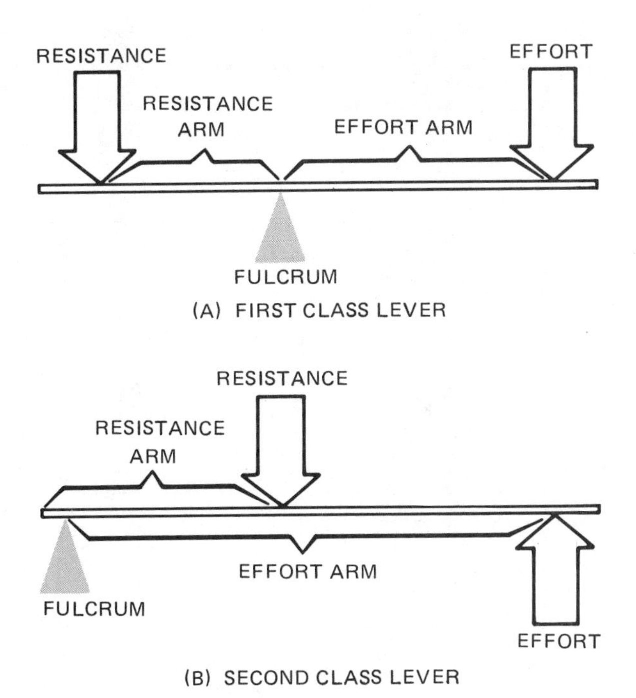

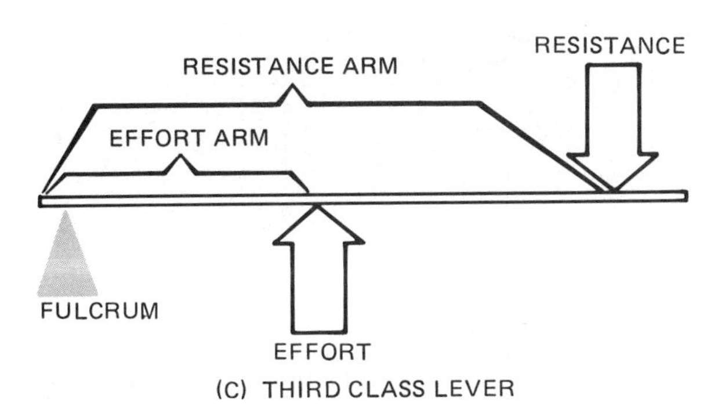

Fig. 1-2 Three types of levers.

equipped car. The brake pedal connection to the master cylinder is an example of a second class lever. A third class lever is found in the design of the accelerator (throttle) pedal of a car.

Figure 1-3 shows how a lever works to provide mechanical advantage. The mechanical advantage of a lever is calculated by finding the ratio of the length of the effort arm to the length of the resistance arm.* In figure 1-3, this is found as follows:

$$\text{Mechanical Advantage} = \frac{\text{length of the effort arm}}{\text{length of the resistance arm}} = \frac{30 \text{ inches}}{3 \text{ inches}} = \frac{10}{1} \text{ or } 10{:}1$$

If the mechanical advantage and the amount of force at the effort arm of a lever are known, the amount of force produced at the resistance arm can be found. For example, suppose that 20 pounds of force is applied to the effort arm of the lever in figure 1-3, which has a mechanical advantage of 10:1. The force at the resistance arm will be 10 times 20, or 200 pounds.

An important point to note is that a mechanical advantage always results in a loss of motion. For example, the lever in figure 1-3 has a mechanical advantage of 10 to 1.

*Note: Mechanical advantage of the lever is expressed here as an effort arm/resistance arm to be consistent with the approach used in finding speed (gear) ratios.

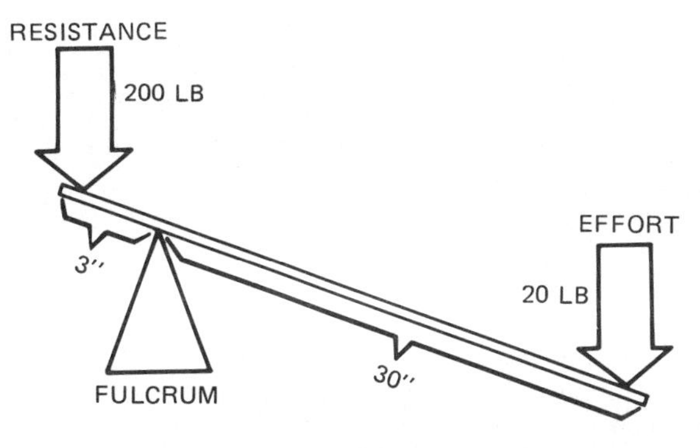

Fig. 1-3

This means that if the effort arm moves a distance of 10 inches, the resistance arm will move only 1 inch.

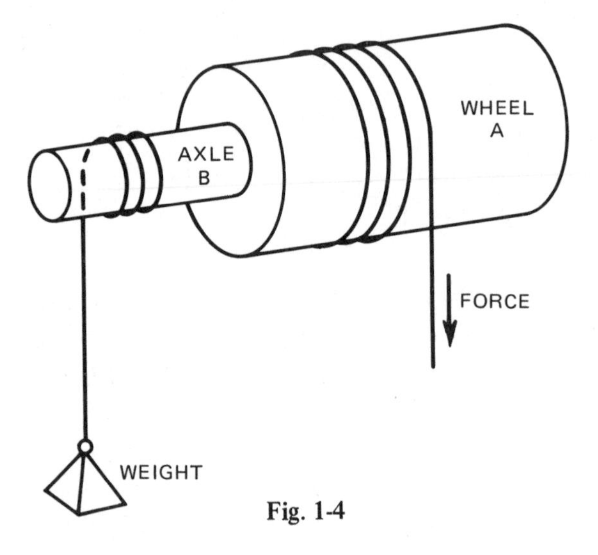

Fig. 1-4

Compound Levers

Compound levers are a series of levers connected together. Two examples of compound levers are the throttle linkage of the carburetor and the shift linkage to the transmission. When levers are used in this way, the ratio of each lever is multiplied by the ratios of the other levers in the series.

Wheel and Axle

Another device that provides a mechanical advantage is the wheel and axle, figure 1-4. This simple machine is made up of two cylinders of different diameters that rotate together on the same axis.

A wheel and axle can be compared to a lever as shown in figure 1-5. The mechanical

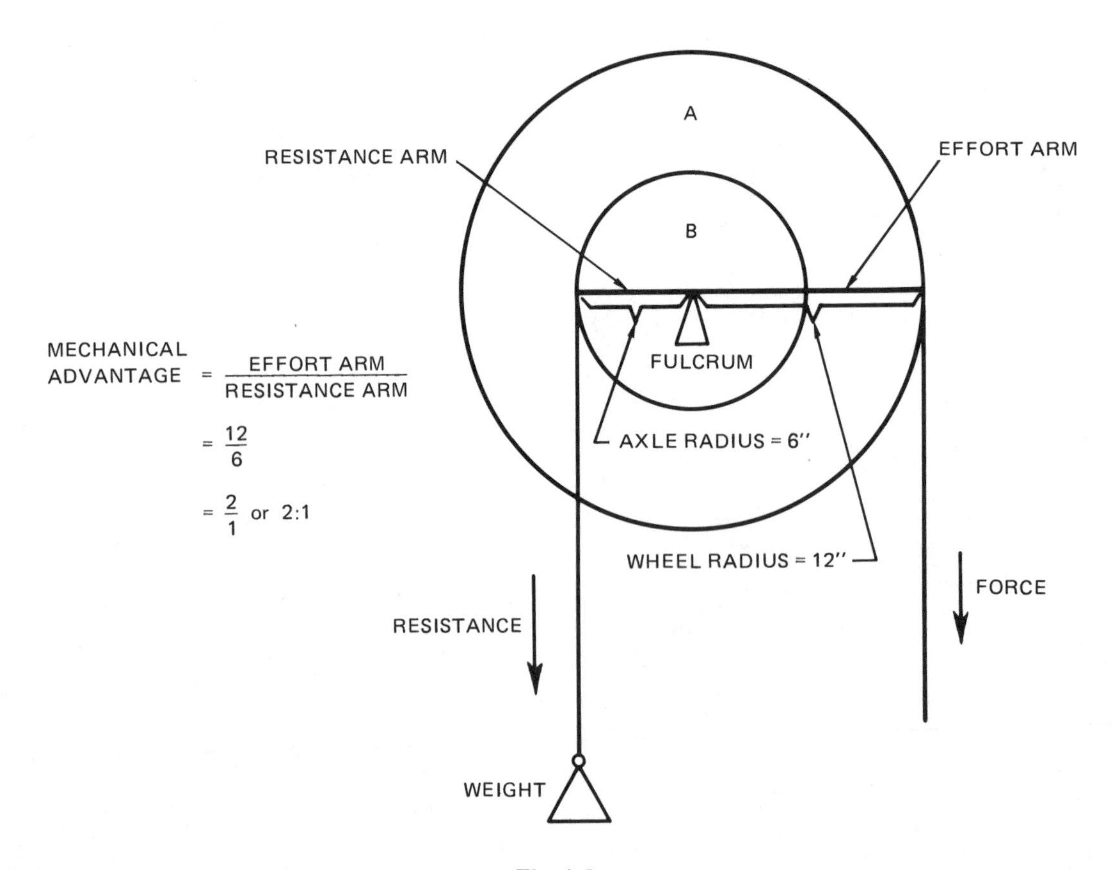

Fig. 1-5

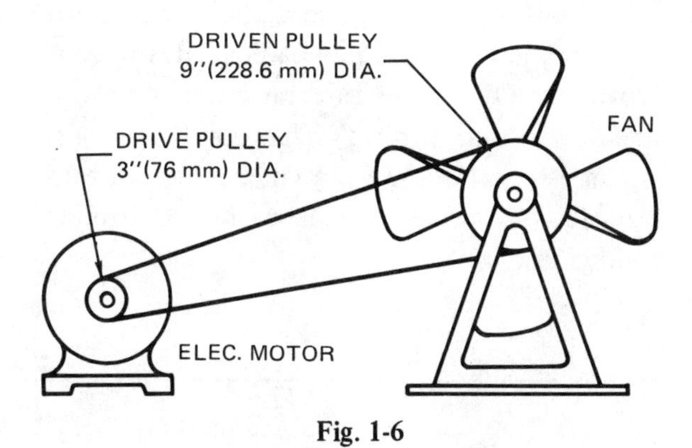

Fig. 1-6

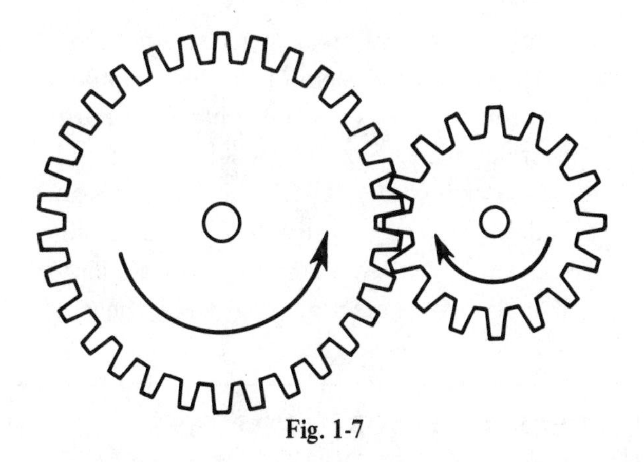

Fig. 1-7

advantage is calculated in the same way as for a lever, and is shown in the illustration.

The wheel and axle does away with some of the disadvantage of loss of motion. The weight can be moved a greater distance because the lever effect works through a full turn of the wheel. However, notice that the movement of the wheel and axle is limited to the number of turns of rope on the wheel.

Pulleys

To decrease the loss of motion that results from a mechanical advantage, a pulley can be used. To provide mechanical advantage, pulleys are used in pairs, figure 1-6. Pulleys of two different sizes are needed. In this case, the small or *drive* pulley is fixed to an electric motor and the large or *driven* pulley is fixed to a fan. The idea here is to turn the fan slower than the motor. The speed of the fan can be determined by dividing the motor speed by the speed ratio of the pulleys. The speed ratio of the pulleys is 9″ : 3″ or 9/3 = 3 : 1. Since most electric motors turn at a speed of 1,725 rotations per minute (r/min), the speed of the fan will be 1,725 ÷ 3 or 575 r/min. Note that there is still motion loss in that the large pulley moves slower than the small pulley.

This also results in a gain in mechanical advantage on the large pulley. For example, assume that the electric motor is capable of producing a twisting force (torque) of 10 lb-ft. Using a ratio of 3:1, the force on the large pulley would be multiplied by 3, resulting in a torque of 30 lb-ft.

With the exception of a crankshaft, the movement of a lever is limited to a maximum arc of somewhat less than 180°. A wheel and axle setup also has limited movement: when the number of turns on the wheel and axle are used up, the object being moved will interfere with the movement of the wheel and axle. The pulleys, however, are kept moving by the motor, and are not limited to the arc through which a lever can be moved or the number of turns of rope on a wheel and axle.

Gears

Gears also provide mechanical advantage and, like pulleys, must work in pairs with a small drive gear turning a larger driven gear, figure 1-7.

Gears are very important in the operation of the automatic transmission. The internal combustion engine depends on relatively high engine speeds to produce the energy that turns the wheels of the car. At low engine speeds, the engine cannot produce the energy needed for a fast and smooth takeoff from a standstill without the use of gears or some other means of mechanical advantage. If an automobile had only a clutch to connect the engine to the drive shaft and rear wheels, the engine would have to be raced and the clutch slipped before enough speed could be reached to allow releasing the clutch pedal. This would be like driving a car with the transmission locked in high gear. By having the engine drive a small gear that is in mesh with a larger gear and by connecting the larger gear to an output shaft, the engine is allowed to speed up. This is because a mechanical advantage is provided. Of course, the transmission is constructed so that different gears can be brought into mesh. This allows the mechanical advantage to be varied to suit road and driving conditions.

MEASURING WORK AND POWER IN THE METRIC SYSTEM

In the metric system, work is the result of multiplying force times distance, just as it is in the English system. However, in the metric system, a special unit, called a *newton,* is used to measure the force of gravity on an object. The newton is used instead of the gram when calculating problems involving work. The unit of measurement for distance is the meter. Work is measured in units called *joules.*

Work = Force X Distance
Joules = Newtons X Meters

For example, if 225 newtons of force are needed to lift a stack of bricks up a ladder measuring 3 meters, 675 joules of work is being done.

Work = 225 newtons X 3 meters
Work = 675 joules

In the metric system, power is measured in joules per second or joules per minute. The metric measurement, joules per second, is commonly called *watts.* For example, if 90 newtons of force are required to lift a weight 1.5 meters in 5 seconds, the amount of power required is found as follows:

Work = Force X Distance
Work = 90 newtons X 1.5 meters
= 135 joules

$$\text{Power} = \frac{\text{Work}}{\text{Time}}$$

$$\text{Power} = \frac{\text{135 joules}}{\text{5 seconds}}$$

= 27 watts

SUMMARY

There are many ways of multiplying force and creating mechanical advantage. The examples given in this unit and the following points provide the basic knowledge needed in the study of automatic transmissions.

- Force is a push or a pull. It may also be called effort.
- To do work, effort or force must cause motion.
- Work is measured in foot-pounds or joules. It is equal to the force times the distance through which the force is exerted.
- Power is the rate of doing work. It is measured in foot-pounds per second (or minute) or joules per second (or minute), also called watts.

- The length of the resistance arm as compared to the length of the effort arm determines the mechanical advantage of speed of a lever.
- The mechanical advantage of force equals the ratio of the resistance to the effort.
- Mechanical advantage always results in a loss of motion. If a pair of pulleys or gears are connected, the larger pulley or gear always turns slower than the smaller pulley or gear.
- The automobile engine is not very efficient at low speeds.
- Gears can be used to move energy from the engine to the rear wheels.

REVIEW

Select the best answer from the choices offered to complete the statement or answer the question.

1. If a small drive pulley is connected by a belt to a larger driven pulley:
   (I) The large pulley will turn slower than the small pulley.
   (II) There will be a mechanical advantage at the large pulley.

   a. I only c. both I and II
   b. II only d. neither I nor II

2. Which of the following statements is (are) correct?
   (I) More mechanical advantage is obtained by using a first class lever than a second class lever.
   (II) The second class lever has its fulcrum located at the end of the resistance arm.

   a. I only c. both I and II
   b. II only d. neither I nor II

3. To be considered work, force must:
   (I) Cause motion.
   (II) Result in a mechanical advantage.

   a. I only c. both I and II
   b. II only d. neither I nor II

4. A lever such as that shown in figure 1-3 has a resistance arm of 2 feet and an effort arm of 8 feet. The ratio of the effort arm to the resistance arm is:

   a. 3:1 c. 1:4
   b. 4:1 d. 10:1

5. A drive pulley with a 3-inch diameter is turning at 1,725 r/min. The drive pulley is turning a driven pulley with a 5-inch diameter. The r/min of the driven pulley would be closest to:

   a. 1,033 c. 2,875
   b. 2,881 d. 1,035

6. An engine pulley measures 6 inches in diameter and the alternator pulley measures 2 1/2 inches in diameter. If the engine is turning 1,500 r/min, how fast is the alternator turning in r/min?

   a. 623
   b. 625
   c. 3,600
   d. 3,750

7. A first class lever has an effort arm of 5 feet and a resistance arm of 2 feet. If the effort arm is moved a distance of 3 feet, the resistance arm will move

   a. 7.5 ft.
   b. 1.66 ft.
   c. 6 ft.
   d. 1.2 ft.

8. Drive pulley A is connected by a belt to driven pulley B. If pulley A is 2 inches in diameter and pulley B is 7 inches in diameter, the ratio between the two pulleys would be

   a. 0.286:1.
   b. 14:1.
   c. 9:1.
   d. 3.5:1.

9. A second class lever has an effort arm of 36 inches and a resistance arm of 6 inches. A force of 25 pounds is applied to the end of the effort arm. The force at the end of the resistance arm would be

   a. 900 lb.
   b. 150 lb.
   c. 216 lb.
   d. 416 lb.

10. A wheel and axle such as the one shown in figure 1-5 can be compared to a lever of:

    (I) The first class.
    (II) The second class.

    a. I only
    b. II only
    c. either I or II
    d. neither I nor II

EXTENDED STUDY PROJECTS

1. A group of levers linked together as shown in figure 1-8, page 9, is called a compound lever. If an effort or force of 50 pounds is applied to the effort arm of lever A, how much weight could be moved by the resistance arm of lever C?
2. What effort would be required to lift a weight of 100 pounds using the setup in figure 1-8?
3. What is the total ratio of the levers in figure 1-8?
4. If the resistance arm were 2 inches, how long would the effort arm of a simple lever (as shown in figure 1-4) have to be to equal the ratio of the compound lever in figure 1-8.
5. A group of pulleys as shown in figure 1-9 multiplies the ratios of pulleys A-B and C-D. The shaft E is known as a countershaft. What would be the speed ratio of pulley D to pulley A?
6. If pulley A is turning at 850 r/min, what is the speed of pulley D?

7. If pulley A is turning 1,000 r/min, how fast is pulley C turning?
8. To equal the ratio of the pulleys in figure 1-9, by using only two pulleys, what must be the diameter of the large pulley if the small pulley is 2 inches in diameter?

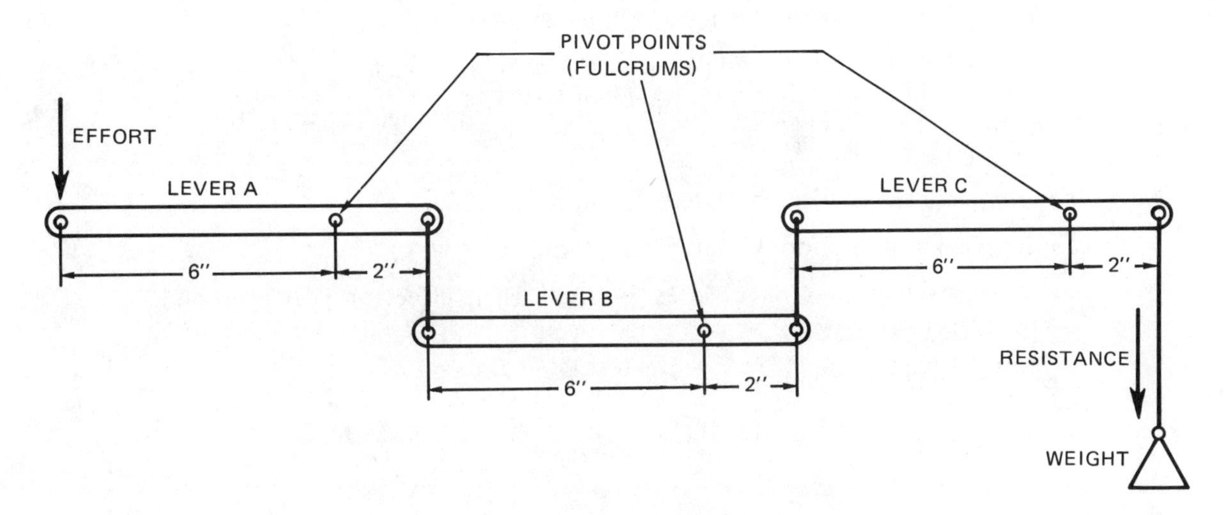

Fig. 1-8

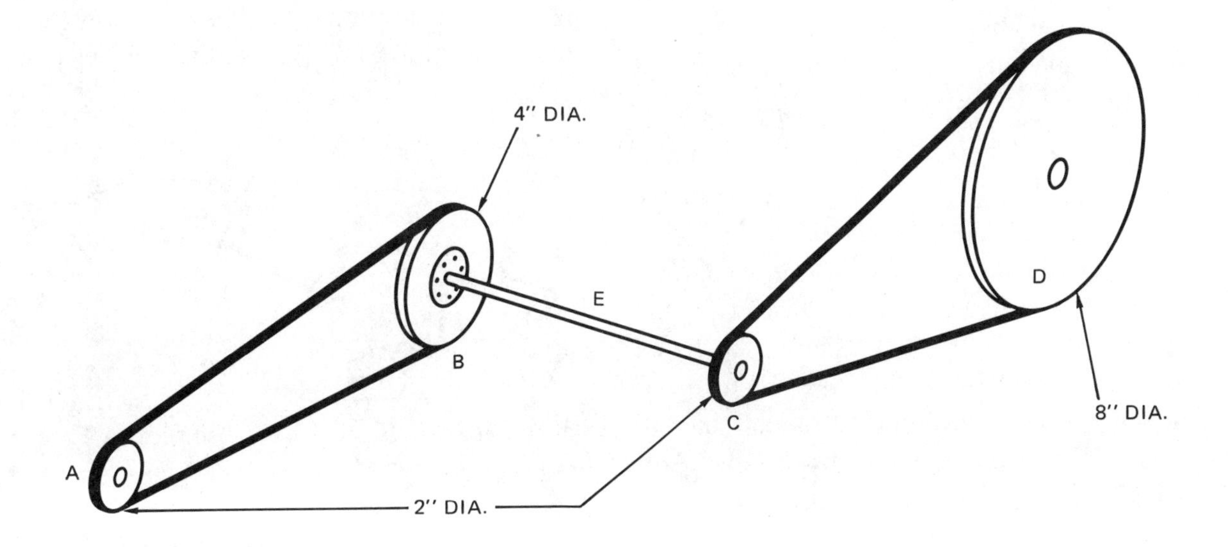

Fig. 1-9

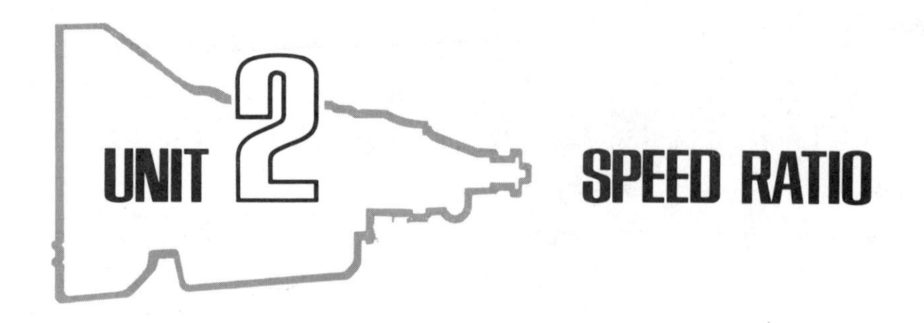

SPEED RATIO

OBJECTIVES

After studying this unit, the student will be able to:

- Determine the speed ratio of a driven gear to a drive gear.
- Explain the relationship of effort, time, and distance to work, torque, and power.
- Explain how friction is related to power loss.
- Explain how friction is related to holding power.
- Describe the function of idler gears and their effect on gear ratio and direction of rotation.
- Determine compound gear ratios.
- Explain how speed ratio effects engine r/min and road speed.

SPEED RATIO

When related to a gear set, *speed ratio* can be defined as the relative speed in number of revolutions that the input (driver) gear makes to one revolution of the output (driven) gear. As an equation, this can be expressed as:

$$\text{Speed ratio} = \frac{\text{\# of turns of (input) driver gear}}{\text{\# of turns of (output) driven gear}}$$

To determine the speed ratio (sometimes called *gear* ratio), count the number of teeth on each gear and set up a ratio in which the teeth of the driven gear are divided by the teeth of the driver. For example, in figure 2-1, drive gear A, which has 12 teeth, is meshed with driven gear B, which has 30 teeth. The ratio for this gear set is B ÷ A or 30/12 = 2.5:1. This means that gear A must make 2.5 revolutions in order for gear B to make one revolution.

This gear set with its speed ratio of 2.5 can be used in a car. For example, if an automobile engine turning 1,000 revolutions per minute (r/min) is connected to a shaft that turns gear A, then gear B will turn at the rate of 1,000 ÷ 2.5 or 400 r/min. This loss of motion between gear A and B results in a mechanical advantage in the same way that loss of motion of a lever results in a mechanical advantage.

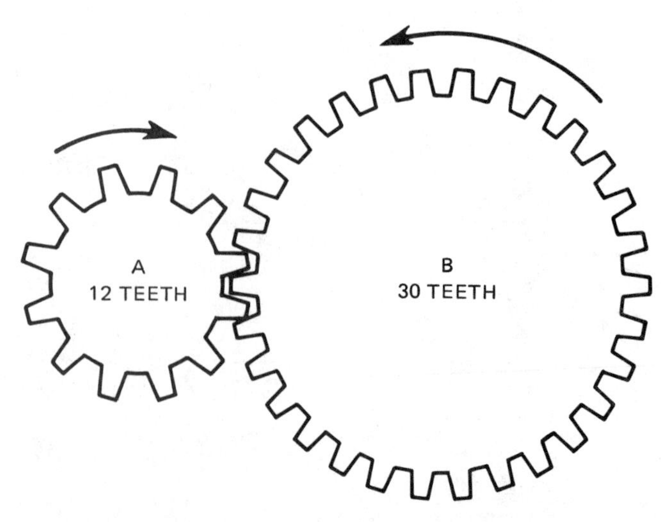

Fig. 2-1 The gear ratio of this set of gears is 2.5:1

TORQUE

Torque is a twisting or turning effort that may or may not cause motion. For example, effort at the rim of a car's steering wheel, figure 2-2, causes torque at the steering gear shaft. If the effort at the rim of the wheel is large enough, the steering gear and linkage will cause the front wheels to turn. If the car is standing still, more effort will be needed to turn the wheel. If the effort is not large enough, the wheel will not turn. However, torque is still being applied to the steering shaft.

Another example of torque is found when tightening or loosening a bolt or nut. In loosening a bolt, torque is applied to the bolt, but no work is performed until the bolt breaks loose. Note this difference between torque and work.

So that torque is not confused with work, torque is measured in pound-feet (lb-ft) rather than foot-pounds. For example, if the effort applied to the rim of the steering wheel were 10 lb, the torque on the steering shaft would be 10 times the radius of the wheel in feet (0.75) or 10 X 0.75 = 7.5 lb-ft (10 N·m). See the end of this unit for an explanation of torque measurement in the metric system.

TORQUE MULTIPLICATION

Torque can be multiplied by the use of a gear set. In the simple gear set shown in figure 2-3, 50 lb-ft of torque is turning the smaller drive gear. Since the ratio of gear B to gear A is 4:1 (60 ÷ 15 = 4) there will be 50 X 4 = 200 lb-ft of torque at the larger driven gear. Note that this does not change the power to gear B; only the torque is changed.

If gear A in figure 2-3 is turning at 1,000 r/min, there would be 50,000 units of power (1,000 r/min X 50 lb-ft of torque). Ignoring power loss due to friction, the units of power at gear B would be found as follows:

$$1{,}000 \text{ r/min} \div 4 = 250 \text{ r/min}$$
$$250 \text{ r/min} \times 200 \text{ lb-ft} = 50{,}000 \text{ units of power}$$

The output power (not counting friction loss) must equal the input power.

HORSEPOWER

The units of power an automobile engine produces are called its *horsepower*. Work, as

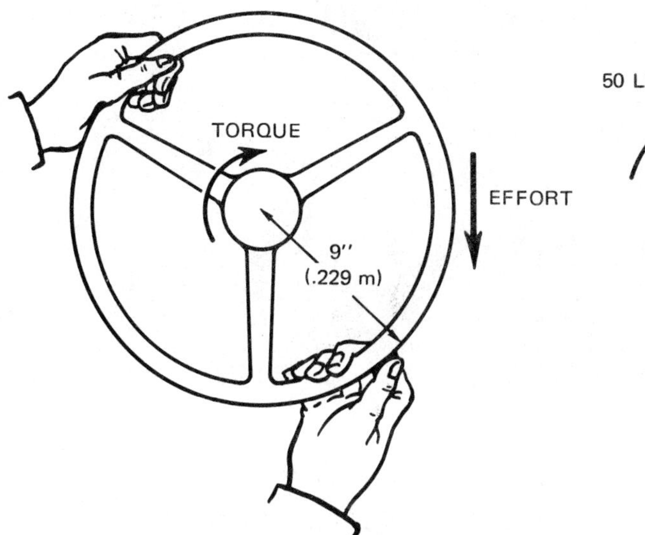

Fig. 2-2 Torque applied to the steering wheel of a car.

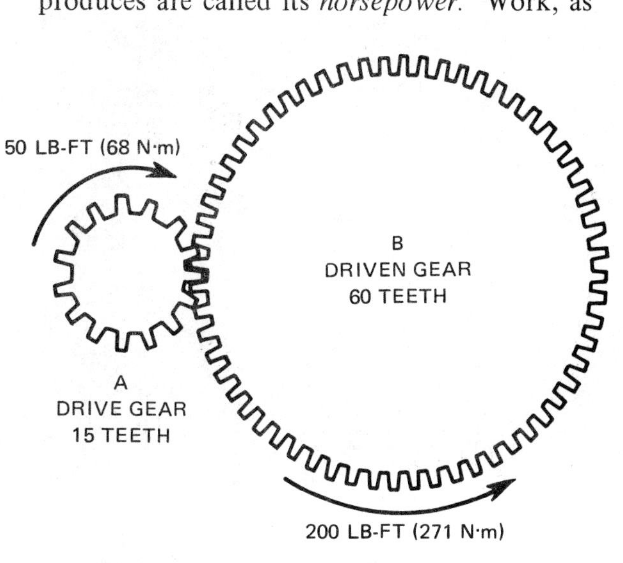

Fig. 2-3

explained in unit 1, can be found by the equation: force × distance moved ÷ time. One horsepower is equal to 33,000 ft-lb of work performed in one minute or 550 ft-lb of work performed in one second. The formula for horsepower is:

$$\text{Horsepower} = \frac{\text{Work}}{\text{Time (in minutes)} \times 33{,}000}$$

or

$$\text{Horsepower} = \frac{\text{Work}}{\text{Time (in seconds)} \times 550}$$

For example, if a truck is able to pull a load having a resistance of 3,000 pounds over a distance of 1 mile (5,280 ft) in 2 minutes time, the horsepower needed to move this load would be:

$$\frac{3{,}000 \times 5{,}280}{2 \times 33{,}000} = 240 \text{ horsepower}$$

Therefore, the horsepower needed to move a load of 3,000 pounds 1 mile in 2 minutes is 240 horsepower.

In the automotive industry, engines are rated according to the total horsepower that they can produce.

BRAKE HORSEPOWER

The horsepower of engines is commonly measured as *brake horsepower.* Brake horsepower is measured on a machine called a *dynamometer.* The dynamometer measures torque and r/min. These figures are then used to determine the horsepower of the engine.

FRICTION

Friction is resistance of two or more parts to motion. Some places where friction occurs are

- between the piston ring and cylinder walls of an automobile engine.
- between the gear teeth of a transmission or the gear teeth of a rear axle.
- between the connecting rod bearing and crankshaft journal (even though the bearing and journal are separated by a film of oil).

Harmful Friction

Friction of the type just described is called *harmful friction* or *unwanted friction.* It results in a loss of horsepower as force is transmitted to the rear wheels. Friction produces heat, which if allowed to build up, could damage an engine or transmission.

Unwanted heat is removed by transferring it to the surrounding air. The automobile engine lubricating and cooling system,

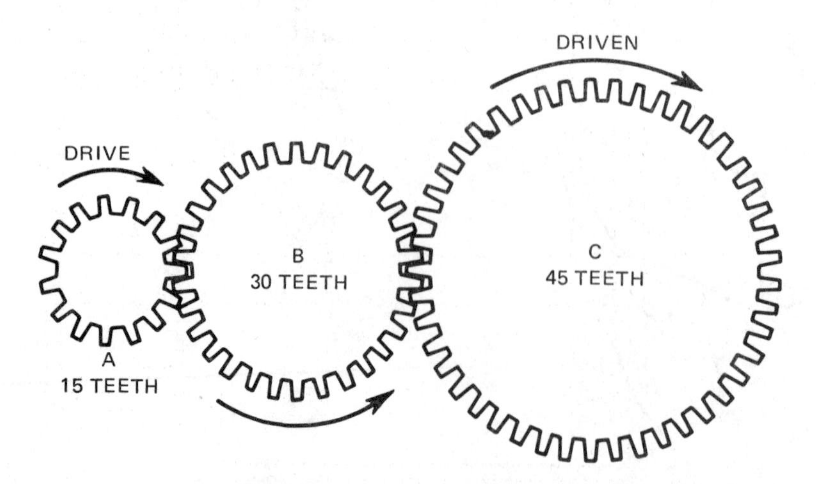

Fig. 2-4 An idler gear (B) allows the drive gear (A) and the driven gear, (C), to move in the same direction.

for example, removes unwanted heat from engine parts. Automatic transmissions have coolers to remove unwanted heat caused by friction.

Useful Friction

While some friction is harmful, other friction may be useful, and even necessary, under certain conditions. For example, an alternator would not turn if it were not for friction between the pulley and the belt. The clutches and bands in an automatic transmission depend on friction for holding power.

IDLER GEARS

In the gear sets shown in figures 2-1 and 2-3, gear A is turning in a clockwise direction. This causes gear B to turn in a counterclockwise direction. In reality, however, it is often necessary to turn the driven gear in the same direction as the drive gear. An idler gear, figure 2-4, makes this possible.

As shown, the direction of rotation of drive gear A and driven gear C are now the same. The idler gear causes a change of direction only; it does not change the ratio of the driven gear to the drive gear. For example, in figure 2-4, the ratio is: 45 ÷ 15 or 3:1. This ratio will be the same regardless of the number of teeth the idler gear has. This can be proven by multiplying the ratio of gear B to gear A by the ratio of gear C to gear B as follows:

$$\frac{B}{A} = \frac{30}{15} = \frac{2}{1} \qquad \frac{C}{B} = \frac{45}{30} = \frac{1.5}{1}$$

and $$\frac{2}{1} \times \frac{1.5}{1} = \frac{3}{1} \text{ or } 3:1$$

Keep this rule in mind when determining gear ratios. Idler gears are often used in automatic transmissions to obtain a change of direction in the gear sets.

COMPOUND GEAR SETS

The simple gear sets studied so far multiply torque in a fixed ratio. Since one ratio would limit the speed of an automobile, transmissions make use of more than one gear

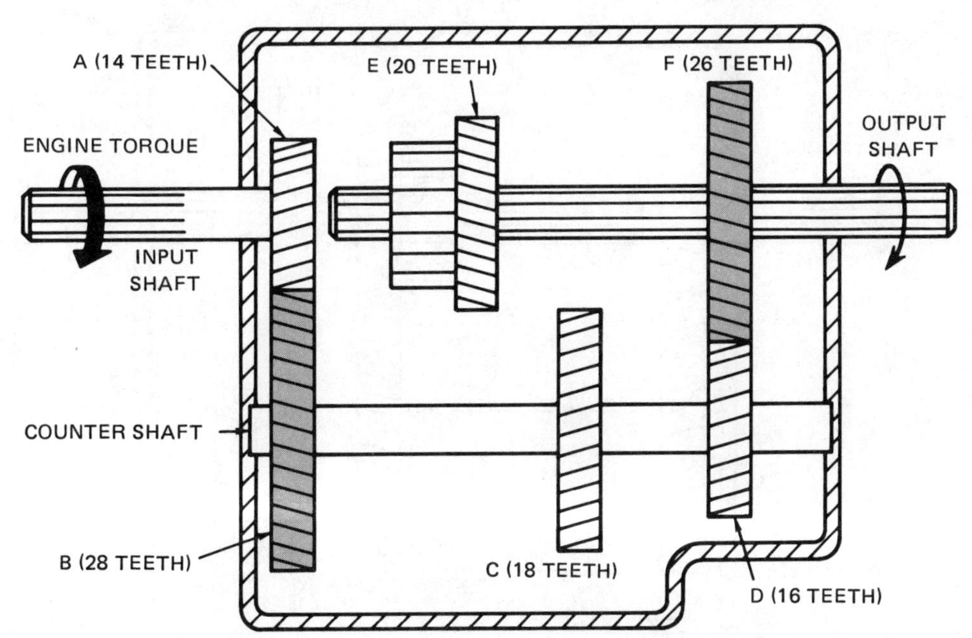

Fig. 2-5 Gears in a standard transmission as positioned for low speed.

set. This is called *compound gearing* and the gear sets are referred to as *compound gear sets.*

As automobile road speed increases, engine r/min also increases. This makes more torque available to drive the rear wheels. To make use of this torque and to increase road speed, a higher ratio is needed. In a standard transmission, this is accomplished when the driver shifts the transmission into second, third or fourth speed.

A diagram of a standard transmission, figure 2-5, shows an input shaft that transfers torque to gear A. Gear A is meshed with larger gear B. Gears B, C, and D are part of the countergear and must turn any time that gear A is turning. Notice that gear D is meshed with gear F and that gear F is larger than gear D. This is the position of the gears for first speed.

Note that none of the gears in figure 2-5 are idlers. This is because in each case one gear is driving another gear of a different size on the same shaft. Compare this gear set to the pulley setup in figure 1-9.

COMPOUND RATIO

Compound gear sets operate in much the same way as the pulleys shown in figure 1-9. A gear set having one ratio is used to drive another gear set having a different ratio. The final ratio then is found by multiplying the ratios of the two sets.

Refer to figure 2-5. Note that there are two gear sets. Gear A is a drive gear and gear B is a driven gear. Gear B turns the counter shaft making gears C and D drive gears and, in this case, gear F a driven gear. The ratio of this compound gear set is determined as follows:

$$\frac{B}{A} \times \frac{F}{D} \quad \text{or} \quad \frac{28}{14} \times \frac{26}{16} = 3.25{:}1$$

When the driver shifts to second speed, gears C and E are in mesh as shown in figure 2-6. The ratio for second speed would be:

$$\frac{28}{14} \times \frac{20}{18} = 2.22{:}1$$

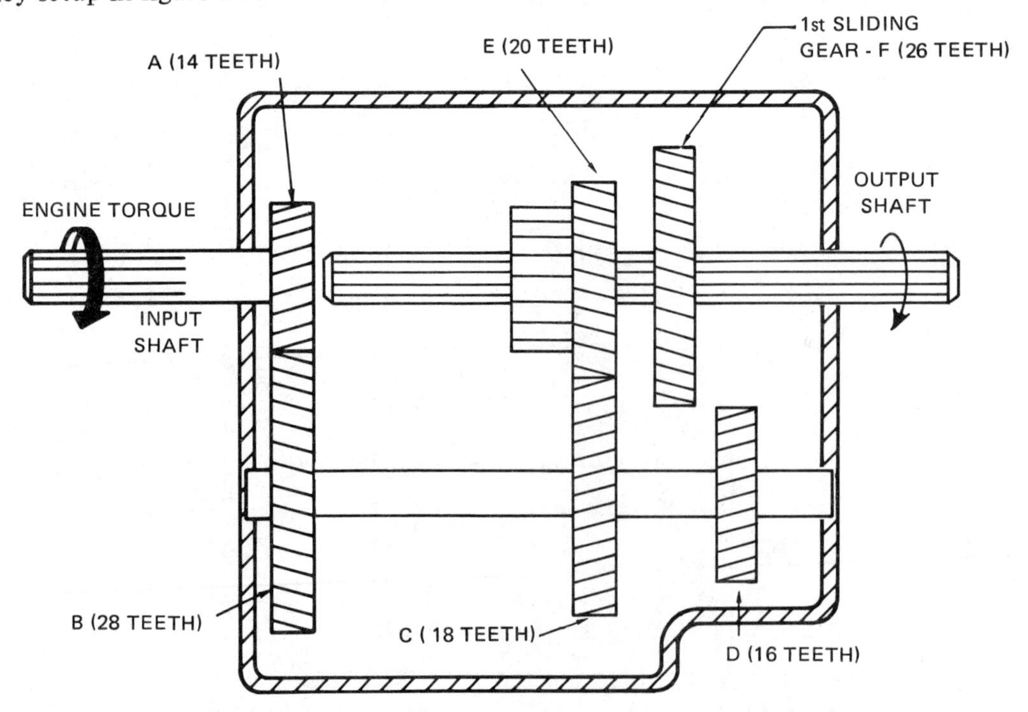

Fig. 2-6 Position of gears in second speed.

Third speed, or direct, is obtained by moving gear E into gear A. Gear A has internal splines that mesh with the splines on the hub of gear E. This locks the input shaft to the output shaft and the transmission is in direct. The ratio for direct is always thought of as 1:1. (Actually, there is no multiplication of torque in direct.)

Additional reduction is gained at the rear axle through the differential. A typical rear axle ratio would be 3.25:1. This means that in high gear (direct) the engine must make 3.25 rotations to every 1 rotation of the rear wheels. Remember, the gear set ratio in direct is 1:1. If this ratio is multiplied by that given by the rear axle, the total ratio is found to be 3.25:1. The speed ratios for first and second speeds are also affected by the rear axle ratio. The overall ratio to the rear wheels would be:

first speed $3.25 \times 3.25 = 10.56:1$
second speed $2.22 \times 3.25 = 7.22:1$

MEASURING TORQUE IN THE METRIC SYSTEM

In the metric system, torque is measured in *newton-meters* (N·m). Remember, the newton is the metric unit used to measure the force of gravity on an object. Torque is found by multiplying the force of gravity on the object by the distance in meters through which the force is applied. For example, refer to the steering wheel shown in figure 2-2. If the effort applied to the rim of the steering wheel is 44 newtons, the torque on the steering shaft would be found by multiplying this force by the radius of the wheel in meters. If the radius of the wheel is 0.229 meters, the torque is found as follows:

$44 \text{ N} \times 0.229 \text{ m} = 10 \text{ N}\cdot\text{m}$ (approximately)

SUMMARY

Torque can be multiplied by using different gear sets or through compound gearing. The following points should be remembered about gear sets and speed ratios.

- As in a set of pulleys, the large gear (the gear with the greatest number of teeth) always turns slower than the small gear (the gear with the least number of teeth).
- To determine the speed ratio of two gears, the number of teeth on the driven gear is divided by the number of teeth on the drive gear.
- Torque is a twisting or turning effort that may or may not cause motion.
- Torque is measured in lb-ft or N·m, and is the product of the effort or force multiplied by the distance through which it is applied.
- Torque can be multiplied by the use of a gear set. However, the power at the output always equals the power at the input.
- Friction is a resistance to motion between two or more parts.
- Friction may be either harmful or useful.

- Idler gears give a change of direction only. They have no effect on the ratio of the driven gear to the drive gear.
- Compound ratios are determined by multiplying together the ratios of two or more gear sets.

REVIEW

Select the best answer from the choices offered to complete the statement or answer the question.

1. If 75 lb-ft of torque is applied to a drive gear that has 22 teeth and is meshed with a driven gear that has 56 teeth, the torque at the driven gear would be approximately

   a. 75 lb-ft.
   b. 255 lb-ft.
   c. 393 lb-ft.
   d. 191 lb-ft.

2. Provided there is no slippage, the fan belt and pulleys on an automobile engine are an example of:

   (I) Harmful friction.
   (II) Useful friction.

   a. I only
   b. II only
   c. both I and II
   d. neither I nor II

3. Taking friction into consideration, if the horsepower of an engine is 200, and the combined ratio of the transmission and rear axle is 4.25:1, what is the horsepower at the rear wheels?

   a. 200
   b. somewhat more than 200
   c. somewhat less than 200
   d. 850

4. A drive gear has 17 teeth and a driven gear has 42 teeth. If the drive gear is turning at the rate of 600 r/min, how fast will the driven gear turn (approximately)?

   a. 247
   b. 1,481
   c. 243
   d. 1,482

5. The amount of torque or twisting effort is usually measured in:

   (I) ft-lb.
   (II) lb-ft.

   a. I only
   b. II only
   c. either I or II
   d. neither I nor II

6. If 7,000 lb is to be lifted 20 ft in 2 minutes, how much horsepower is required?

   a. 2.12
   b. 0.212
   c. 4.24
   d. 0.106

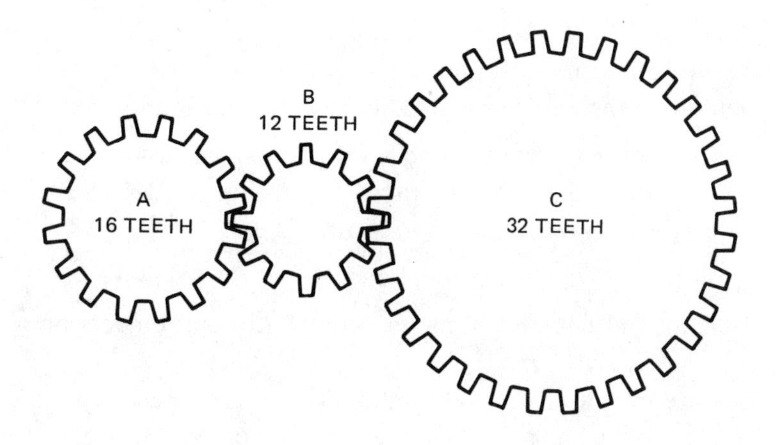

Fig. 2-7

7. If 85,000 ft-lb of work is performed in 1 minute, how much horsepower is produced (approximately)?

   a. 0.388
   b. 1.00
   c. 1.258
   d. 2.58

8. For the gear set shown in figure 2-7, which of the following statements apply to gear C?

   (I) The ratio of gear C to gear A is 2:1.
   (II) Gear C would be turning in the same direction as gear A.

   a. I only
   b. II only
   c. both I and II
   d. neither I nor II

9. If a drive gear has 17 teeth and is turning a driven gear with 52 teeth, the speed ratio would be approximately

   a. 3.06:1.
   b. .327:1.
   c. 3.27:1.
   d. 8.84:1.

10. If an automobile engine is producing 175 horsepower with the transmission in direct, what is the horsepower to the rear wheels if the rear axle ratio is 3.27:1 (not counting friction loss)?

    a. 175
    b. 535
    c. 572
    d. 327

11. In a gear set such as that shown in figure 2-4, if gear A has 20 teeth, gear B has 32 teeth, and gear C has 50 teeth, what is the final ratio of the gear set?

    a. 1.6:1
    b. 2.5:1
    c. 1.56:1
    d. 31.25:1

EXTENDED STUDY PROJECTS

1. a. Disassemble a standard transmission far enough to count the number of teeth on the gears.
   b. Determine the gear ratios for the forward speeds and reverse.
2. It takes a hiker 2 hours to climb a mountain that has a vertical height of 3,000 feet. If the hiker weighs 175 pounds and carries a pack that weighs 60 pounds, how much horsepower is used?
3. Perform the following experiment and record the results:
   a. Place an old cylinder head (or other flat metal object) on a clean, dry, metal, workbench.
   b. Use a spring scale to determine 1) the force needed to start the object moving, and 2) the force needed to keep the object moving.
   c. Put a thin film of chassis lube on the bench and object, and repeat step b.
   d. Write the answers to the following questions:
      1. What is the difference between the force needed to start the object moving and the force needed to keep the object moving?
      2. Why is there a difference between these two forces?
      3. What is the difference between the forces needed to move the object on a dry surface and a lubricated surface?
      4. Why is there a difference in these forces?
      5. What is the purpose of using a lubricant?

SIMPLE PLANETARY GEAR SETS

OBJECTIVES

After studying this unit, the student will be able to:

- Identify the parts of a planetary gear set.
- Determine the drive, driven, and holding members of a simple planetary gear set for the ratios used in automatic transmissions.
- Find the gear ratios of a simple planetary gear set for low, intermediate, reverse, and reverse-overdrive.

In the standard transmission, the gear ratios must be changed by the driver to obtain the different speeds and directions necessary to operate the automobile. If the gears are not brought into mesh properly, they will clash or grind. In automatic transmissions, a different type of gear set is used. In this gear set, called a planetary gear set, the gears are always in mesh. Thus, they cannot clash or grind. A planetary gear set is shown both disassembled and assembled in figure 3-1.

Planetary gear sets get their name from the way in which the units of the gear set work together, much like the planets in the solar system. The planetary gear set is made up of a sun gear, a planet carrier assembly, and a ring gear. These parts are called its *members*. The gears rotate about a common

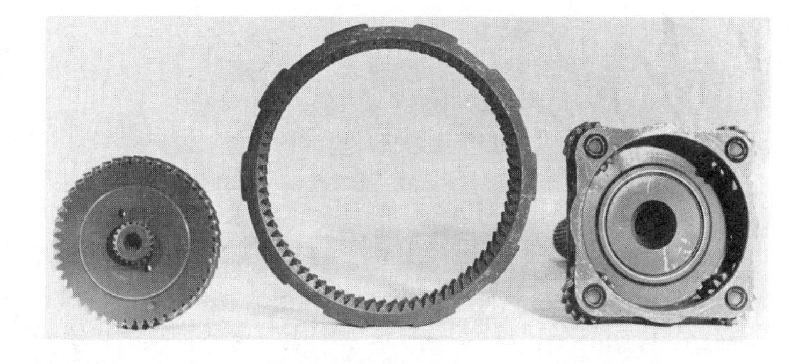

Fig. 3-1(A) Members of a planetary gear set (disassembled).

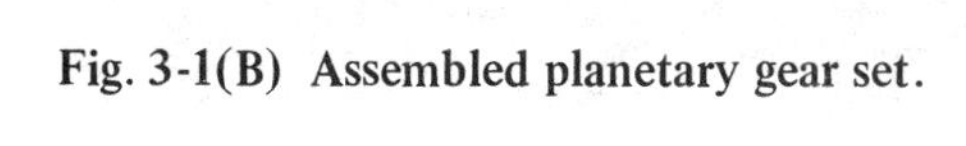

Fig. 3-1(B) Assembled planetary gear set.

axis with the sun gear at the center, figure 3-2. The planet gears, called *pinions,* mesh with the sun gear and are pinned to the *carrier assembly* (sometimes called the *reaction carrier* or *reaction member*). The ring gear is in mesh with the planet pinions. Notice that the ring gear has internal teeth. In other words, the outside of the gear is smooth and the teeth are facing toward its open center. Gears of this type are called internal gears or annulus gears. (*Annulus* means ring).

ADVANTAGES OF PLANETARY GEAR SETS

Planetary gear sets have many advantages over standard gear sets. They are stronger, because more teeth are in mesh than in standard gear sets. A planetary gear set may have two, three, or four pinions in mesh with the sun and ring gear. Because they are stronger and rotate around a common axis, planetary gears can be made smaller and more compact. In addition, they are always in mesh and cannot clash or grind.

OPERATION

The speed or gear ratio of a planetary gear set is changed by holding and driving different members. This is accomplished in the transmission by using brake bands and clutches. The actual operation of these parts will be discussed in a later unit. At this point, it is only important to know which gears are held and which are drivers and driven for each speed.

In the gear set shown in figure 3-2, eight different combinations are possible. They are: neutral; forward speeds of low, second, or intermediate; direct; overdrive 1; overdrive 2; and reverse speeds of low and overdrive. The chart in figure 3-3 gives information about the status of the gears in each one of these combinations. Combinations 2, 3, 6, 7, and 8 in the chart are used in automatic transmissions.

Study the chart in figure 3-3 and note the following points:

- When the carrier is the driven or output member, there is a lower gear ratio, or a *speed reduction* (conditions 1 and 2).
- When any two members are driven at the same speed in the same direction or locked together, there is *direct drive* (condition 3).
- When the carrier is the drive or input member, there is a *speed increase* or *overdrive* (conditions 4 and 5).
- When the carrier is held, the pinions act as idler gears and the output member turns in a *reverse direction* to the input member (conditions 6 and 7).
- When there is no input or no member being held the gear set is in *neutral* (condition 8).

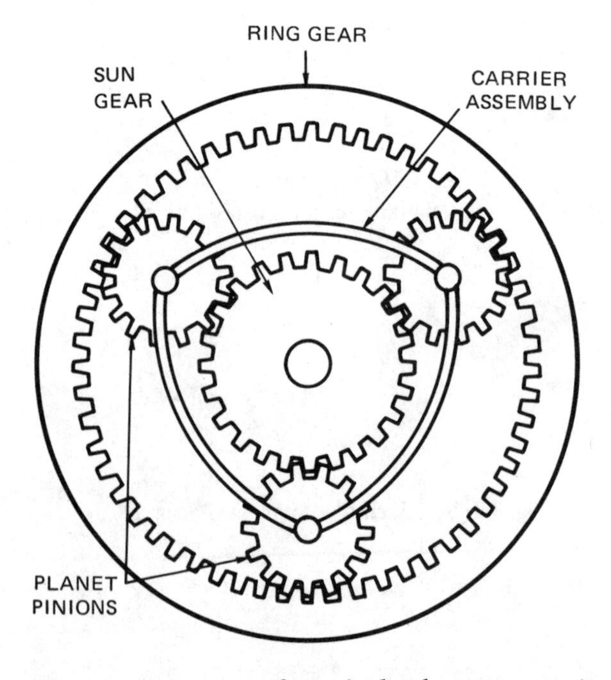

Fig. 3-2 Arrangement of gears in the planetary gear set.

Condition	Drive	Hold	Driven	Direction	Speed Ratio
1	sun	ring	carrier	forward	low
2	ring	sun	carrier	forward	intermediate
3	any 2	(or) any 2	unit locked	forward	direct
4	carrier	sun	ring	forward	1st overdrive
5	carrier	ring	sun	forward	2nd overdrive
6	sun	carrier	ring	reverse	low
7	ring	carrier	sun	reverse	overdrive
8	none	none	none	none	neutral

Fig. 3-3 Combinations possible with a simple planetary gear set.

Low

The chart shows that low (first speed) is obtained by driving the sun gear and holding the ring gear as shown in figure 3-4. The sun gear, as the drive or input gear, is turning in a clockwise direction. This turns the pinions in the opposite, or a counterclockwise, direction. Since the ring gear is being held, the pinions are forced to *walk* around the ring gear. As the pinions walk around the ring gear which is held, the carrier, or driven member, turns in a clockwise direction at a speed slower than the sun gear. This condition is **not** used to obtain low speed in modern automatic transmissions, but an understanding of this setup is helpful in the study of combinations that are used.

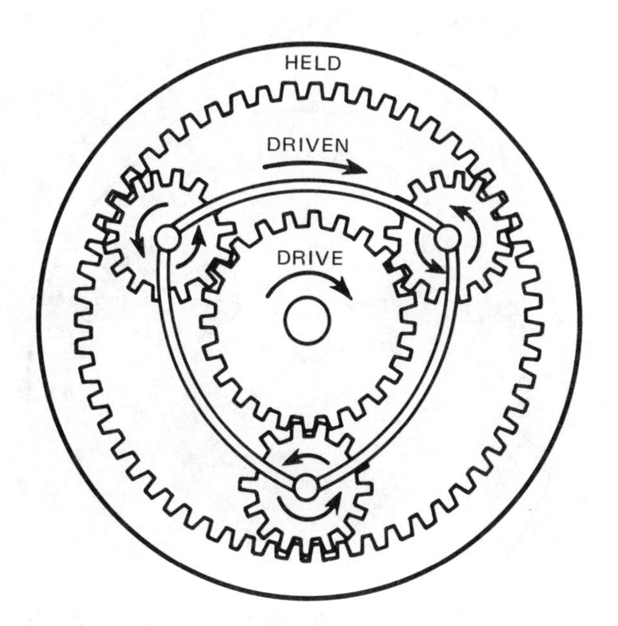

Fig. 3-4 The lowest possible speed ratio with a simple planetary gear set is obtained by driving the sun gear and holding the ring gear.

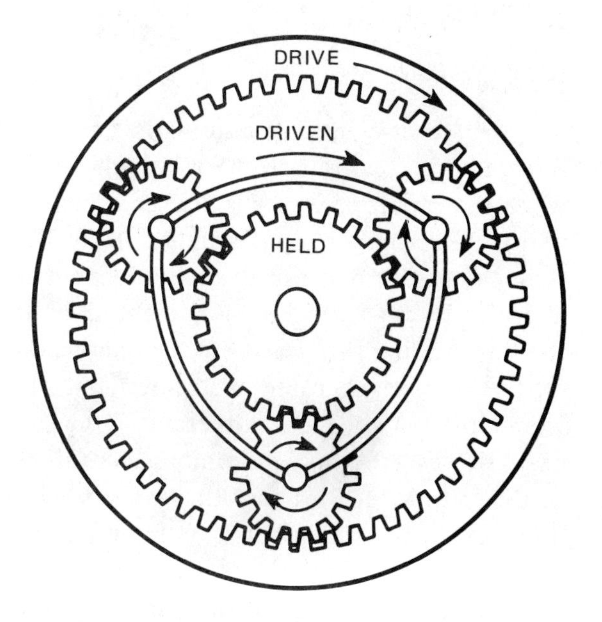

Fig. 3-5 Intermediate or second speed in a simple planetary gear set.

Intermediate

The setup or condition for intermediate (second speed) shown in figure 3-5 is used in most American-made automatic transmissions. In this case, the ring gear is the drive gear. The sun gear is being held, and the carrier is the driven member. The ring gear, turning in a clockwise direction, causes the pinions to turn in the opposite direction. This forces the pinions to walk around the stopped sun gear, thus moving the carrier in the same direction as the ring gear. However, the carrier moves at a slower speed. This results in the speed ratio of intermediate or second speed.

Direct

In most automatic transmissions, direct, or third, speed is obtained by locking the sun gear and the ring gear to the input shaft of the transmission, figure 3-6. This causes the sun gear and ring gear to turn at the same speed in the same direction. With this setup, the planet pinions cannot rotate. The gear set is locked together and must turn as a unit.

Reverse

For reverse in most automatic transmissions, the sun gear is the drive or input member while the carrier assembly is held and the ring gear is the driven or output member, figure 3-7. With the carrier being held, the pinions act as idlers, and reverse the direction of rotation. If the sun gear and ring gear were meshed together, the ring gear, because of its internal teeth, would turn in the same direction as the sun gear. However, because the pinions act as idlers in this setup, the ring gear turns in a reverse direction to the sun gear. Note that there is a gear reduction in this condition. This reduction is equal to the ratio of the ring gear to the sun gear or ring gear divided by the sun gear.

$$\frac{\text{ring gear}}{\text{sun gear}}$$

Overdrive

Overdrive is created when the output or driven member is turning faster than the input or drive member. A simple planetary gear set is used to obtain overdrive on standard transmissions. It is not currently in use in automatic transmissions. The setup

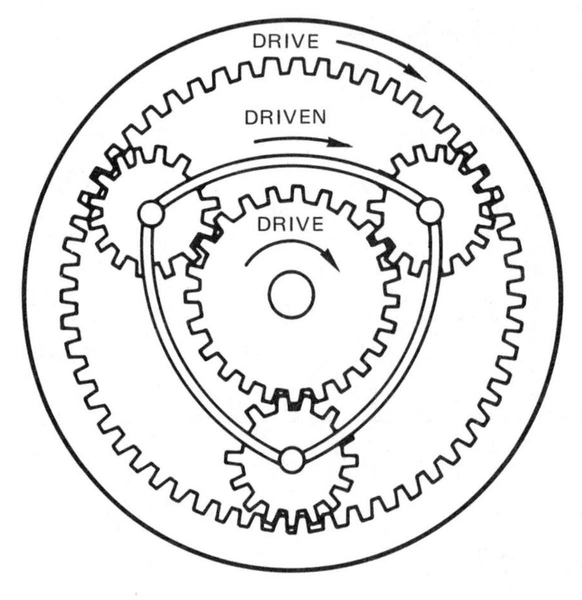

Fig. 3-6 Direct, or third, speed in a planetary gear set.

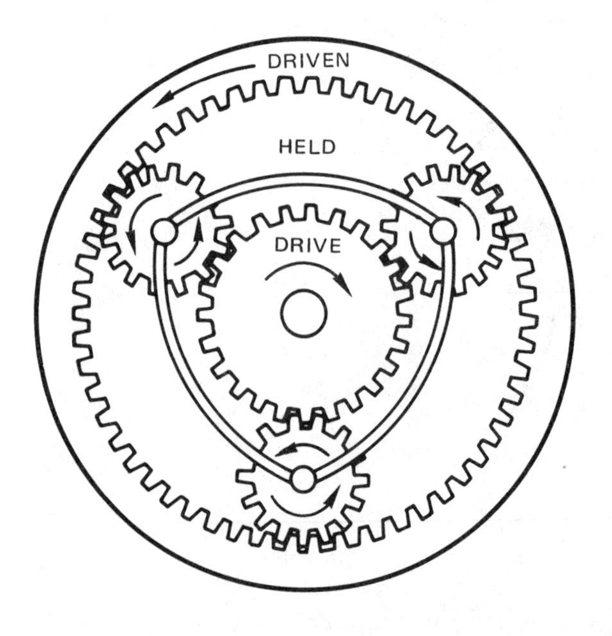

Fig. 3-7 The planetary gear system in reverse reduction.

of the gears for overdrive is shown in figure 3-8.

In this case, the input or drive member is the carrier; the sun gear is stopped, and the output or driven member is the ring gear. This combination causes the ring gear to turn faster than the carrier. The ratio for overdrive is less than one, and is expressed as a decimal. For example, if the ratio of the gear set shown in figure 3-8 is 0.7:1. This means that 7/10 of a turn on the driver produces one (1) full turn on the driven gear. Compare this condition with that shown in figure 3-5. Note that with the sun gear stopped, the carrier must turn slower than the ring gear.

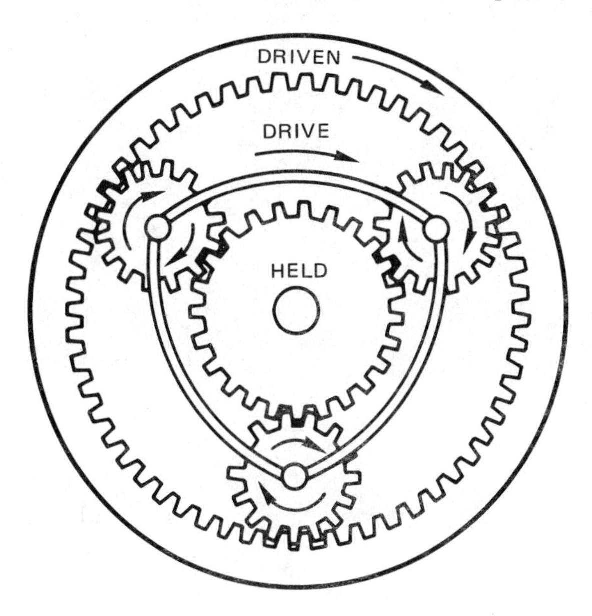

Fig. 3-8 The planetary gear system in overdrive.

Another overdrive, that is not in use at this time, can be obtained by driving the carrier and holding the ring gear. This gives an overdrive on the sun gear. In the set in figure 3-8 this would give a ratio of about 0.36:1.

Reverse Overdrive

In some transmissions, reverse overdrive is used with a compound planetary gear set to obtain a low range ratio. Compound gear sets will be discussed in greater detail in a later unit.

For reverse overdrive, as in reverse reduction, the planet carrier is held, causing the pinions to act as idlers, figure 3-9. However, in this condition, the ring gear is the drive gear, and the sun gear is driven. This means that the sun gear must turn faster than the ring gear in the ratio of the number of teeth on the sun gear divided by the number of teeth on the ring gear. This ratio is always less than one, and is expressed as a decimal.

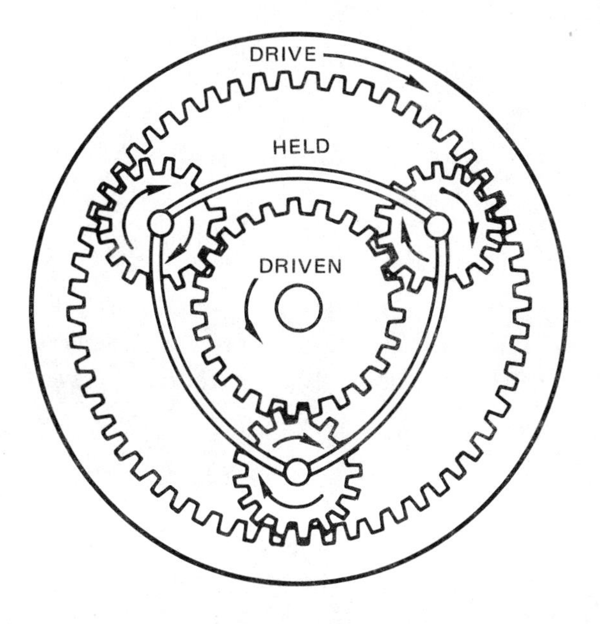

Fig. 3-9 Reverse overdrive.

SPEED RATIO

Except for the reverse condition, the speed ratio for planetary gear sets must be determined in a different way than it is for standard gear sets. This is because of the way in which the pinions and carrier assembly walk around the stopped gear. Even though the walking of the carrier assembly affects the ratio, the pinions act as idlers and need not be used to determine the ratio.

When the carrier walks around the stopped gear, the pinions must mesh with all of the teeth of that gear in order to make the carrier rotate one complete turn. The first step then, is to add the number of teeth on the drive gear to the number of teeth on the gear being held. This sum is then divided by the number of teeth on the drive gear. This ratio can be stated as follows:

$$\frac{\text{driver teeth} + \text{driven teeth}}{\text{driver teeth}}$$

Finding the Ratio for Low

For example, in the gear set in figure 3-10, the sun gear A has 22 teeth and the ring gear B has 70 teeth. If the sun gear is driving (the condition for low or first speed), the ratio of the carrier to the sun gear is:

$$\frac{A + B}{A} \text{ or } \frac{22 + 70}{22} = \frac{4.18}{1} \text{ or } 4.18{:}1$$

or 4.18 rotations of the sun gear to 1 rotation of the carrier.

Finding the Ratio for Intermediate

The ratio for intermediate is found the same way as for low. In this case, the drive gear is the ring gear. For example, the ratio in intermediate (second speed) for the gear set in figure 3-10 would be:

$$\frac{A + B}{B} \text{ or } \frac{22 + 70}{70} = \frac{1.31}{1} \text{ (approx.) or } 1.31{:}1$$

Finding the Ratio for Overdrive

The ratio for overdrive is found by using the sun gear and the ring gear (gears A and B in figure 3-10) as the drive gear.

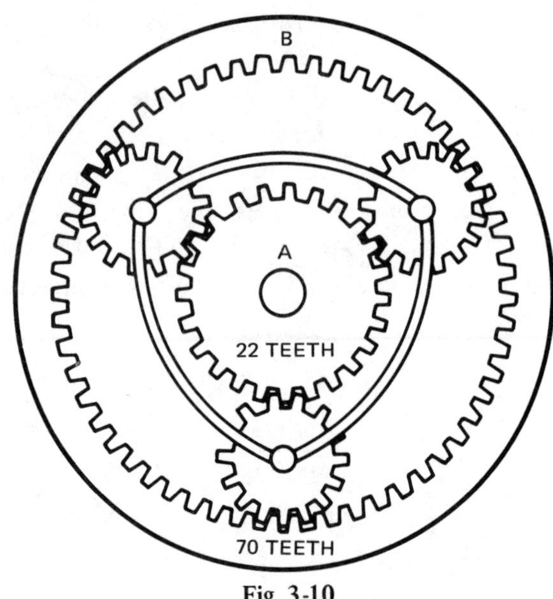

Fig. 3-10

If gear B is the driven gear, the ratio would be:

$$\frac{B}{A + B}$$

If A is the driven gear, the ratio would be:

$$\frac{A}{A + B}$$

Using the gear set in figure 3-10, the ratios for overdrive would be:

$$\frac{70}{22 + 70} = .76{:}1$$

or

$$\frac{22}{22 + 70} = .24{:}1$$

Finding the Ratio for Reverse

The ratio for reverse is found in the same way as for standard gear sets. In the gear set shown in figure 3-10, the ratio for reverse would be:

$$\frac{70}{22} = 3.18{:}1$$

Note: When the gear set is in direct, the unit is locked and turns as one member. Although there actually is no ratio in direct, it is sometimes considered to be 1:1.

SUMMARY

In an automatic transmission, planetary gears are the means of moving or transmitting engine torque to the drive shaft and rear wheels. The operation of the planetary gear set can be summarized by the following points.

- In a planetary gear set, the gears are always in mesh.
- Changes in the gear ratio are made by driving and holding different members of the gear set.
- When the planet carrier is held, the condition is reverse. The speed ratio for reverse is found in the same way as for standard gears.
- Except for in direct, the carrier always rotates slower than either the sun gear or the ring gear.
- To find the speed ratio of the carrier to either the sun or ring gear, divide the total number of teeth on the sun gear and ring gear by the number of teeth on the drive gear.

REVIEW

I. Figure 3-11 represents a planetary gear set. Write the letter of the member in the diagram which represents each of the following parts.

1. carrier
2. planet pinion
3. sun gear
4. ring gear

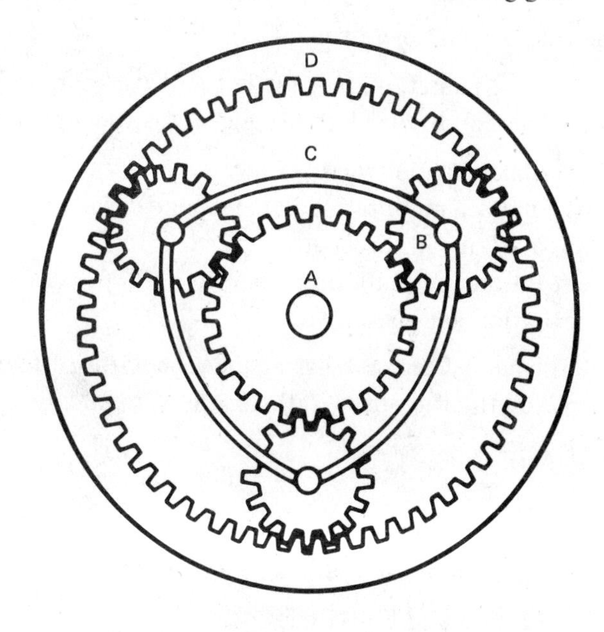

Fig. 3-11

II. Select the best answer from the choices offered to complete the statement or answer the question.

5. The ring gear of a planetary gear set is the input or drive gear, and the sun gear is being held. The condition at the carrier (as to rotation and speed) is
   a. forward rotation and low (first) speed.
   b. forward rotation and intermediate (second) speed.
   c. forward rotation and direct speed.
   d. reverse rotation and reduction.
   e. reverse rotation and overdrive.

6. If a simple planetary gear set has a sun gear with 18 teeth and a ring gear with 52 teeth, the approximate speed ratio of reverse-overdrive would be
   a. 2.89 : 1.
   b. 0.26 : 1.
   c. 0.74 : 1.
   d. 0.35 : 1.

7. If a simple planetary gear set has a sun gear with 17 teeth, and a ring gear with 51 teeth, the approximate speed ratio for reverse (reduction) would be
   a. 4.00 : 1.
   b. 1.33 : 1.
   c. 3.00 : 1.
   d. 0.75 : 1.

8. The sun gear and ring gear of a planetary gear set are both acting as drive gears. The condition at the carrier (as to rotation and speed) is
   a. forward rotation and low (first) speed.
   b. forward rotation and intermediate (second) speed.
   c. forward rotation and direct speed.
   d. reverse rotation and reduction.
   e. reverse rotation and overdrive.

9. The sun gear of a planetary gear set is the drive gear, and the carrier is being held. The condition at the ring gear (as to rotation and speed) is
   a. forward rotation and low (first) speed.
   b. forward rotation and intermediate (second) speed.
   c. forward rotation and direct speed.
   d. reverse rotation and reduction.
   e. reverse rotation and overdrive.

10. If a simple planetary gear set has a sun gear with 20 teeth, and a ring gear with 52 teeth, the speed ratio for low would be approximately
   a. 1.38 : 1.
   b. 3.60 : 1.
   c. 0.72 : 1.
   d. 2.60 : 1.

11. The ring gear of a planetary gear set is the drive gear, and the carrier is being held. The condition at the sun gear (as to rotation and speed) is
   a. forward rotation and low (first) speed.
   b. forward rotation and intermediate (second) speed.

c. forward rotation and direct speed.
d. reverse rotation and reduction.
e. reverse rotation and overdrive.

12. If a simple planetary gear set has a sun gear with 18 teeth and a ring gear with 48 teeth, the speed ratio for intermediate would be approximately

    a. 1.38 : 1.
    b. 2.66 : 1.
    c. 3.66 : 1.
    d. 2.48 : 1.

13. The sun gear of a planetary gear set is acting as the drive gear, and the ring gear is being held. The condition at the carrier (as to rotation and speed) is

    a. forward rotation and low (first) speed.
    b. forward rotation and intermediate (second) speed.
    c. forward rotation and direct speed.
    d. reverse rotation and reduction.
    e. reverse rotation and overdrive.

EXTENDED STUDY PROJECTS

1. Count the number of teeth on the sun gear and ring gear of a simple planetary gear set, and find the ratio for low, intermediate, reverse, and overdrive.
2. Prove that the ratios determined in number 1 are correct by placing chalk marks on the different members and counting the rotations of the drive and driven member for each of the conditions.

FLUID COUPLINGS AND TORQUE CONVERTERS

OBJECTIVES

After studying this unit the student will be able to:

- Explain the operation of a fluid coupling.
- Identify the parts of a torque converter, and describe their functions.
- Explain how fluids can be used to move mechanical parts.
- Describe the use of hydraulics to obtain torque multiplication.

An automotive transmission should have many different speed (or gear) ratios. The ideal transmission would have an unlimited number of ratios that could move the automobile from a standstill to cruising speed at the best ratio for throttle opening and road speed. The standard transmission, as described in unit 2, has three forward speeds and reverse. At best, the standard transmission has four, five or six forward speed ratios and reverse.

The modern automatic transmission has only three forward speeds and reverse, but it has an additional unit that makes it almost an ideal transmission. In the modern automobile, this special unit is called a *torque converter.*

Manual transmissions have a mechanical unit, called a *clutch,* that connects and disconnects the engine from the transmission. With the exception of a few vehicles such as race cars, tractors, trucks, and some recreational vehicles, a fluid coupling or torque converter connects the automatic transmission to the engine. The torque converter, figure 4-1, has the ability to multiply engine torque as well as serve as a connection between the engine and transmission. The torque converter, therefore, acts as a transmission, itself. A torque converter can provide ratios of approximately 2.5 : 1 to 1.11 : 1, depending on engine torque and road speed. A torque converter also makes it possible to stop the car with the transmission in gear.

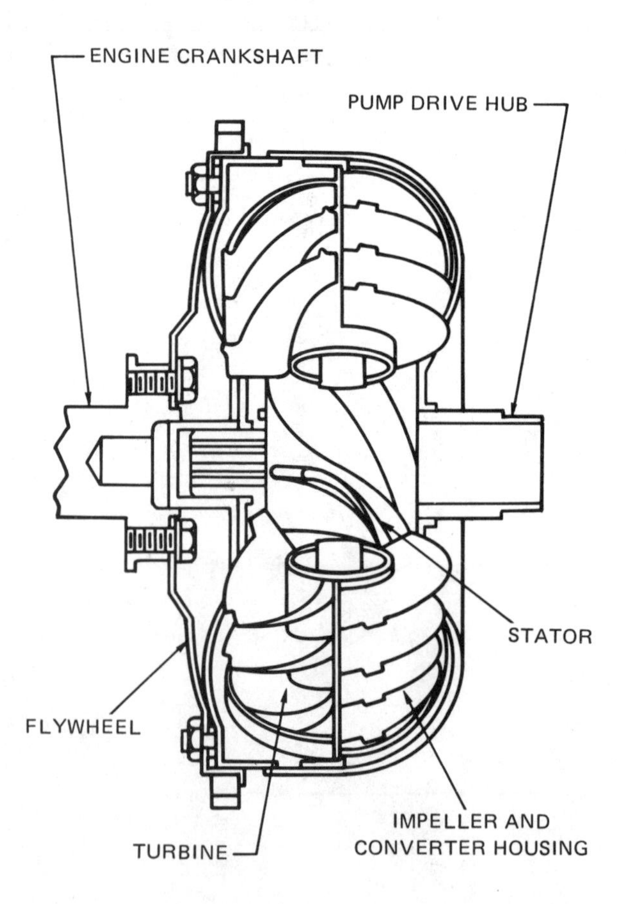

Fig. 4-1 A torque converter for use with an automatic transmission.

Fig. 4-2 The windmill is one type of fluid coupling.

FLUID COUPLINGS

To understand how a torque converter multiplies torque, it is necessary to know something about fluids and fluid couplings. A windmill, figure 4-2, is a simple fluid coupling. In the case of the windmill, the fluid is air. The air pushing against the paddles of the wheel causes it to move. The hub of the windmill can be connected by a shaft to some form of machinery and made to do work, such as pumping water.

A water wheel is also a type of fluid coupling. The fluid in this case is water. The water, pushing on the blades of the wheel, causes the wheel to rotate. This rotating force can be used to do work.

Another type of water wheel is called a *tub wheel* or *drop wheel,* figure 4-3. Water dropping from above pushes against the curved blades and turns the wheel. This type of wheel is often called a turbine. A turbine is used as part of the fluid coupling between the automatic transmission and the engine.

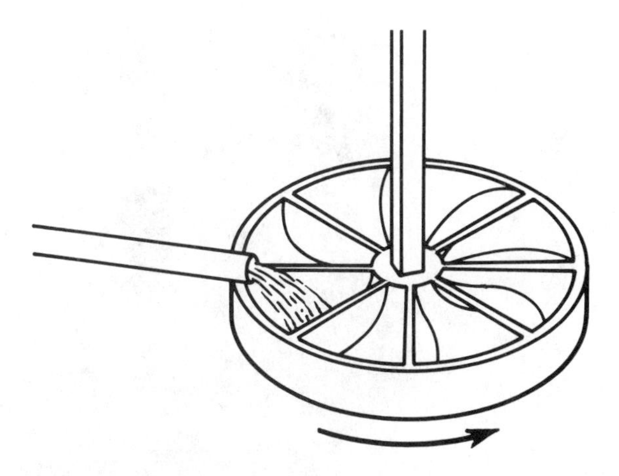

Fig. 4-3 This type of water wheel works in the same way as a turbine.

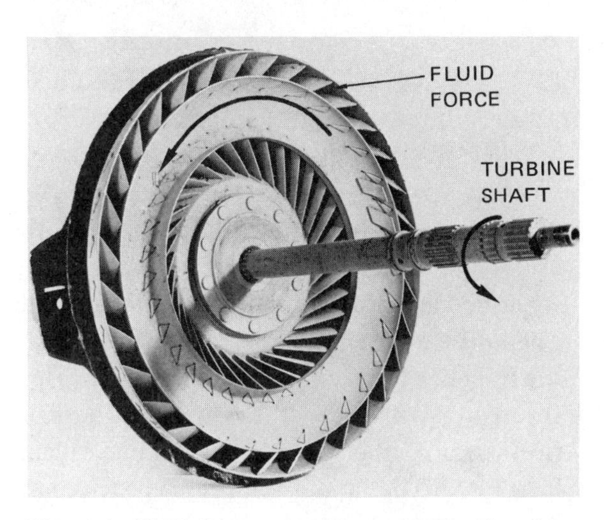

Fig. 4-4 Fluid force turns the turbine and turbine shaft, and this forms a connection between engine and transmission.

The Turbine

The input shaft of an automatic transmission is splined to fit the hub of a turbine encased in the housing of the fluid coupling. For this reason, it is also called a *turbine shaft,* figure 4-4. The turbine is supported by a bearing in the housing, and it is free to turn. The blades of a turbine are known as *vanes.* The force of the transmission fluid pushing on the vanes rotates the turbine and input shaft of the transmission.

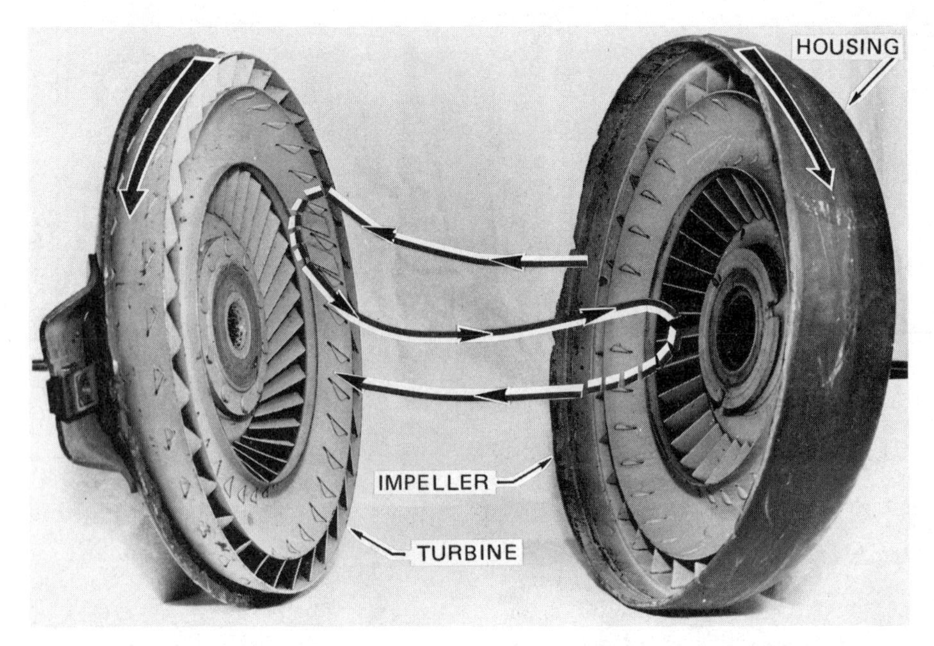

Fig. 4-5 The impeller pumps fluid through the vanes of the turbine, causing it to turn.

The Impeller and Housing

To make the turbine turn, fluid must be forced between the vanes. This is done by the use of an *impeller,* figure 4-5. The impeller has vanes that are attached to the converter housing. The housing turns with the engine, thus causing the impeller to act as a pump and forcing fluid through the vanes in the turbine. The fluid flows around the vanes in the turbine causing it to turn. The fluid is then forced back to the impeller, against the direction of rotation of the impeller and engine. This reduces the amount of work that is performed. This combination of impeller and turbine is a simple fluid coupling. Because of the direction of fluid flow, it cannot multiply torque.

TORQUE CONVERTERS

The Stator

To multiply torque, the direction of the fluid returning from the turbine must be turned so that it pushes against the impeller vanes in a helping direction. In other words, it must push the impeller in the same direction as it is already moving. This is accomplished by adding a third member called a *stator*, figure 4-6.

As shown in figure 4-7, the stator turns, or redirects, the fluid to the impeller. Since the returning fluid is now helping the impeller and engine to turn, torque is increased at the turbine and transmission input shaft. This torque increase can be compared to the torque increase obtained by using a gear train. However, in the case of the torque converter, the amount of torque increase depends on the amount of *slippage* (or speed difference)

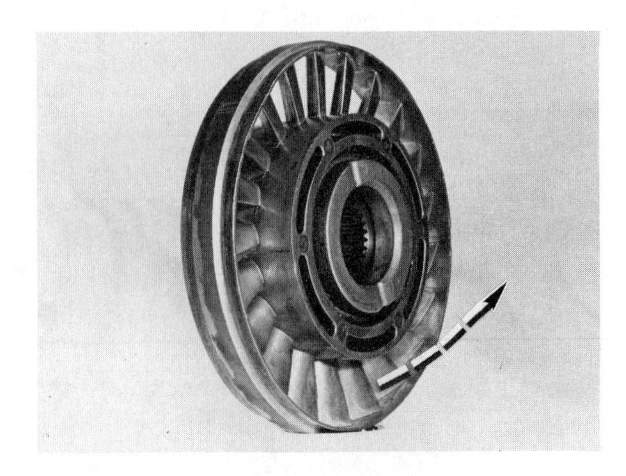

Fig. 4-6 The stator reverses the flow of fluid.

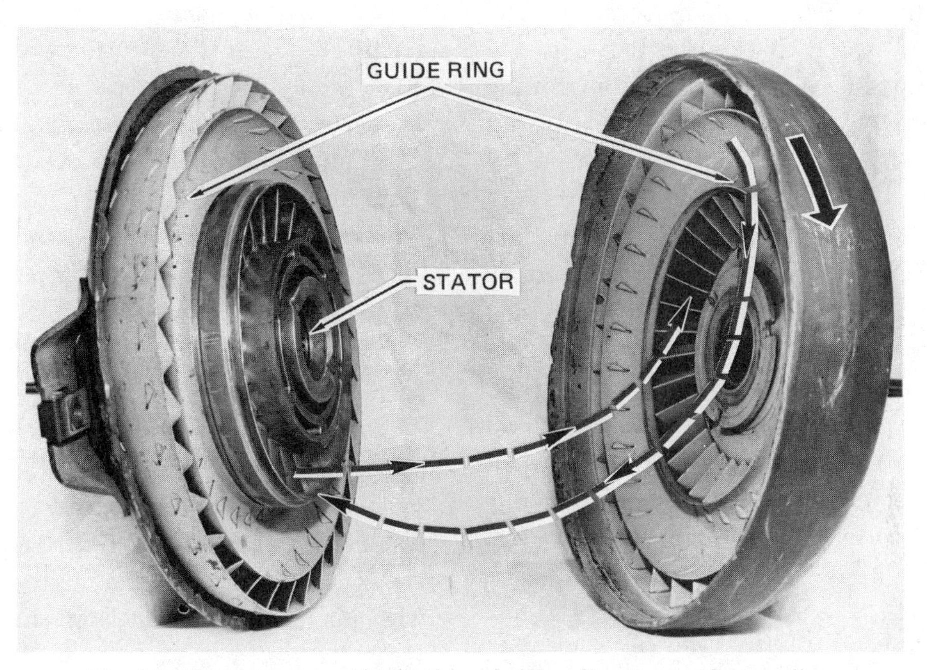

Fig. 4-7 The stator turns the fluid in a helping direction to the impeller.

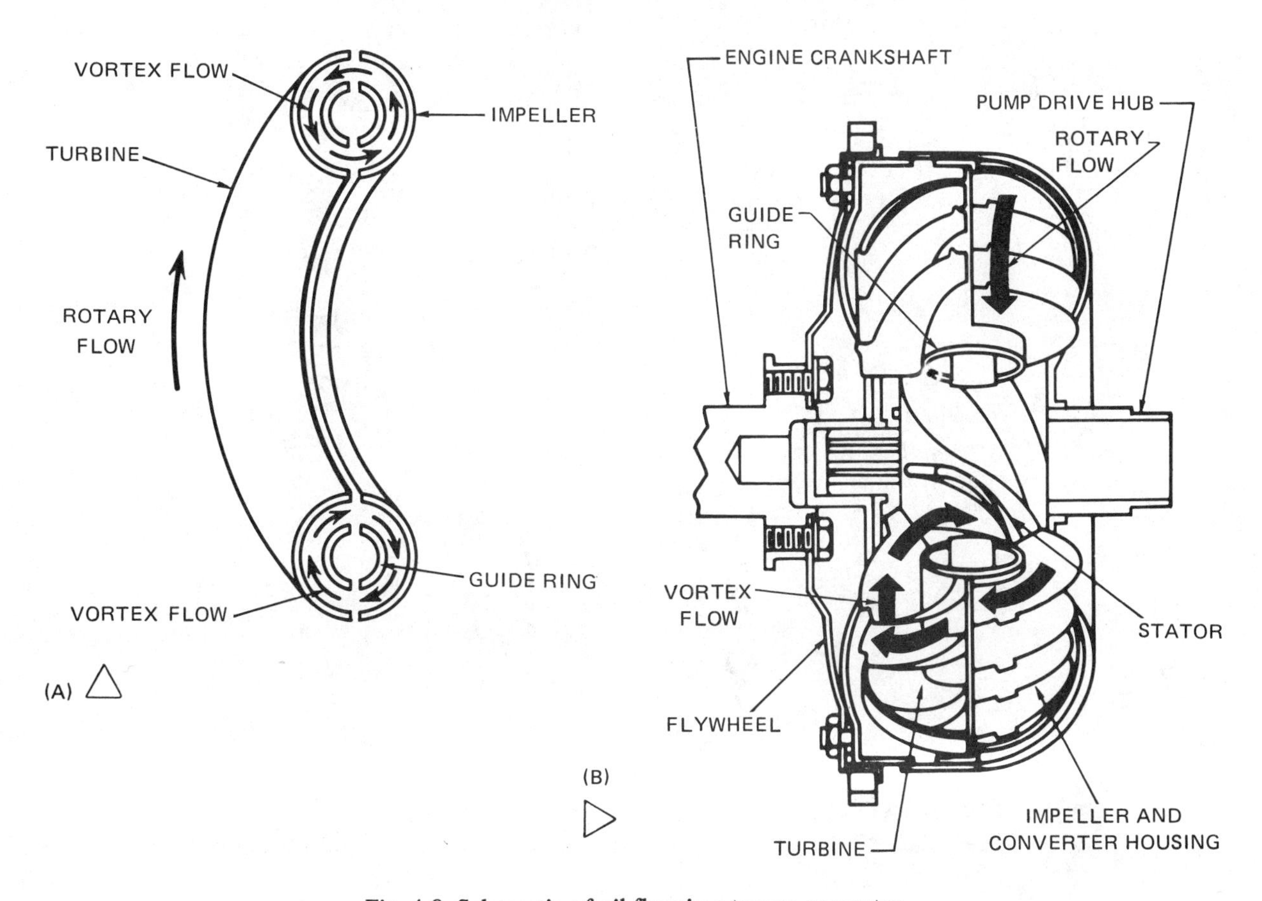

Fig. 4-8 Schematic of oil flow in a torque converter.

between the impeller and the turbine. The greater the speed difference between the turbine and the impeller, the more torque is increased.

Vortex and Rotary Flow

The impeller, stator, and turbine are placed very close together, but do not touch one another. As shown in figure 4-7, the *guide ring* causes the fluid to flow between the vanes of the impeller, turbine, stator, and finally, back to the impeller. This spiral-like flow of the fluid is called *vortex flow.* Fluid also tends to flow around the circumference of the torque converter. This is called *rotary flow,* figure 4-8.

Coupling Stage

Vortex flow is greatest at maximum torque multiplication or acceleration but falls off as cruising speed is reached. At cruising speed there is very little speed difference between the impeller and turbine, and therefore, almost no vortex flow. This is called the *coupling stage.* In this stage, fluid flow is almost entirely rotary, figure 4-9.

Note that there is no mechanical connection between the impeller and the turbine. Also, there is some speed difference or slippage. However, the slippage at cruising speed is very small and results in a ratio of about 1.11 : 1. This means that the power loss to the rear wheels is very small. For example, at cruising speed, the turbine will make nine turns to every ten turns of the impeller.

One-way Clutch

In the coupling stage, the vanes of the stator would tend to slow rotary flow. To avoid this problem, a *one-way roller clutch* is built into the stator hub. This allows the stator to turn with the unit, figure 4-10. The inner race (channel) is smooth, like a track for a bearing, and is splined to fit the stator support at the front of the transmission. The stator support is part of the transmission front pump. However, the support is stationary, and this keeps the inner race from turning.

The outer race is splined to the stator. It is separated from the inner race by roller bearings. In figure 4-10, notice that ramps

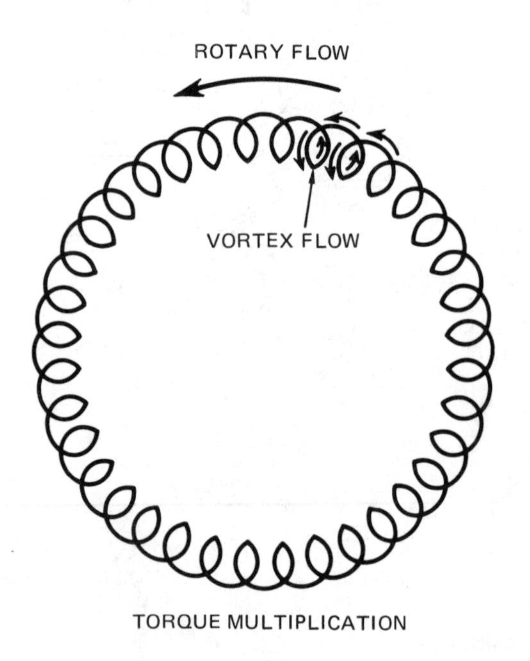

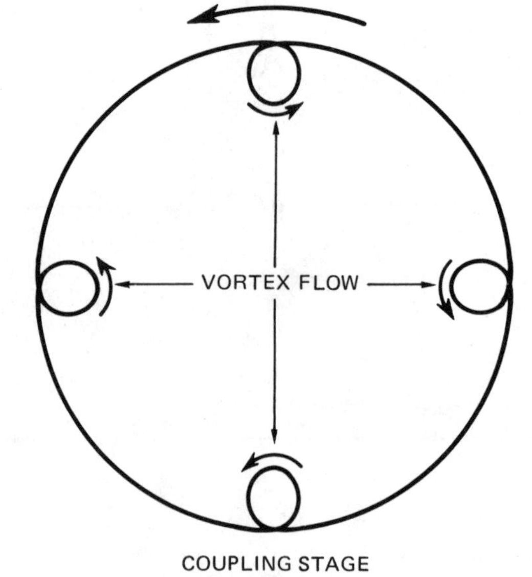

Fig. 4-9

or notches are cut into the outer race. Also, each roller has a spring that tends to push the roller into the small space made by the ramp. These springs are called *energizer springs.* They eliminate any slack or free play between the inner race, rollers, and outer race.

During vortex flow, the returning fluid from the turbine strikes the vanes of the stator with a counterclockwise force. The counterclockwise force tends to wedge the rollers into the small space between the inner and outer race, locking the stator so that it cannot turn in a counterclockwise direction.

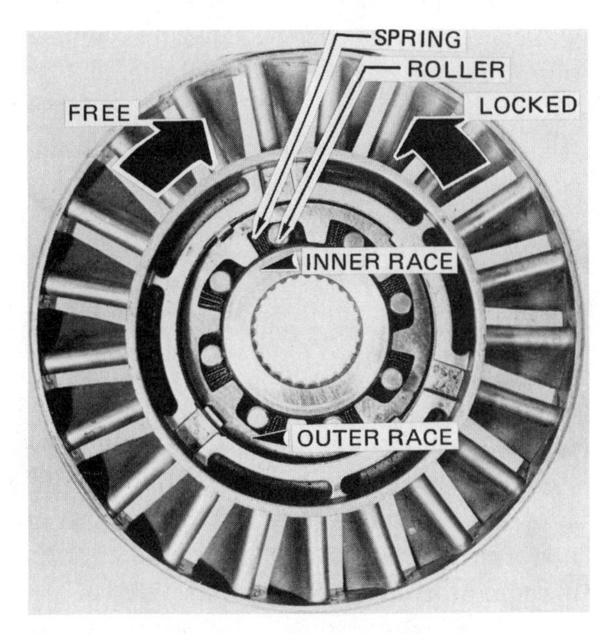

Fig. 4-10 The stator assembly locks up in counterclockwise direction, but runs free in clockwise direction.

In the coupling stage, the fluid flow in the converter is almost entirely rotary. This sets up a clockwise force on the stator, and since the rollers are not wedged in place, the stator is free to turn in a clockwise direction. At this point, the roller clutch is acting as a ball bearing. However, any time the driver of the automobile opens the throttle enough to cause vortex flow, the stator locks up and torque is multiplied.

If the driver releases the throttle at cruising speed (when coming to a stop or going down a steep hill, for instance) the rear wheels will tend to drive the turbine. This makes the turbine act as a pump forcing fluid into the impeller. Since the impeller and housing assembly are connected to the crankshaft, the engine helps to brake the vehicle. At idle speed, the impeller does not pump enough fluid to turn the turbine. This allows the torque converter to act as a clutch which disconnects the engine from the transmission.

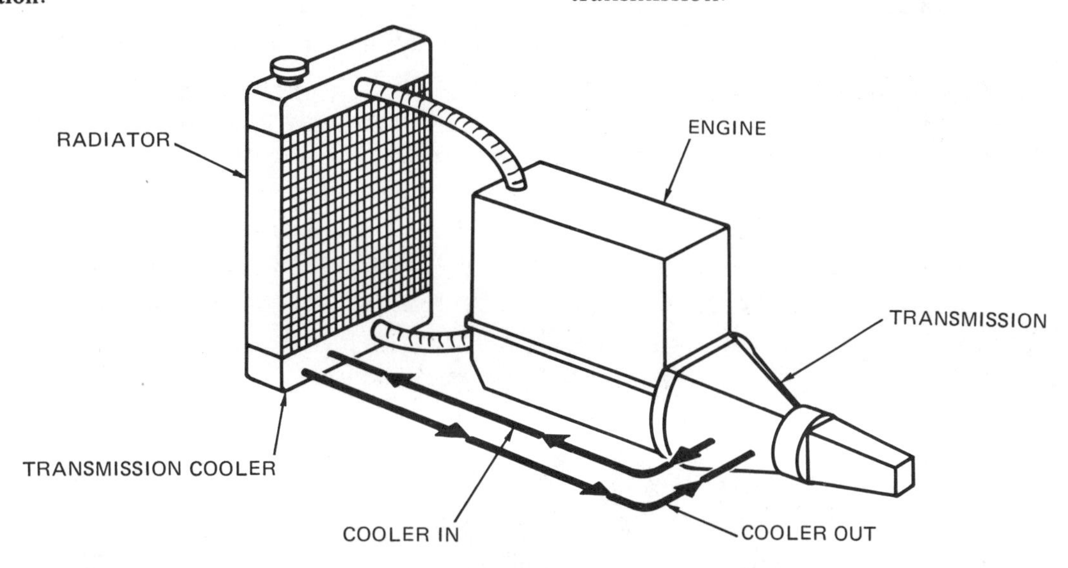

Fig. 4-11 Operation of the transmission cooler.

Cooling

The impeller, stator, and turbine are spaced very close together, but there is no metal-to-metal contact. However, the speed difference between the impeller and the turbine tends to pull the oil molecules apart, causing friction. This type of friction is called *viscous friction*, and, like any form of friction, it causes heat. To prevent harmful heat buildup, most modern automatic transmissions use a cooler, figure 4-11.

The cooler is usually a small tank or tubing inside the radiator of the car. In this location the engine coolant comes into contact with the cooler tank. Since the transmission fluid is hotter than the engine coolant, the fluid gives up its heat to the coolant in the radiator.

Stall Speed

For best performance from a vehicle equipped with an automatic transmission, the torque converter should allow the engine to reach the speed of maximum torque. At a standstill, with the brakes locked, the transmission in one of the drive ranges, and the throttle wide open, engine r/min for most passenger cars should be between 1,300 and 2,900 r/min, depending on engine size and converter design. The converter limits the engine to a certain r/min, known as *converter stall speed.*

Stall speed varies with different engine-transmission combinations. For example, Chrysler Corporation lists a stall speed of 1,400 to 1,700 r/min for the six-cylinder, 225-cubic inch engine. The high performance 440-cubic inch engine/transmission combination, however, has a stall speed of 2,600 to 2,900 r/min.

Transmissions equipped with high stall converters need extra cooling to prevent harmful heat buildup. This subject is covered in more detail in a later unit.

Caution: Stall testing can be dangerous if the automobile brakes are not set properly. Also, some manufacturers do not recommend stall testing because it may damage the transmission. For these reasons, stall testing should last only a few seconds and be done under the direct supervision of an instructor. Stall testing for the purpose of troubleshooting is discussed in later units. The proper steps and precautions must be known before doing a stall test.

SUMMARY

A torque converter is a type of fluid coupling capable of multiplying torque. Thus, a torque converter can be thought of as a transmission in itself, having forward speeds, but no reverse. In the modern automobile the torque converter provides many different ratios for the transmission of power to the rear wheels.

The greatest torque multiplication of a torque converter takes place at full stall speed. For example, one type of automatic transmission/engine combination has a low ratio of 2.46 : 1. At full stall, the ratio of the engine speed to the drive shaft speed is equal to:

$$\text{converter ratio} \times \text{transmission ratio}$$

If the stall ratio of the converter is 2.2 : 1, then the overall ratio is:

$$2.46 \times 2.2 = 5.41 : 1$$

At the coupling stage the ratio is:

$$2.46 \times 1.11 = 2.73 : 1$$

In addition, any ratio between stall and coupling is multiplied by the transmission ratio.

The following points concerning torque converters are important in the further study of automatic transmissions:

- The torque converter acts as a clutch to connect and disconnect the engine from the transmission.
- Vortex flow is the flow of fluid in a circular path through the vanes of the impeller, turbine, and stator.
- Vortex flow is greatest at a maximum torque multiplication.
- Rotary flow is the flow of the fluid around the circumference of the converter.
- Rotary flow is greatest at the coupling stage.
- When the throttle is closed at cruising speed, the turbine pumps oil to the impeller. This allows the engine to help in braking the vehicle.

REVIEW

Select the best answer from the choices offered to complete the statement or answer the question.

1. In figure 4-12, part A is called the

   a. impeller. c. turbine.
   b. stator. d. guide ring.

2. Part B in figure 4-12 is the

   a. impeller. c. turbine.
   b. stator. d. guide ring.

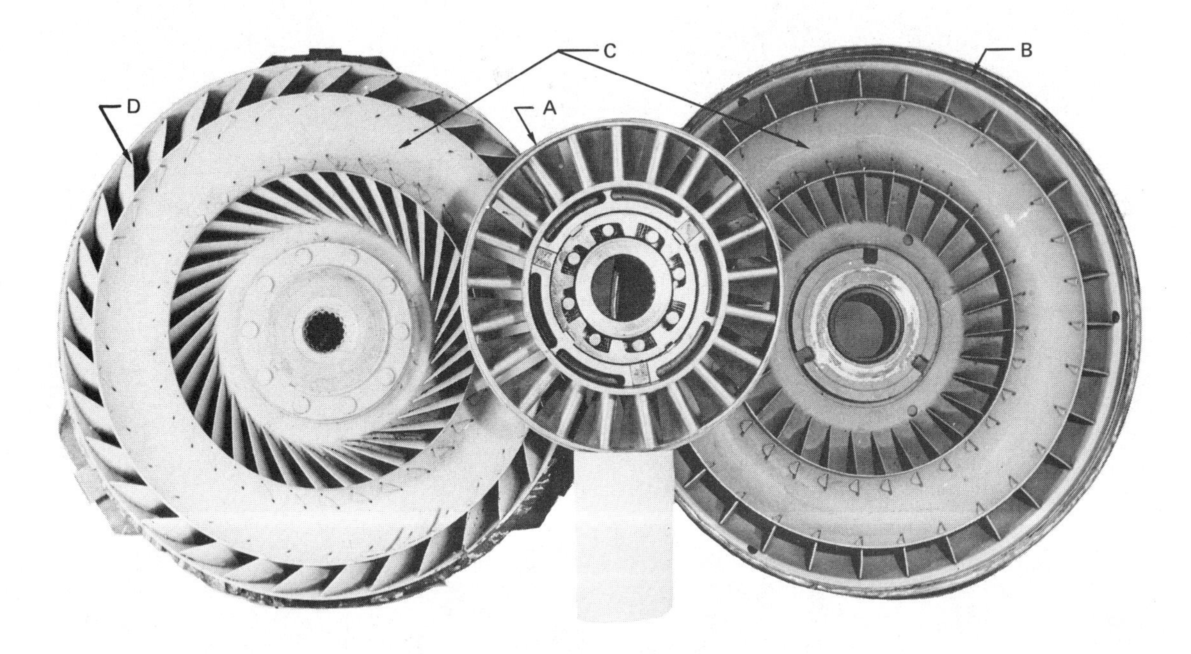

Fig. 4-12

3. Part C is called the

   a. impeller.
   b. stator.
   c. turbine.
   d. guide ring.

4. Part D is called the

   a. impeller.
   b. stator.
   c. turbine.
   d. guide ring.

5. The stator in a torque converter:

   (I) Changes vortex flow to rotary flow.
   (II) Is locked in all speed ranges.

   a. I only
   b. II only
   c. both I and II
   d. neither I nor II

6. In a torque converter, the most torque multiplication is achieved at

   a. the time of greatest vortex flow.
   b. idle.
   c. cruising speed.
   d. the time of greatest rotary flow.

7. The driven member of a torque converter is called the

   a. stator.
   b. impeller.
   c. turbine.
   d. vane.

8. The driving member of a torque converter is called the

   a. stator.
   b. impeller.
   c. turbine.
   d. vane.

9. Slippage in a torque converter occurs at:

   (I) Cruising speed.
   (II) Stall speed.

   a. I only
   b. II only
   c. both I and II
   d. neither I nor II

10. The greatest torque multiplication in a torque converter occurs at

    a. cruising speed.
    b. stall speed.
    c. half throttle.
    d. idle speed.

11. To multiply torque, the fluid in a fluid coupling must be turned so that it pushes the pumping member in a helping direction. The member that does this is called the

    a. stator.
    b. impeller.
    c. turbine.
    d. guide ring.

EXTENDED STUDY PROJECTS

1. Place two electric fans facing each other on a work bench or table, about four feet apart. Mark one blade of one fan with chalk. Do not plug this fan into an electrical outlet. Plug in the second fan and turn it on at high speed. Observe the effect on the unconnected fan. Move the fans closer together a little at a time, until the guards touch. Observe the effect that this has on the fan that is not plugged in.

   Caution: Do not touch the moving blades with your fingers or other objects.

2. Write the answers to the following questions.

   a. What effect does the running fan have on the other fan when they were located four feet apart?
   b. As the fans are moved closer together, what happens?
   c. How could this experiment be compared to (1) a fluid coupling, and (2) a torque converter?
   d. Compare the fan that is running to a member in a fluid coupling.
   e. Compare the fan that is not plugged in to a member in a fluid coupling.
   f. Would the two fans be able to multiply torque? Why or why not?

BASIC HYDRAULICS AND PRESSURE REGULATION

OBJECTIVES

After studying this unit, the student will be able to:

- Define force, pressure, and area as they apply to hydraulic systems.
- Discuss the relationship between pressure, motion, and mechanical advantage in hydraulic systems.
- Explain the function of fluid pumps.
- Identify bypass valves and restriction valves and explain their function in pressure regulation.
- Identify valves having differential force areas and explain the effect of differential force on valve movement.
- Describe how clutch and band application are affected by orificing.
- Explain the use of clutch and band application to drive or hold planetary gear members.

Hydraulics is a branch of physics that deals with fluids under motion and pressure. A fluid may be a liquid or a gas. Gases are fairly easy to compress, but it would take many tons of force to compress a liquid even a small amount. Therefore, liquids are usually considered incompressible. This property of liquids is put to use in the hydraulic brake system of a car. In the brake system, force from the driver's foot is transferred to the brake shoes through the use of brake fluid. Hydraulic principles are used in much the same way to control the operation of automatic transmissions.

CLOSED HYDRAULIC SYSTEMS

One principle of hydraulics states that pressure on a confined liquid is transmitted equally in all directions and acts with equal force on equal areas. Thus, in a *closed hydraulic system* such as that shown in figure 5-1, all of the pressure gauges read the same. This is because under normal conditions liquids are incompressible, and the pressure is the same in all parts of the system. This is true of any closed system, regardless of the shape of the container or the length of the connecting pipes. If the faucet in figure 5-1 were opened and the fluid could flow, the pressure would drop.

It is often necessary to know the pressure in a certain system. A specific formula is used for this purpose:

$$\text{Pressure} = \frac{\text{Force}}{\text{Area}}$$

For example, in figure 5-1, a force of 100 pounds is pushing on a piston with 1 square inch of surface area. This results in a pressure of 100 pounds per square inch (psi) in the container and pipes.

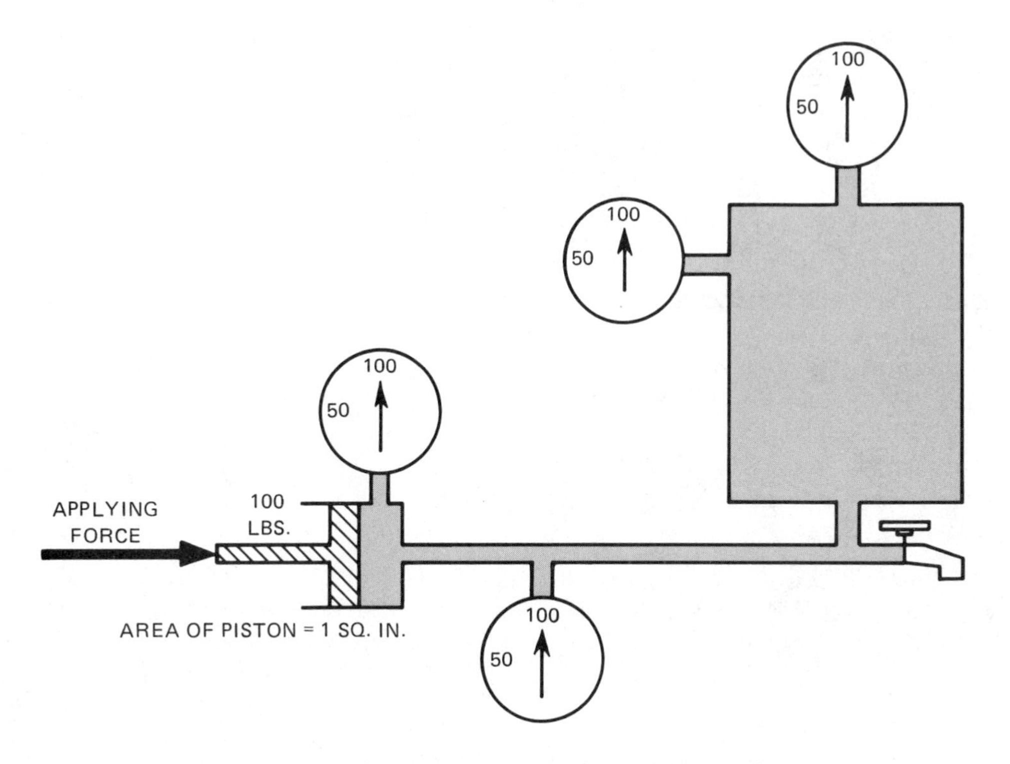

Fig. 5-1 Pressure is the same in all parts of a closed hydraulic system.

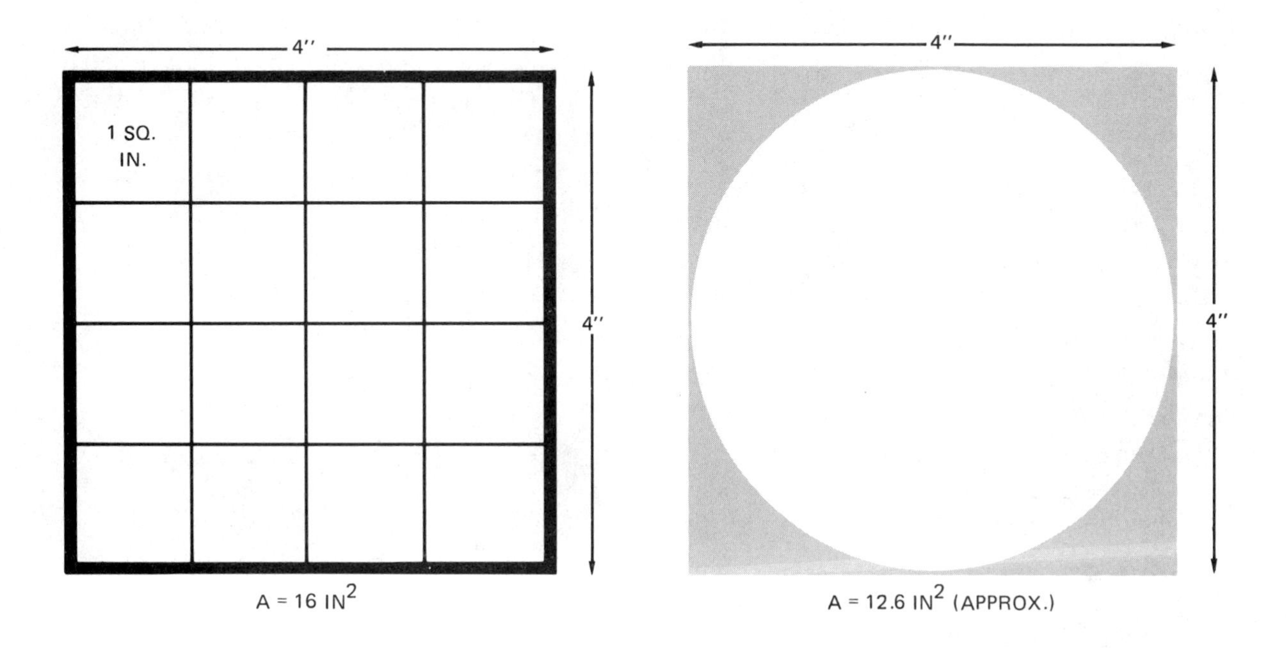

Fig. 5-2 Area of a square and a circle.

Force

Force is any effort that causes or tends to cause motion. In the formula, *F* stands for force in pounds. For example, if a person lifts a 10-pound weight, a force of 10-pounds is necessary to move it. However, force does not always produce motion. If a person tries to push a car and cannot move it, a force is still used in trying to move it.

Surface Area

Area is the measurement of any plane surface having bounds. In the formula for pressure, *A* stands for surface area, measured in square inches. To understand the idea of area, study figure 5-2. If the 4-inch square shown in the figure is divided into 1-inch squares, there are 16 squares in all, or an area of 16 square inches. A circle that fits into the 4-inch square would have an area of about 12.6 square inches. Notice that the circle has less area because parts of the square are not inside the circle. However, the area is still measured in square inches. This is true for any plane (two dimensional) surface such as a diamond, triangle, or rectangle.

Pressure

Pressure is defined as force divided by area or force per unit area. Keep in mind that pressure is evenly distributed on the entire surface area to which it is applied. This supports what has been said of closed hydraulic systems. That is, in a closed hydraulic system, the pressure is the same in the entire system regardless of the size and shape of the container.

In the formula, *P* stands for pressure in pounds per square inch. For example, in figure 5-3 a force of 160 pounds exerted by a fluid is evenly distributed on a surface area of 16 square inches. The pressure is found by using the formula, P = F/A. Thus, 160 pounds divided by the area, 16 square inches equals 10 pounds per square inch (psi).

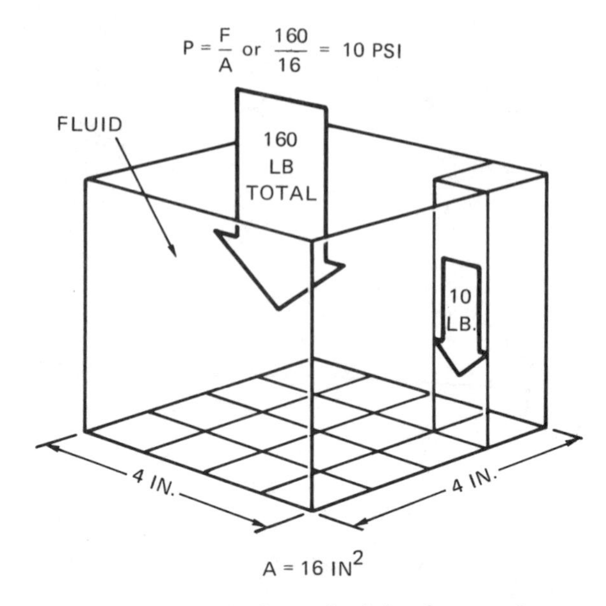

Fig. 5-3 Pressure is determined by force and area.

CALCULATING PRESSURE IN HYDRAULIC SYSTEMS

Using the formula for pressure, the pressure in any hydraulic system can be calculated. For example, the hydraulic system illustrated in figure 5-4 shows a force of 50 pounds pushing on a piston with a surface area of 2 square inches. Substituting these numbers in the formula P = F/A, gives the following equation:

$$P = \frac{50}{2} \qquad P = 25 \text{ psi}$$

Fig. 5-4

Thus, the pressure in this system is 25 pounds per square inch.

An interesting thing happens if the piston area is changed and the force remains the same. This is illustrated in figure 5-5. Notice that the pressure is not as great when a larger piston (5 square inches) is used with the same amount of force. This is because the force is spread over a larger area.

Figure 5-5 also shows that if the piston size is reduced, the pressure increases. This is because the force is not spread over as great an area; the push is harder against a smaller area. Of course, if more or less force is applied, the pressure will increase or decrease. That is, for a given piston area, greater force results in greater pressure, and less force results in less pressure.

CALCULATING FORCE IN HYDRAULIC SYSTEMS

It has been shown that if the force and area of a system are known, the pressure can be calculated. The force can also be calculated, if the pressure and area are known. To do this, the formula for pressure is used in a different form: F = P X A. The letters in the formula have the same meaning as before. However, now the formula states that force equals pressure multiplied by area.

For example, study the system shown in figure 5-6. The force on an input piston having an area of 1 square inch is 100 pounds. Thus, the pressure at the input end is 100 psi (100 pounds/1 square inch). The force on the output piston can be found by using the formula, F = P X A. In this case, the output piston also has an area of 1 square inch, and the pressure being exerted by the fluid is 100 psi. These numbers are substituted in the formula as follows:

F = 100 psi X 1 square inch

F = 100 pounds per square inch

The force at the output piston is 100 psi.

TRANSMITTING MOTION

In a system such as that shown in figure 5-6, any movement of the input piston will

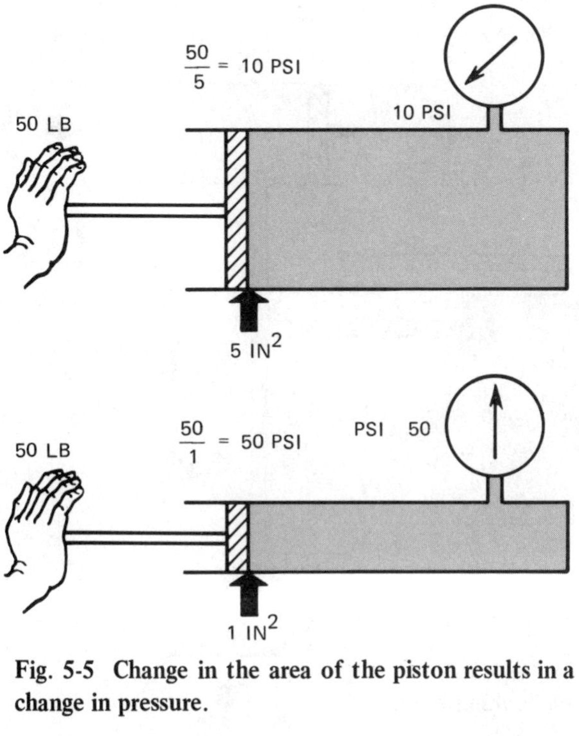

Fig. 5-5 Change in the area of the piston results in a change in pressure.

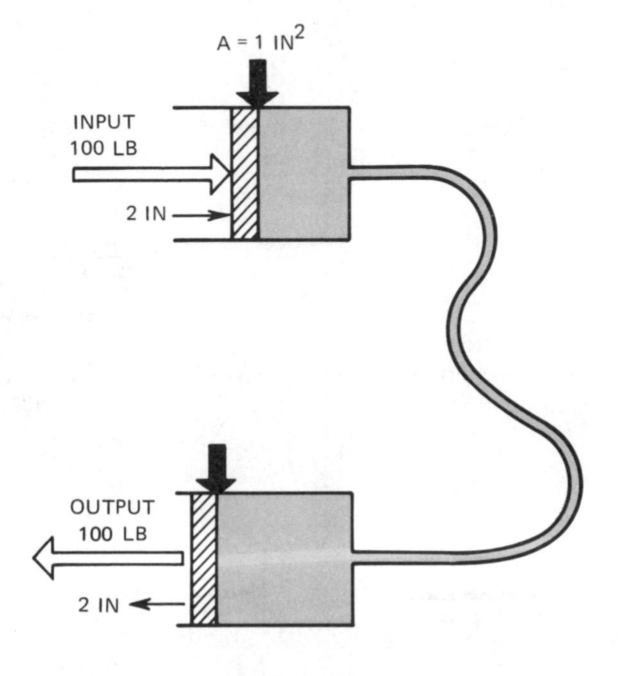

Fig. 5-6

cause the output piston to move the same amount. Remember, fluids cannot be compressed. Therefore, when the input piston in figure 5-6 is pushed in 2 inches, the output piston is pushed out 2 inches. When this occurs, it is said that motion is transmitted.

Note that the input and output piston do not have to be part of the same cylinder to transmit motion. In figure 5-6, two cylinders of the same size are connected by a length of tubing. The motion is transferred through the tubing, regardless of its length or the number of bends it may have.

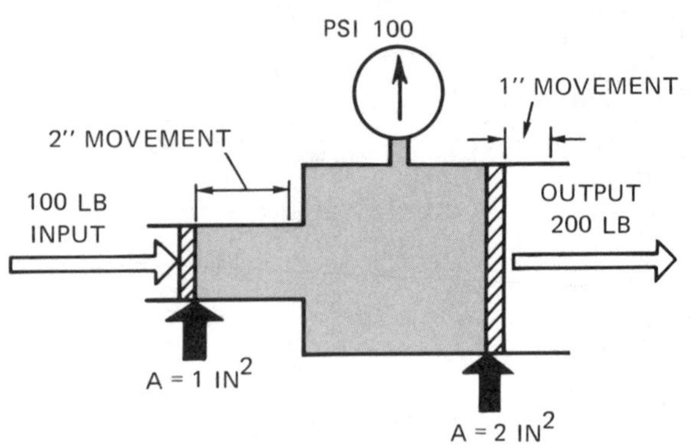

Fig. 5-7 A hydraulic system used to provide mechanical advantage.

MECHANICAL ADVANTAGE OF HYDRAULIC SYSTEMS

The fluid in a hydraulic system can provide a mechanical advantage in much the same way as a lever does. Changing the area of either the input piston or the output piston changes the output force. In figure 5-6, an input force of 100 pounds and piston area of 1 square inch still result in 100 psi in the system. However, when an output piston of 2 square inches is used with an input piston of 1 square inch, figure 5-7, the output force that results is 200 pounds (100 psi X 2 square inches = 200 pounds of force). Thus, the force at the output piston is twice as great as that at the input piston, giving a mechanical advantage. In this case, the mechanical advantage is 2:1. To obtain a mechanical advantage, the output piston must be larger than the input piston.

The use of a smaller input piston and a larger output piston can be compared to the braking system of a car. In the braking system, the wheel cylinders are larger than

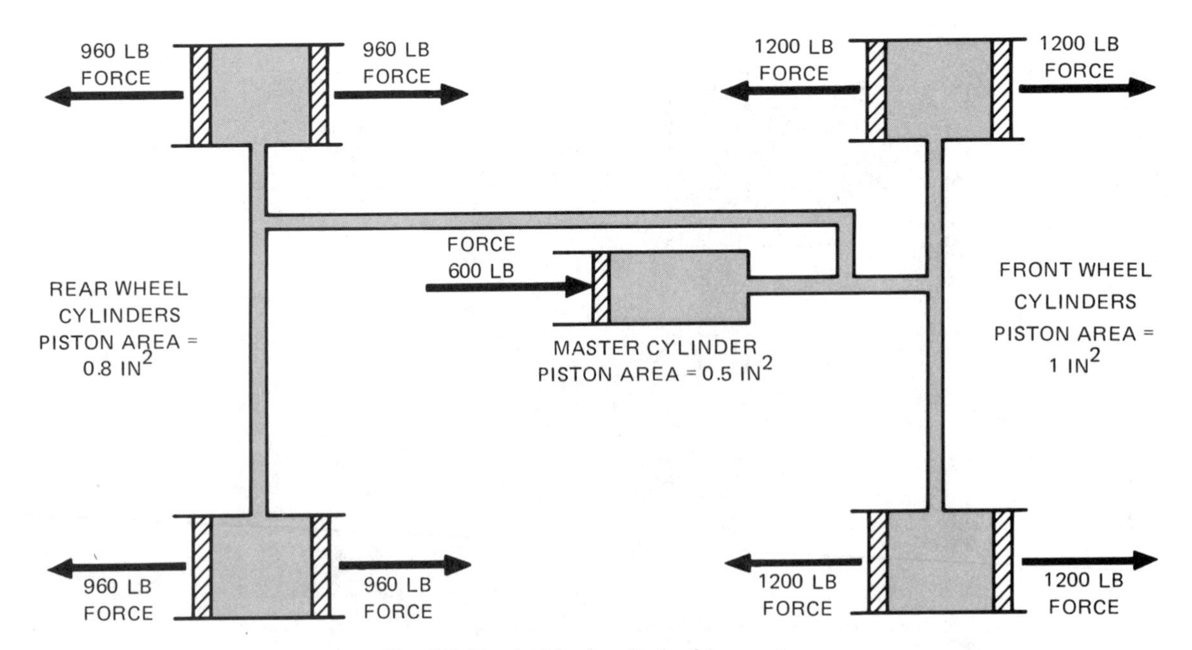

Fig. 5-8 Typical hydraulic braking system.

the master cylinder. In addition, the front wheel cylinders are larger than the rear wheel cylinders, because more braking force is needed on the front wheels. A typical braking system is shown in figure 5-8. Notice that the length of the tubing and the number of bends in it do not affect the mechanical advantage.

Recall from unit 1 that mechanical advantage results in a loss of motion. In a hydraulic system, when the output force is increased, the travel of the output piston is reduced. This motion loss is in direct proportion to the difference in size between the input and output piston. For example, in figure 5-7, if the input piston moves 2 inches the output piston will move only 1 inch.

THE HYDRAULIC SYSTEM IN THE AUTOMATIC TRANSMISSION

In the automatic transmission, the hydraulic system controls the clutches and bands which make it possible to change the gear ratio of the transmission. Several special parts have been developed which help the system to carry out this function.

Fluid Pump

In automatic transmissions, a pump is used to pressurize the fluid in the hydraulic system. The pump

- eliminates the loss of motion that results from mechanical advantage, and
- assures that the system is supplied with enough fluid to operate.

Modern transmissions use a gear-type pump such as that shown in figure 5-9. The pump consists of a drive gear, an outer driven gear, and a half-moon or crescent to separate the two gears, figure 5-10. The drive gear has lugs that are keyed to the torque converter hub. Whenever the engine runs, the pump gears turn. This pulls fluid from the sump which is like a reservoir or holding tank. As the gears turn, the fluid is moved past the crescent area of the pump to where the gear teeth

Fig. 5-9 Automatic pumps; the pump on the right has been disassembled to show inner and outer gears and crescent.

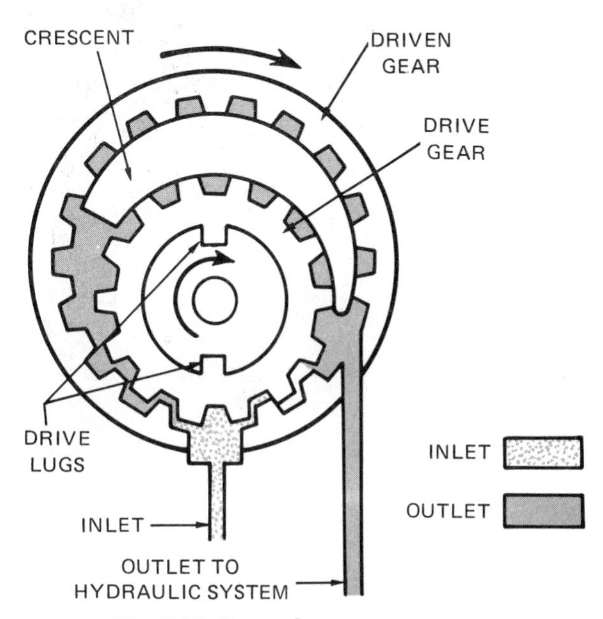

Fig. 5-10 Parts of a gear-type pump.

begin to come together. The fluid is forced between the gear teeth and into the hydraulic system. Since the transmission hydraulic system is a closed system (except for very small leaks at the bushings in the lubrication systems and at certain parts that use cast iron seal rings) the pressure increases as the system is filled. The use of a pump does not change the relationship between force and pressure. As shown in figure 5-11, a pump supplying 100 psi provides 200 pounds of output force on a piston area of 2 square inches.

Regulating Pump Pressure. In the system shown in figure 5-11, the pump is connected directly to the hydraulic cylinder. In an automatic transmission, however, the pump pressure must be regulated to avoid possible damage to the transmission. Pressure regulation is also necessary because the pressure needs of the automatic transmission change with driving conditions and speed range. For example, more pressure is needed when climbing a steep hill than when driving on a level road.

The pump must be able to supply the transmission's hydraulic system with enough fluid, and at high enough pressure, to meet the greatest torque needs. At low speeds and low torque, unregulated pressure would cause harsh, jerky shifting and could cause damage to the transmission. To avoid this problem, the pressure is reduced or regulated at the pump outlet before the fluid enters the hydraulic system.

Keep in mind that pressure may vary if the fluid in the system is in motion. This concept is illustrated by the plumbing system in a home, figure 5-12. As shown by gauge B, the pressure drops when the faucet is opened. The pressure can be regulated by changing the amount the faucet is opened or closed. Automatic transmission hydraulic systems make use of this principle to regulate pressure. In an automatic transmission, pressure is regulated automatically to handle different driving conditions.

Spool Valves

Spool valves, figure 5-13, are used to regulate pressure and control fluid flow in automatic transmissions. The lands of the spool valve and the valve body fit each other exactly, figure 5-14. Passages allow the fluid to enter and leave the valleys. Since the fluid is under pressure, it exerts a force on the valve faces. Spool valves are also used to send fluid to the servos, clutches, and other controls.

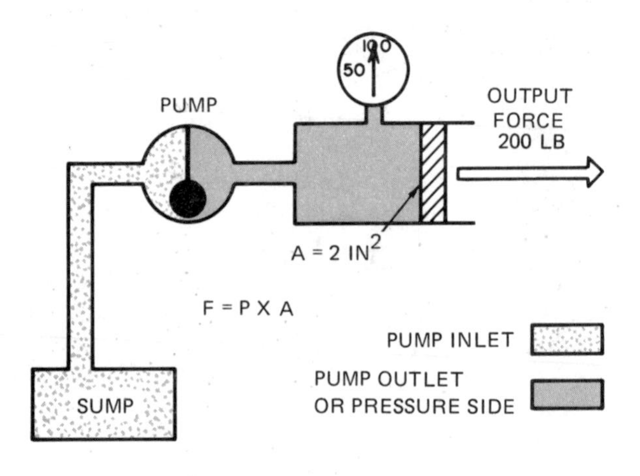

Fig. 5-11

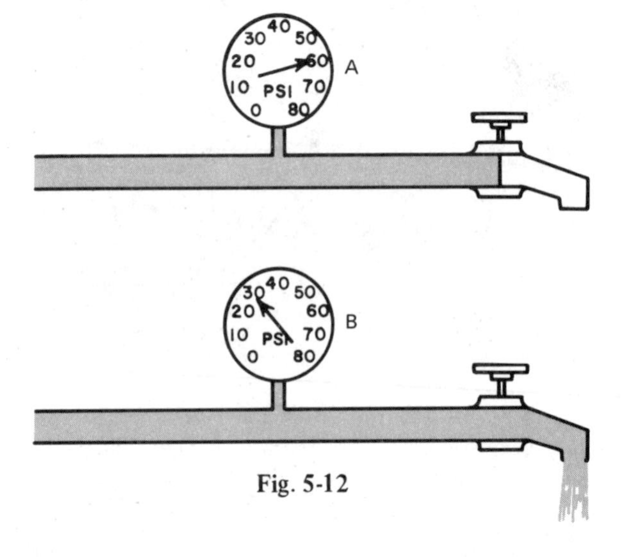

Fig. 5-12

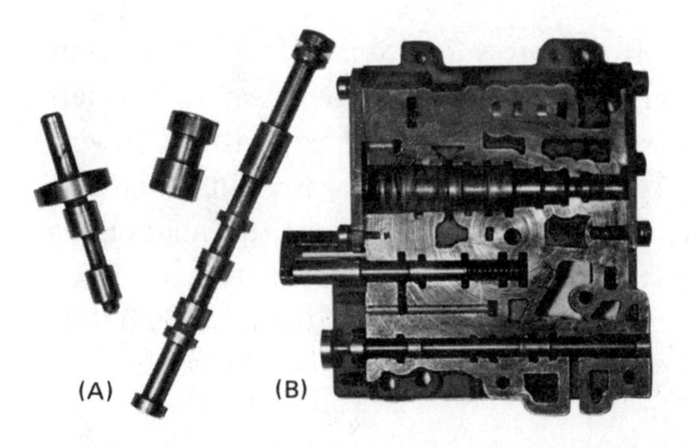

Fig. 5-13 (A) spool valves, (B) A valve body that has been cut down to show how the valves fit in the bores.

Simple Bypass Regulator

The spool valve in figure 5-15 has only one face and one land. It is being used as a simple bypass regulator. The pressure from the pump exerts a force on the face of the valve. This causes the valve to move to the right against the spring force, figure 5-16. As the valve moves to the right, an exhaust port is uncovered by the land of the valve. The fluid returns through this port to the sump or inlet side of the pump, and by-passes the system.

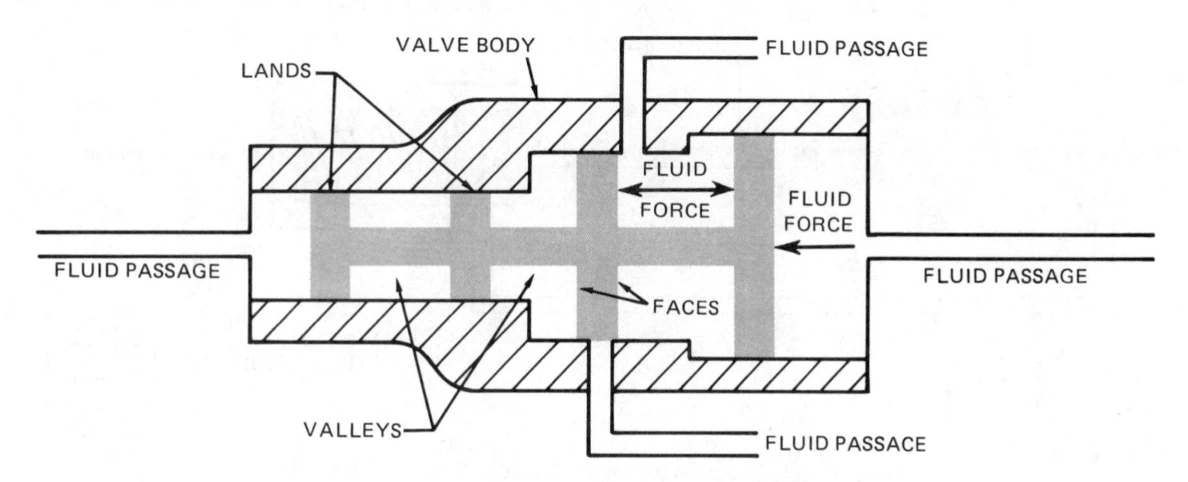

Fig. 5-14 Parts of a spool valve.

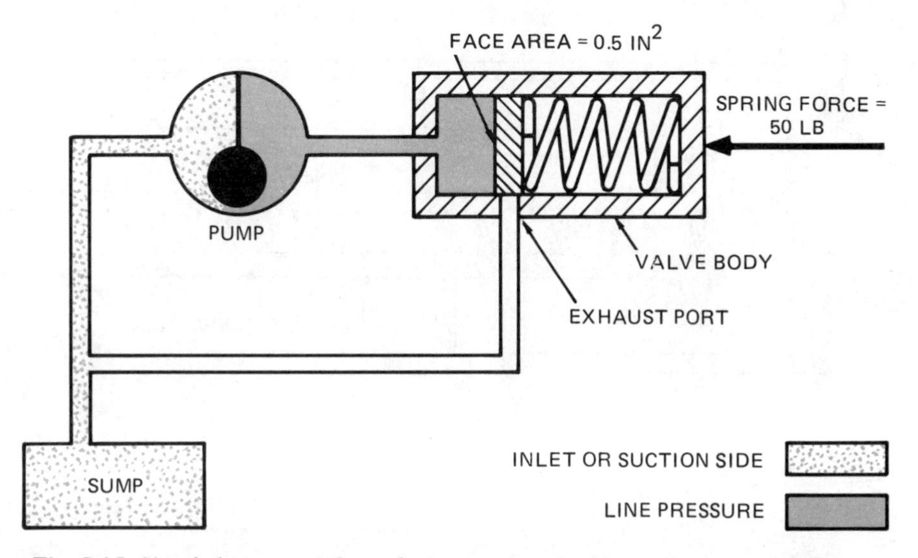

Fig. 5-15 Simple bypass regulator during pressure buildup, valve not regulating.

Since the sump and inlet side of the system are not under pressure, the fluid has nothing more to push against, and the pressure is reduced or regulated. The formula P = F/A applies to this system. The spring force of 50 pounds divided by the valve face area of 0.5 square inches is equal to a pressure of 100 psi.

If the pressure drops below 100 psi, the spring force will push the vaive to the left and close off the exhaust port, thus allowing the pressure to build up again. In actual use, the valve moves just enough to maintain a pressure of 100 psi in the system at all times.

Regulated pressure (usually called *line* or *control pressure*) will be the same any-

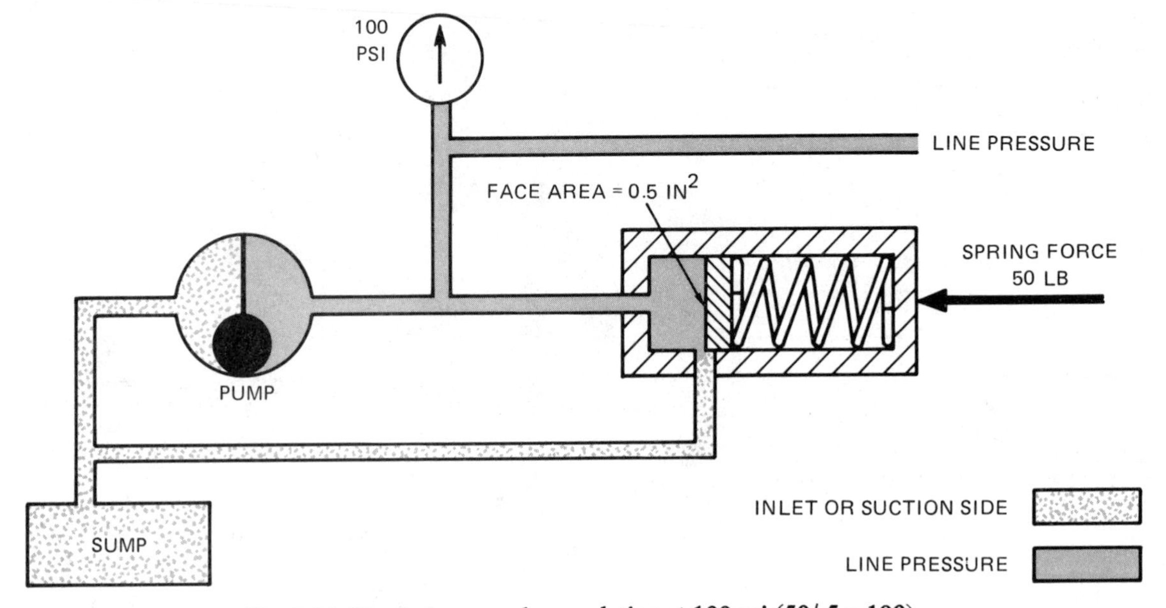

Fig. 5-16 Simple bypass valve regulating at 100 psi (50/.5 = 100).

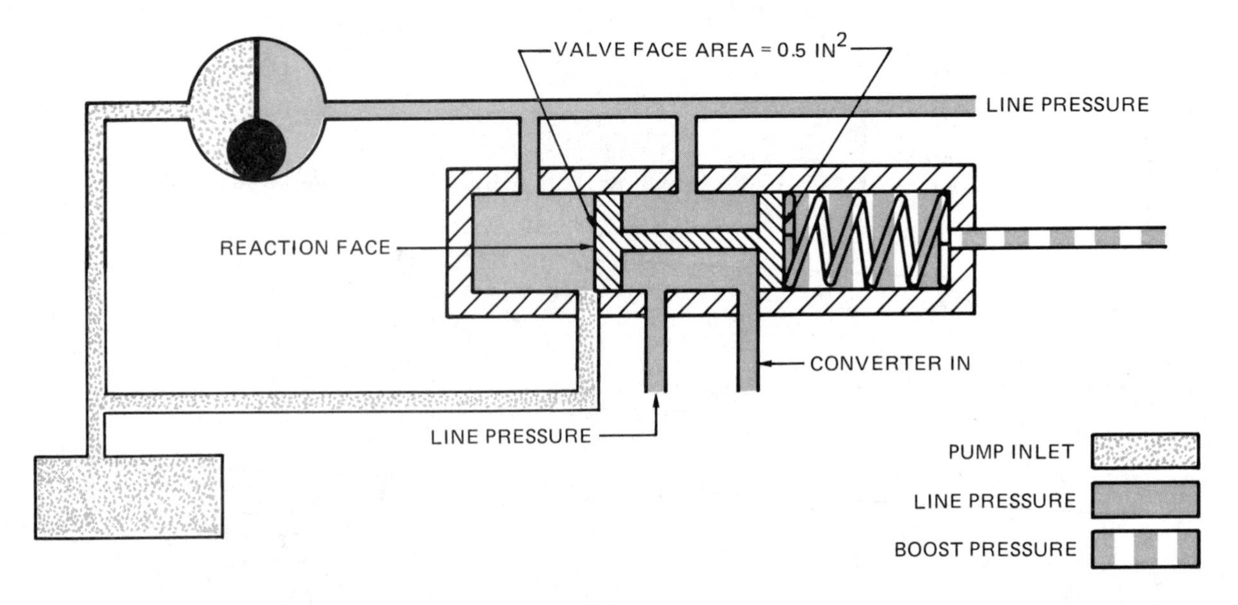

Fig. 5-17

where in the main line. The pressure remains even in the entire main line system except for slight surges which occur when a servo or other hydraulic control is filled or exhausted.

Balanced Valve

The type of regulator valve used in automatic transmissions is the balanced valve, figure 5-17. As pump pressure builds up, fluid moves to the valley of the valve and the reaction face. Pressure on the reaction face moves the valve to the right to open the exhaust port. In this way, pressure is regulated in the same way as by the simple bypass valve. As the valve moves to the right, the other land of the valve opens a port to feed the torque converter. Note that converter pressure is also line pressure, and the converter port will not be open unless the valve is regulating. Since both valve faces are of the same size, the pressure in the valley of the valve does not affect line pressure. The reaction (line) pressure acts as a balancing force on the reaction face of the valve.

Boost Pressure

Under certain conditions it is necessary to increase line pressure. This can be accomplished by using a force called *boost pressure.* In figure 5-17, two forces act against reaction (line) pressure: a spring force, and a force set up by boost pressure. Reaction pressure must balance both the spring force and the force set up by boost pressure. In this way, line pressure is changed to suit driving conditions. If there is no boost pressure, line pressure is regulated by the spring force. For example, if the spring force is 25 pounds, then line pressure is found as follows:

$$P = \frac{F}{A} = \frac{25}{0.5} = 50 \text{ psi}$$

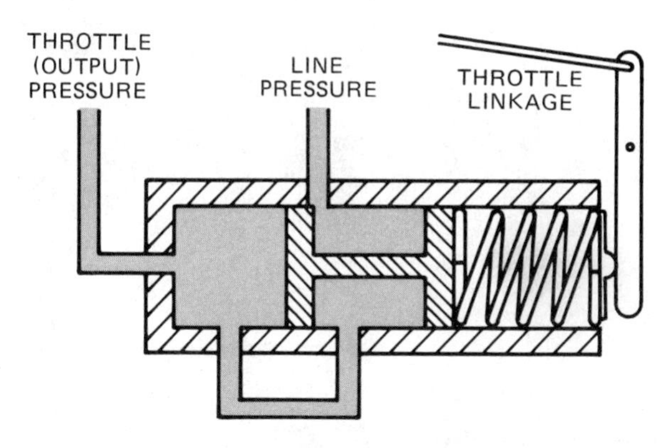

Fig. 5-18 Restriction-type regulator.

If a boost pressure of 25 psi is added, the force on the end of the valve is found as follows:

$25 \times 0.5 = 12.5$, and

$12.5 + 25$ pounds spring force $= 37.5$ total force

Line pressure would then be:

$$\frac{37.5}{0.5} = 75 \text{ psi}$$

If more line pressure is needed, boost pressure can be increased. This, in turn, results in an increase in line pressure. Boost pressure usually comes from a restriction-type regulator.

Restriction-Type Regulators

All of the regulating valves described thus far are of the bypass type. Another type of regulator used in automatic transmissions is the restriction type, figure 5-18. The restriction-type regulator is similar to the balanced bypass regulator in appearance and operation. However, when output or reaction pressure builds up in the restriction-type valve, (causing it to move to the right), the input, or line, pressure is cut off, or *restricted,* by the land of the valve. Since this reduces the flow of input fluid, the output pressure is regulated to the value of the spring force.

Note that mechanical linkage has been added to this valve. The pressure can be increased or decreased by putting more or less spring force on the end of the valve. On some vehicles, this type of valve is connected to the throttle linkage. In this setup, the output pressure is known as *throttle* pressure and is used to control shift speeds and line pressure.

The restriction-type regulator is used as a *secondary pressure regulator.* That is, the input pressure to the valve has already been regulated. It is usually main-line pressure from a bypass-type valve. If a restriction-type regulator were directly connected to the pump, dangerous pressures could build up and cause damage to the transmission.

DIFFERENTIAL FORCE AREAS

The discussion of balanced valves stated that the pressure in the valley of the valve makes no difference because both valve faces in the valley are the same size. However, there are valves in which the faces are different sizes. A spool valve that has a valley with two different face areas is said to have a *differential force area.* A valve with a differential force area is shown in figure 5-19. As pressure builds up in the valley, more force is exerted on the larger face. This causes the valve to move to the left with 100 pounds of force. This can be calculated as follows:

force on larger face = P × A = 100 psi × 2 in² = 200 lb

force on smaller face = P × A = 100 psi × 1 in² = 100 lb

200 lb – 100 lb = 100 lb differential force

In some cases, the differential force area may be located on the outside faces of the valve, figure 5-20. Again, more force is exerted on the larger face, and the valve moves to the right. This principle is used quite often in automatic transmissions. Main pressure regulators, shift valves, and servos are a few of the controls that use differential force areas.

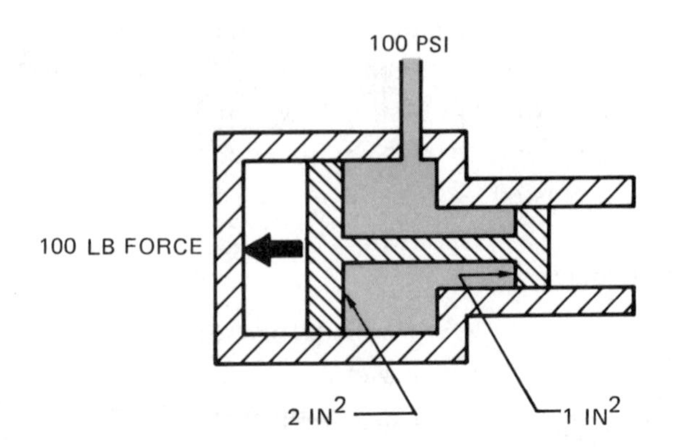

Fig. 5-19 Valve with differential force area.

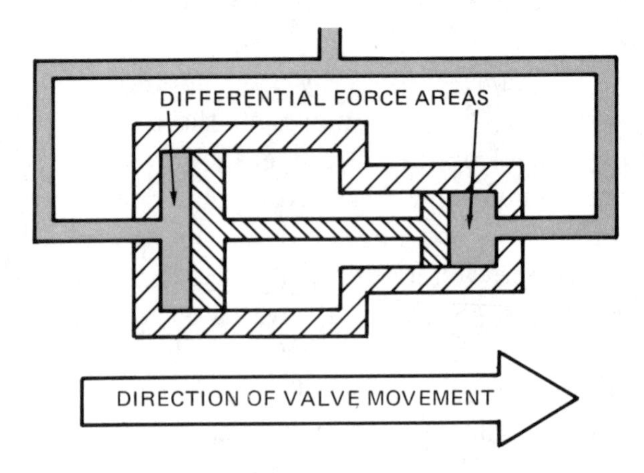

Fig. 5-20 Valve with differential force areas outside of the valve faces.

SERVOS, CLUTCHES, AND ORIFICING

A *servo* is a piston and cylinder which is usually built into the transmission case, figure 5-21, page 49. The servo applies a brake band that holds a member of the planetary gear set. When the control valve (inside the valve body) is moved, fluid at line pressure passes through the valley of

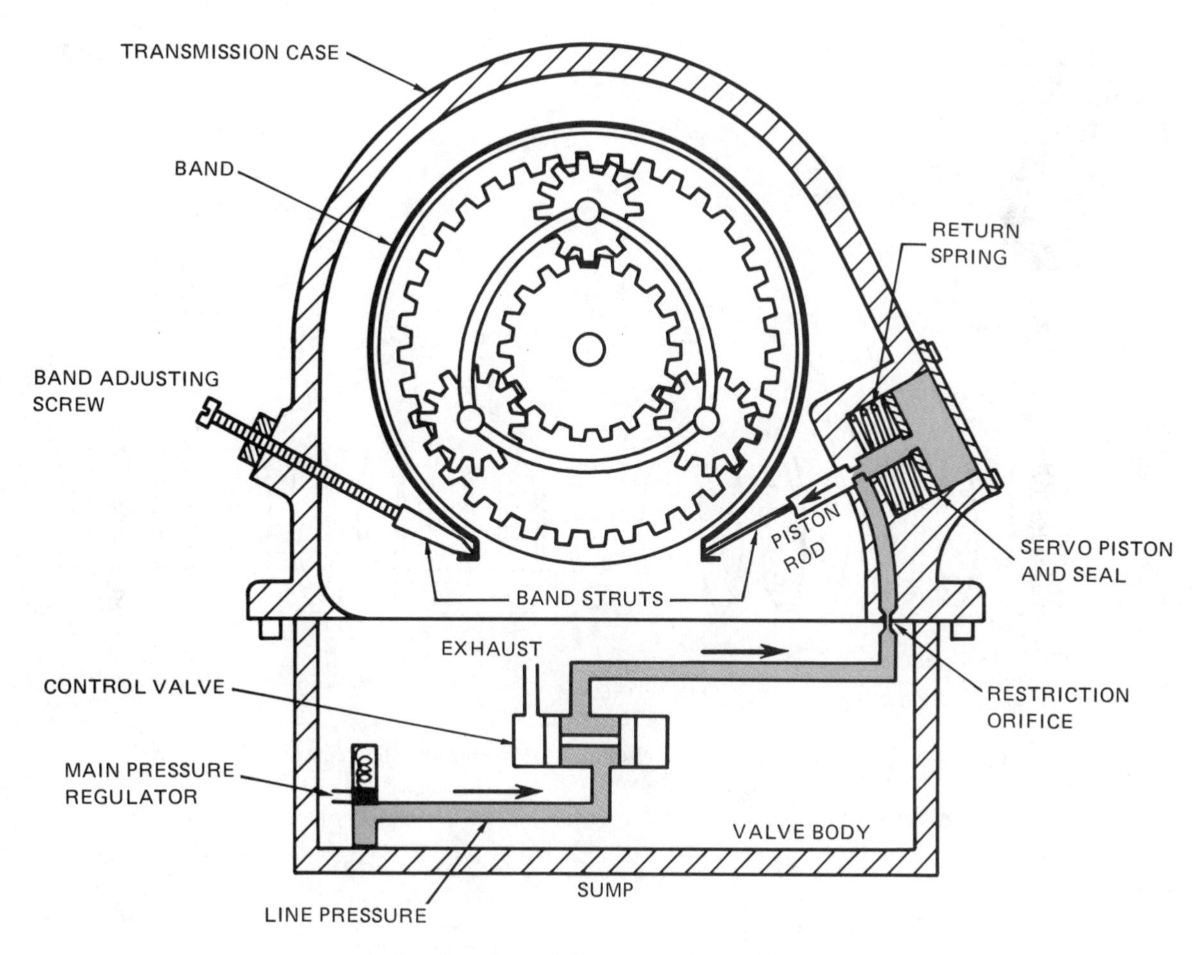

Fig. 5-21 Function of the servo in band application.

the valve. A passage in the valve body sends the fluid to a hole drilled, in the case, through the hollow piston rod, to the top of the piston. Here, line pressure forces the piston and rod to apply the band.

A small, drilled hole, called a *restriction orifice,* is used to control servo application. Until the servo is filled all the way, the pressure on the servo side of the orifice is less than line pressure. This slows up and smooths out band application. When the servo is completely filled, full line pressure is applying the band.

When the control valve is moved to the left, the land on the right side of the valve blocks line pressure. The pressure in the servo is then exhausted, and the return spring moves the servo to the OFF position to release the band.

Clutches are used in automatic transmissions to drive or hold different members of the planetary gear set. Because it has several steel drive discs and lined driven discs, this type of clutch is known as a *multidisc clutch,* figures 5-22 and 5-23. The number of discs depend on the make and model of the transmission. The steel drive discs are splined to the clutch drum and the lined driven discs are splined to the clutch hub. In this case, the clutch hub is part of the ring gear of a planetary gear set.

When the clutch is released the drive discs are free to rotate and cannot turn the driven discs. To apply the clutch, line

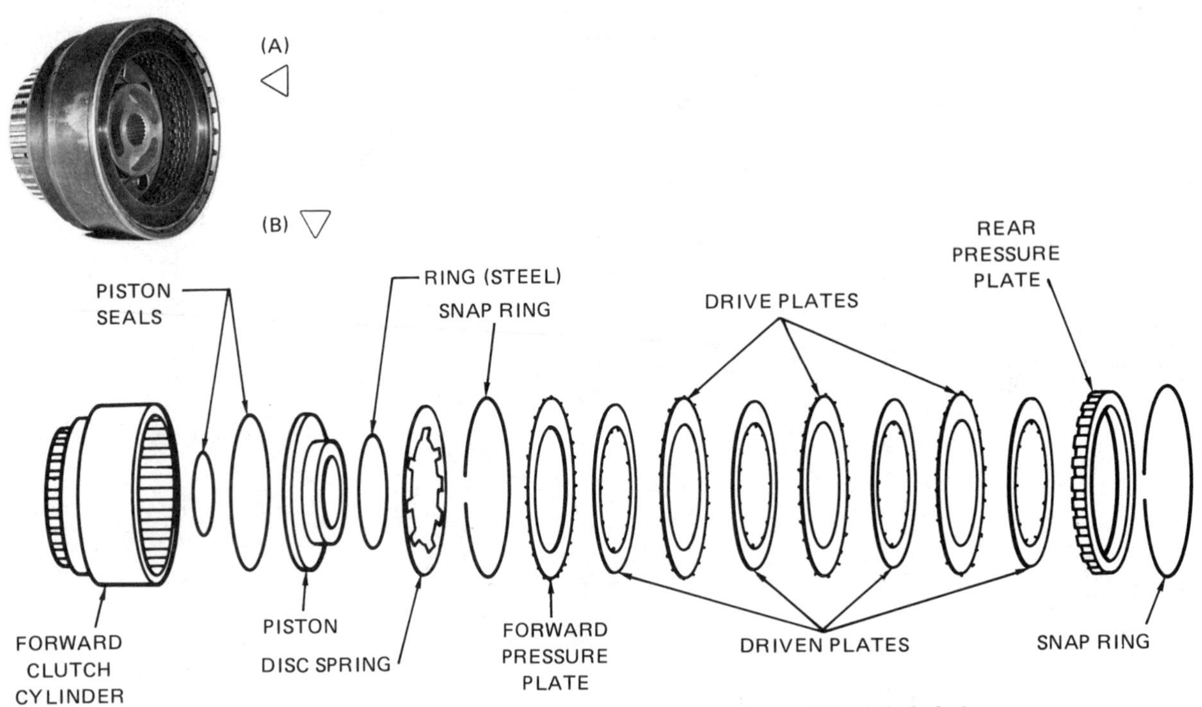

Fig. 5-22 Automatic transmission clutch. (A) Assembled, (B) exploded view.

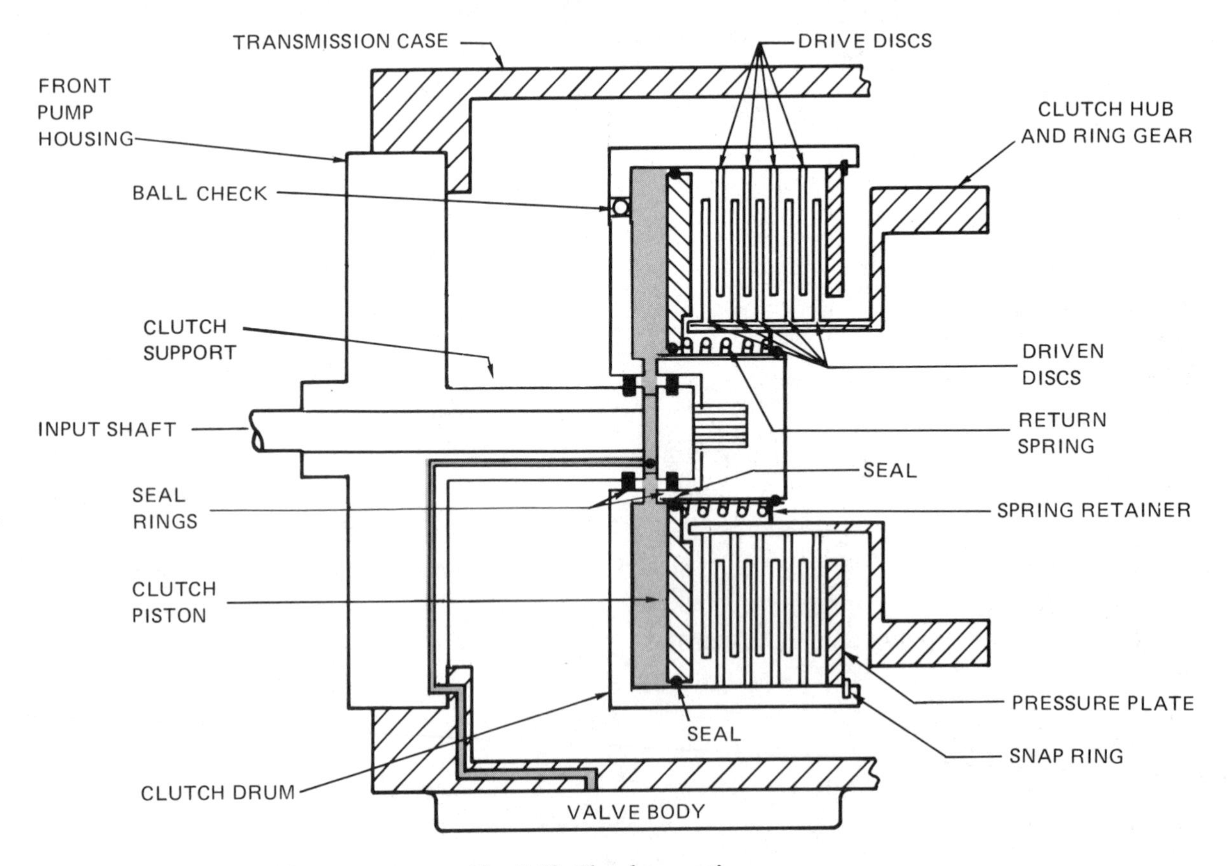

Fig. 5-23 Clutch operation.

pressure from the valve body flows through drilled holes in the case, pump housing and clutch support. Drilled holes in the clutch drum allow the fluid to fill the clutch. As pressure builds up, it forces the piston toward the pressure plate. This squeezes the drive and driven discs together, and in this way the input shaft is locked into the ring gear.

A *ball check* in the clutch drum or piston allows trapped air to be bled off until pressure builds up. The ball check then seats, which stops the fluid from flowing through the ball check hole. To control clutch apply pressure, an orifice is sometimes used in the same way as in a servo.

To release the clutch, pressure is exhausted and a return spring pushes the piston away from the discs and pressure plate. This allows the drive and driven discs to rotate freely again.

When a clutch is used to hold a member of a planetary gear set, the steel clutch discs are splined to the transmission case. The lined discs are splined to a hub that is a part of a planetary gear member. When the clutch is applied, the gear member is locked into the transmission case and cannot turn. Releasing the clutch allows the member to turn again.

CALCULATING PRESSURE IN THE METRIC SYSTEM

In the metric system, the following units are used to measure force, area, and pressure:

force – newtons (N)

area – square meters (m^2)

pressure – pascals (Pa) or, more commonly, kilopascals (kPa) [1 kPa = 1 000 Pa]

For example, if a liquid is exerting a force of 667 newtons on an area of .0007 m^2, the pressure is found as follows:

$$P = \frac{F}{A} \qquad P = \frac{667N}{.0007\ m^2}$$

P = 952 857 Pa or 953 kPa (approx.)

SUMMARY

The hydraulic principles discussed in this unit and summarized below apply to all automatic transmissions.

- Fluids are not compressible.
- In a closed system, the pressure of a fluid is equal in all directions. Pressure applied at one part of the system can be transferred to another part of the system.
- Motion in one part of the transmission can be transferred through hydraulic lines to another part of the transmission.
- A fixed relationship exists between force, pressure, and surface area. It can be expressed by either of the following formulas: $P = F/A$ or $F = P \times A$.
- Fluid pumps are used in automatic transmissions to supply fluid under pressure to the controls.

- In automatic transmissions, pressure is regulated by the use of spool valves.
- Servos are used to apply a band that holds a member of a planetary gear set.
- Clutches are used to drive or hold members of planetary gear sets.
- An orifice is used in some transmissions to cushion or soften the application of a clutch or band.

REVIEW

Select the best answer from the choices offered to complete the statement or answer the question.

1. If fluid with a pressure of 10 psi is pumped into one end of a closed cylinder with a diameter of 6 inches until the cylinder is full, what will a gauge at the opposite end read?
   a. 10 psi
   b. 60 psi
   c. 283 psi
   d. It would depend on the length of the container.
2. In a hydraulic system such as that shown in figure 5-24, if the input piston has a force of 100 lb and an area of 0.5 inches, what is the pressure in the system?
   a. 20 psi
   b. 50 psi
   c. 200 psi
   d. 2,000 psi
3. If the output piston in system described in question 2 has an area of 2 inches, what is the output force of the system?
   a. 50 lb
   b. 200 lb
   c. 250 lb
   d. 400 lb
4. What is the mechanical advantage of the system described in questions 2 and 3?
   a. 0.5:1
   b. 1:1
   c. 2:1
   d. 4:1

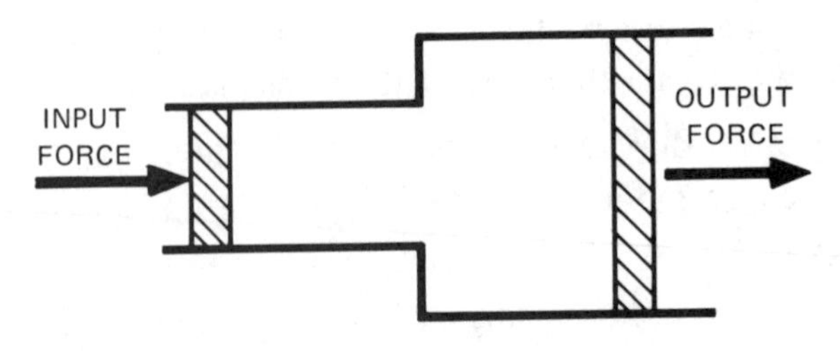

Fig. 5-24

5. If the input piston in the same system is moved into the cylinder 3 inches, how far will the output piston move?

   a. 0.75 in.
   b. 1.5 in.
   c. 2 in.
   d. 4 in.

6. Pump pressure must be regulated in order to

   a. increase its pressure.
   b. reduce its pressure.
   c. increase the flow of fluid.
   d. reduce horsepower losses.

7. The valve shown in figure 5-17 is of the:

   (I) Bypass type.
   (II) Balanced type.

   a. I only
   b. II only
   c. both I and II
   d. neither I nor II

8. The type of valve shown in figure 5-18 is used as:

   (I) A variable pressure control.
   (II) A main pressure regulating valve.

   a. I only
   b. II only
   c. both I and II
   d. neither I nor II

9. Hydraulic pressure on the inside faces of a valve having a differential force area, would result in a force that would

   a. hold the valve in the same position.
   b. move the valve in the direction of the small end.
   c. move the valve in the direction of the large end.
   d. increase the pressure, in psi, on the large end.

10. If the face areas of a differential force area valve are 3 inches and 2 inches, and the pressure on the valve faces is 75 psi, what is the differential force?

   a. 75 lb
   b. 150 lb
   c. 225 lb
   d. 375 lb

11. A clutch in an automatic transmission:

   (I) Drives a planetary gear member.
   (II) Holds a planetary gear member.

   a. I only
   b. II only
   c. both I and II
   d. neither I nor II

12. A ball check in the clutch drum or piston is used to:

    (I) Prevent fluid from coming in contact with the clutch discs.
    (II) Allow air trapped between the clutch and drum to bleed off.

    a. I only
    b. II only
    c. both I and II
    d. neither I nor II

13. In a hydraulic system, an orifice

    (I) Slows down clutch or band apply.
    (II) Prevents dangerous pressure buildup.

    a. I only
    b. II only
    c. both I and II
    d. neither I nor II

14. A band in an automatic transmission

    (I) Drives a planetary gear member.
    (II) Holds a planetary gear member.

    a. I only
    b. II only
    c. both I and II
    d. neither I nor II

EXTENDED STUDY PROJECTS

Using an oil circuit diagram of a Chrysler six-cylinder TorqueFlite transmission, answer the following questions:

1. How many pressure regulators are there? Name them.
2. Is the main pressure regulator of the bypass or restriction type?
3. Is the throttle pressure regulator of the bypass or restriction type?
4. Is the pressure to the torque converter regulated? Explain.

AUTOMATIC TRANSMISSION PRESSURE REGULATION & CONTROL DEVICES

OBJECTIVES

After studying this unit, the student will be able to:

- Describe how the main pressure regulator controls line pressure under different driving conditions and speed ranges.
- Explain the relationship between shift timing, road speed, and throttle opening.
- Explain how accumulator action and orificing affect shift quality or feel.
- Describe the control of the planetary gear set by the transmission hydraulic system.

Unit 5 described the use of a simple bypass or restriction valve to control pressure. In the automatic transmission, pressure must be regulated to meet a number of different driving conditions. For example, more pressure is required at high torque loads, such as when going up a steep hill. Less pressure is needed at cruising speeds, or with small throttle openings. To enable the automatic transmission to meet these different driving conditions, a slightly more complicated pressure regulating system is required.

Pressure regulation is also an important factor in *shift feel* or *quality*. At low speeds, the shift should be soft and barely noticeable. At high speeds, or large throttle openings, the shift should be fast and firm. This prevents band or clutch wear caused by slipping at high torque loads.

This unit explains the use of hydraulic control devices to obtain automatic shifting, to control shift timing and feel, and to supply different speed ranges in the planetary gear set.

MAIN PRESSURE REGULATION

The main pressure regulator of an automatic transmission is a balanced bypass valve, figure 6-1. Note that the valve has two differential force areas and a face area on the spring end of the valve. This design makes it possible to control pressure for different driving conditions.

Increased Pressure for Reverse

Automatic transmissions require an increase in line pressure for reverse operation. Greater pressure is needed because of the increased torque needs placed on the band or clutch holding the planetary carrier assembly in reverse. Normal line pressure in neutral is about 60 psi as shown in figure 6-1. When the driver selects reverse, the pressure is increased, figure 6-2.

The shift linkage of an automatic transmission is connected to a valve in the transmission called the *manual valve*. When the driver moves the shift lever to reverse, the manual valve moves and opens the reverse

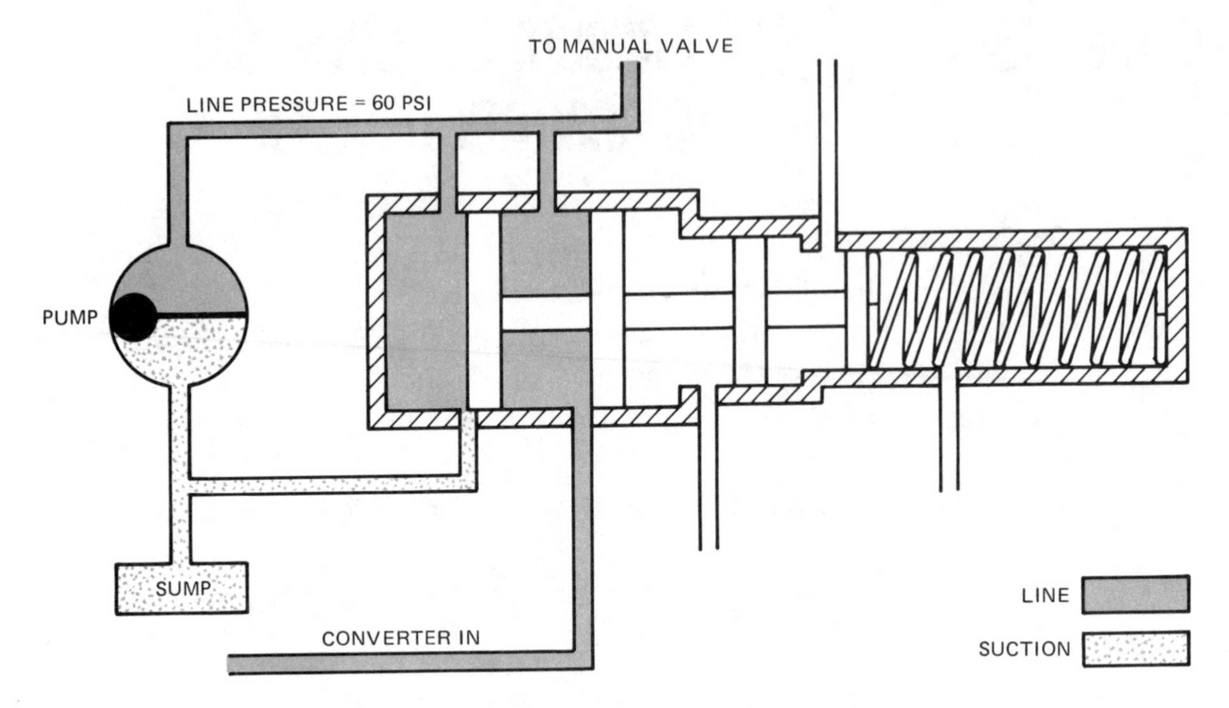

Fig. 6-1 Main pressure regulation control by spring force only.

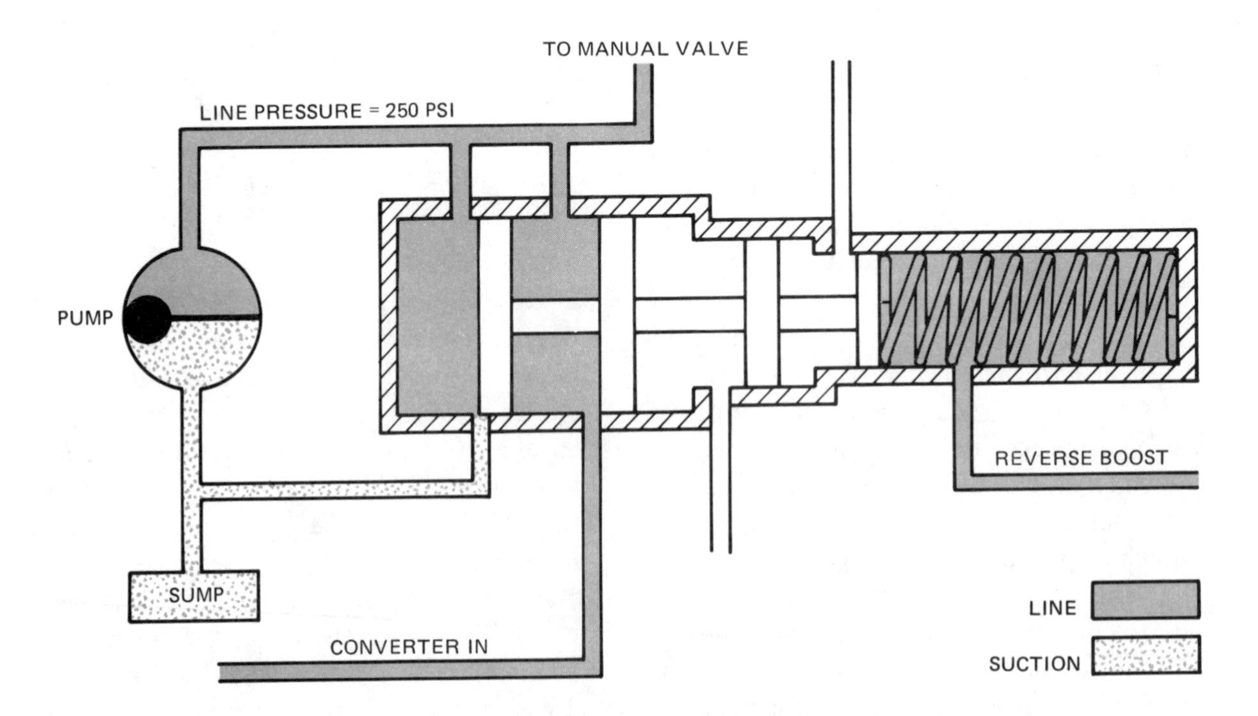

Fig. 6-2 Main pressure regulation, reverse.

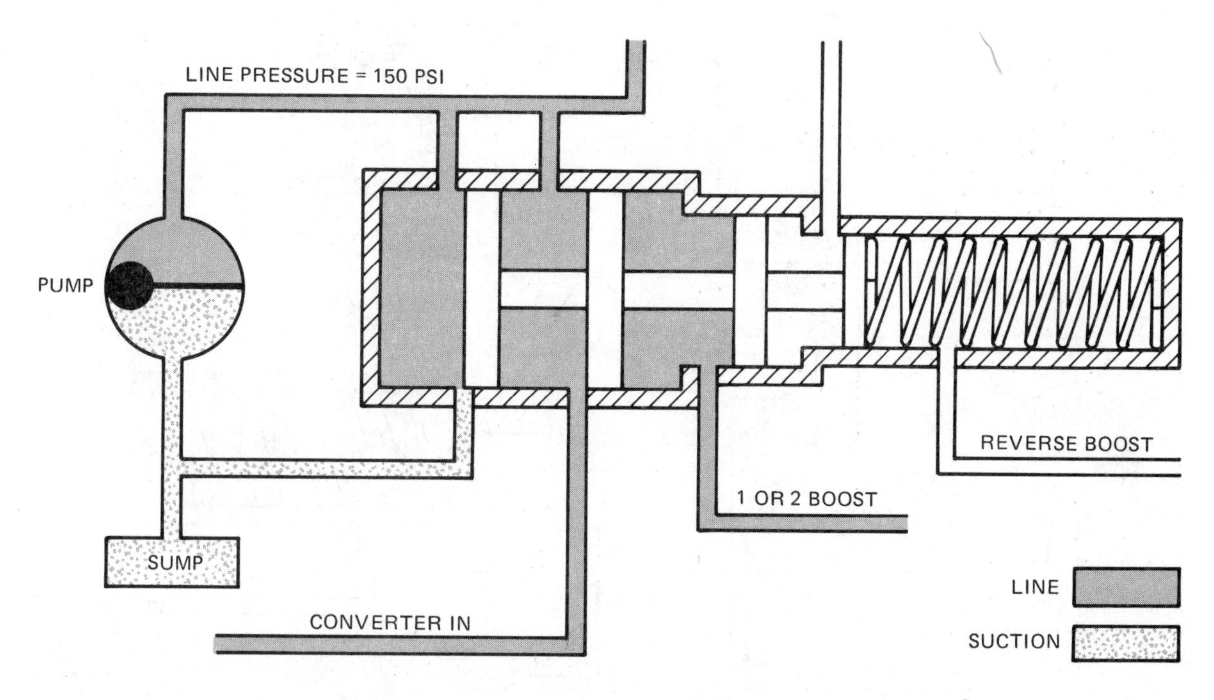

Fig. 6-3 Main pressure regulation system, manual 1 or 2.

lines or passages to line pressure. The *reverse boost passage* is part of this system. The line pressure supplied to the boost system adds to the spring force, resulting in a line pressure of about 250 psi.

Increased Pressure for Manual Second and First

Some transmissions require line pressure boost in manual second (intermediate) and first (low). The same system is used as for reverse, but in this case, line pressure from the manual valve is sent to a differential force area of the valve, figure 6-3. The differential force, added to the spring force, increases line pressure to about 150 psi.

INCREASING PRESSURE WITH INCREASING TORQUE

When engine torque loads are high, more pressure is needed to supply the torque converter and to keep the clutches and bands from slipping. High torque loads occur when a vehicle is climbing steep grades or pulling heavy loads. Additional pressure gives faster transmission clutch or band apply and a firm shift. This helps to prevent slipping and wear. At cruising speeds, or small throttle openings, the shift should be soft. This means smooth clutch and band apply.

The Throttle Valve

One way to change line pressure to meet torque requirements is through the use of a *throttle valve,* figure 6-4. *Throttle valve pressure,* called *TV pressure,* is present whenever the engine is running. As shown in the diagram, TV pressure acts on a differential force area on the main pressure regulator. Note that the throttle valve is of the balanced restriction type, and that spring force on the end of the valve can be changed by throttle opening.

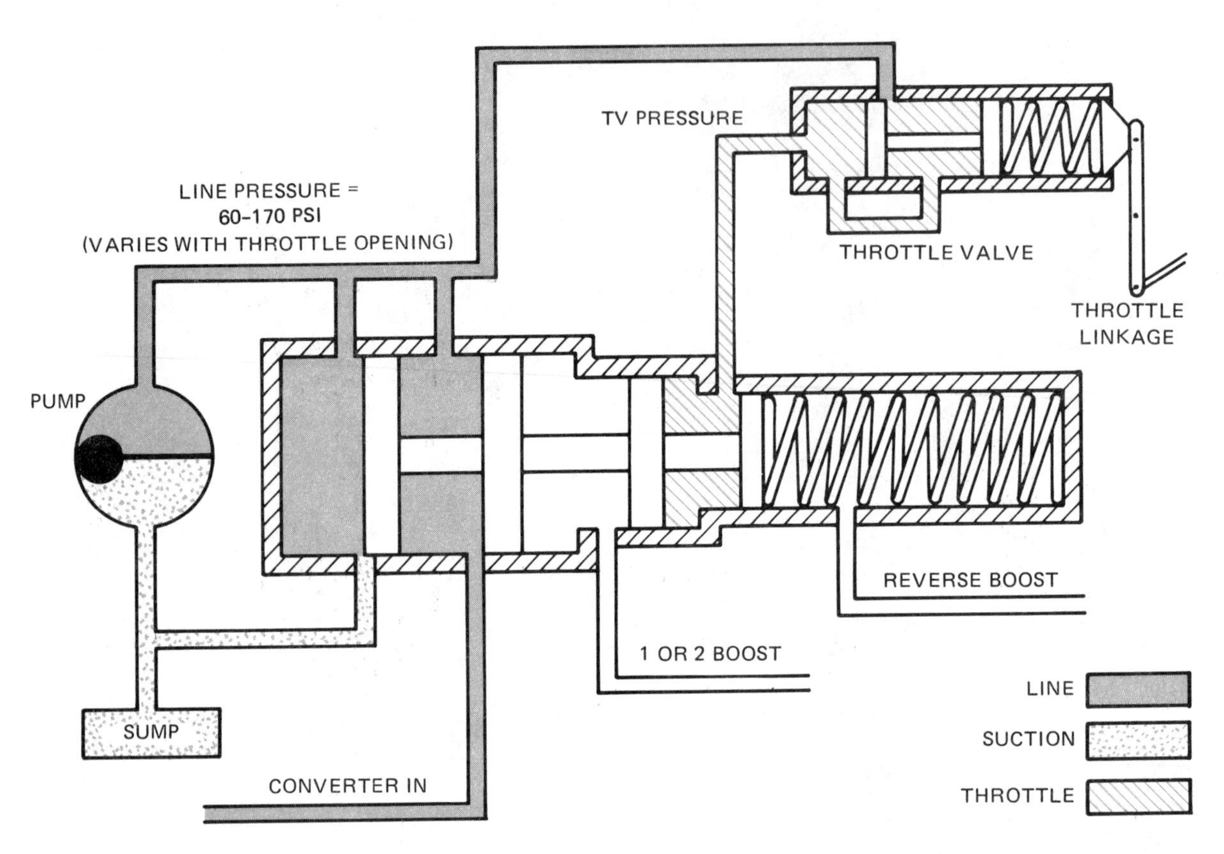

Fig. 6-4 Main pressure regulation controlled by spring force and throttle valve pressure.

At idle speed there is very little TV pressure. Hence, line pressure is low – about 60 psi. As the throttle is opened, spring force on the valve is increased. This increases TV pressure. TV pressure then acts to increase line pressure. The increase in line pressure is in proportion to throttle opening. With the throttle wide open, the pressure is about 170 psi, depending on the make and model of the transmission.

Vacuum Control

Instead of a linkage-controlled throttle valve, many transmissions use a vacuum diaphragm unit to control throttle pressure and line pressure, figure 6-5. (*Note:* General Motors calls their vacuum unit a *modulator*, and the pressure it controls, *modulator pressure.* In this text, the terms *vacuum control unit* and *throttle* or *TV pressure* are used.)

As shown in figure 6-5, the throttle valve is a balanced, restriction-type regulator. Two forces control throttle pressure. A spring force acts to increase throttle pressure. The force of atmospheric pressure acts to decrease throttle pressure.

Remember, fluids can be either liquids or gases. The air in the atmosphere acts according to the same principles as the fluid in an automatic transmission. Air is constantly exerting pressure on everything on the earth's surface. This pressure, called *atmospheric pressure*, varies with altitude (height of the land above sea level). Atmospheric pressure can be measured in either pounds per square inch (psi) or inches of mercury (in Hg). Normal atmospheric pressure at sea level is approximately 14.7 psi or 29.9 in Hg.

At small throttle openings, engine manifold vacuum is high, and the force of atmospheric pressure pushes on the diaphragm and compresses the spring. With the spring compressed, line pressure is restricted, and throttle pressure is low. For example, at engine idle, manifold vacuum is about 18 in Hg and throttle pressure is about 10 psi. (*Vacuum*, or pressure that is less than atmospheric pressure, is measured in inches of mercury.)

At wide-open throttle, engine manifold vacuum falls off to less than 1 in Hg. This allows the spring force to move the valve to the left, and throttle pressure increases to about 80 psi.

When diagnosing transmission problems, the following points should be kept in mind:

- High engine manifold vacuum (small throttle opening) = low throttle pressure.
- Low engine manifold vacuum (large throttle opening) = high throttle pressure.

For example, if engine vacuum were low (due to a leak or poor engine tuning), throttle pressure and line pressure would be high. This, in turn, could be a cause of harsh clutch or band application.

At high altitudes, both atmospheric pressure and engine manifold vacuum are less than at sea level. At sea level, atmospheric pressure is about 29.9 in Hg and engine manifold vacuum is about 18 in Hg. At 5,000 feet, atmospheric pressure is about 24.5 in Hg and engine manifold vacuum is about 15 in Hg. At idle speed, this results in an increase in line pressure of approximately 10 psi over that at sea level – about 70 psi at 5,000 feet as compared to 55 to 60 psi at sea level.

Altitude-Compensated Vacuum Unit

The unit shown in figure 6-5 is a non-compensated vacuum unit. This means that it cannot adjust to changes in altitude. At high altitudes, engine output is less than at sea level. For this reason, some transmissions are made to work at reduced throttle and line pressure at high altitudes. An altitude-compensated vacuum unit, figure 6-6, is used for this purpose. As shown, this unit has three forces that act to control throttle pressure:

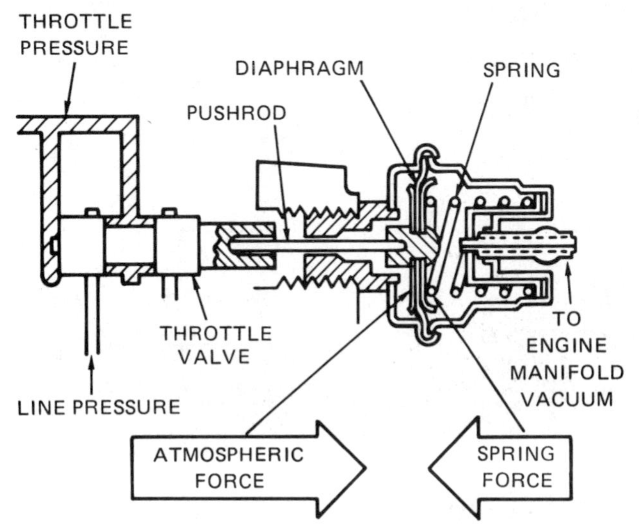

Fig. 6-5 Non-compensated vacuum unit.

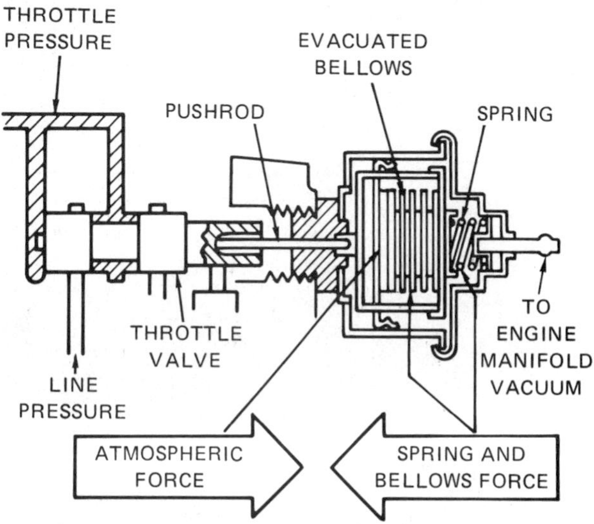

Fig. 6-6 Altitude-compensated vacuum unit.

- a spring force (increases throttle pressure).
- a bellows force (increases throttle pressure).
- atmospheric force (decreases throttle pressure).

The bellows is made of pleated metal and resembles the pleats of an accordion. Air is removed from the bellows, and it is sealed off from atmospheric pressure. In other words, the inside of the bellows is under a vacuum. Because the pressure inside the bellows is less than atmospheric pressure, atmospheric pressure tends to collapse or squeeze the bellows. The bellows is stretched and is inserted into the unit in such a way that it pulls on the diaphragm with what is called *collapsing force.*

At sea level, the spring force and the bellows' collapsing force on the diaphragm are equal to the spring force on the diaphragm of the non-compensated unit. This means that, at sea level, the throttle pressure will be the same with either unit. At high altitudes, the compensated unit reduces throttle pressure. Atmospheric pressure is less, and therefore, the bellows does not have as much collapsing force. For example, at sea level and 18 in Hg manifold vacuum, line pressure in either unit will be about 60 psi. At 5,000 feet, the non-compensated unit will regulate line pressure to about 70 psi. At the same altitude, the compensated unit will regulate line pressure to about 60 psi, or to about the same pressure as at sea level.

MANUAL VALVE

An automatic transmission allows the driver to select from several drive ranges including, park, reverse, neutral, drive, second, and first. The driver puts the transmission in different ranges by moving the shift lever. The shift lever is connected by linkage to the

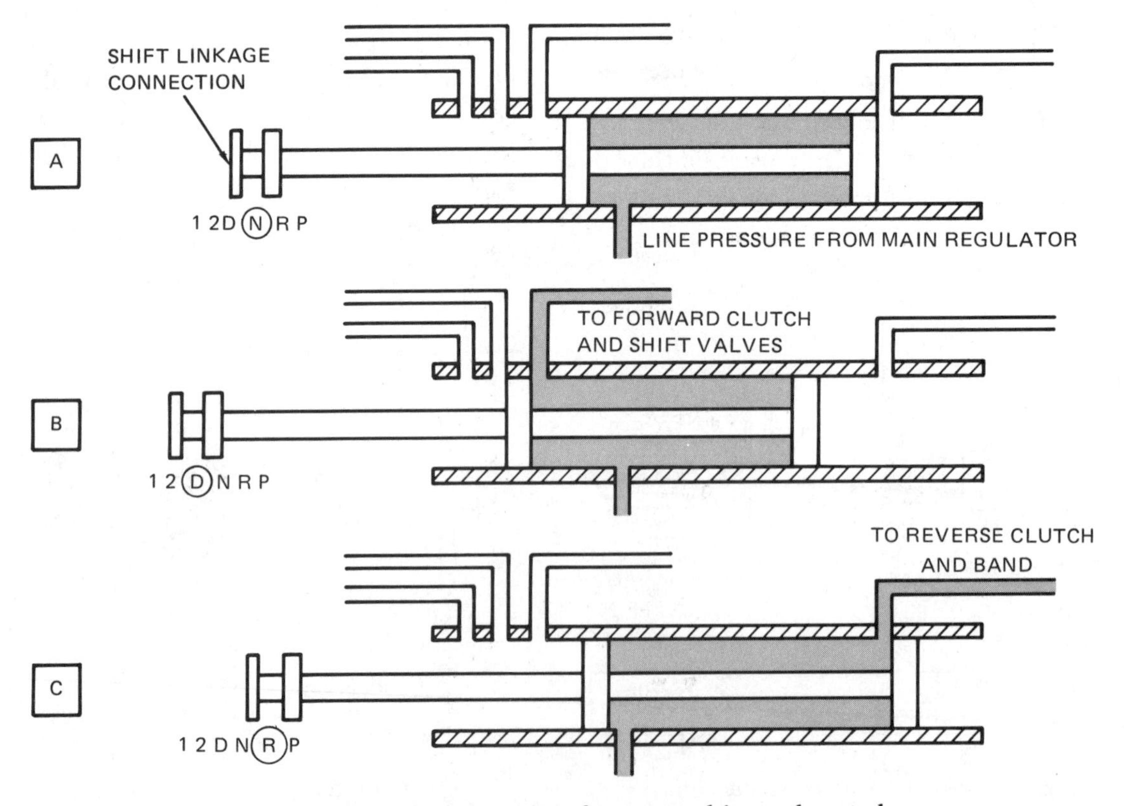

Fig. 6-7 Manual valve positions for reverse, drive, and neutral.

manual valve in the transmission valve body, figure 6-7. As shown, line pressure can be directed to different passages in the valve body. For example, in view A, the valve is positioned for neutral. Line pressure is trapped in the valley of the valve and cannot leave the valve body. In view B, the valve has been moved to the left to the drive position. In this position, the fluid can flow to the forward clutch and shift valves. For reverse, as shown in view C, the valve is moved to the right. This allows fluid to flow to the reverse clutch and band.

The valves in an actual transmission do not operate exactly as shown here. In general, they are somewhat more complicated. However, the diagrams shown here present the basic principles of valve action. In later units, these principles are applied to actual valves of different makes of transmissions.

SHIFT VALVES

When the driver selects the drive or D range, fluid flows to a clutch or servo and a shift valve. The shift valve or valves move by spring force and hydraulic pressure, figure 6-8. They are held in the downshift position by the spring force on the right side of the valve. To move the valve to the upshift position, force must be applied at the left side of the valve. The force that is used is *governor pressure.*

There are two main types of governors, figure 6-9. The governor made by General Motors uses separate flyweights to move a

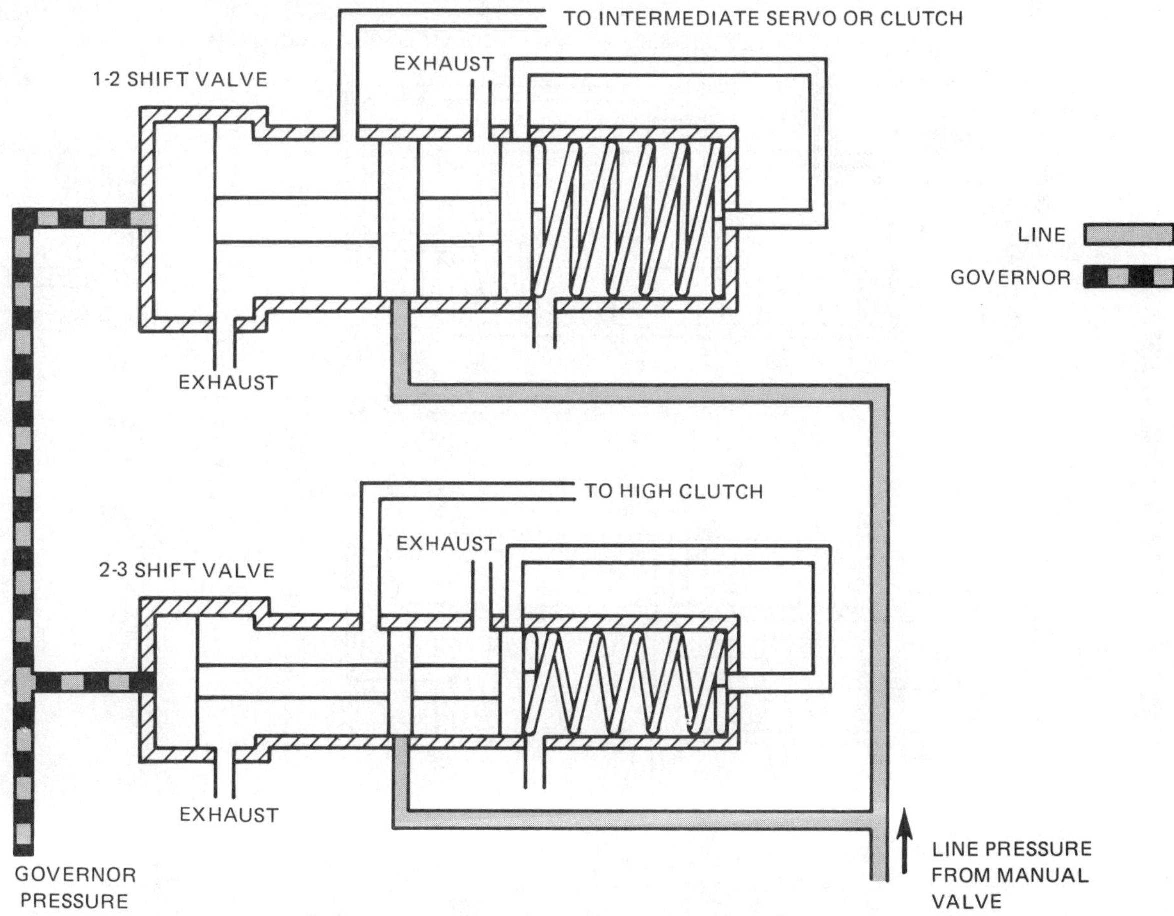

Fig. 6-8 Shift valves in the downshift position.

restriction-type valve. A centrifugal force is set up when the weights (valves) revolve about the axis of the shaft. This force tends to move the weights away from the center of the shaft. This can be compared to a car rounding a curve in the road: the faster the curve is taken, the more force must be used at the steering wheel to keep the car into the curve. Centrifugal force tends to force the car to the outside of the turn.

The Ford and Chrysler governors have valves that act as their own weights. Both types of governors are driven by the output shaft of the transmission. Centrifugal force works on the weights or valves. This causes governor pressure to increase as the speed of the car increases. Governor action is covered in greater depth in the units of this text dealing with specific transmissions.

Fig. 6-9 Two types of governors.

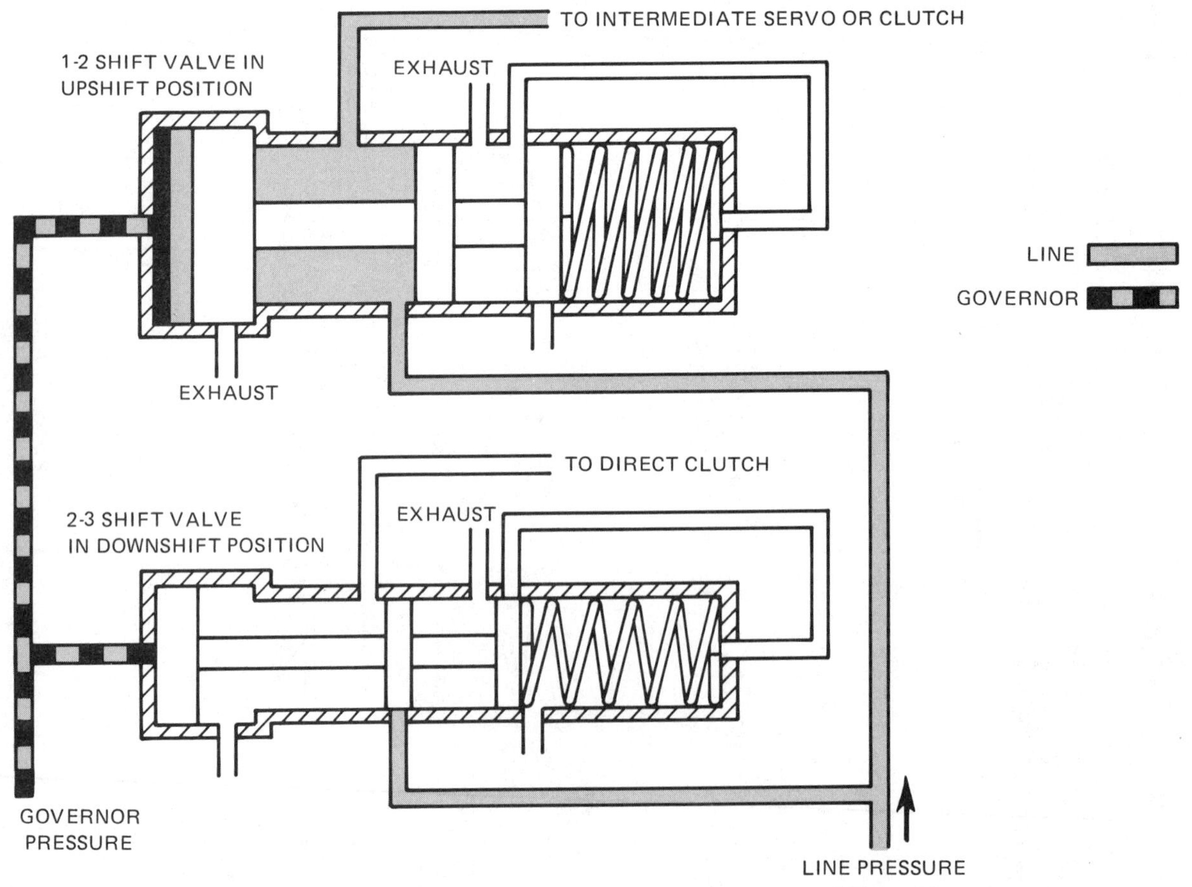

Fig. 6-10 Shift valves showing the 1-2 shift valve in the upshift position.

Upshifting

When governor pressure is high enough, it overcomes the spring force. The shift valve then moves to the right, and line pressure flows through the valley of the valve, figure 6-10. This line pressure is then used to apply an intermediate band or clutch. Note that the 2-3 shift valve is still in the downshift position. The 2-3 shift valve has less face area at the governor end than the 1-2 shift valve. Therefore, less force is applied at this face ($F = P \times A$) and the valve remains stationary.

As road speed increases, governor pressure also increases. When governor pressure reaches a certain point, the 2-3 shift valve moves to the right. This causes line pressure to apply the direct clutch. As the direct clutch is applied, two members of the planetary gear set are locked. The gear set then turns as a unit. (The operation of the shifting mechanism is discussed in greater detail in relation to specific transmissions.)

Downshifting

As the driver slows the car, governor pressure decreases, and the spring force moves the valve to the left. When this happens, line pressure exhausts from the clutch or servo, and the transmission downshifts. Note that to this point, only two forces have been working on the shift valve:

- **spring force**, which works to keep the valve in the downshift position, and
- **governor pressure**, which works against the spring force to move the valve to the upshift position.

With this setup, all shifts would take place at the same road speed.

Shift Delay System

In the modern transmission, shifts take place at different road speeds depending on engine torque or throttle opening. For example, if the car is going up a steep hill or the throttle is opened more for a fast takeoff, the shifts take place at a higher road speed. To make this possible, TV pressure is used on the downshift side of the valve, figure 6-11. As shown, TV pressure works with the spring force on the right side of the valve. With this arrangement, the shifts can take place at a higher road speed, depending on throttle opening.

At small throttle openings, TV pressure is low, and governor pressure is able to move the valve to upshift the transmission at low road speeds. However, if the driver opens the throttle wider for a fast takeoff or to climb a steep hill, TV pressure will be high. This means that the car must reach a higher speed to build enough governor pressure to work

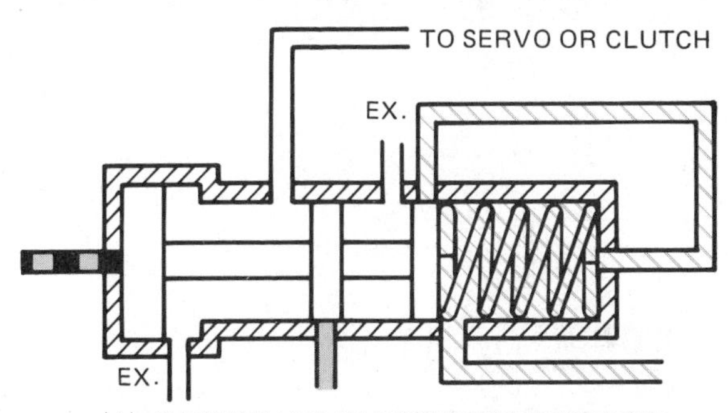

(A) SHIFT VALVE IN DOWNSHIFT POSITION:
TV PRESSURE ACTING ON VALVE

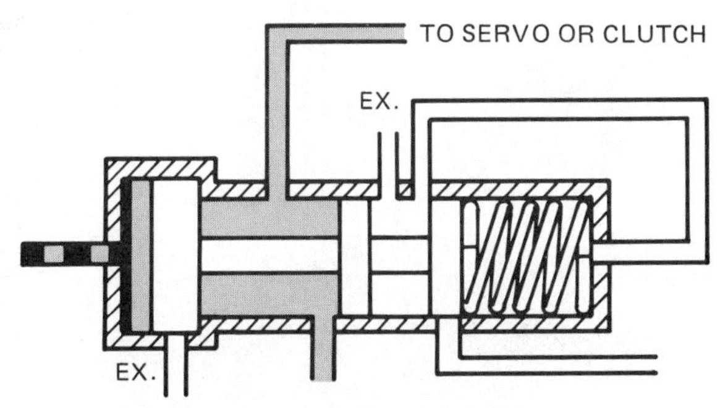

(B) SHIFT VALVE IN UPSHIFT POSITION:
TV PRESSURE CUT OFF.

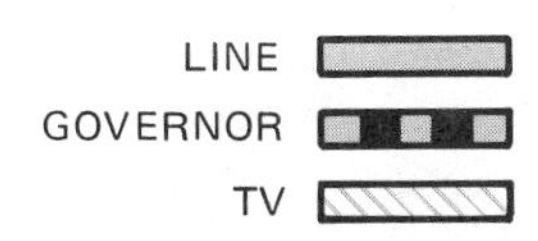

Fig. 6-11

against TV pressure and the spring force.

At small throttle openings, a 1-2 shift may take place at 8 to 15 miles per hour (mph), and a 2-3 shift at 15 to 23 mph. For the same engine-transmission setup, shift speeds at wide-open throttle would be 38 to 53 mph for a 1-2 shift, and 77 to 93 mph for a 2-3 shift.

In figure 6-11, note that TV pressure has been cut off and exhausted in the upshift position of the valve. In this way, the car is prevented from downshifting if the driver uses the throttle to increase speed. This allows full use to be made of the torque converter's ability to multiply torque. At slower speeds, if the driver moves the throttle to almost wide open, line pressure will increase and work on the differential force area to help move the valve to the downshift position.

Snap Action

Another reason for cutting off TV pressure is to give the valve what is called *snap action*. The shift valve, when it starts to move, must snap open. This fast opening of the shift valve is necessary to prevent shifts that are too soft, which in turn, may cause burned clutches and bands.

Forced Downshift

Below road speeds of about 60 to 90 mph (depending on the engine-transmission combination), the driver can force a 3-2 downshift by pushing the throttle wide open. When the throttle is opened wide, a downshift valve opens the TV passages to almost full line pressure. This causes the shift valve to move to the downshift position. Also,

ACCUMULATOR PISTON
ACCUMULATOR CYLINDER
LINE PRESSURE FROM MANUAL VALVE

Fig. 6-12 Accumulator before upshift.

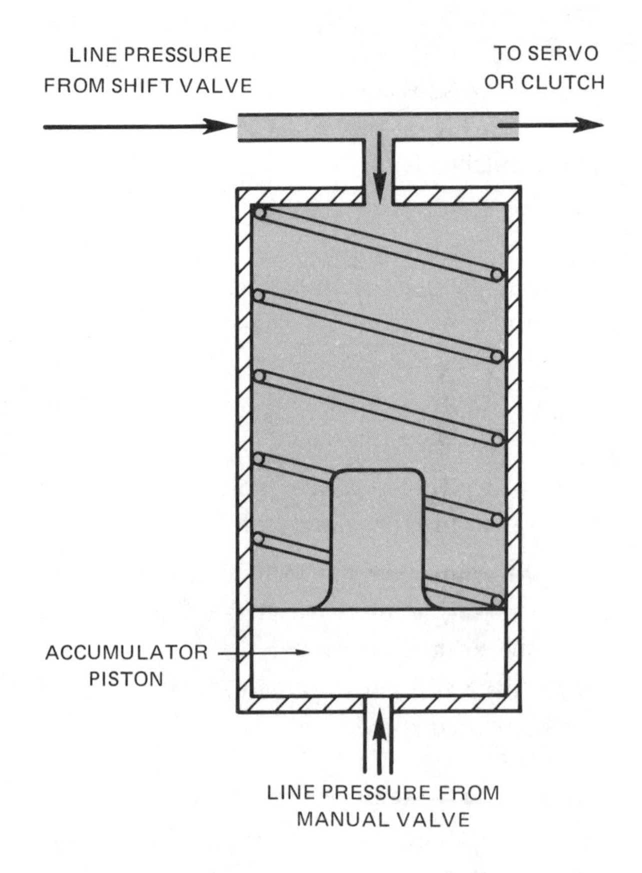

Fig. 6-13 Accumulator after upshift.

below some road speeds a 2-1 or 3-1 shift can take place. Various companies accomplish this type of shift in different ways.

CONTROLLING SHIFT FEEL

Changing line pressure with throttle opening has an important effect on shift quality. In addition, most transmissions use accumulators and orificing to help control shift feel. Orificing is explained in unit 5, and therefore, will not be covered here. Review unit 5 for details.

Accumulators

In an automatic transmission, an *accumulator* is a hydraulic control that collects or holds hydraulic fluid, figure 6-12. When the driver moves the shift lever to drive range, line pressure is also sent to the accumulator. Line pressure strokes or moves the piston up against the spring force.

When the shift valve moves to send line pressure to a clutch or servo, a passage or path in the valve body sends line pressure to the top of the accumulator piston, figure 6-13. Since the pressure is now the same on both the top and bottom of the accumulator piston, the spring force strokes the accumulator piston down to the bottom of the accumulator cylinder. As the accumulator piston moves down, it must take on, or *accumulate,* fluid. Full clutch or band application cannot take place until the accumulator is completely filled. In this way, the application of the clutch or band is softened.

SUMMARY

- Main line pressure is regulated by a balanced bypass valve. The forces acting on the valve to control pressure are: (1) a spring force, and (2) a control or boost pressure in the different ranges. Line pressure senses engine torque through the use of TV pressure that changes with throttle opening. The throttle valve may be controlled by linkage, or by a vacuum unit.
- The manual valve allows the driver to choose different driving ranges. It sends line pressure to other valves in the valve body and to the clutch or band needed for the range that has been chosen.
- The shift valves send line pressure to the clutch or band necessary for automatic shifts. The forces acting on the shift valves are: (1) governor pressure, which tends to move the valve to the upshift position; and (2) spring force and TV pressure, which tend to keep or move the valve to the downshift position.
- The governor is connected to the output shaft of the transmission. Governor pressure is increased as road speed is increased.
- Orificing and accumulators are used to control automatic shift feel, or quality, which varies with road speed and engine torque.

REVIEW

Select the best answer from the choices offered to complete the statement or answer the question.

1. When engine torque loads are high, line pressure must be increased. The force acting on the main pressure regulator to increase line pressure for high torque loads is

   a. governor pressure.
   b. converter stator pressure.
   c. accumulator pressure.
   d. throttle pressure.

2. The force or forces acting on a shift valve to move it to the upshift position are:

   (I) Throttle pressure.
   (II) Governor pressure.

   a. I only
   b. II only
   c. both I and II
   d. neither I nor II

3. In an automatic transmission, line pressure can be used to

   a. fill the torque converter.
   b. lubricate the transmission.
   c. apply a clutch or band.
   d. all of the above

4. For automatic upshifts, the shift valves are moved by

   a. modulator pressure.
   b. exhaust pressure.
   c. governor pressure.
   d. main line pressure.

5. The throttle valve can be made to sense torque by use of:

   a. a spring force.
   b. governor pressure.
   c. engine manifold vacuum.
   d. main line pressure.

6. When going up a steep hill, line pressure is increased. This increase in pressure is caused by

   a. road speed.
   b. throttle opening.
   c. 1-2 boost pressure.
   d. governor pressure.

7. The force or forces acting on a shift valve to move it to the downshift position are:

   (I) Throttle pressure.
   (II) Governor pressure.

   a. I only
   b. II only
   c. both I and II
   d. neither I nor II

8. The part of an automatic transmission that takes on clutch or band application fluid in order to control shift feel is called an:

   (I) Accumulator.
   (II) Orifice.

   a. I only
   b. II only
   c. both I and II
   d. neither I nor II

9. A clutch or band in an automatic transmission is applied by

   a. governor pressure.
   b. throttle pressure.
   c. boost pressure.
   d. line pressure.

10. A 1-2 upshift takes place when

    a. governor pressure overcomes throttle pressure.
    b. throttle pressure overcomes governor pressure.
    c. main line pressure overcomes throttle pressure.
    d. none of the above

11. The manual valve is controlled by

    a. governor pressure.
    b. the driver.
    c. throttle pressure.
    d. main line pressure.

12. To obtain good shift quality, clutch or band application is controlled by an:

    (I) Accumulator.
    (II) Orifice.

    a. I only
    b. II only
    c. both I and II
    d. neither I nor II

13. The reverse boost system is used to:

    (I) Raise line pressure for reverse operation.
    (II) Release a clutch or band.

    a. I only
    b. II only
    c. both I and II
    d. neither I nor II

14. The low and intermediate boost system is effective in:

    (I) Drive ranges 1 and 2.
    (II) Manual 1 and manual 2.

    a. I only
    b. II only
    c. both I and II
    d. neither I nor II

15. Low engine vacuum supplies:

    (I) High throttle or modulator pressure.
    (II) Higher main line pressure.

    a. I only
    b. II only
    c. both I and II
    d. neither I nor II

EXTENDED STUDY PROJECTS

The following projects are an introduction to diagnosis or troubleshooting. This is an important part of transmission service. A review of units 5 and 6 is suggested before these projects are attempted. Consult with the instructor for further guidance.

1. State two problems in a hydraulic system that could cause late upshifts.
2. Name three hydraulic system problems that could cause no upshift.
3. List three faults in a transmission that could cause upshifts that are too soft.
4. List two reasons for harsh upshifts in a transmission.
5. State five reasons for lack of the drive range in a transmission when the driver chooses the D shift position.

FLUIDS AND SEALS

OBJECTIVES

After studying this unit, the student will be able to:

- Differentiate between various types of transmission fluids, and select the correct fluid for a specific transmission.
- Describe the function of additives in transmission fluid.
- List the properties of transmission fluid and explain how they affect transmission performance.

AUTOMATIC TRANSMISSION FLUIDS

When automatic transmissions were first marketed, ordinary motor oil was used as a transmission fluid. However, it was found that motor oil did not work well in transmissions. Then in 1949, a special automatic transmission fluid was developed. It was called AQ–ATF (Armour Qualification–Automatic Transmission Fluid), and was developed for General Motors by the Armour Research Foundation.

Since that time, changes in transmission fluid have kept pace with modern transmissions and loads. For example, in older cars having a horsepower rating of about 100 horsepower, about 14 quarts of fluid circulated in the transmission. This meant that each quart of fluid had to handle a heat load created by the friction elements and torque converter of about 7 hp. Today, horsepower may be over 300, while fluid capacity has been cut to about nine quarts. Each quart now handles about 33 hp. Fluid must be specially made to handle this large a load.

Fig. 7-1 Two types of transmission fluid.

TYPES OF FLUID

There are four basic types of transmission fluid for American cars today: Dexron, and Dexron II for use in General Motors, Chrysler, and American Motors transmissions; Type F, for use in some Ford Motor Company transmissions from 1964 on; and CJ fluid, for use in Ford Jatco transmissions and Ford C–6 transmissions from 1977 on. Both type F and Dexron are dyed red, while CJ fluid has an amber-orange color. Since motor oil has a golden brown color, this helps the mechanic tell if an oil leak is from the engine or transmission.

Coefficient of Friction

The coefficient of friction is the force needed to stop movement between two parts.

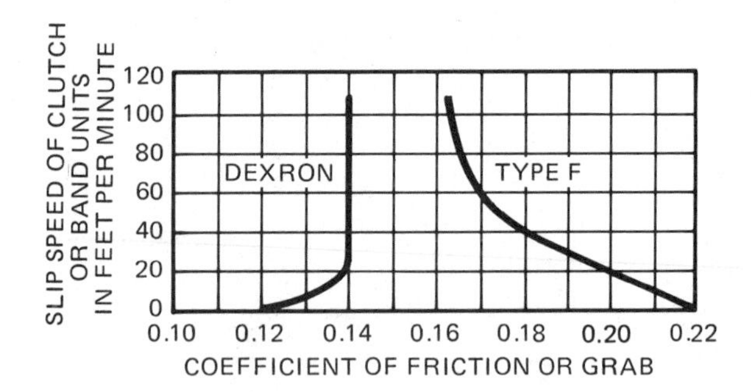

Fig. 7-2

The two parts may be a drum inside of a band, or adjacent clutch discs. The lining materials used by Chrysler, Ford, and GM in making clutches and bands have different coefficient of friction values. Fluids are also made to have different coefficient of friction values, and should only be used with the type of lining for which they are designed. The difference in the coefficient of friction values for Dexron and Type F, for example, is shown in figure 7-2. As the chart shows, the coefficient of friction or *grab* of Dexron is less as the clutch or band becomes tighter. This slows down the slip speed. With Type F, there is more grab as the clutch or band gets tighter. For this reason, fluids should be used only in the transmissions for which they are designed. Use of the wrong fluid or mixing fluids may cause clutch or band failure and serious transmission damage over a period of time.

PROPERTIES OF TRANSMISSION FLUID

Automatic transmission fluid has certain properties or traits that set it apart from other oils. These properties help the fluid do the following jobs:

- Transmit pressure and motion through the hydraulic control system.
- Transmit power from the engine to the transmission through the torque converter.
- Move heat from the transmission to the cooler.
- Help seal the transmission parts and keep them clean.
- Lubricate the bushings, bearings, thrust washers, gears, clutches, and bands.

In addition to these functions, the fluid must be *compatible,* or get along with, the materials in the clutches, bands, seals, and the different metal parts used in the transmission.

ADDITIVES

To make a fluid which meets all of these requirements, a high quality mineral oil base is blended with many additives. An *additive* is a small amount of a material that is added to another substance to give it special properties. Additives are used in transmission fluid to increase its ability to prevent rust and corrosion, and to give it other qualities it needs to do its job. Some of these additives are described here.

Viscosity Index Improvers

Temperature affects viscosity in that mineral oil tends to become thin at high temperatures and thick at low temperatures. It is important that viscosity index is not confused with viscosity. *Viscosity* is a measure of an oil's body and fluidity and is usually referred to as the *weight* of the oil. Motor oils, for example, have viscosity ratings of SAE (Society of Automotive Engineers) 10W, 20W, and so forth. (The W stands for winter grade.)

Some motor oils cover more than one grade – SAE 10W-40 – for instance. Such an oil is said to have a high *viscosity index.* That is, it will keep its body and fluidity over a wide temperature range. An oil with a high viscosity index will stay at a more constant viscosity. That is, it will not become too thin to lubricate at high temperatures, or become too thick to flow at low temperatures.

Pour Point Depressants

The *pour point* of an oil is the temperature at which the oil becomes too thick to pour. A pour point depressant helps the oil to flow at lower temperatures.

Oxidation Inhibitors

The temperature of band or clutch discs, at the time they are applied, can be as high as 600°F. At this temperature, oil combines with oxygen from the surrounding air. This reaction, called *oxidation,* tends to form gum and varnish which build up on transmission parts. The addition of oxidation inhibitors helps stop the oxidation of the fluid.

Corrosion and Rust Inhibitors

At high temperatures, water, oxygen and oil tend to cause acids to form that rust and corrode parts. Corrosion inhibitors help to neutralize the acids, while rust inhibitors keep water from damaging the transmission parts.

Foam Inhibitors

The gear train of a transmission has an "egg-beater effect" on the fluid. It whips air into the fluid causing it to foam. This process is called *aeration.* It is undesirable in a transmission fluid because air can be compressed, and would keep the clutches or bands from locking up. Foam inhibitors help prevent air from being churned into the fluid. However, incorrect fluid level (too high or too low) can also cause aeration.

However, incorrect fluid level (too high or too low) can also cause aeration.

Detergents – Dispersants

Detergents and dispersants work together in a transmission fluid to clean the parts and valves in the same way a laundry detergent cleans clothes. These additives also stop the cleaned off sludge and varnish from forming large clots that could plug the pickup screen or jam a shift valve.

Extreme Pressure Agents

Extreme pressure agents work chemically with metal parts to form a strong, tough surface film. This surface film helps stop metal to metal contact. One place this is important is between gear teeth.

Friction Modifiers

Mineral oils have a low coefficient of friction, while clutch and band facings depend on a high coefficient of friction for lockup. For this reason, friction modifiers are added to the fluid. Different types are needed for different fluids.

Seal Swell Agents

To keep seals soft and pliable, a seal swell agent is added to the fluid. High temperatures tend to dry out the seals and make them hard. The seal swell agent helps the fluid to pass into the seals and keep them soft.

Fluidity Modifiers

A transmission fluid must be able to flow readily through small openings in the valve body. Fluidity modifiers are added to the fluid to make this possible. The operation of fluidity modifiers can be compared to gasoline and water. Gasoline is more fluid and is able to pass through certain filters that would trap water.

All of the additives discussed except the last three, are also used in high quality motor oil, but in different amounts. Friction modifiers, seal swell agents, and fluidity modifiers are the additives that set transmission fluid apart from motor oil.

AUTOMATIC TRANSMISSION SEALS

Seals in an automatic transmission serve to stop outside leaks at the output shaft, torque convertor, and manual linkage connections. Seals also prevent internal leaks at servo and clutch pistons and help to control leaks at accumulators and at the clutch support. The small amount of leakage which does occur at the clutch support serves to lubricate the support and clutch bore.

Lip-type Seals

Lip-type seals are made from synthetic rubber and are used to stop outside leaks (those visible on the outside of the transmission). Seal A in figure 7-3 is a transmission output shaft seal and has a boot that stops dirt from entering the transmission from the drive shaft slip joint. Seal B is a front pump seal and has been cut to show the small spring that goes around the circumference of the lip. This spring is called a *garter spring* and is used to keep the seal lip tight on the pump drive hub of the torque converter.

One important point about this type of seal is that the lip of the seal must face the fluid. In other words, the lip of the seal must face the inside of the transmission, figure 7-4.

Other lip-type seals are used on some clutch and servo pistons. On the servo piston shown in figure 7-5, the seal has been partially removed to show how it fits into a machined groove on the piston. Notice that this type of seal does not have a garter spring. The lip of the seal is larger in diameter than

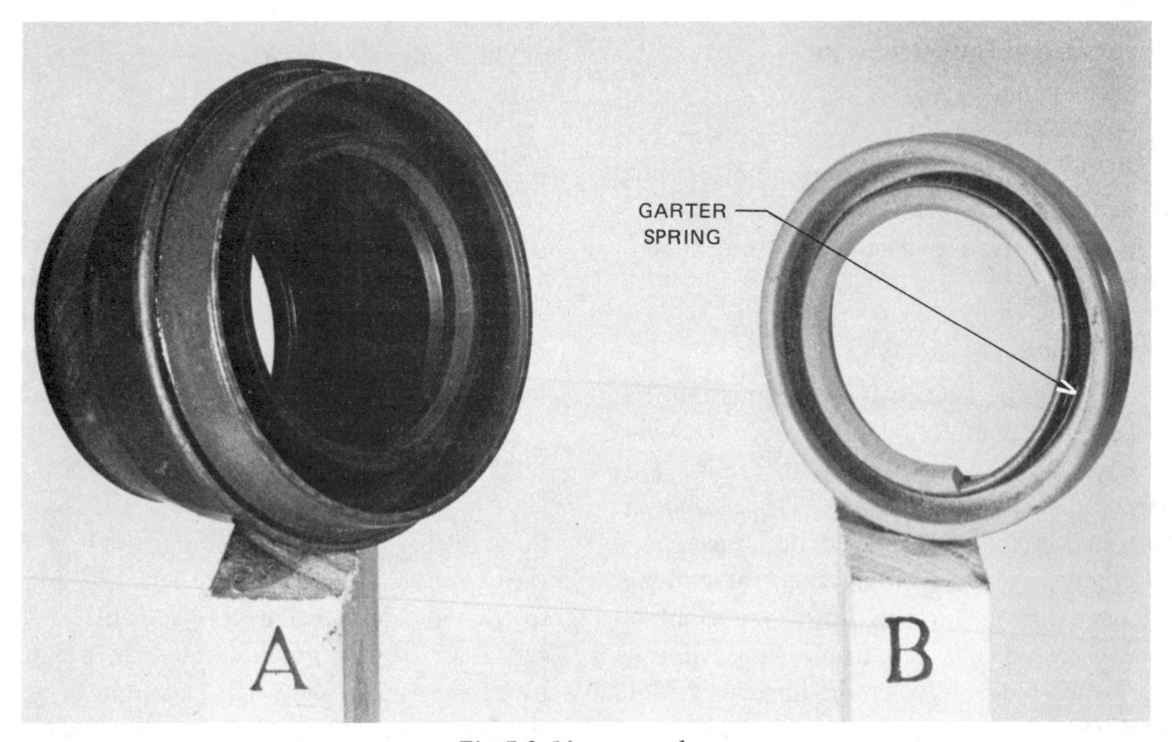

Fig. 7-3 Lip-type seals.

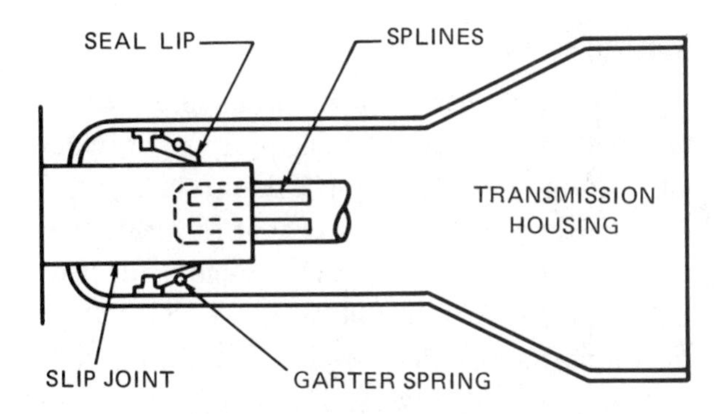

Fig. 7-4 Placement of lip-type seal.

Fig. 7-5 Lip-type servo piston seal (seal partly removed to show groove).

the cylinder. This keeps lip tension on the cylinder wall. When the servo is applied, fluid pressure forces the lip tight against the cylinder wall for an even better seal.

O-rings

O-rings are used to seal certain shafts in automatic transmission. As shown in figure 7-6, fluid pressure crowds the seal to one side of the groove, forcing the O-ring to expand and form a tight seal. Notice that, unlike the lip-type seal, the O-ring can seal from both sides.

If O-ring seals were used on clutch or servo pistons, they would tend to roll as the piston moved back and forth in the cylinder. For this reason, some clutch and servo pistons are sealed with lathe cut seals, figure 7-7. On the clutch piston shown, the seal has been removed to show how the seal fits in a machined groove in the piston. The lathe cut seal is square or oblong in cross section, and seals in the same way as an O-ring. However, because of its square shape, the lathe cut seal cannot roll.

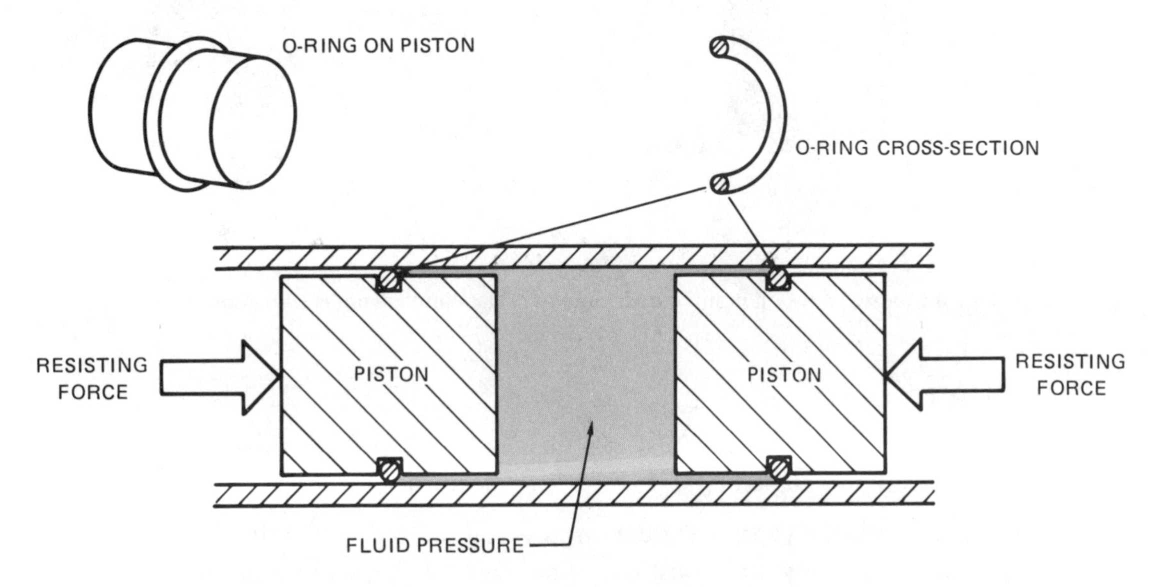

Fig. 7-6 Use of O-rings to seal shafts in an automatic transmission.

Fig. 7-7 Lathe cut seal and clutch piston.

Controlled-Leak Seals

In some cases small leaks are not a problem, or may even be necessary to provide lubrication. In figure 7-8, the pump clutch support has cast iron seal rings that fit inside the clutch hub. Fluid flows through a drilled passage in the clutch hub to move the clutch piston and apply the clutch. The slight fluid leak at the seal rings serves to lubricate the clutch support and hub. This small leakage is provided for in the design of the transmission and does not affect clutch lockup.

Fig. 7-8 Cast iron seal rings on pump clutch support. The clutch drum is shown on the right.

SUMMARY

- Automatic transmission fluid is a blend of high quality mineral oil and additives. It is formulated to meet the specifications of the car makers.
- The two main types of fluid in use today are: Type F, for Ford Motor Company transmissions; and Dexron, for American Motors, Chrysler Corporation, and General Motors Corporation transmissions.

- The blend of additives helps the fluid to (1) transmit pressure and motion; (2) transmit power through the torque converter; (3) help cool the transmission; and (4) lubricate the transmission.
- There are four types of seals used in automatic transmissions: lip-type, O-ring, lathe cut, and cast iron.
- The lip-type seal must be installed so that the lip faces the fluid to be sealed.

REVIEW

Select the best answer from the choices offered to complete the statement or answer the question.

1. Cast iron seal rings are designed to:

   (I) Be leak tight.
   (II) Control leakage.

   a. I only
   b. II only
   c. both I and II
   d. neither I nor II

2. O-ring seals

   a. are used on clutch and servo pistons.
   b. have garter springs.
   c. have two lips.
   d. depend on fluid pressure to help form a tight seal.

3. The lip of a seal must:

   (I) Face the fluid.
   (II) Have a garter spring.

   a. I only
   b. II only
   c. both I and II
   d. neither I nor II

4. Which of the following additives are found only in transmission fluid?

   a. viscosity index improver, pour point depressants, oxidation inhibitors
   b. friction modifiers, seal swell agents, fluidity modifiers
   c. rust inhibitors, corrosion inhibitors, foam inhibitors
   d. extreme pressure agents, detergents, dispersants.

5. The ability of an oil to keep a more constant viscosity over a wide temperature range is called

   a. viscosity scale.
   b. pour point.
   c. viscosity index.
   d. fluidity.

6. Fluidity refers to the ability of an oil to

   (I) Flow through small openings.
   (II) Thicken at sub-zero temperatures.

   a. I only
   b. II only
   c. both I and II
   d. neither I nor II

7. The pour point of an oil is determined by

   (I) Measuring the amount of oil that will flow through a metered orifice at 0°F.
   (II) The ability of an oil to flow under a wider range of temperature conditions.

   a. I only
   b. II only
   c. both I and II
   d. neither I nor II

8. In an oil, extreme pressure agents:

   (I) Tend to increase oil pressure.
   (II) Increase film strength.

   a. I only
   b. II only
   c. both I and II
   d. neither I nor II

9. Oxidation of an oil would take place **most readily**

   a. at abnormally high temperatures.
   b. at abnormally low temperatures.
   c. at normal operating temperatures.
   d. during warm-up.

10. An **increase** in temperature tends to:

    (I) Decrease the fluidity of an oil.
    (II) Increase the pour point of an oil.

    a. I only
    b. II only
    c. both I and II
    d. neither I nor II

11. A detergent-dispersant is used in an oil to:

    (I) Prevent varnish buildup.
    (II) Prevent impurities from forming large clots or particles.

    a. I only
    b. II only
    c. both I and II
    d. neither I nor II

EXTENDED STUDY PROJECTS

1. What effect might there be on the fluid if the fluid level is too high?
2. What effect might a fluid level that is too low have on the fluid?
3. What effect would the conditions in questions 1 and 2 have on transmission operation?
4. Why are lathe cut seals used on clutch and servo pistons instead of O-rings?
5. What effect would a leaking clutch or servo seal have on main line pressure?
6. How can a mechanic tell if an oil leak is from the transmission or the engine?
7. Why shouldn't Dexron and Type F fluids be mixed?

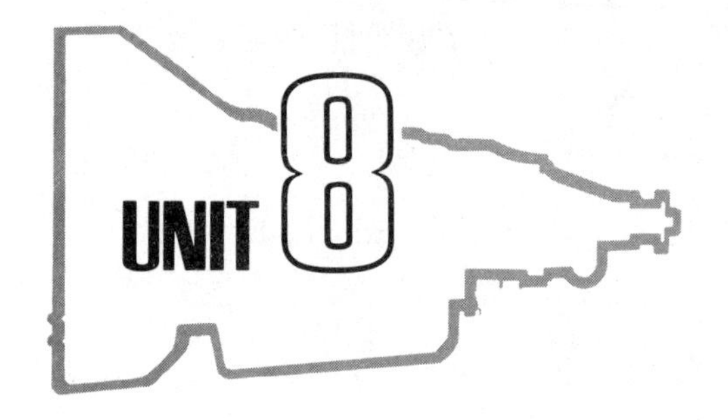

GENERAL MOTORS TYPE 300 AUTOMATIC TRANSMISSION

OBJECTIVES

After studying this unit, the student will be able to:

- Identify the bands and clutches in the General Motors Type 300 transmission.
- Trace the power flow in low, drive, and reverse.
- Explain the application of the bands and clutches to obtain the necessary power flow in the planetary gear set.
- Explain how the hydraulic system controls clutch and band application.
- Explain how fluid is directed in the hydraulic system for automatic and manual shifts.
- Diagnose transmission problems.
- Trace fluid flow on oil circuit diagrams for each speed range.

APPLICATION

The basic principles of the flow of power in a simple planetary gear set were introduced in unit 3. This unit illustrates the flow of power through a compound planetary gear set in a two-speed transmission. The General Motors Type 300 transmission is used as an example because of its popularity. It has been used in many Chevrolet vehicles, where it is called the Aluminum Powerglide. Other forms of this transmission have been used by Buick (Super Turbine 300), Oldsmobile (Jetaway) and Pontiac (M-35). Although the Type 300 is no longer being manufactured, it is found in many vehicles on the road today.

SHIFT POSITIONS OF THE TYPE 300

The transmission is coupled to the engine by a torque converter and has five shift lever positions: P(park), R(reverse), N(neutral), D(drive), and L(low). In park, the lubrication system is working, but the output shaft is locked to the transmission housing so the rear wheels cannot turn. In reverse, the reverse clutch is applied and this provides for reverse operation. In neutral, the rear wheels are free to turn, no clutches or bands are applied, but the lubrication system is working. Drive is the normal operating position and gives automatic upshifts and downshifts. This is accomplished by the application and release of a low band and high clutch. In low, the transmission will start and stay in low regardless of road speed.

In a variation of this transmission, called the Torque Drive, the hydraulic system is designed so that the transmission has no automatic shifts. With this transmission, the driver must shift manually from low to direct.

COMPOUND PLANETARY GEAR SETS

As described in unit 3, a simple planetary gear set has one sun gear, one set of planetary pinions, and one ring gear. A compound gear

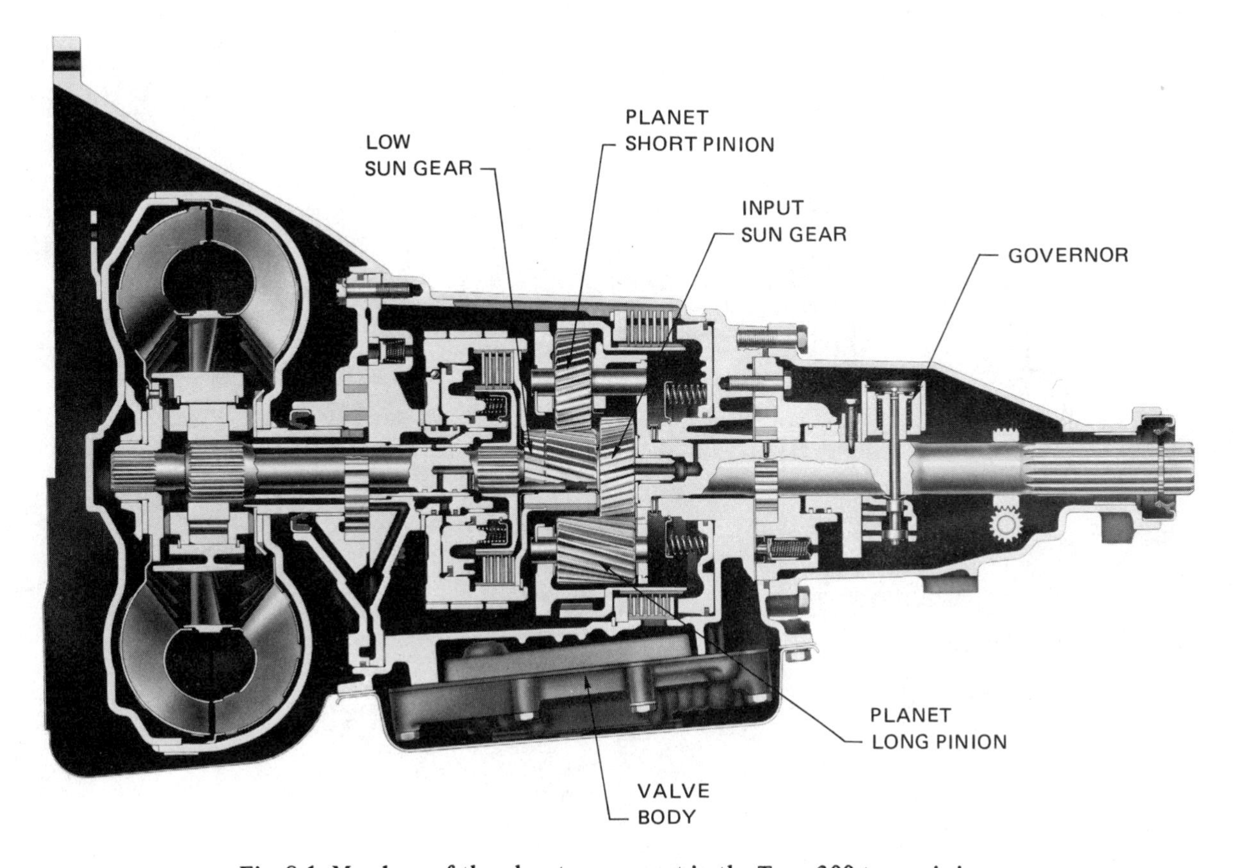

Fig. 8-1 Members of the planetary gear set in the Type 300 transmission.

set has more than one of some of these members. The Type 300 has two sun gears, two sets of planet pinions, and one ring gear, figure 8-1. This type of gear set is known as the *Ravingeau gear train* and is easy to control. In this case, it is controlled by one band and two clutches. The control required for a simple planetary gear set would be too complicated for dependable operation. Hence, a compound gear set is used.

POWER FLOW

When diagnosing problems, it is very important for the mechanic to know what is happening inside the transmission. Specifically, he or she must know which clutch or band is being used for each speed range and what effect each clutch or band has on the planetary gear set.

Notes on the Illustrations. In the illustrations that follow, the power flow through the gear train is indicated by shading. The illustrations show only two long pinions and two short pinions. The actual gear set has three long pinions and three short pinions. Only two are shown in the illustrations to make it easier to follow the power flow.

By referring to figure 8-1, it can be seen that the actual gear train is more compact than it appears to be in the illustrations. Also, note that the complete gear train is shown only in figure 8-5. This is also done to make it easier to follow the power flow from one range to the next.

Neutral or Park

Power flow in neutral or park is through the torque converter, the turbine or input shaft, and to the input sun gear, figure 8-2. This means that whenever the engine is running, the input shaft and input sun gear are turning. Since no other member of the gear set is being held, the condition is neutral.

Drive Range Low and Low Range

When the shift lever is moved to D, or drive range, the low band is applied, figure

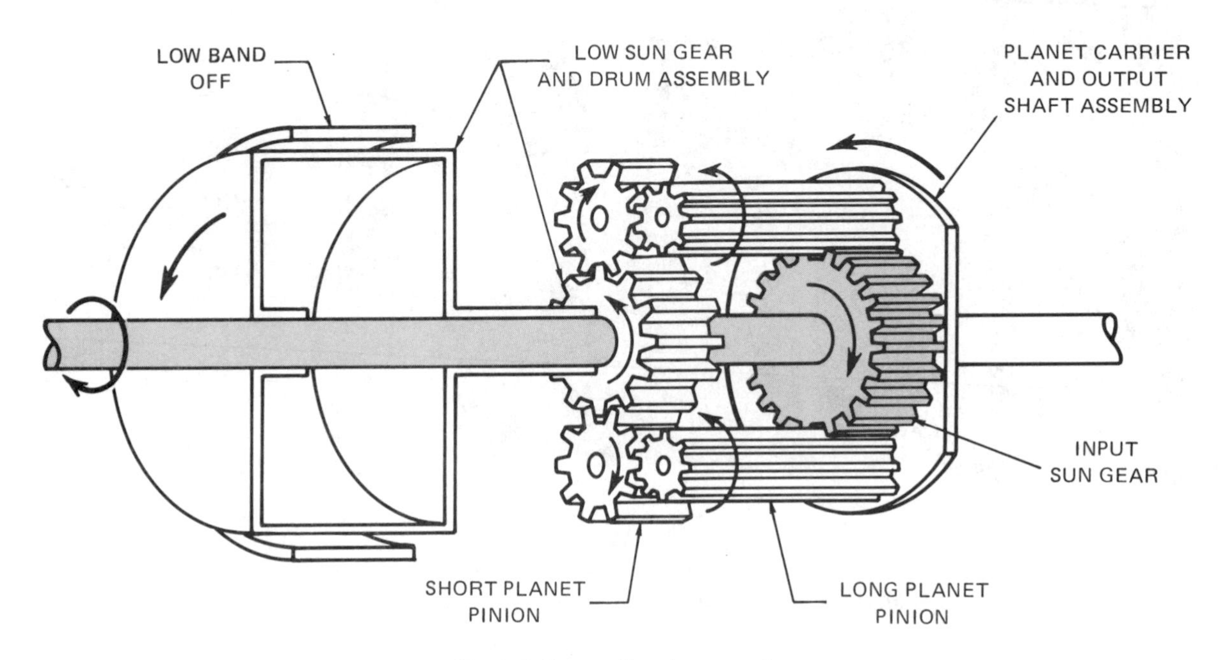

Fig. 8-2 Power flow in neutral.

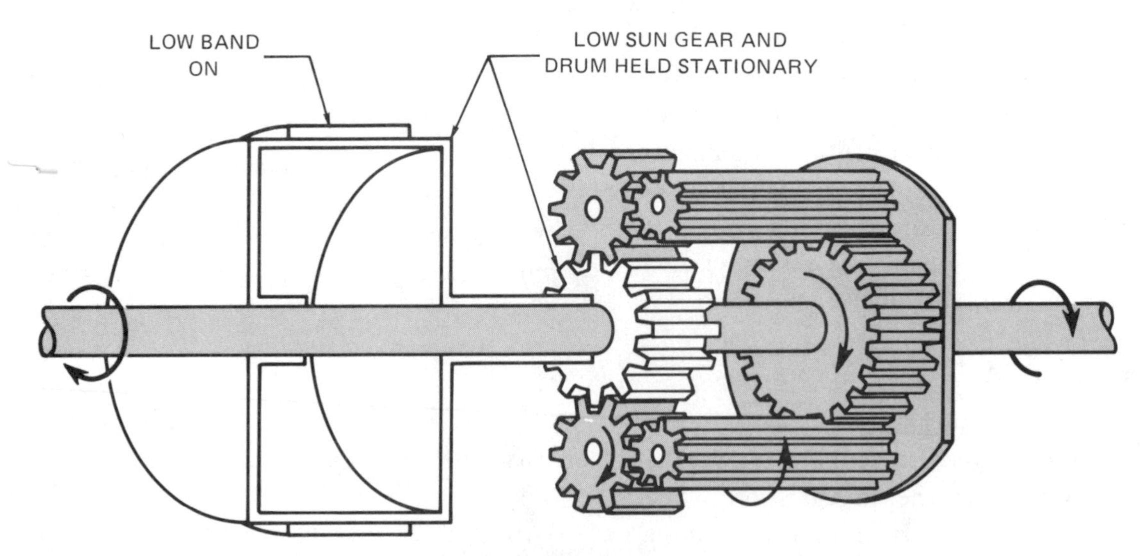

Fig. 8-3 Power flow in low.

8-3. Since the low band holds the low sun gear stationary, the power flow is through the input shaft or input sun gear, and to the long planet pinions. Because the long pinions are in mesh with both the input sun gear and the short pinions, the short pinions must also turn. The low sun gear is being held stationary by the low band. This means that the short pinions must walk around the low sun gear. Both sets of pinions are pinned to the carrier and output shaft assembly. Therefore, the walking of the short pinions turns the carrier and output shaft. This causes a reduction in speed.

Depending on the engine/transmission combination used, the reduction is 1.82:1 or 1.76:1, and stays in effect as long as the low band is on. In manual low, or L range, the transmission will not shift out of low and power flow will be as shown in figure 8-3. In drive range low, however, the transmission will shift to high according to road speed and throttle opening.

Drive Range High

When the transmission shifts to high gear, two things happen: the low band is released and, at the same time, the high clutch is applied, figure 8-4. The high clutch hub is splined to the input shaft of the transmission, and the clutch drive plates are splined to the clutch hub. The clutch-driven plates are splined to the clutch drum and sun gear assembly. Thus, both the input sun gear and the low sun gear must turn with the input shaft. With the two sun gears driving, the gear set must turn as a unit in direct drive. This is the same as driving two members of a simple planetary gear set, with the result being direct drive.

Reverse

In reverse operation, both the high clutch and low band are off, and the reverse clutch is applied, figure 8-5. Clutch plates are splined to both the ring gear and the transmission housing as shown. With the reverse clutch on,

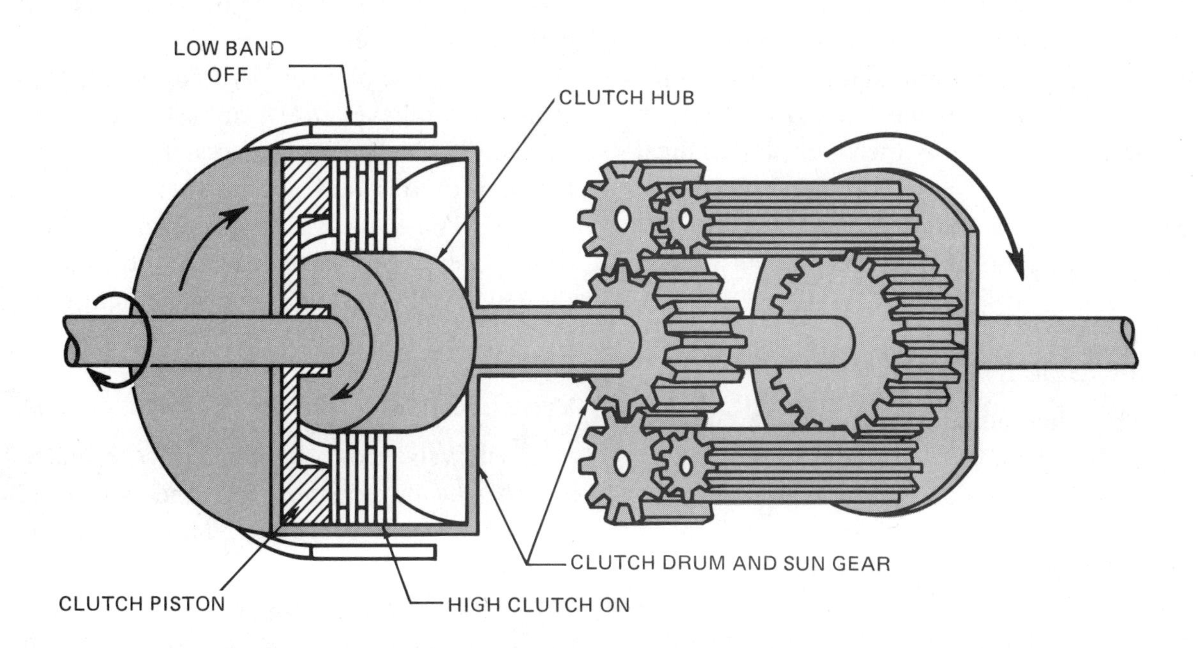

Fig. 8-4 Power flow in drive range high.

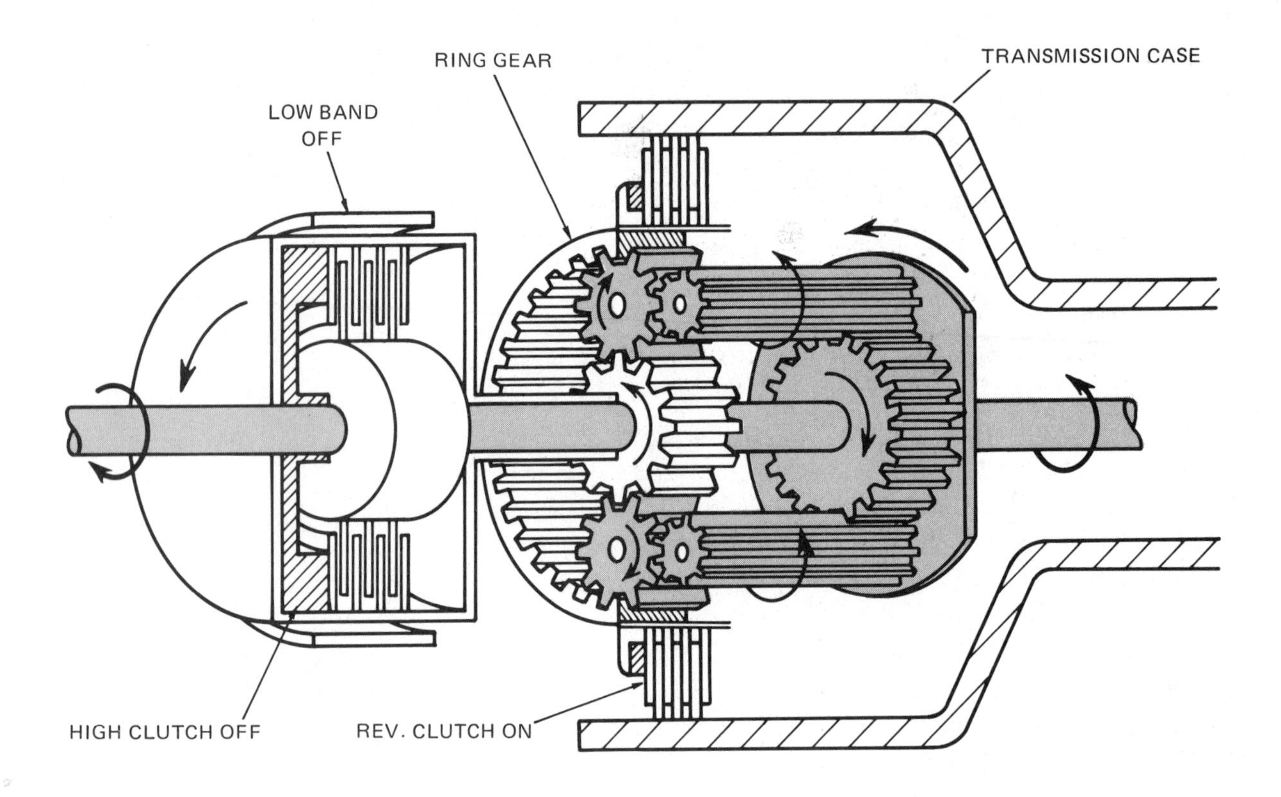

Fig. 8-5 Power flow for reverse.

the ring gear is held stationary, and the power flow is through the input shaft and input sun gear to the long pinions, and to the short pinions. To this point, the setup is the same as for low, but in reverse, the short pinions must walk around the inside of the stopped ring gear. This turns the carrier and output shaft in the same ratio as that for low, but in a reverse direction.

HYDRAULIC SYSTEM

The hydraulic system can be thought of as a computer that does the thinking for the transmission. The valve body is the "brain," the governor is the speed sensor, and the throttle valve and vacuum systems are torque sensors. The information from the sensors is fed into the valve body. Using this information, the valve body decides when to cause a shift either up or down and when to raise or lower main line pressure.

The principles of hydraulics are presented in units 5 and 6 and should be reviewed at this point if necessary. The remainder of this unit applies these principles to the GM Type 300 transmission. Many transmission problems can be traced to the hydraulic system. Hence, it must be thoroughly understood.

Valve Locations

The valve body is bolted to the underside of the transmission case and is covered by the oil pan or sump. Most valves are located in the valve body with the exception of the governor valve and three valves in the oil pump body. The governor valve is located on the output shaft. The downshift timing valve, the cooler bypass valve, and the

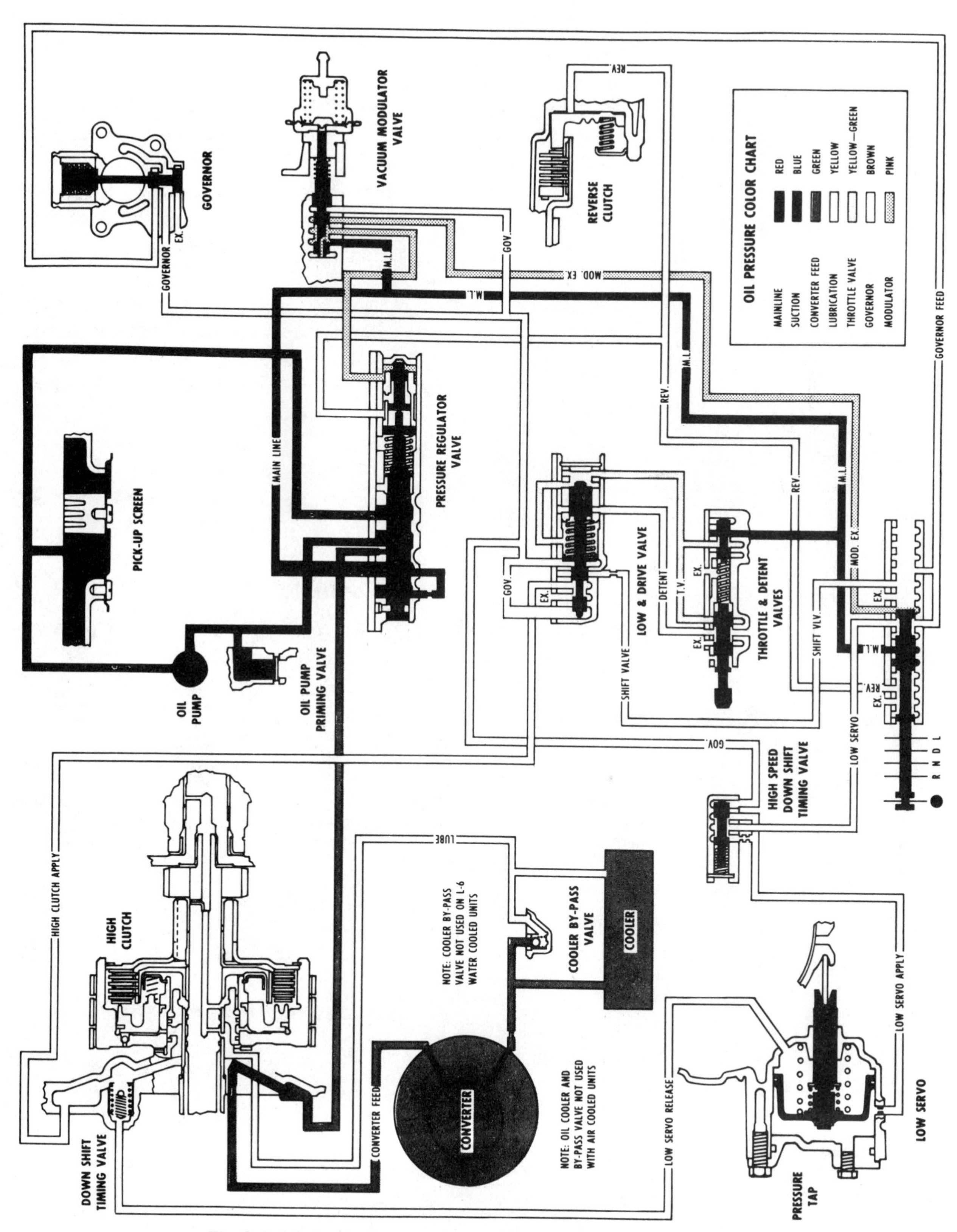

Fig. 8-6 Oil circuit diagram. (Neutral or park with engine running.)
[For color diagram, see page 299.]

oil pump priming valve are found in the oil pump body. There are usually several things going on at once in the hydraulic system. To give a clear explanation, each valve and its use will be discussed separately.

Converter and Lubrication System

The oil pump is of the internal-external gear type and is driven by the converter hub. This means that whenever the engine runs, fluid is pumped to the hydraulic system. *Note:* Oil circuit diagrams are used here to show the operation of the lubrication system. Study each of the diagrams presented and trace the fluid flow through the system. Being able to read an oil circuit diagram is a must for quick, accurate, diagnosis of transmission problems.

As shown in figure 8-6, fluid is drawn through the pickup screen by the pump and flows under pressure to the pressure regulator valve. Note the oil pump priming valve in this line. The purpose of this valve is to bleed off any air that may be trapped in the system. After the air is bled, the rising pressure closes the valve.

At the pressure regulator valve, fluid passes through a valley in the valve to the reaction face. As the valve moves to the right against the spring force, the converter feed port is opened. Fluid is forced through the converter feed circuit to a drilled passage in the pump body. It is then forced into the converter between the converter hub and pump stator support.

The return fluid passes between the input shaft and inside diameter of the stator support. From here the fluid moves through the cooler and back to the lubrication system. Note the cooler bypass valve. This valve will open and bypass the cooler should the cooler become blocked or plugged. Air-cooled models do not use a cooler or by-pass valve.

Main Line Circuit

As the converter is filled and pressure builds, the regulator valve moves further to the right and oil bypasses to the suction side of the pump. This regulates main line pressure according to the spring force at the right of the valve. The passage marked *main line* or

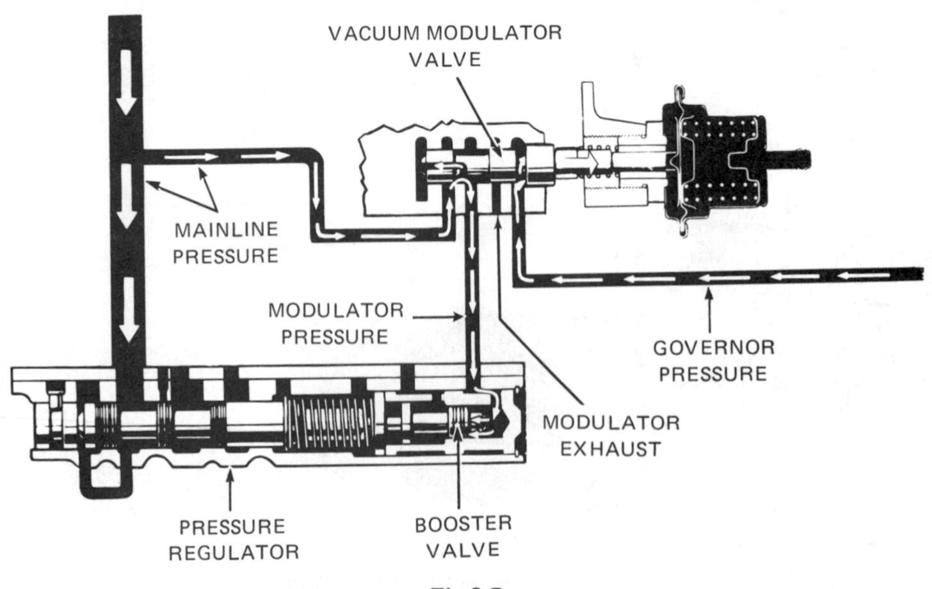

Fig 8-7

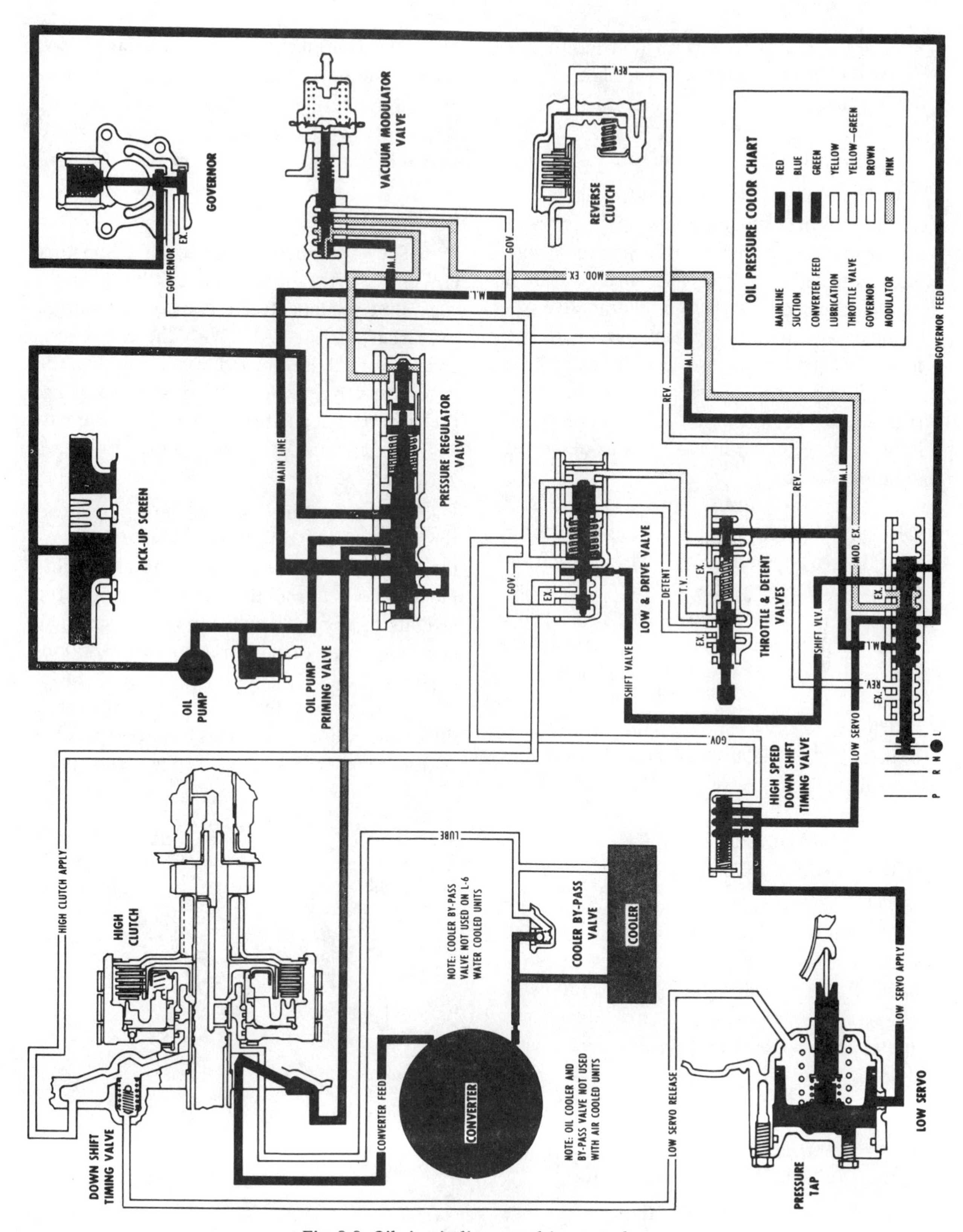

Fig. 8-8 Oil circuit diagram - drive range low.

[For color diagram, see page 300.]

ml can be traced to the vacuum modulator valve, the manual valve, and the throttle and detent valves.

Modulator Circuit

The vacuum modulator valve, figure 8-7, is made up of a vacuum unit which screws to the rear face of the transmission case and works on the modulator valve that is located in the valve body. The modulator valve is a balanced, restriction-type regulator. Main line pressure enters between the first two valleys of the valve and flows through a drilled passage in the valve to the reaction face. Pressure is regulated in the valve as described in units 5 and 6.

Modulator pressure flows to the booster valve at the right of the pressure regulator (also shown in figure 8-7). The force from modulator pressure adds to the spring force, and main line pressure now depends on spring force and throttle opening. Remember, low engine vacuum results in high modulator pressure.

Thus far, two separate circuits at different pressures have been considered: main line pressure and modulator pressure.

THE HYDRAULIC SYSTEM IN DRIVE RANGE LOW AND HIGH

In drive range low, main line pressure and modulator pressure are controlled the same as in park or neutral. In drive range low, however, the manual valve opens the following passages to main line pressure: low servo apply, shift valve, and governor feed, figure 8-8, page 85.

Low servo oil from the manual valve flows through the valley of the high speed downshift timing valve to the apply side of the low servo. (The use of this valve will be discussed in detail later in the unit.) Since the low sun gear is now being held, the transmission is in drive range low. Shift valve oil is blocked by a land of the low-drive valve. The low-drive valve is a shift valve, and works in the same way as those shown in unit 6.

Governor Circuit

The governor is made up of a hub that is locked to the output shaft, a body, and a valve that is connected to two counterweights by a shaft, figure 8-9. With the car at rest, governor feed is blocked by a land on the governor valve. As the car starts moving, the governor regulates main line pressure to governor pressure. Governor pressure changes with road speed.

Counterweights control governor pressure through centrifugal force, figure 8-10. As road speed increases, the governor weights move outward, and the shaft connecting the weights to the valve pulls the valve open. The governor valve then operates as a restriction-type regulator.

Notice that the governor valve has a differential force area. This balances the force of the weights and helps keep governor pres-

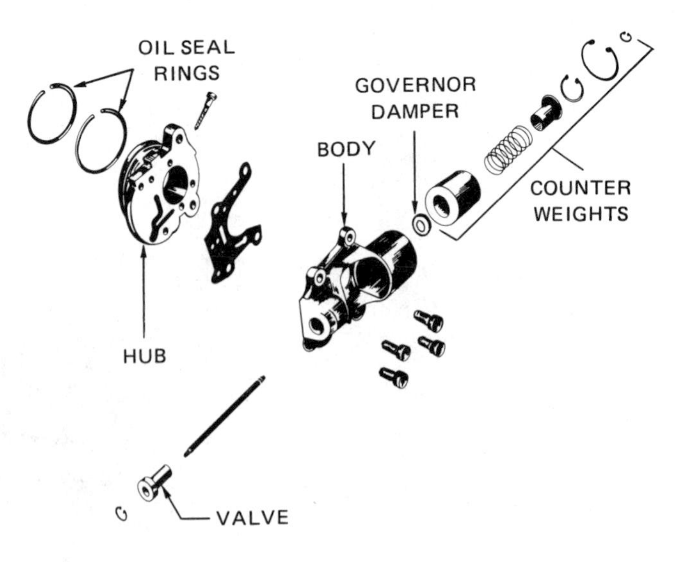

Fig. 8-9 Parts of a governor valve.

sure balanced to road speed. The large counterweight controls governor pressure at low road speeds, while the small weight controls governor pressure at high road speeds.

Governor pressure (GOV) passes from this valve to the modulator valve, the low-drive valve, and the high speed downshift timing valve. Governor pressure works on a differential force area of the modulator valve. This tends to reduce modulator pressure and main line pressure. Lower main line pressure helps to smooth out part-throttle upshifts.

Low-drive Valve

Governor pressure at the low-drive valve works to move the valve to the upshift position, while throttle pressure and spring force work to keep the valve in the downshift position. Throttle or TV pressure is regulated by throttle opening as explained in units 5 and 6.

As road speed increases, governor pressure overcomes the spring force and TV pressure on the low-drive valve. When the low-drive valve moves to the right, main line pressure passes through a differential force area of the valve to the high clutch apply passage, figure 8-11. This helps snap the valve to the right. Notice that this snap action is also helped by the exhaust of TV pressure from a differential force area of the valve. Note that TV pressure exhausts through the detent passage of the throttle and detent valve.

Timing Clutch Apply and Band Release

With the low-drive valve moved to the right, main line (shift valve) oil takes two paths:

- through the valve body to a drilled passage in the front pump housing, to the clutch apply piston.

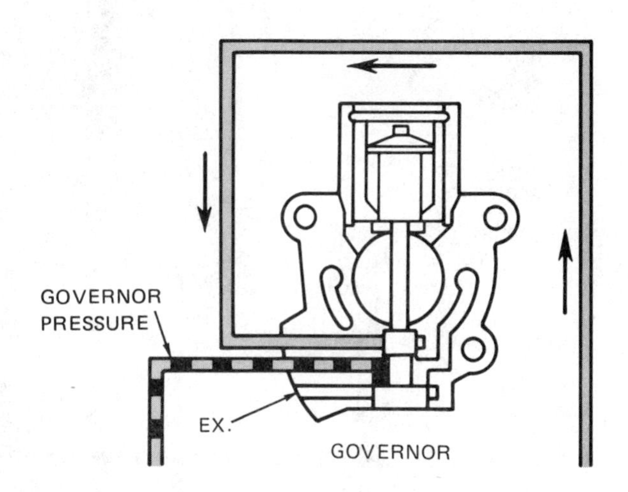

Fig. 8-10

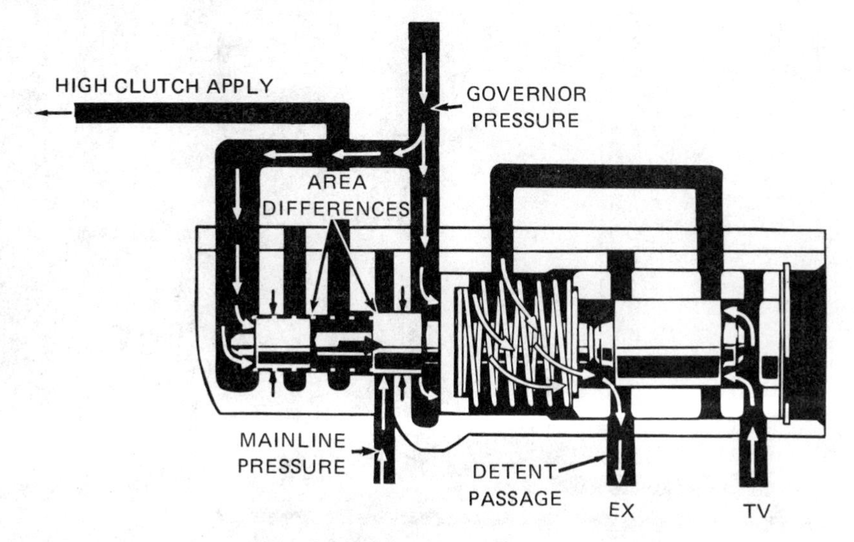

Fig. 8-11 Operation of low-drive valve.

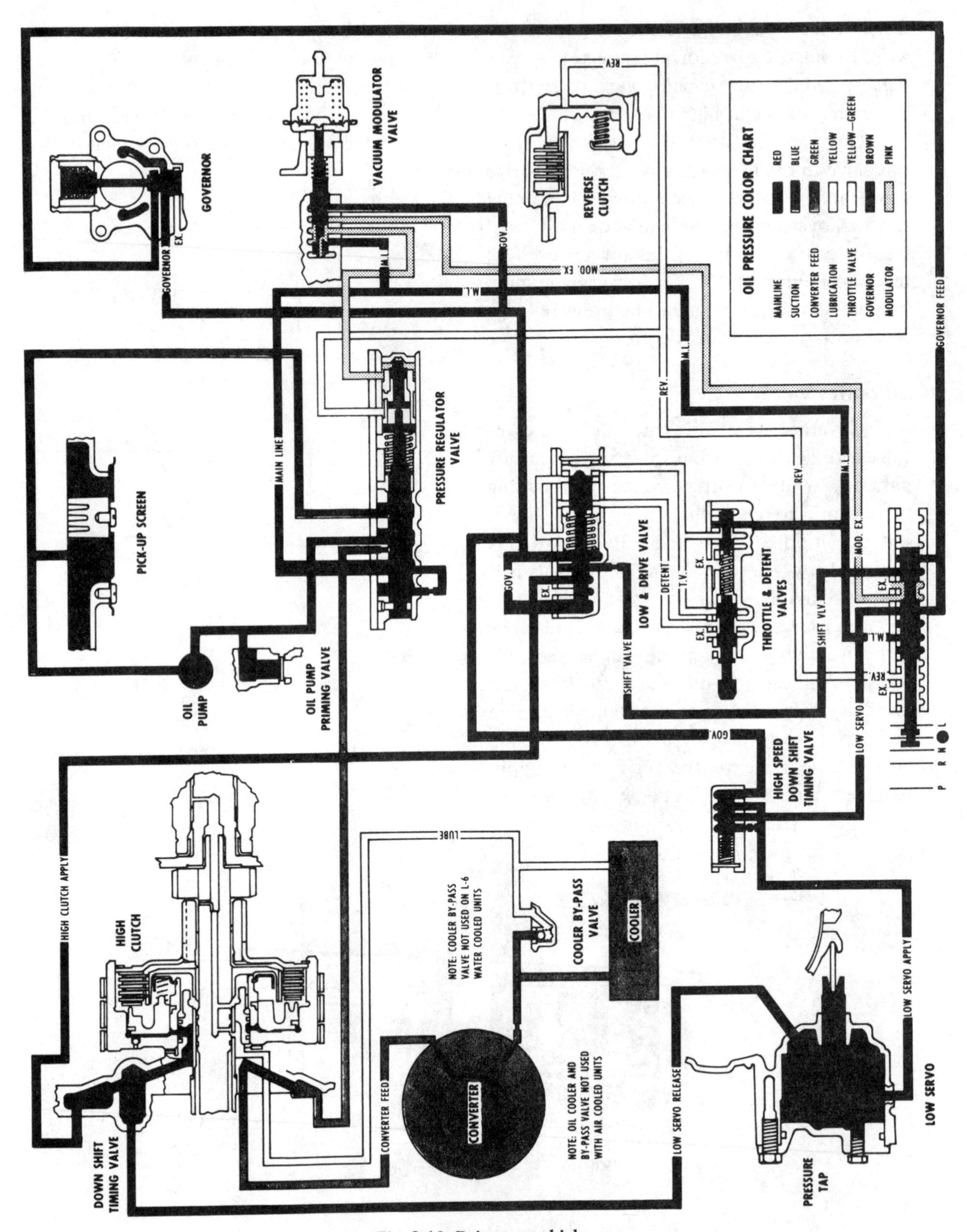

Fig. 8-12 Drive range high.

[For color diagram, see page 301.]

- unseating the downshift timing valve and passing through the low servo release passage to the release side of the servo.

Since servo apply oil and servo release oil are both in the main line system, the pressure is the same on both sides of the servo piston. This allows the servo return spring to stroke the piston to the OFF position. With the low servo off and the high clutch on, both sun gears are driven, and the transmission is in high or direct drive, figure 8-12.

Controlling Shift Quality

Shift quality or feel is controlled by:

- modulator pressure acting on main line pressure.
- the accumulator action of the servo release.

At small throttle openings, modulator and main line pressure are low. This gives a smooth, soft shift. As the throttle is opened more, the shift becomes firmer, until at wide-open throttle, it is very fast and firm.

An orifice, located at the low-drive valve, also helps to control clutch apply and band release. In addition, because high clutch apply and low servo release are part of the same passage, the servo acts as accumulator for the high clutch as the piston is stroked to the OFF position.

TYPES OF SHIFT

As stated earlier, the shift point depends on road speed and throttle opening. Three types of shifts are used in road testing: closed throttle, detent touch, and full detent. The following shift speeds are typical of only one engine/transmission use. Actual shift speeds vary with tire size, axle ratio, and engine size.

Closed Throttle

On a level road, using the least throttle opening to cause a speed increase, a shift from low to high should take place at 14 to 17 mph (25 to 27 km/h). If the shift does not take place at these speeds, something is wrong. For instance, if the shift is late, three things should be recognized immediately as possible causes of the problem: a bad governor, stuck low-drive valve, or a problem in the TV system such as adjustment or a stuck valve.

Detent Touch

If the driver pushes the throttle toward the floor board, a resistance will be felt just before wide-open throttle is reached. This is known as *detent touch.* Since TV pressure is very high at this point, upshift should take place at 44 to 53 mph (71 to 85 km/h).

Full Detent

Full detent is wide-open throttle, and under this condition, upshift should take place at 53 to 60 mph (85 to 97 km/h). At full detent, full TV pressure works on the low-drive valve, and high road speeds are

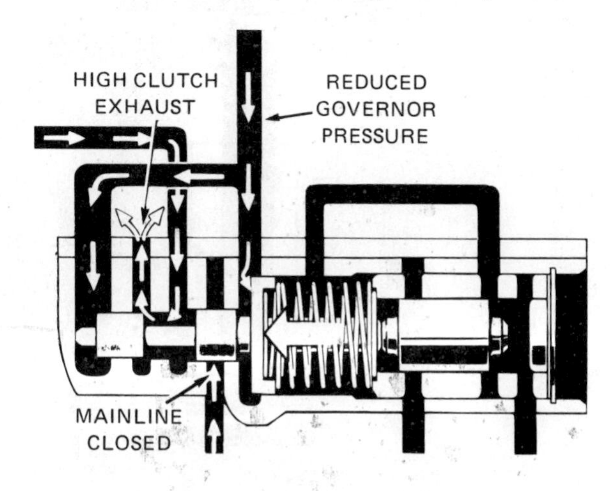

Fig. 8-13 Closed throttle downshift.

needed to cause enough governor pressure to move the low-drive valve.

AUTOMATIC DOWNSHIFTS

The types of downshift are the same as for upshift: closed throttle, detent touch, and full detent. Recall that on upshift, TV pressure is exhausted from a differential force area of the low-drive valve. This helps keep the transmission in high even at quite large throttle openings, allowing full use to be made of the torque converter.

Closed-Throttle Downshift

At coast, with the throttle closed, a downshift should take place between 12 and 16 mph (19 to 26 km/h). This occurs when governor pressure, due to the slower road speed, falls off, and TV pressure and spring force move the low-drive valve to the left, figure 8-13, page 89. This allows the high clutch to exhaust as shown. To provide a smooth downshift, servo release oil must pass through two orifices at the downshift timing valve. This gives the high clutch time to release before the low band is fully on. The transmission is again in low gear and the oil circuit is as shown in figure 8-9.

Detent Touch Downshift

At speeds below about 15 to 23 mph (24 to 37 km/h) a forced downshift can be made by opening the throttle enough to cause high TV pressure. Governor pressure is overcome and the transmission shifts to low. As long as the throttle is held steady, the transmission will stay in low until governor pressure is again high enough to overcome TV pressure. This type of shift is also called a *torque-demand shift.*

Full Detent

At road speeds below 49 to 57 mph (79 to 92 km/h), the driver can force a shift to low by pushing the throttle wide open. This allows full TV pressure to act on the low-drive valve through both the TV and detent passages, figure 8-14. This type of shift is known as a *detent downshift.* As long as the throttle

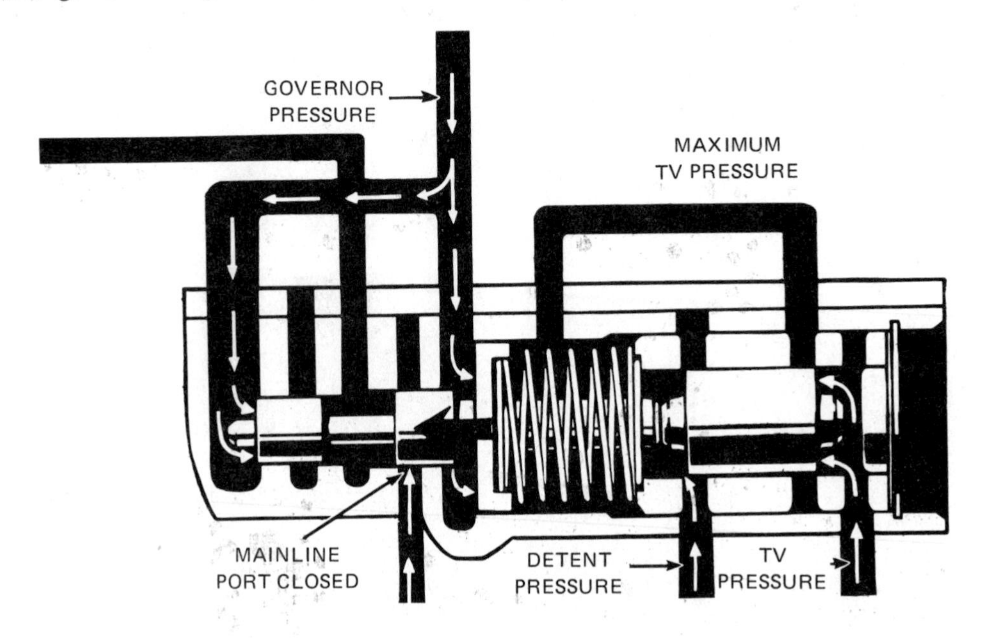

Fig. 8-14 Detent downshift.

is held wide open, the transmission will stay in low until road speeds of 53 to 60 mph (85 to 97 km/h) are reached.

HIGH-SPEED DOWNSHIFT TIMING VALVE

At high road speeds, the engine r/min should increase during a downshift. This provides a smooth downshift in much the same way as does increasing engine speed to match road speed during a downshift in a standard transmission. To obtain this effect, a high-speed downshift timing valve is used, figure 8-15. At speeds of about 30 mph or less (48 km/h), the spring force holds the valve to the right. Low servo apply oil can then follow two paths:

- through the valley of the valve to the open path on the right, or
- through the passage at the left that has a restriction orifice.

In actual use, with the valve in the closed position, most of the fluid flows through the open path, and servo apply is quite fast.

At speeds of 30 mph (48 km/h) and above, governor pressure moves the valve to the left. When this happens, the open path is blocked by the right-hand land of the valve. All apply oil must then pass through the restriction orifice. This action slows up and times the band application to the release of the clutch, giving a smooth downshift.

MANUAL LOW CIRCUIT

The transmission can be locked in low by moving the shift lever to the L position. In this range, main line pressure is sent to the low servo and modulator exhaust, figure 8-16, page 92. Main line pressure at the modulator exhaust stops the modulator from regulating and allows full main line pressure to act on the pressure regulator boost valve. This provides the additional main line pressure needed for higher torque loads on the band (caused by pulling heavy loads) or for engine braking on steep hills. Since line pressure is not sent to the shift valve, the transmission will stay in low as long as the manual valve is in the L position.

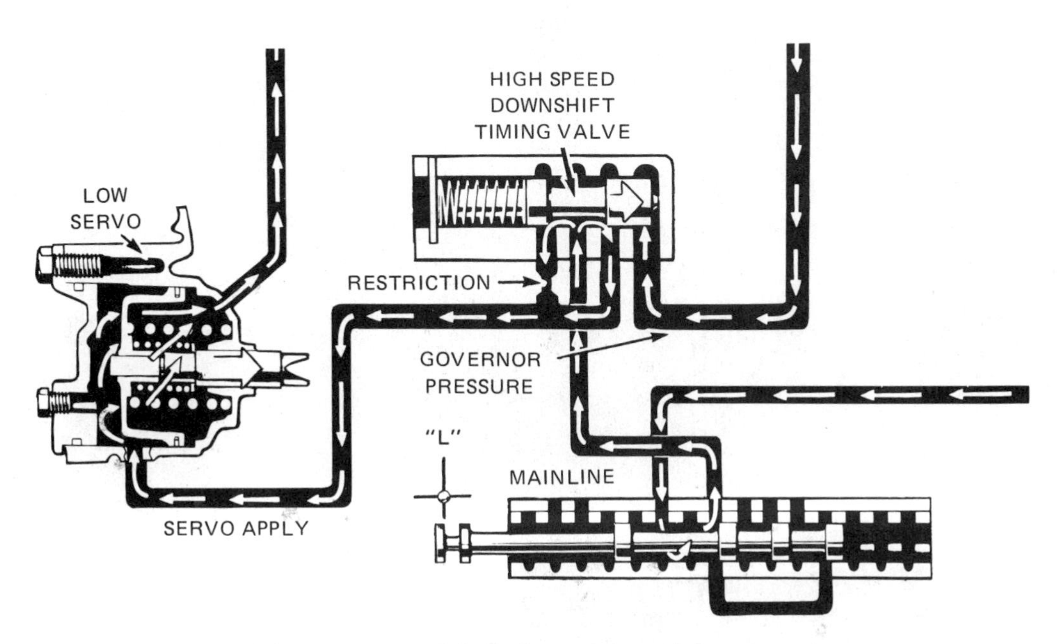

Fig. 8-15 Operation of the high speed downshift timing valve.

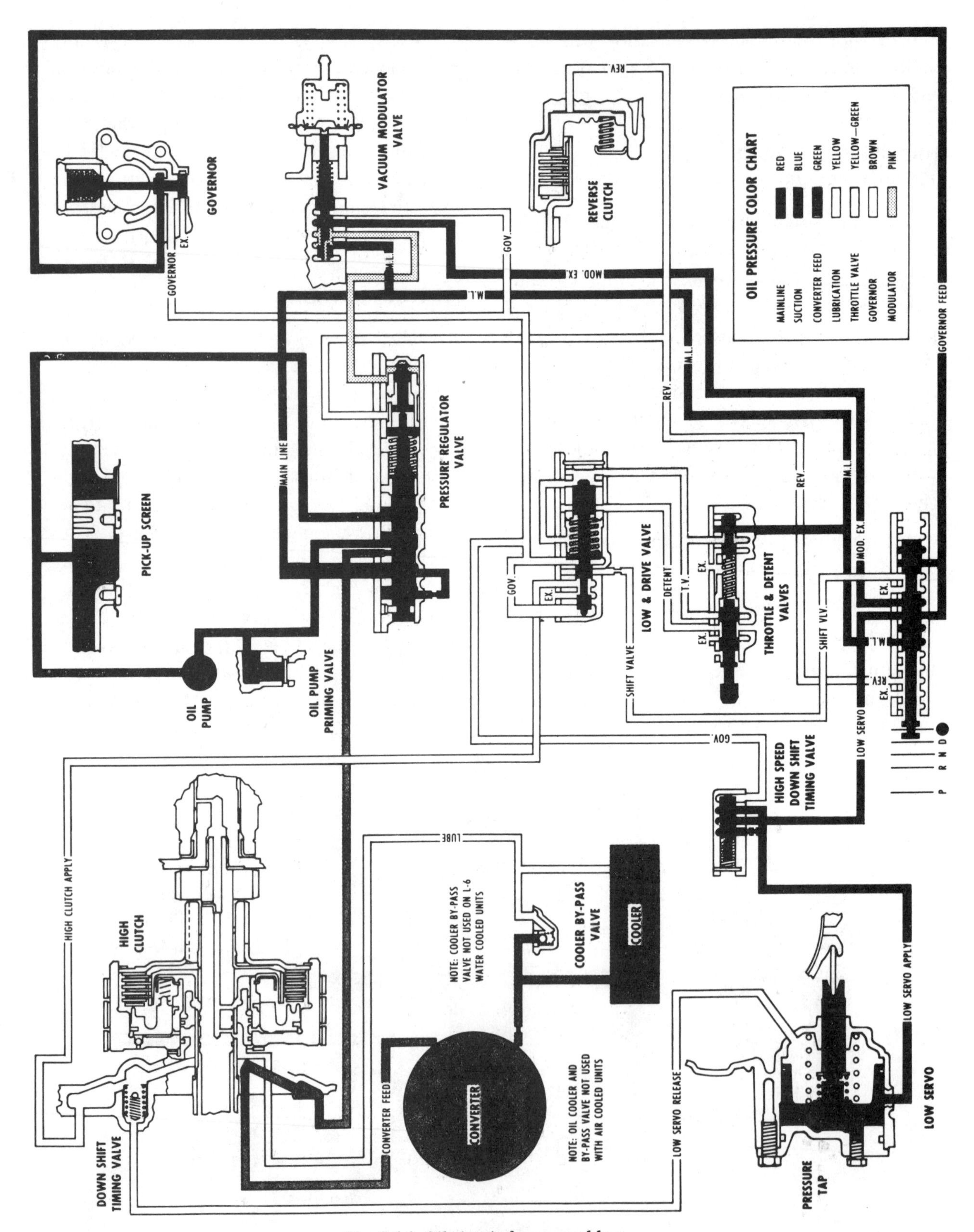

Fig. 8-16 Oil circuit for manual low.

[For color diagram, see page 302.]

Downshift to Manual Low

If the transmission is in high and the manual valve is moved to L, high clutch oil will pass through the low-drive valve and exhaust at the manual valve. This will shift the transmission to low at any road speed. The oil circuit for this condition is the same as that in figure 8-16.

Caution: Since, in manual low, the transmission is locked in low or will downshift at any speed, the driver should not use manual low at speeds above the full detent shift points. To do so could cause serious damage to the engine, transmission, or drive line.

REVERSE

With the manual valve in R, main line pressure passes only through the reverse circuits. The high and low circuits are open to exhaust at the manual valve, figure 8-17. With the reverse clutch holding the ring gear, the pinions walk around the stopped ring gear, rotating the carrier and output shaft in a reverse direction.

Due to the increase in torque loads in reverse, main line pressure is sent to a differential force area of the pressure regulator booster valve. This boost pressure, plus modulator pressure, increases main line pressure to its maximum. High main line pressure gives a tighter lockup to the reverse clutch and prevents slipping due to the increased torque load.

DIAGNOSIS

Knowing how the clutches, bands, and hydraulic system work with one another is an important tool in diagnosing transmission problems. All of the steps used in diagnosis, or troubleshooting, are discussed in unit 16. At this point, note and keep in mind what takes place when the transmission is in the different ranges.

Knowledge of the system is used in the following way. If the problem is no reverse, the mechanic remembers, or traces on an oil circuit diagram, the various things that could cause this problem. In this case, it could be a manual valve out of

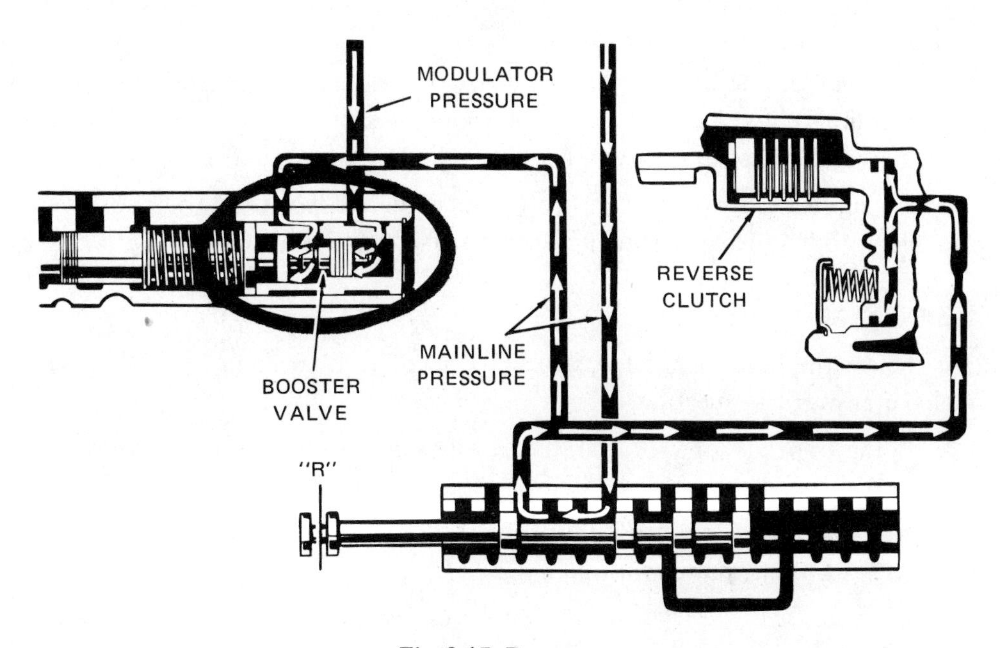

Fig. 8-17 Reverse.

place (linkage adjustment), a blown seal in the reverse clutch, worn or burned reverse clutch plates, or a blocked reverse passage. Knowing these possibilities, the mechanic makes other tests to close in on the problem. If the problem is a blocked reverse passage, the valve body can be removed and repaired without removing the transmission; this saves both time and money.

SUMMARY

The General Motors Type 300 transmission is a fully automatic transmission having two forward speeds and reverse. Following is a summary of the power flow and hydraulic system in this transmission.

- In D range when first starting off, or in L range, the low band is applied to hold the low sun gear stationary.
- When the shift is made to high, the low band is released and the high clutch is applied. In this range, both sun gears are driving.
- In R, the reverse clutch is applied to hold the ring gear, and both the low band and high clutch are released.
- The hydraulic system has four separate systems: main line, TV, governor, and modulator.
- The main line system serves to charge the converter; cool and lubricate the transmission; feed the governor, modulator, and TV system; and apply the band and clutches.
- The governor system provides pressure that increases with road speed. It works on the low-drive valve to produce upshifts, on the modulator to decrease modulator pressure, and on the high-speed downshift timing valve.
- The modulator provides pressure that increases with increasing engine torque. It exerts a force on the main pressure regulator to increase main line pressure as torque is increased.
- TV pressure increases with throttle opening, and works on the low-drive valve to help balance shift points with governor pressure.

REVIEW

Select the best answer from the choices offered to complete the statement or answer the question.

1. If the transmission will not shift until high road speeds are obtained, a probable cause is:

   (I) Improper TV linkage adjustment.
   (II) A leaking modulator diaphragm.

   a. I only
   b. II only
   c. both I and II
   d. neither I nor II

2. To control the planetary gear system, the transmission makes use of

   a. a high clutch, a low band, and a reverse band.
   b. a high clutch, a reverse clutch, a low band, and a one-way clutch.
   c. a high clutch, a reverse clutch, and a high band.
   d. a high clutch, a reverse clutch, and a low band.

3. A harsh closed-throttle downshift could be caused by:

   (I) A stuck high-speed downshift timing valve.
   (II) A leak in the modulator diaphragm.

   a. I only
   b. II only
   c. both I and II
   d. neither I nor II

4. Which of the following would not cause a harsh upshift?

   a. TV rod adjustment
   b. high clutch not applying
   c. improper band adjustment
   d. stuck throttle valve

5. When the band is applied the:

   (I) Input sun gear is allowed to turn.
   (II) Low sun gear is held.

   a. I only
   b. II only
   c. both I and II
   d. neither I nor II

6. Governor pressure acts on the modulator valve to:

   (I) Decrease modulator pressure.
   (II) Help smooth out part throttle upshift.

   a. I only
   b. II only
   c. both I and II
   d. neither I nor II

7. If the high clutch and the band is being applied, the transmission is in

   a. intermediate.
   b. direct.
   c. reverse.
   d. none of these

8. The input sun gear is splined to the

   a. reverse clutch drum.
   b. output shaft.
   c. input shaft.
   d. high clutch drum.

9. The low sun gear is meshed with the

   a. output shaft.
   b. planet short pinions.
   c. output sun gear.
   d. planet long pinions.

10. In the Torque Drive transmission,

    a. the driver must change speed ranges manually.
    b. there is no main pressure regulator valve.
    c. the manual valve has been removed.
    d. a clutch is used instead of a torque converter.

11. The upshift is controlled by
    a. control pressure.
    b. throttle valve pressure and governor pressure.
    c. modulator pressure and governor pressure.
    d. modulator pressure and TV pressure.

12. Creep in neutral could be caused by:
    (I) Improper manual linkage adjustment.
    (II) The servo piston not releasing.
    a. I only
    b. II only
    c. both I and II
    d. neither I nor II

13. Engine flare or slipping under large throttle openings could be caused by a:
    (I) Clogged oil suction screen.
    (II) Vacuum leak.
    a. I only
    b. II only
    c. both I and II
    d. neither I nor II

14. Which of the following would **not** be a cause of no upshift?
    a. defective governor
    b. stuck low-drive valve
    c. stuck throttle valve
    d. clutch not releasing.

15. The input sun gear is meshed with the
    a. planet short pinions.
    b. planet long pinions.
    c. output sun gear.
    d. high clutch.

16. When the high clutch is off and the band is on, the transmission is in
    a. low.
    b. direct.
    c. reverse.
    d. none of these

17. No drive in any range could be caused by:
    (I) TV linkage out of adjustment.
    (II) A defective pump.
    a. I only
    b. II only
    c. both I and II
    d. neither I nor II

18. On a detent downshift, the low-drive valve is moved by:
    (I) governor pressure.
    (II) modulator pressure.
    a. I only
    b. II only
    c. both I and II
    d. neither I nor II

19. When the high clutch is on and the band is off, the transmission is in
    a. low
    b. direct
    c. reverse
    d. none of these

20. In a Powerglide-equipped car, no upshift could be caused by
    (I) The clutch not releasing.
    (II) A defective governor.
    a. I only
    b. II only
    c. both I and II
    d. neither I nor II

EXTENDED STUDY PROJECTS

On blank oil circuit diagrams, trace the circuits listed below. The following color codes are suggested for use with colored pencils or markers:

Main line – red
Suction – blue
Converter feed – green
Lubrication – yellow
TV – yellow-green
Governor – brown
Modulator – pink

1. Indicate the oil circuit for park or neutral.
2. Indicate the oil circuit for drive range low. On the back of the circuit, list three possible causes for no drive in D.
3. Indicate the circuit for drive range high. On the back of the circuit, list three possible causes for no upshift.
4. Indicate the oil circuit for low.
5. Indicate the oil circuit for reverse. On the back of the circuit, list one possible cause for no reverse.

Try to complete the diagrams as far as possible without referring to the text. A helpful method is to trace the line back from the clutch or band that is in use. For example, if the circuit being traced is drive range low, start at the low servo and trace the low servo apply passage through the high speed downshift timing valve to the manual valve.

THE BORG-WARNER AUTOMATIC TRANSMISSION

OBJECTIVES

After studying this unit, the student will be able to:

- Identify the bands and clutches in the Borg-Warner transmission.
- Trace the power flow in all forward speeds and reverse.
- Compare band and clutch application to power flow through the planetary gear set.
- Describe how the hydraulic system controls clutch and band application in this model.
- Explain how the hydraulic system controls automatic and manual shifts.
- Diagnose transmission problems in this model.
- Trace fluid flow on oil circuit diagrams for each speed range.

APPLICATION

The Borg-Warner automatic transmission is in use today in both American and foreign cars. Variations of this transmission are being or have been used by American Motors, Ford Motor Company, and many imported auto makers. The transmission is coupled to the engine by a torque converter, and the gear train is much like that of the GM Type 300. The Borg-Warner, however, has three forward speeds and reverse.

To simplify the description of this transmission, one particular model is discussed here. The model chosen is that used in the Plymouth Cricket. There are many variations from the model described here, but the major points are the same in all models. For example, the transmission discussed in this unit uses a throttle-linkage-controlled TV system, while other models use a vacuum-controlled TV. This is not an important difference however, since all forms of this transmission are related. Study of the Plymouth Cricket provides basic knowledge that can be applied to all models of the Borg-Warner transmission.

OPERATION

The shift selector on this model has five positions: P (park), R (reverse), N (neutral), D (drive), 2 (manual 2), and 1 (manual 1). Park, reverse, and neutral serve the same function as in the GM Type 300 and other transmissions. Drive range in the Cricket gives automatic shifts starting from low, and moving to intermediate and high.

In 2 range, the transmission starts off in low, shifts to intermediate, but will not shift to high. If the transmission is in high and the driver moves the shift lever to 2, the transmission will shift to intermediate and, as car speed decreases, to low.

Caution: To avoid damage to the engine from overspeeding, 2 range should not be used at speeds over 60 mph (97 km/h).

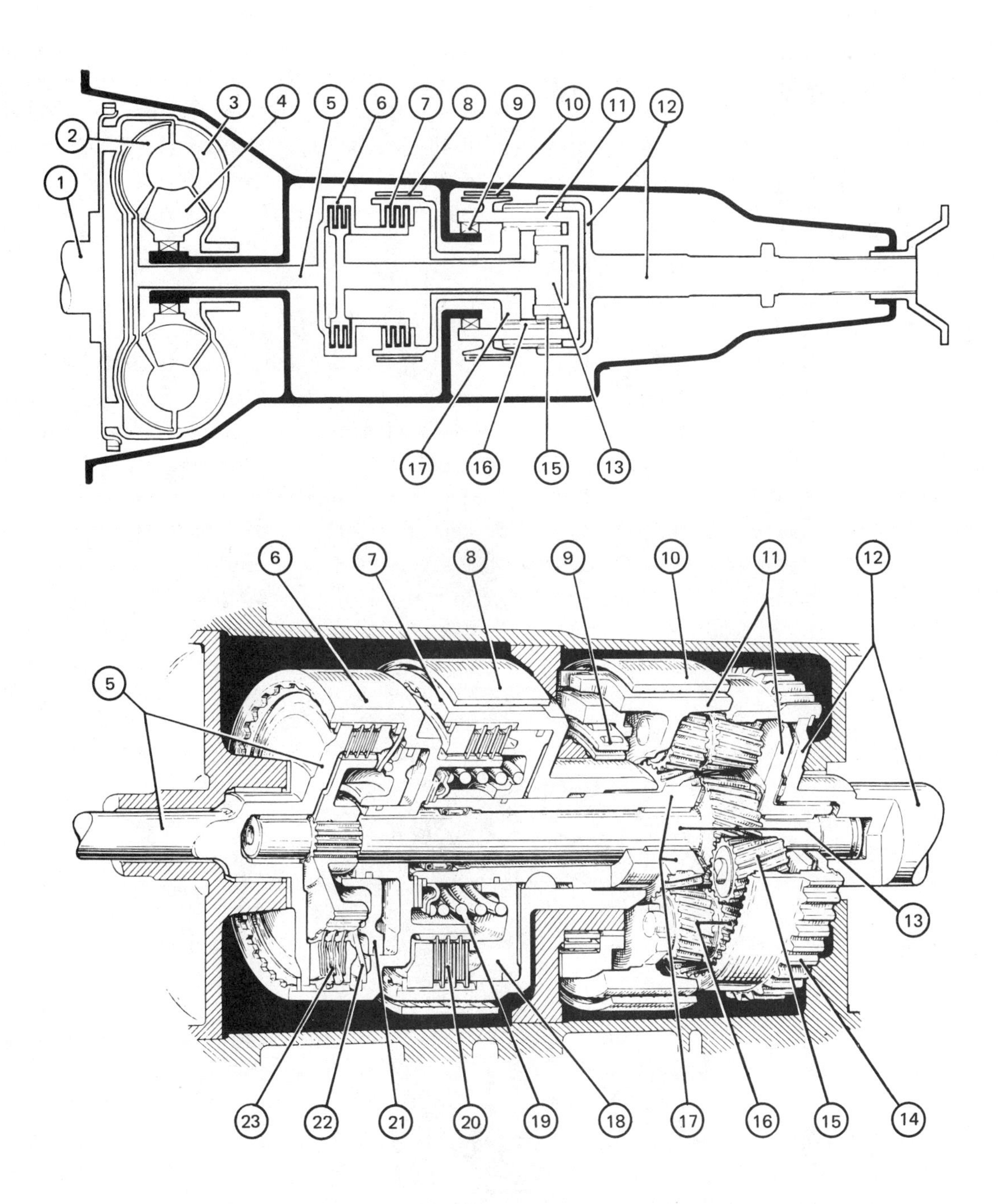

1. Engine Crankshaft
2. Turbine
3. Impellor Torque Converter
4. Stator
5. Input Shaft
6. Front Clutch
7. Rear Clutch
8. Front Brake Band
9. One-way Clutch (Sprag Type)
10. Rear Brake Band
11. Planet Pinion Carrier
12. Ring Gear (Annulus) and Output Shaft
13. Forward Sun Gear and Shaft
14. Parking Pawl Teeth
15. Short Planet Pinion
16. Long Planet Pinion
17. Reverse Sun Gear
18. Rear Clutch Operating Piston
19. Rear Clutch Piston Return Spring
20. Rear Clutch Plates
21. Front Clutch Operating Piston
22. Front Clutch Piston Return Spring (Disc Type)
23. Front Clutch Plates

Fig. 9-1 Schematic view of transmission and cutaway view of clutches and planetary gears.

In 1 range, the transmission starts off in low and will stay in low, no matter what road speed is reached. The shift lever can be moved to 1 at higher road speeds, and the transmission will shift to intermediate. Then, when road speed is low enough, the transmission will shift to low and remain there as long as the shift lever is in the 1 position. Below certain road speeds, the driver can force downshifts by pushing the throttle to the wide-open position.

POWER FLOW AND HYDRAULIC CONTROL

To control the gear train, this transmission makes use of a front clutch, a rear clutch, a front band, a rear band, and a one-way clutch, figure 9-1. The gear set has two sun gears, two sets of pinions, a pinion carrier, and a ring gear, figure 9-2. The difference between this gear set and the GM Type 300 gear set is that the ring gear of this set is part of the output shaft. Recall that in the Type 300, the carrier is part of the output shaft.

Neutral and Park

In neutral and park, power flow is from the converter to the input shaft. Since the front clutch is off, power cannot reach the gear set. In park, the rear band is applied. This is due to the design of this valve body in this particular transmission. However, since there is no power input, the gear set is still in neutral.

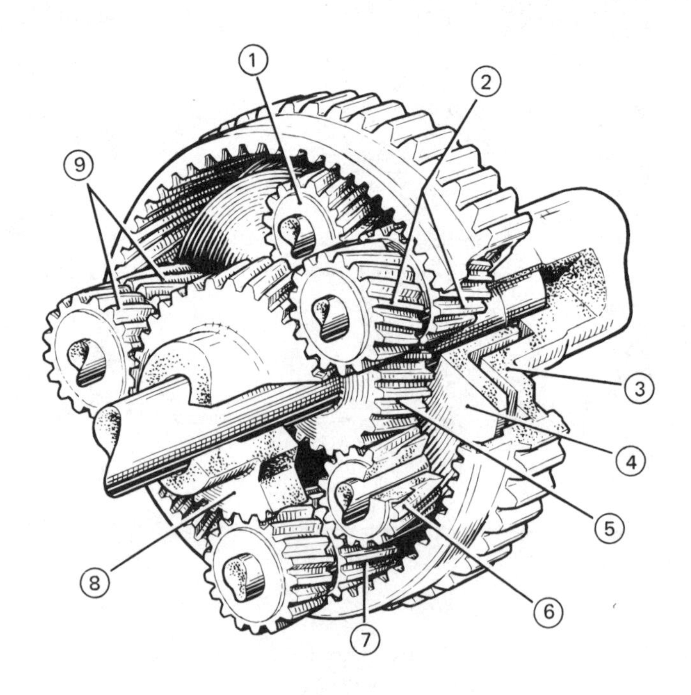

1 & 6	Short Planet Pinion (Third pinion not shown)
2, 7 & 9	Long Planet Pinion
3	Ring Gear (Annulus) and Output Shaft
4	Planet Carrier
5	Forward Sun Gear
8	Reverse Sun Gear

Fig. 9-2 Planetary set, cutaway view.

Whenever the engine is running, oil circulates in the transmission. The oil passages are shown in figure 9-3. Two passages (marked *21*) serve as converter feed and converter return lines. Oil is also moved through passages (marked 23) to lubricate the bushings and thrust washers, and through line 1, to the throttle and manual valves. As each range is discussed, the passage numbers on this illustration will be given. Refer to figure 9-3 as needed.

Drive Range Low

When the manual valve is placed in D range, oil flows through line 5 (figure 9-3) to the front clutch, the 1-2 shift valve, and the governor. Oil also flows to the 2-3 shift valve. With the front clutch applied, power flows from the input shaft through the front clutch to the forward sun gear (5) and shaft, figures 9-4 and 9-5. This sets up a force that tends to walk the carrier (4) in a reverse direction. However,

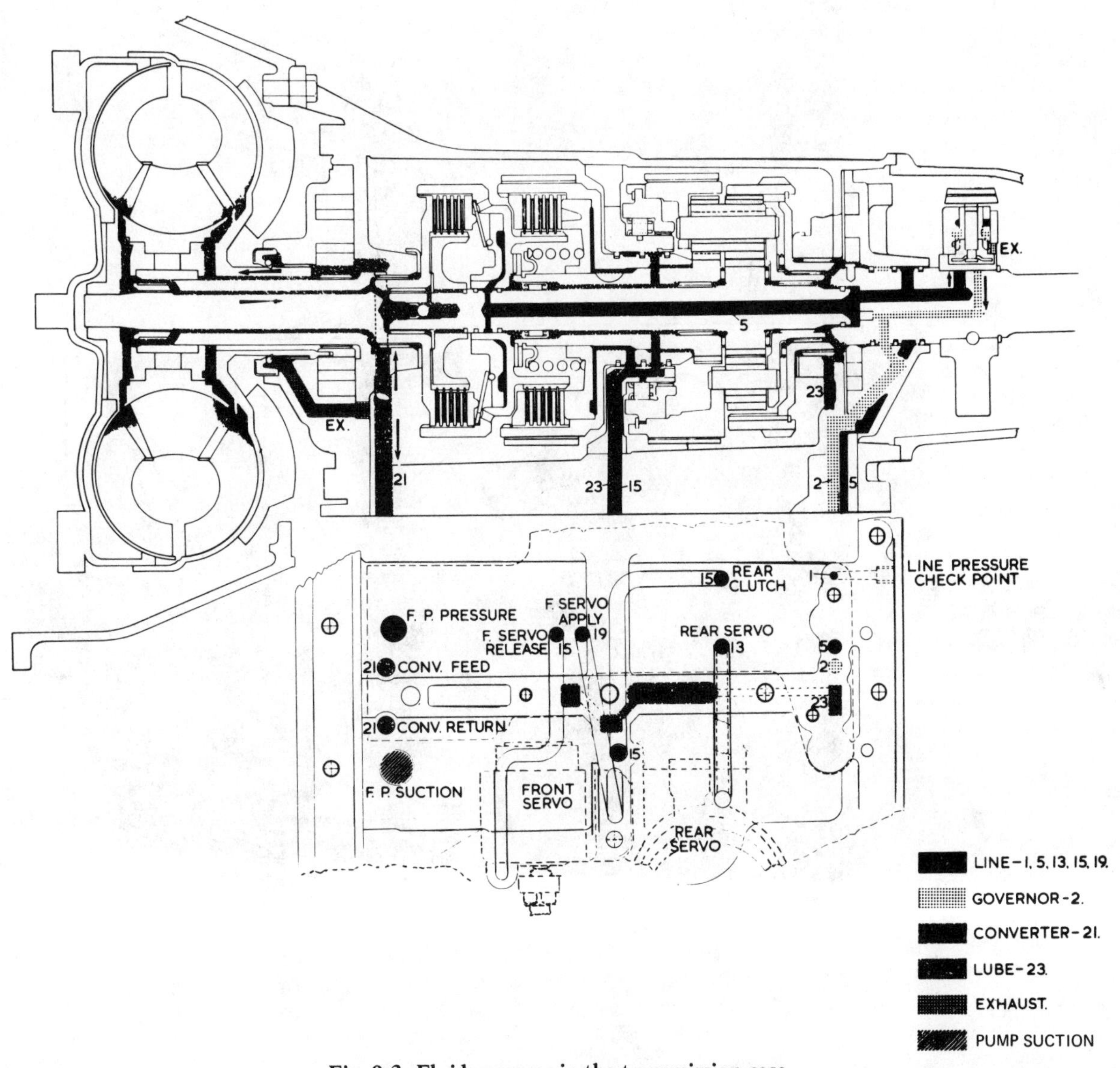

Fig. 9-3 Fluid passages in the transmission case.

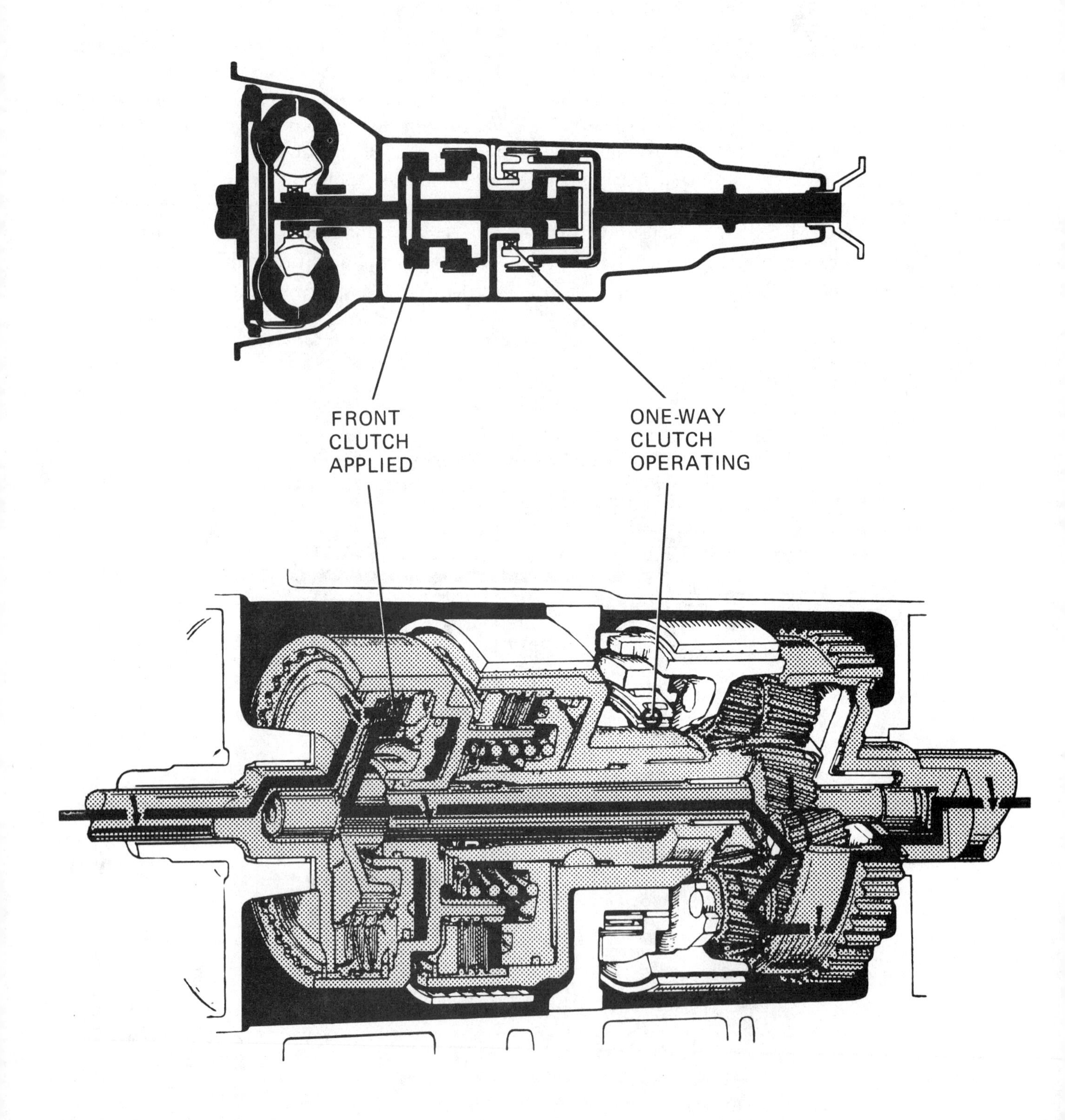

Fig. 9-4 Schematic for drive range low.

[For color diagram, see page 303.]

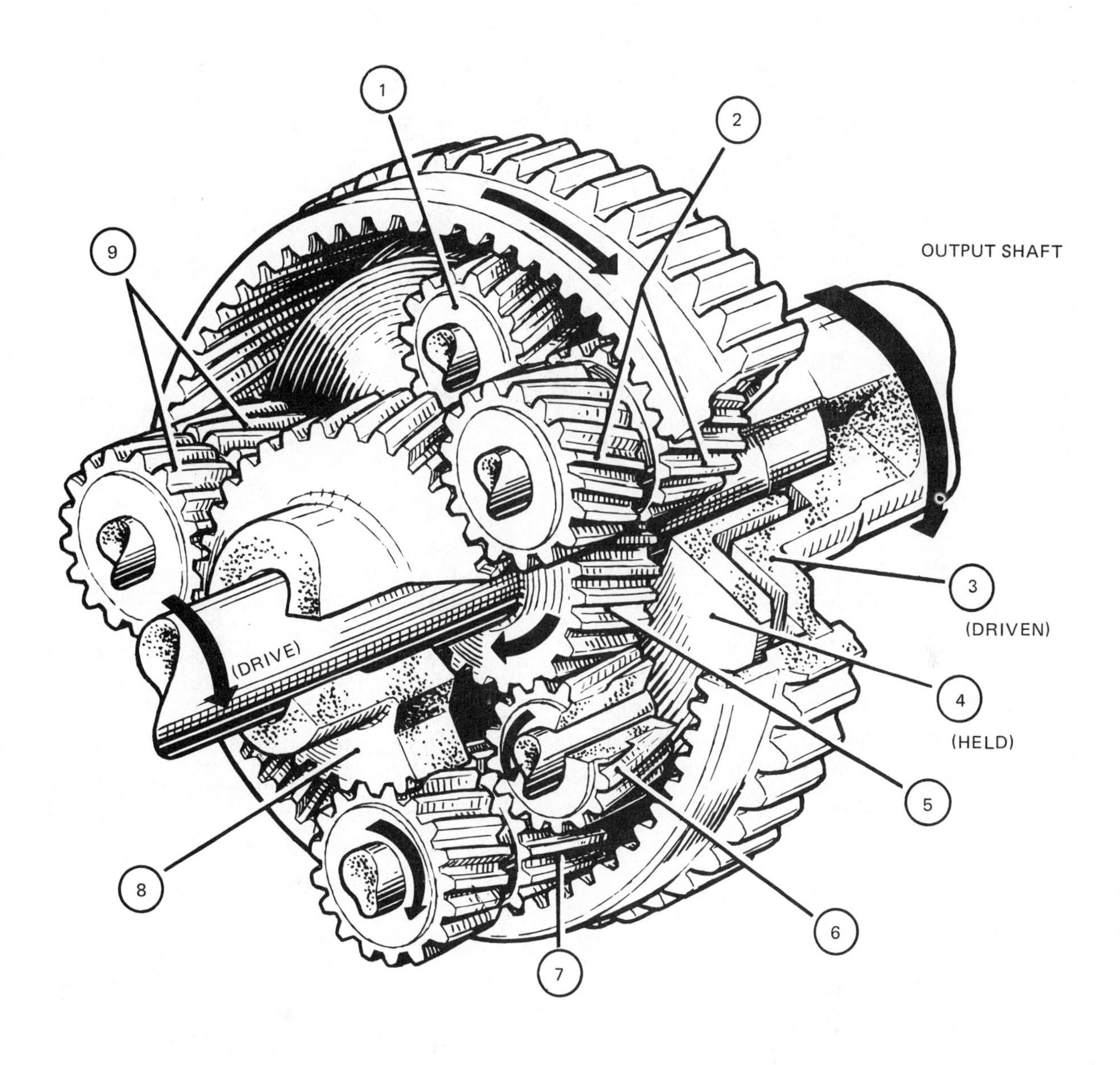

1 & 6	Short Planet Pinion (Third pinion not shown)
2, 7 & 9	Long Planet Pinion
3	Ring Gear (Annulus) and Output Shaft
4	Planet Carrier
5	Forward Sun Gear
8	Reverse Sun Gear

Fig. 9-5 Gear set in drive range low.

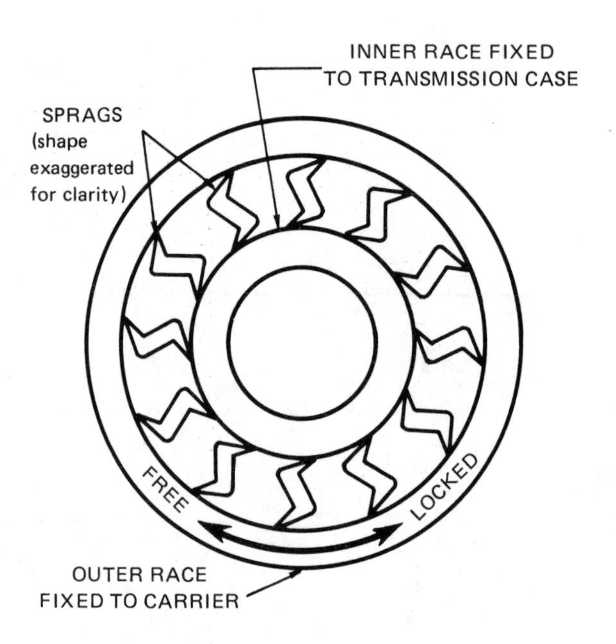

Fig. 9-6 One-way clutch (Sprag type).

the one-way clutch (figure 9-4) acts on the carrier to prevent reverse rotation. Since the carrier cannot turn, the ring gear and output shaft (3) are turned in a forward direction at a ratio of 2.39:1. Note that the reverse sun gear (8) merely rotates free in a reverse direction.

Sprag Clutch

The type of one-way clutch used in this transmission is called a *sprag clutch*, figure 9-6. The inner race of this clutch is fixed to the transmission case, and the outer race is fixed to the carrier. The sprags are placed between the races and are shaped in such a way that they tilt to lock the outer race for reverse rotation, or tilt for free rotation in a forward direction.

Use of a one-way clutch for drive range low leads to a smooth shift to D range 2. The band is not needed to hold the carrier, and there is no need to time the release of the band with clutch application.

Drive Range 2 or Intermediate

As road speed is increased, governor pressure moves the 1-2 shift valve against spring force and throttle pressure. When this happens, line pressure moves to the front servo apply through passage 19 (figure 9-3). As shown in figures 9-7 and 9-8, when the band comes on, the reverse sun gear (8) is brought to a stop. The forward sun gear is still being driven by the front clutch, but, with the reverse sun gear stopped, the one-way clutch will release. This causes the planet carrier and pinions (4, 6 and 7) to walk in a forward direction around the reverse sun gear. Through the action of the short and long pinions, the ring gear and output shaft (3) are turned in a forward direction at a ratio of 1.45:1.

Drive Range High

As road speed increases, governor pressure moves the 2-3 shift valve. When the shift valve moves, line pressure is sent

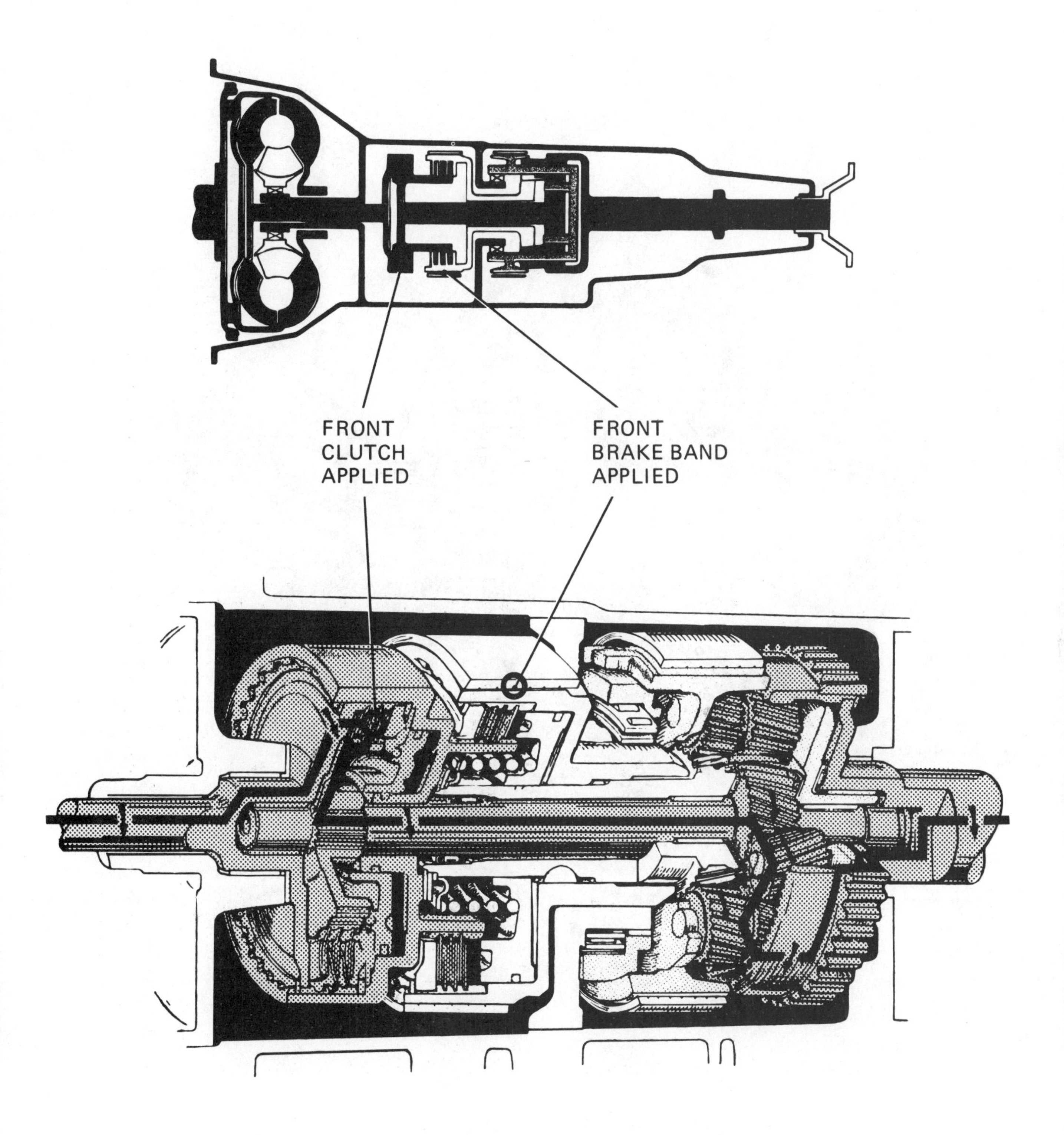

Fig. 9-7 Schematic for second gear.

[For color diagram, see page 304.]

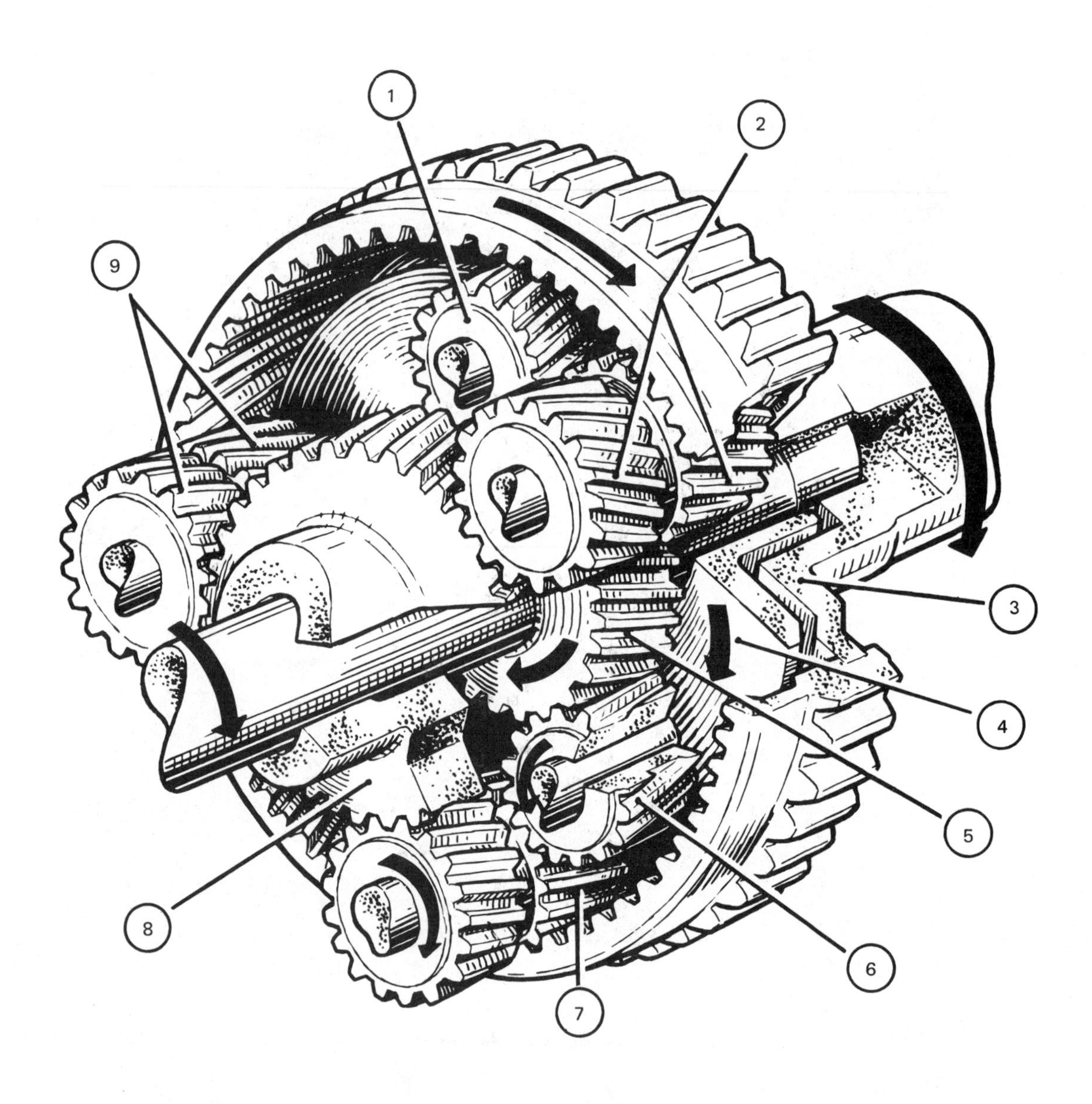

1 & 6	Short Planet Pinion (Third pinion not shown)
2, 7 & 9	Long Planet Pinion
3	Ring Gear (Annulus) and Output Shaft
4	Planet Carrier
5	Forward Sun Gear
8	Reverse Sun Gear

Fig. 9-8 Gear set in second gear. Planet carrier must walk around stopped reverse sun gear.

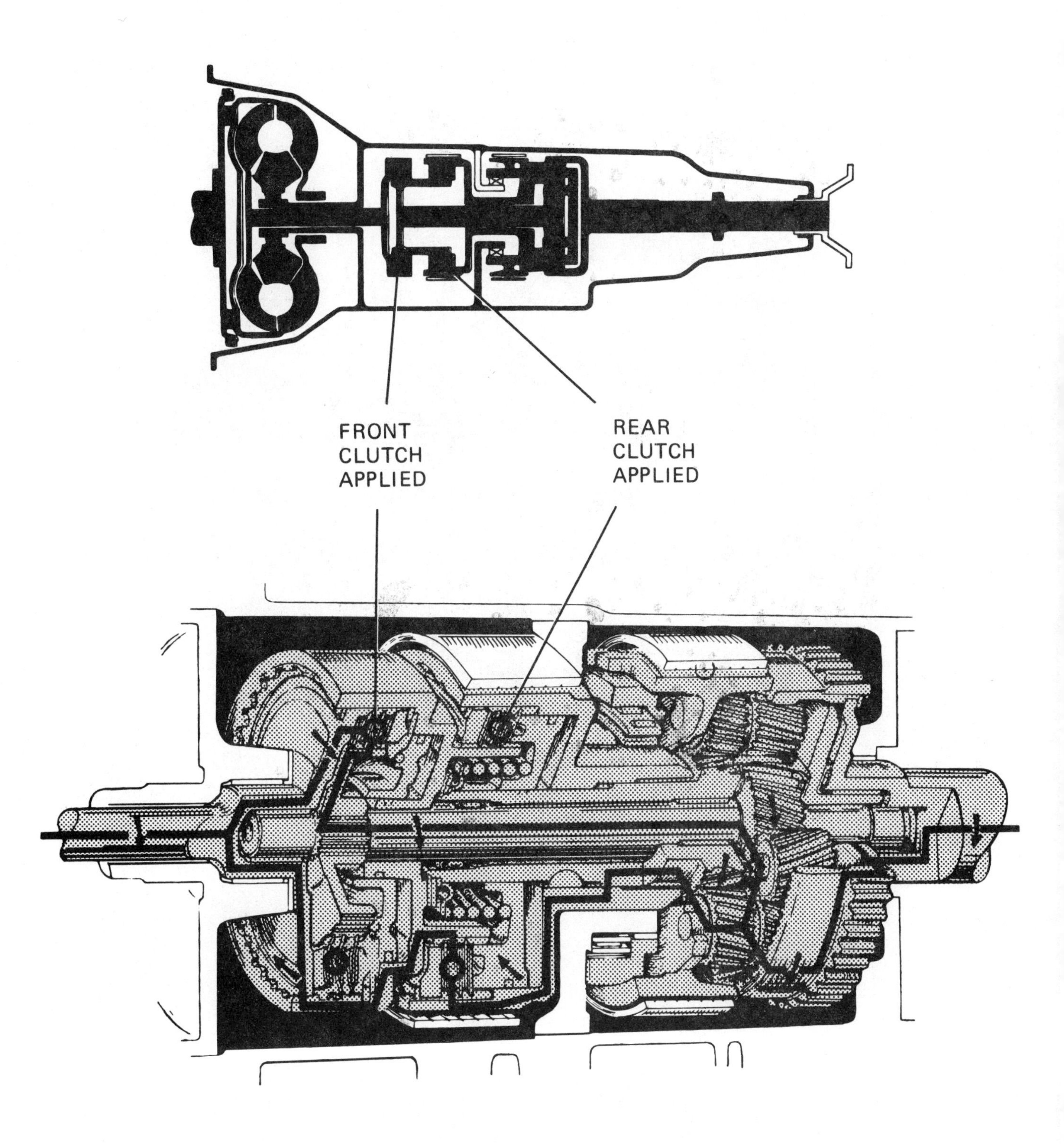

Fig. 9-9 Schematic for third gear or direct.

[For color diagram, see page 305.]

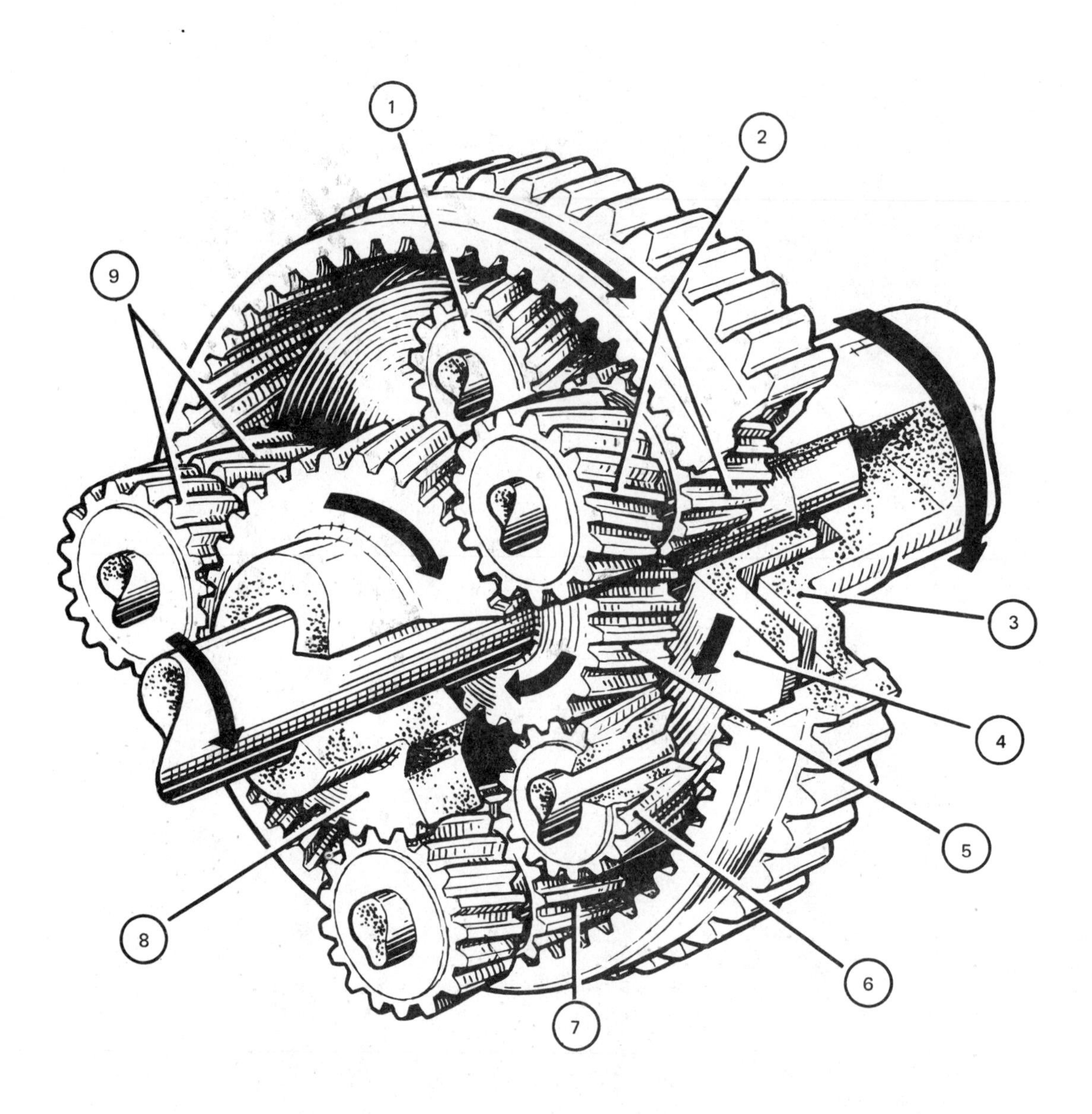

1 & 6	Short Planet Pinion (Third pinion not shown)
2, 7 & 9	Long Planet Pinion
3	Ring Gear (Annulus) and Output Shaft
4	Planet Carrier
5	Forward Sun Gear
8	Reverse Sun Gear

Fig. 9-10 Gear set in third or direct.

through line 15 (figure 9-3) to the rear clutch apply, and through the servo orifice control valve to the release side of the front servo. With the front band off and the rear clutch on, power flows to both sun gears, figures 9-9 and 9-10. As shown, both sun gears are driving. The pinions cannot turn or walk, and the whole gear set must turn as one unit.

At light throttle, a 1-2 upshift will take place at 7 to 11 mph (11 to 18 km/h). A 2-3 upshift will take place at 10 to 15 mph (16 to 24 km/h). At full throttle, shift speeds will be 27 to 39 mph (43 to 63 km/h) for a 1-2 shift, and 55 to 65 mph (89 to 105 km/h) for a 2-3 shift.

AUTOMATIC DOWNSHIFTS

Automatic downshifts take place as the car slows down. Part-throttle downshifts also take place at road speeds below about 28 mph (45 km/h). In addition, the driver may force a downshift by pushing the throttle to the wide-open position. This is called a *kickdown shift*. With the throttle wide open, a 3-2 shift will take place at road speeds below 48 to 61 mph (77 to 98 km/h). A 2-1 shift will occur at road speeds below 17 to 28 mph (27 to 45 km/h).

MANUAL LOW

In manual low (or *1 lockup* as it is also called) the power flow through the gear set is the same as in D range low. The difference here is that the rear band is on, line pressure keeps the 1-2 shift valve in the downshift position, and line pressure is cut off from the 2-3 valve. The rear band is applied in this range to give extra holding force to the carrier for heavy torque loads, and to provide engine braking when going down steep hills. Without this provision, the rear wheels would tend to drive the engine when the vehicle was going down steep hills. This would set up a force turning the carrier in a forward direction, causing the one-way clutch to unlock. The transmission would be allowed to freewheel, and there would be no engine braking.

If the manual valve is moved to 1 when the transmission is in high, rear clutch oil is exhausted, and the transmission shifts to 2. Line pressure to downshift the 1-2 valve is blocked by a land of the valve, and the shift to 1 will not take place until a low speed is reached – about 15 to 22 mph (24 to 35 km/h).

MANUAL INTERMEDIATE

Manual intermediate is also called *2 lockup*. The power flow in this range is the same as in drive range intermediate. With the manual valve in the 2 position, line pressure is exhausted from the rear clutch and the release side of the front servo. As noted before, the transmission cannot shift to high, but can downshift to low.

REVERSE

When the manual valve is moved to the R position, line pressure is sent through line 15 (figure 9-3) to the rear clutch and the release side of the front servo. Line pressure also moves through line 13 (figure 9-3) to the rear servo to apply the band, figure 9-11.

With the band applied, power moves through the rear clutch to the reverse sun gear, figure 9-12. The reverse sun gear (8) is in mesh only with the long pinions (7) and the carrier (4) is being held. Therefore, the pinions act as idlers and turn the ring gear and output shaft (3) in a reverse direction, at a ratio of 2.09:1.

HYDRAULIC SYSTEM

All of the valves used to control the hydraulic system, except the governor, are located in the valve body, figure 9-13.

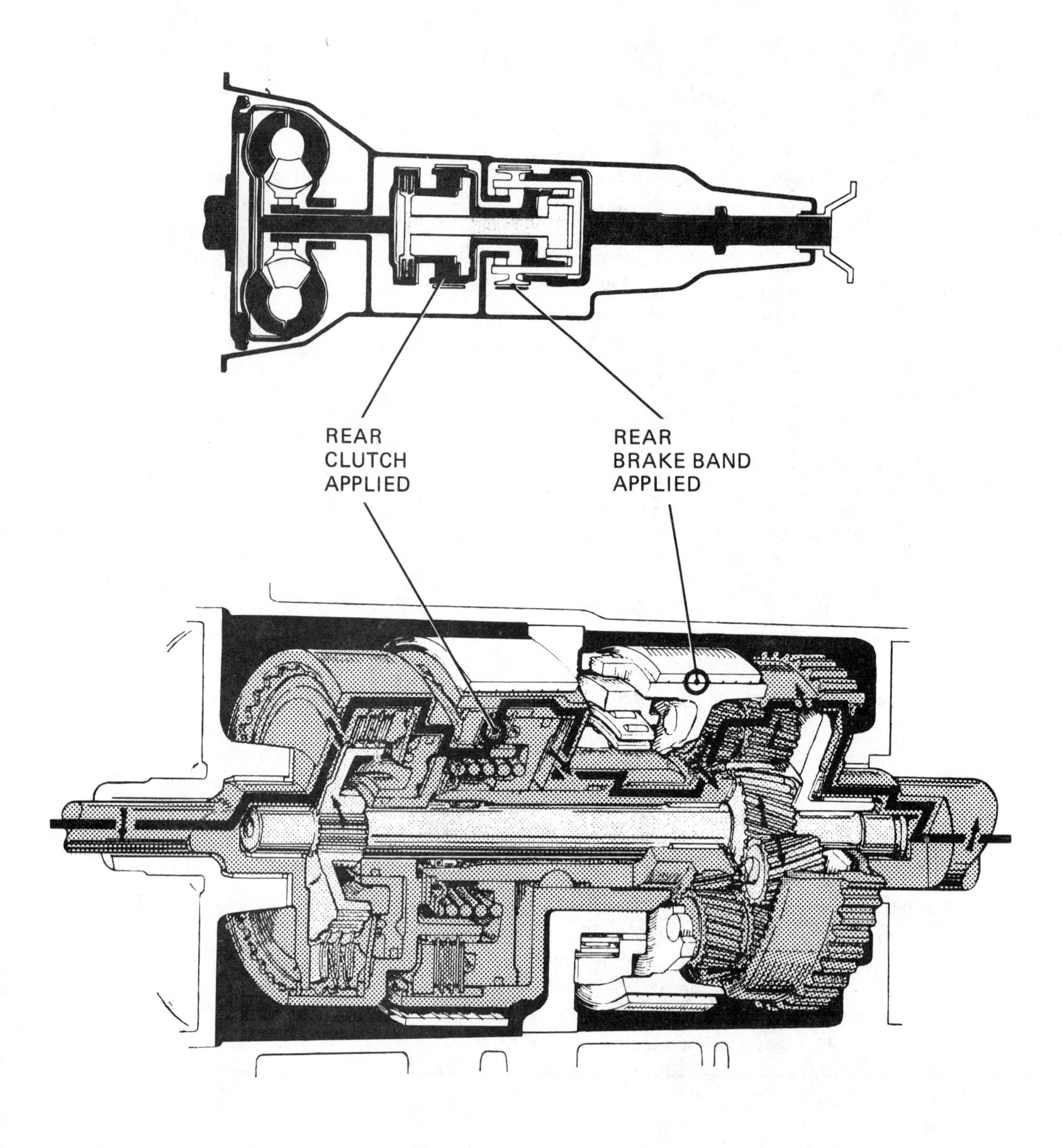

Fig. 9-11 Schematic for reverse.
[For color diagram, see page 306.]

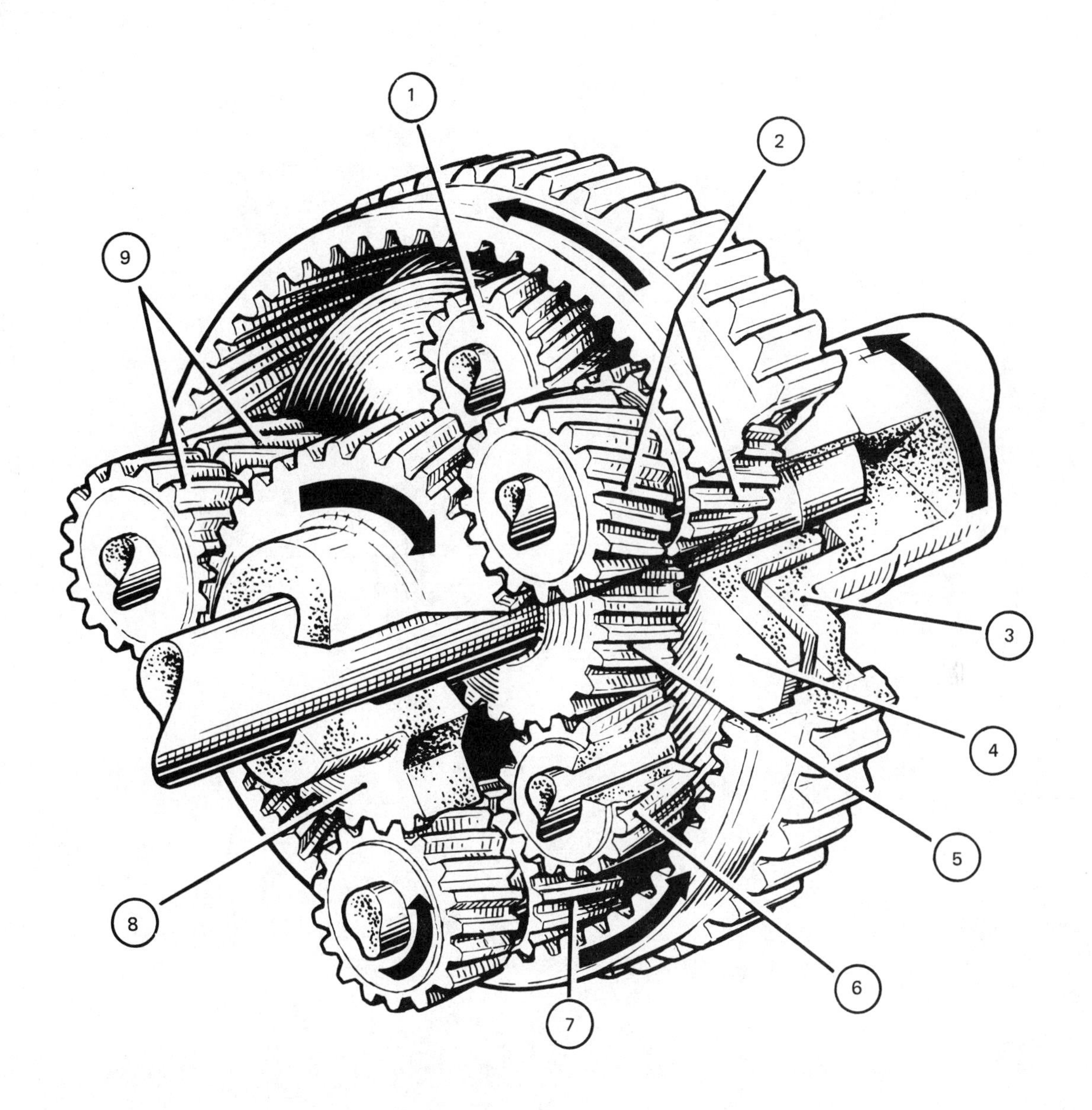

1 & 6	Short Planet Pinion (Third pinion not shown)
2, 7 & 9	Long Planet Pinion
3	Ring Gear (Annulus) and Output Shaft
4	Planet Carrier
5	Forward Sun Gear
8	Reverse Sun Gear

Fig. 9-12 Gear set in reverse.

Pump and Pressure Regulation

The front pump is of the internal-external gear type. The action of this type of pump is described in unit 5. Pressure regulation is handled by two systems: a primary regulation system, and a separate secondary regulator system.

Both the primary and secondary regulators are of the balanced bypass type. The primary pressure regulator regulates pump pressure to line pressure. Fluid at line pressure is passed to the manual valve and the throttle valve. (As line pressure leaves the manual valve it is also called *directed line pressure.*) As the throttle is opened, throttle pressure works on the top or spring end of the primary pressure regulator to increase line pressure. In this way, line pressure will vary from 57 to 120 psi (393 to 1 103 kPa).

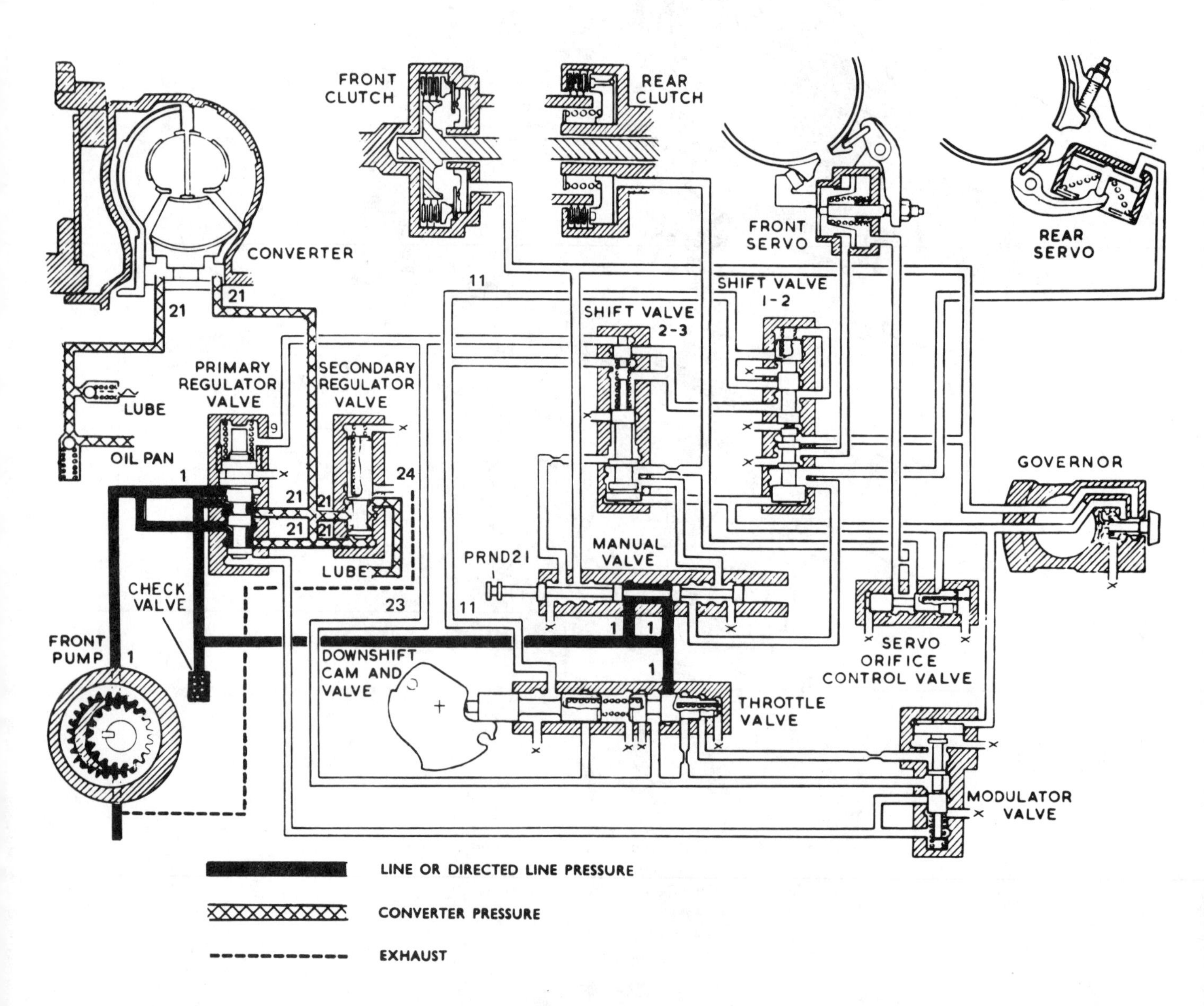

Fig. 9-13 Hydraulic circuit in neutral.
[For color diagram, see page 307.]

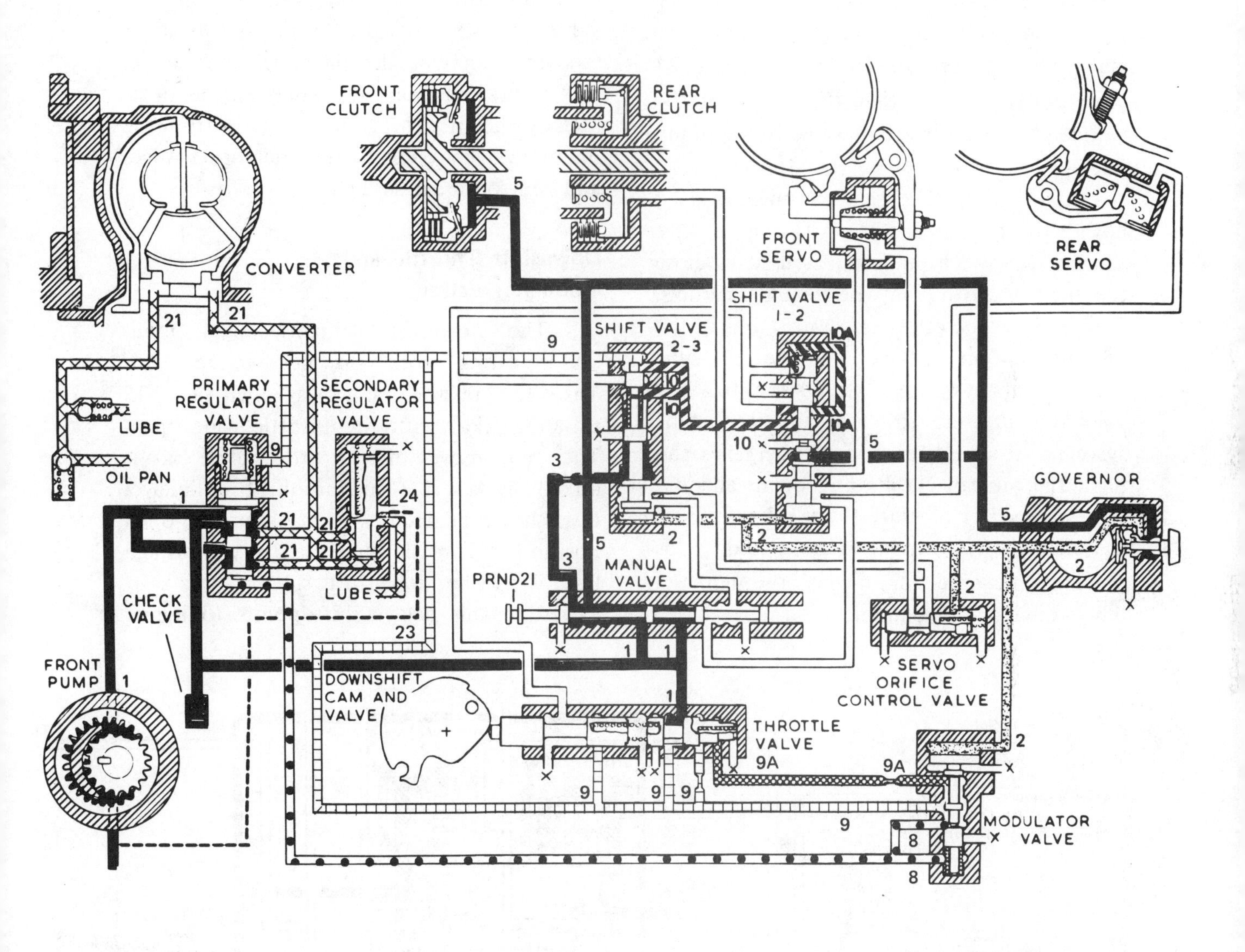

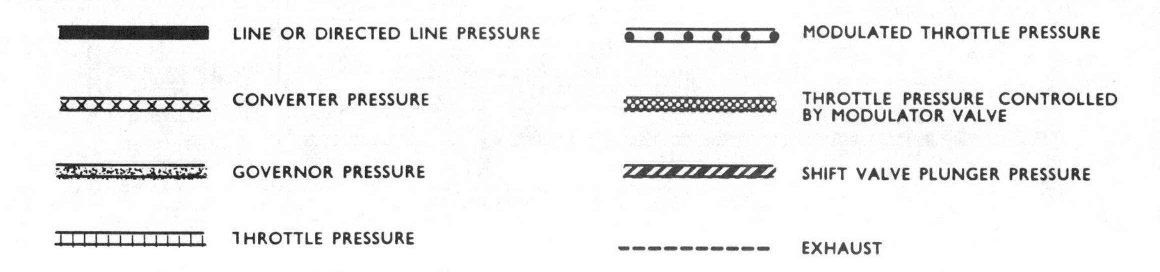

Fig. 9-14 TV at half-open throttle, modulator operating.

[For color diagram, see page 308.]

As line pressure is bypassed to the secondary pressure regulator, it is decreased to converter and lubrication pressure of 18 to 25 psi (124 to 172 kPa). Note that the bypass of the secondary regulator is exhausted to the pump inlet.

Older models of this transmission used both a front and rear pump. The check valve shown near the front pump is used to close the line to the rear pump used on older models. Two other check valves are shown: one in the lube line and one in the oil pan line. The lube valve unseats at about 5 psi (34 kPa) to allow converter pressure to enter the front lube system. If converter pressure becomes too high, the oil pan check valve opens and bypasses converter pressure to the oil pan. With the engine off, both valves are closed. This prevents converter drain-back through the lube system and oil pan line.

As shown in figure 9-13, the manual valve has six oil paths and two exhausts (shown as an *x* on the diagram). The manual valve works in much the same way as those already shown. Therefore, the paths from the manual valve are dealt with when other valves and controls are described.

Downshift, Throttle, and Modulator Valves

The downshift, throttle, and modulator valves work together, figure 9-14. Throttle pressure ranges as high as 135 psi (931 kPa) with the throttle wide open. The cam moves the throttle valve and is linked by a cable to the throttle linkage. It is shown in figure 9-15 at about half-open position. Throttle pressure passes through line 9 to the downshift side of the shift valves. This delays shift points to higher

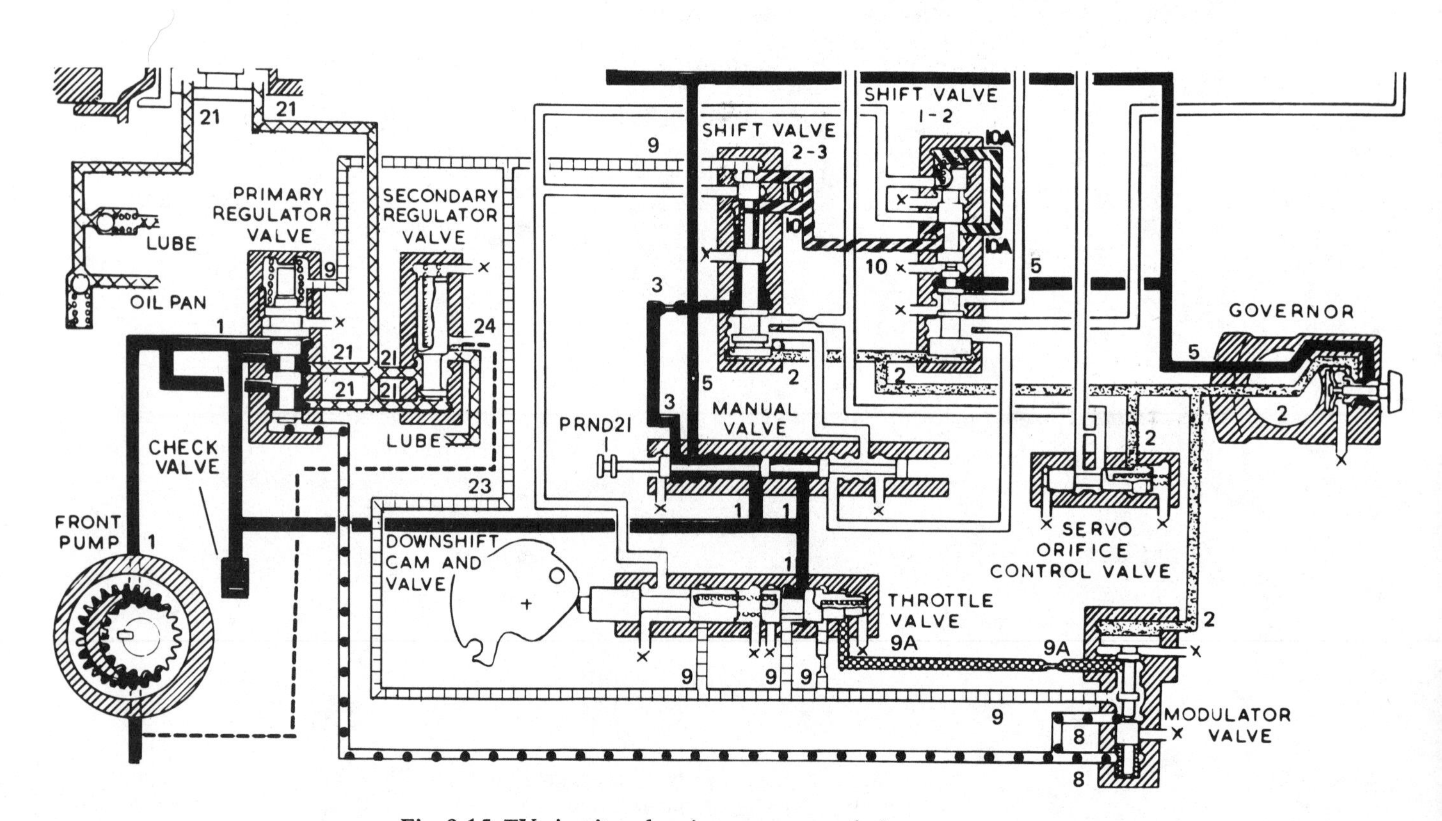

Fig. 9-15 TV circuit and main pressure regulation.

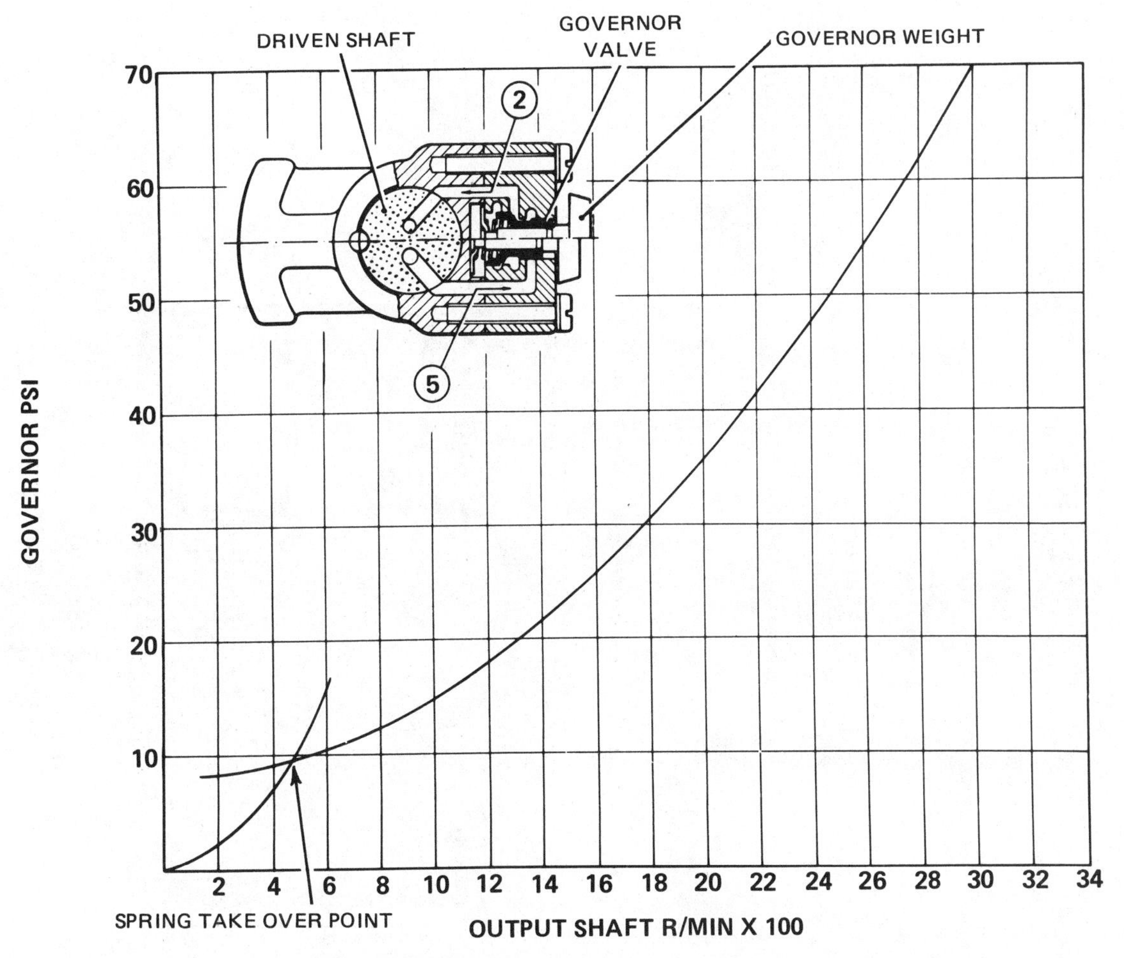

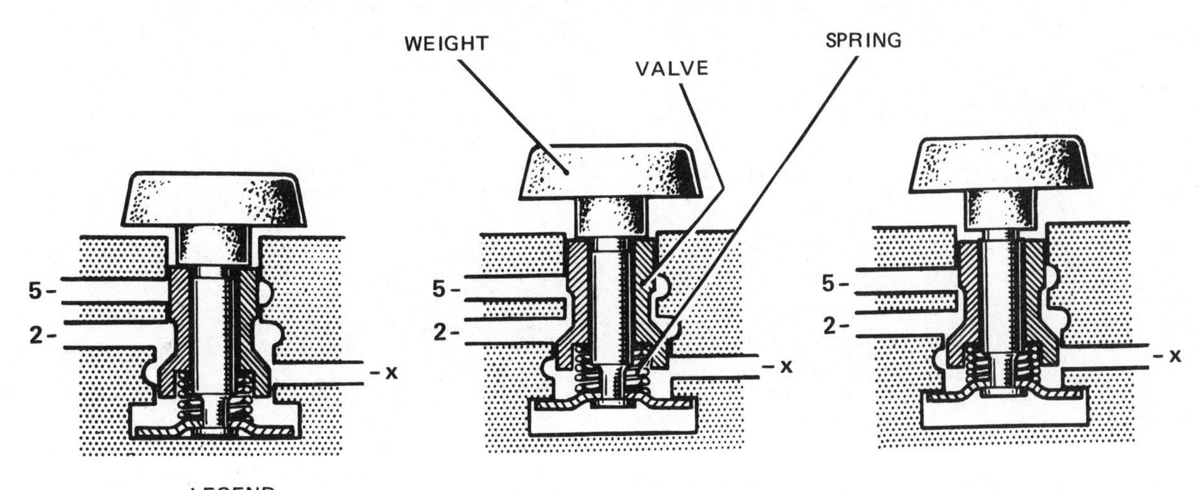

LEGEND
5 DIRECTED LINE PRESSURE FROM MANUAL CONTROL VALVE
2 GOVERNOR PRESSURE
X EXHAUST

Fig. 9-16 Governor operation.

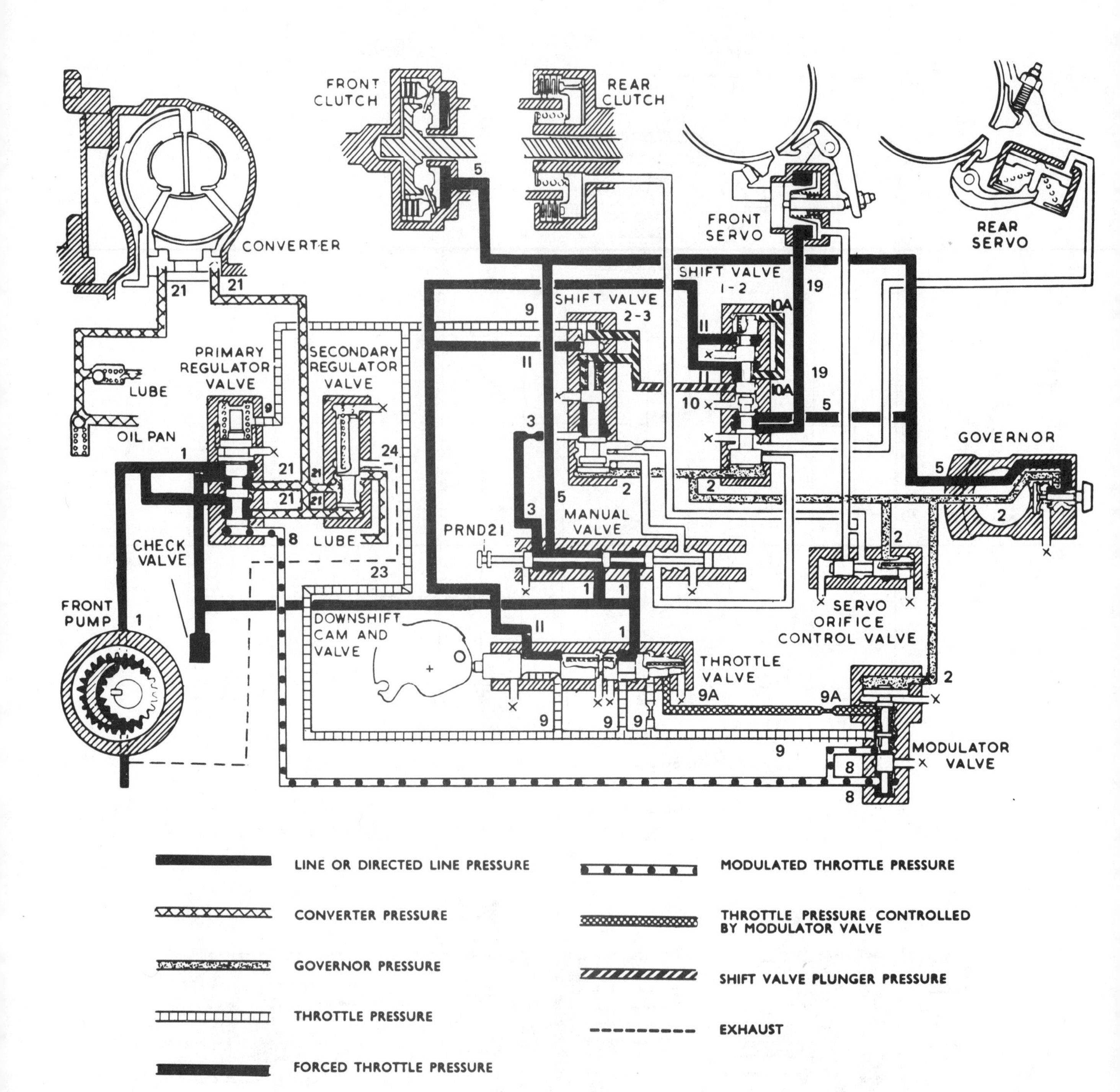

Fig. 9-17 Hydraulic circuit for D intermediate or 2nd gear.

[For color diagram, see page 309.]

road speeds, or downshifts the transmission to intermediate at speeds below about 28 mph (45 km/h) and to low at speeds below about 12 mph (19 km/h).

The throttle valve is a balanced restriction-type regulator and is separated from the downshift valve by a spring. Movement of the downshift valve puts more or less spring force on the throttle valve. Line pressure coming to the throttle valve through line 1 is changed to throttle pressure at line 9.

Throttle pressure is also sent through line 9 to a differential force area in the modulator valve. At small throttle openings, this changes throttle pressure to modulated throttle pressure. The fluid then passes through line 8 to the primary pressure regulator. Modulated throttle pressure works on the primary pressure regulator to reduce line pressure for smooth part-throttle shifts.

As road speed increases, governor pressure works on the large face of the modulator valve. This moves the valve wide open and stops it from regulating. In this position, modulated throttle pressure is the same as throttle pressure, and a decrease in line pressure takes place. Throttle pressure, controlled by the modulator valve, is sent through line 9a to the throttle valve to further reduce throttle pressure.

Most automatic transmissions reduce line pressure for part-throttle and cruising speeds. The means by which this is done varies and is discussed separately for each transmission.

Governor Action

The governor, figure 9-16, is a restriction-type regulator that is keyed to, and turns with, the output shaft of the transmission. Forces acting on the governor are: (1) spring force, (2) hydraulic force, and (3) centrifugal force. With the manual valve in D, 2, or 1, line pressure enters the governor through port 5. When the car is standing still, line pressure is blocked by a land of the valve. As the car moves, centrifugal force causes the weight to swing out. This, in turn, exerts force on the spring, and the valve moves regulating line pressure to governor pressure in port 2.

In port 2, governor pressure works on the differential force area to balance out centrifugal and spring force. As the speed of the car increases, the weight moves out to a stop. At this point, governor pressure depends on the centrifugal force on the valve, the spring force, and the balancing force of governor pressure. Since centrifugal force varies with the square of the speed, this design is used to make governor pressure rise evenly from low to high speed. Governor pressure varies from 0 to 70 psi (0 to 483 kPa). It is sent through line 2 to work on the shift valves and modulator valve. As the car slows to a stop, the valve moves back to the closed position, and governor pressure is exhausted at X.

Shift Valves

The forces acting on the shift valves are:

- governor pressure, which tends to move the valves to the upshift position; and
- throttle pressure, spring force, and line pressure on differential force areas, which tend to keep the valves in the downshift position.

As car speed increases, governor pressure overcomes the spring, line, and throttle pressure forces on the 1-2 shift valve. This causes the valve to move to the upshift position, figure 9-17. Line pressure moves through line 19 to apply the front servo. This stops the sun gear, causing the transmission to move into intermediate. (*Note:* Due to area and spring force differences, the 1-2 valve will move before the 2-3 valve.)

As road speed increases, governor pressure builds and moves the 2-3 valve, figure 9-18. Line pressure now moves through line 15 and a restriction orifice to apply the rear clutch. Fluid also flows through the servo orifice control valve to the release side of the front servo. Now, with both sun gears driving, the transmission is in high.

Note that as each valve moves, line pressure is exhausted from the differential force areas. The result is to give the valves snap action, and to reduce the force on the downshift side of the valve. This gives shift speed differences between upshift and downshift points and allows quite large throttle openings without downshifting the transmission.

In manual 1, figure 9-19, page 120, line pressure is sent to the front clutch through line 5, and to the rear servo through lines 6 and 13. Note that line pressure works on the large differential force area of the 1-2 shift valve so that the transmission cannot shift out of low.

In manual 2, line pressure moves through line 5 to apply the front clutch and to the apply side of the front servo when the 1-2 shift valve moves. Line pressure to the 2-3 valve, the rear clutch, and the release side of the front servo is exhausted at the manual valve. This means that the transmission will start in low and shift to 2, but cannot shift to high. Also, by exhausting line pressure in this way, the transmission can shift from high to 2 at any road speed.

Servo Orifice Control Valve

The *servo orifice control valve* is a shift timing valve that helps time 2-3 and 3-2 shifts. During a 2-3 shift at low road speeds, line pressure passes through line 15, without restriction, to the release side of the front servo. This gives a fast band release that is timed to clutch application for a smooth shift. At higher road speeds, governor pressure (at line 2) moves the servo orifice control valve to the right against a spring force. As shown in figure 9-18, the unrestricted path through line 15 is now blocked by a land of the valve and line pressure must pass through the restriction orifice to release the band. This slows up band release and prevents a *run-up* or *slipping shift* to high.

During a 3-2 downshift, the orifice in the line slows up band application so that the clutch has time to release. This prevents tie-up on a 3-2 shift.

SUMMARY

The Borg-Warner transmission has three speeds. It can be used for full automatic shifts or, if the driver chooses, can be shifted by hand. A summary of band and clutch use in this transmission is shown in figure 9-20. To review how the different shifts work and the effect that band or clutch application has on the gear set, restudy the section of this unit that covers these points. Proper understanding of the power flow and hydraulic system is essential for diagnosis.

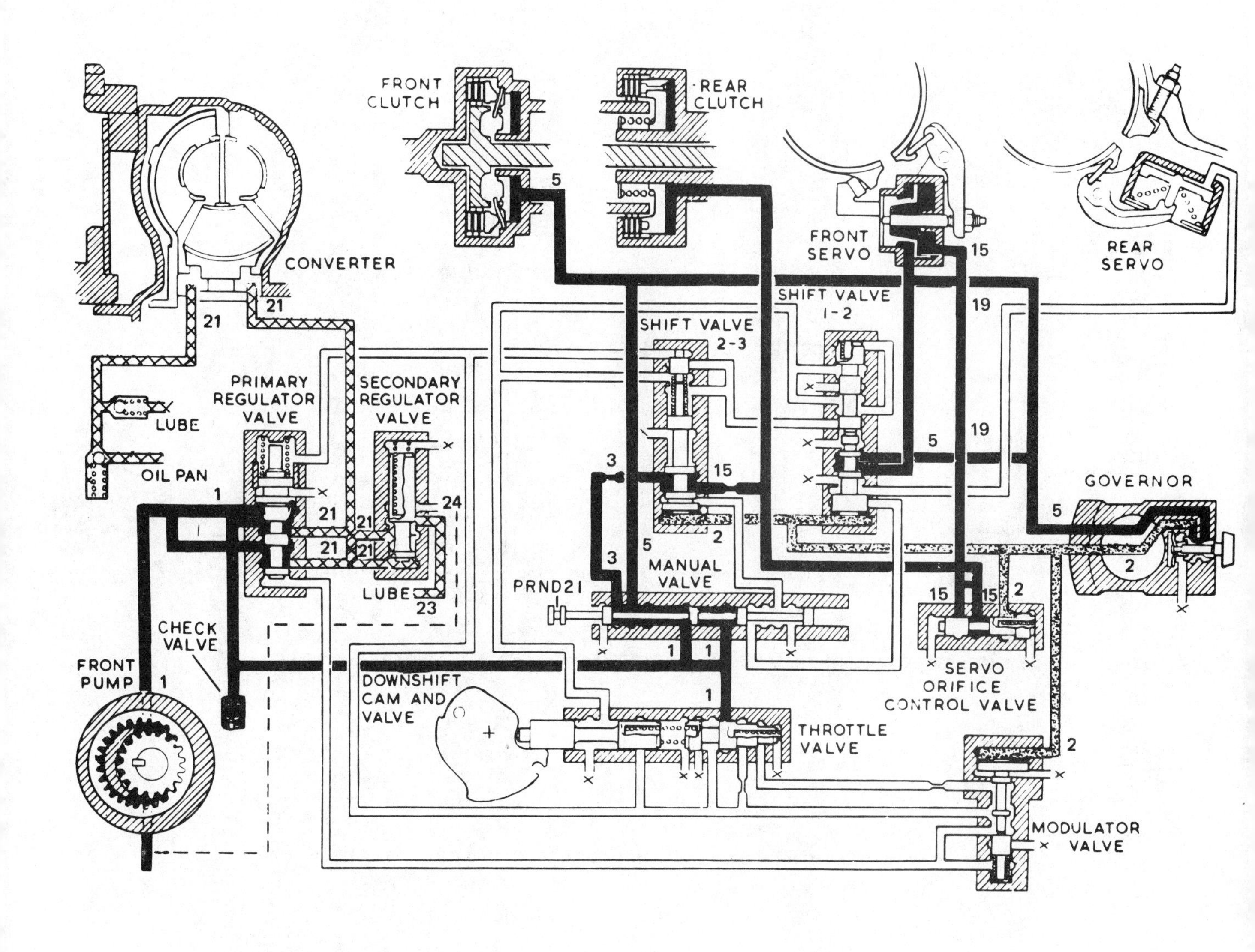

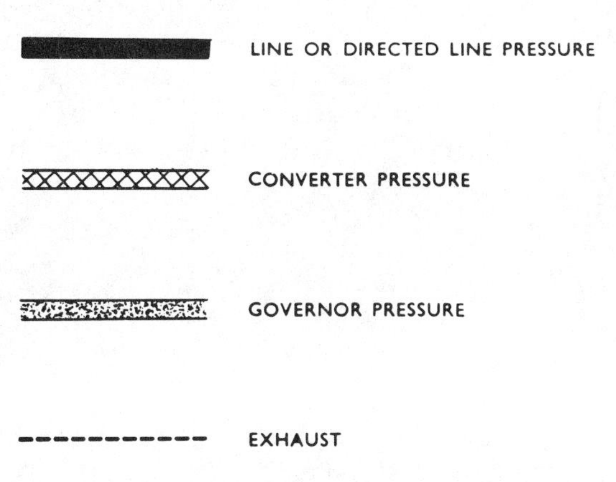

Fig. 9-18 Hydraulic circuit in D high.

[For color diagram, see page 310.]

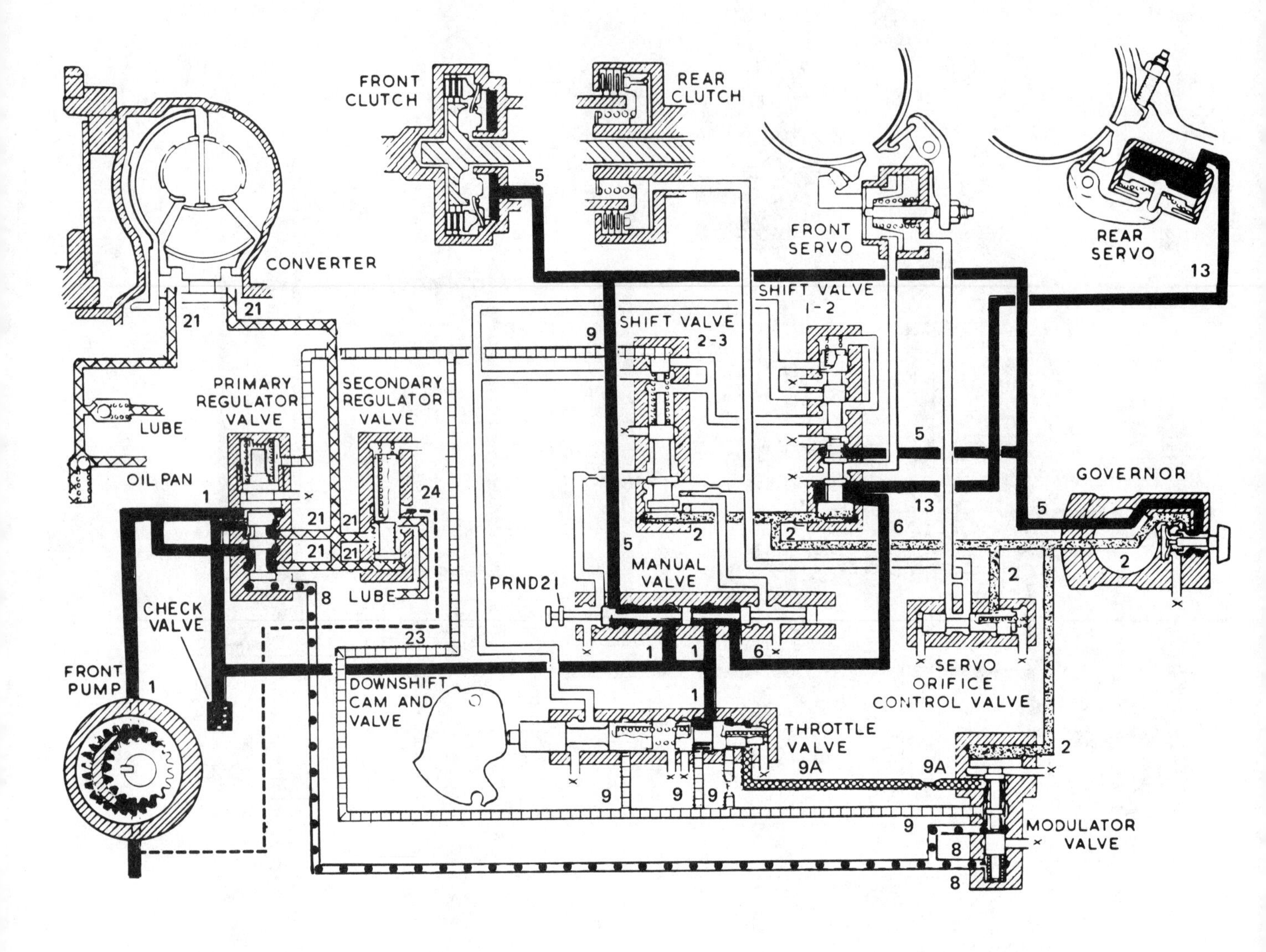

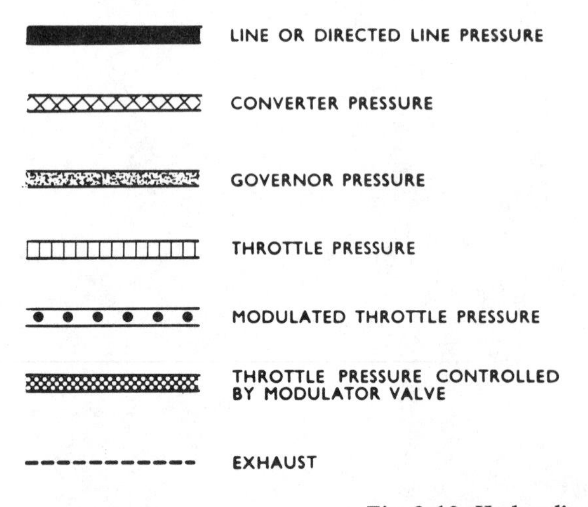

Fig. 9-19 Hydraulic circuit, Manual 1.

[For color diagram, see page 311.]

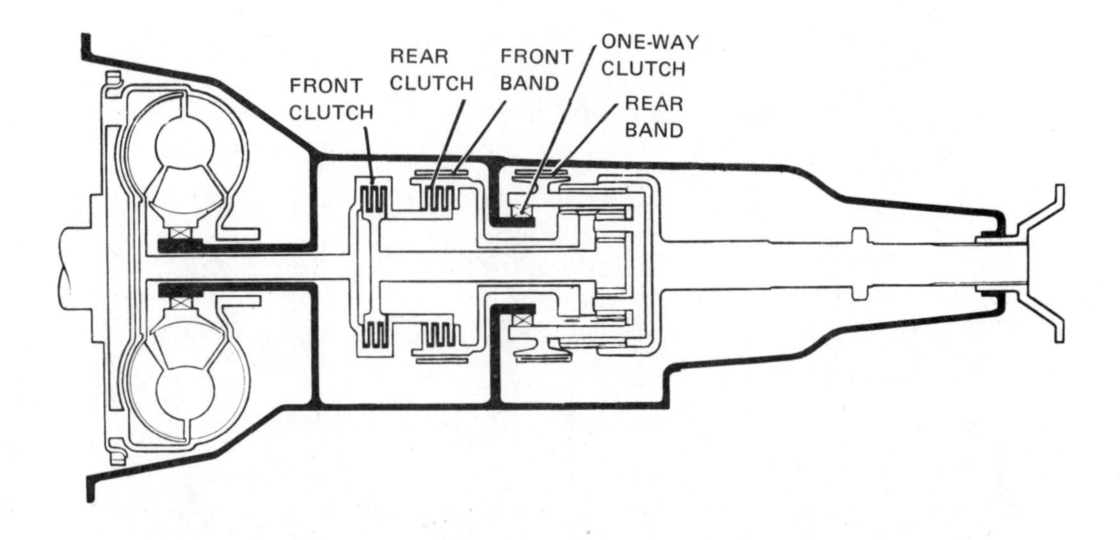

MANUAL VALVE POSITION	FRONT CLUTCH	REAR CLUTCH	FRONT BAND	REAR BAND	ONE-WAY CLUTCH
MANUAL 1	X			X	
DRIVE 1st	X				X
DRIVE 2nd OR MANUAL 2	X		X		
DRIVE 3rd	X	X			
NEUTRAL					
REVERSE		X		X	
PARK				X	

X = APPLIED

Fig. 9-20 Power flow summary.

REVIEW

Select the best answer from the choices offered to complete the statement or answer the question.

(*Note:* In answering the questions dealing with diagnosis, refer to figure 9-20, and those parts of the text that deal with the range or shift in question.)

1. With only the front clutch applied, the forward sun gear

   a. turns in a reverse direction.
   b. turns in a forward direction.
   c. is held.
   d. freewheels.

2. Modulated throttle pressure:

   (I) Helps in line pressure control.
   (II) Moves the shift valves.

   a. I only
   b. II only
   c. both I and II
   d. neither I nor II

3. A customer complains that his transmission slips in both D range high and in R. This would most likely be caused by a faulty:

   (I) Rear servo.
   (II) Rear clutch.

   a. I only
   b. II only
   c. both I and II
   d. neither I nor II

4. A cause of no drive in 1, could be caused by a faulty:

   (I) One-way clutch.
   (II) Front band.

   a. I only
   b. II only
   c. both I and II
   d. neither I nor II

5. The unit of the gear set that is part of the output shaft is the

   a. forward sun gear.
   b. reverse sun gear.
   c. carrier.
   d. ring gear.

6. When the front clutch and front band are applied, the forward sun gear

   a. turns in a reverse direction.
   b. turns in a forward direction.
   c. is held.
   d. freewheels.

7. When the one-way clutch is locked the carrier

   a. turns in a reverse direction.
   b. turns in a forward direction.
   c. is held.
   d. freewheels.

8. In manual 1, line pressure is exhausted from the:

   (I) Rear clutch.
   (II) Front servo.

   a. I only
   b. II only
   c. both I and II
   d. neither I nor II

9. In manual 2, line pressure is exhausted from the:

   (I) Front clutch.
   (II) Rear clutch.

   a. I only
   b. II only
   c. both I and II
   d. neither I nor II

10. In D range low, there is no engine braking because the carrier

    a. turns at a ratio of 2.39:1.
    b. turns in a forward direction.
    c. is held.
    d. freewheels.

11. The clutch and servo pistons are moved by:

    (I) Line pressure.
    (II) Governor pressure.

    a. I only　　c. both I and II
    b. II only　　d. neither I nor II

12. For automatic shifts, the shift valves are moved by:

    (I) Governor pressure.
    (II) Throttle pressure.

    a. I only　　c. both I and II
    b. II only　　d. neither I nor II

13. A customer complains of a slipping 2-3 shift at high road speeds. The most likely cause of this would be a:

    (I) Sticking servo orifice control valve.
    (II) Faulty front clutch.

    a. I only　　c. both I and II
    b. II only　　d. neither I nor II

14. When the rear clutch and rear band are applied, the transmission is in

    a. reverse.　　c. intermediate.
    b. low.　　d. high.

15. In D range low, the manual valve sends line pressure to the

    a. rear clutch.　　c. front band.
    b. front clutch.　　d. rear band.

16. Line pressure to apply the front servo must pass through the:

    (I) 1-2 shift valve.
    (II) Servo orifice control valve.

    a. I only　　c. both I and II
    b. II only　　d. neither I nor II

17. A cause of no drive in D could be a:

    (I) Faulty one-way clutch.
    (II) Faulty front servo.

    a. I only　　c. both I and II
    b. II only　　d. neither I nor II

18. In R, line pressure is exhausted from the:

    (I) Rear clutch.
    (II) Rear servo.

    a. I only
    b. II only
    c. both I and II
    d. neither I nor II

19. No drive in any range could be caused by a faulty:

    (I) Front pump
    (II) Primary pressure regulator.

    a. I only
    b. II only
    c. both I and II
    d. neither I nor II

20. When the front and rear clutches are on, the transmission is in

    a. reverse.
    b. low.
    c. intermediate.
    d. high.

EXTENDED STUDY PROJECTS

On blank oil circuit diagrams, complete the following exercises.

1. Indicate the oil circuit for park.
2. Indicate the oil circuit for neutral.
3. Indicate the oil circuit for D-low. On the back of the circuit, list three possible causes of no drive in D.
4. Indicate the oil circuit for D-intermediate. On the back of the circuit, list two causes of no 1-2 shift.
5. Indicate the oil circuit for D-high. On the back of the circuit, list two possible causes of no 2-3 shift.
6. Indicate the oil circuit for reverse. On the back of the circuit, list three possible causes of no reverse.
7. Indicate the oil circuit for manual 1.
8. Indicate the oil circuit for manual 2.

THE SIMPSON GEAR TRAIN

OBJECTIVES

After studying this unit, the student will be able to:

- Identify the bands and clutches used to control the gear train.
- Trace the power flow in all forward speeds and reverse.
- Relate band and clutch application to power flow.
- Diagnose gear train problems.

APPLICATION

The Simpson gear train is accepted throughout the industry as the best gear train for use in modern automatic transmissions. It is used in transmissions made by all of the major auto manufacturers including the Chrysler TorqueFlite; American Motors Torque-Command; Ford C-3, C-4, and C-6, and General Motors Turbo Hydra-Matic series. Made up of two simple planetary gear sets joined by one common sun gear, it is both simple and strong. As shown in figure 10-1, it

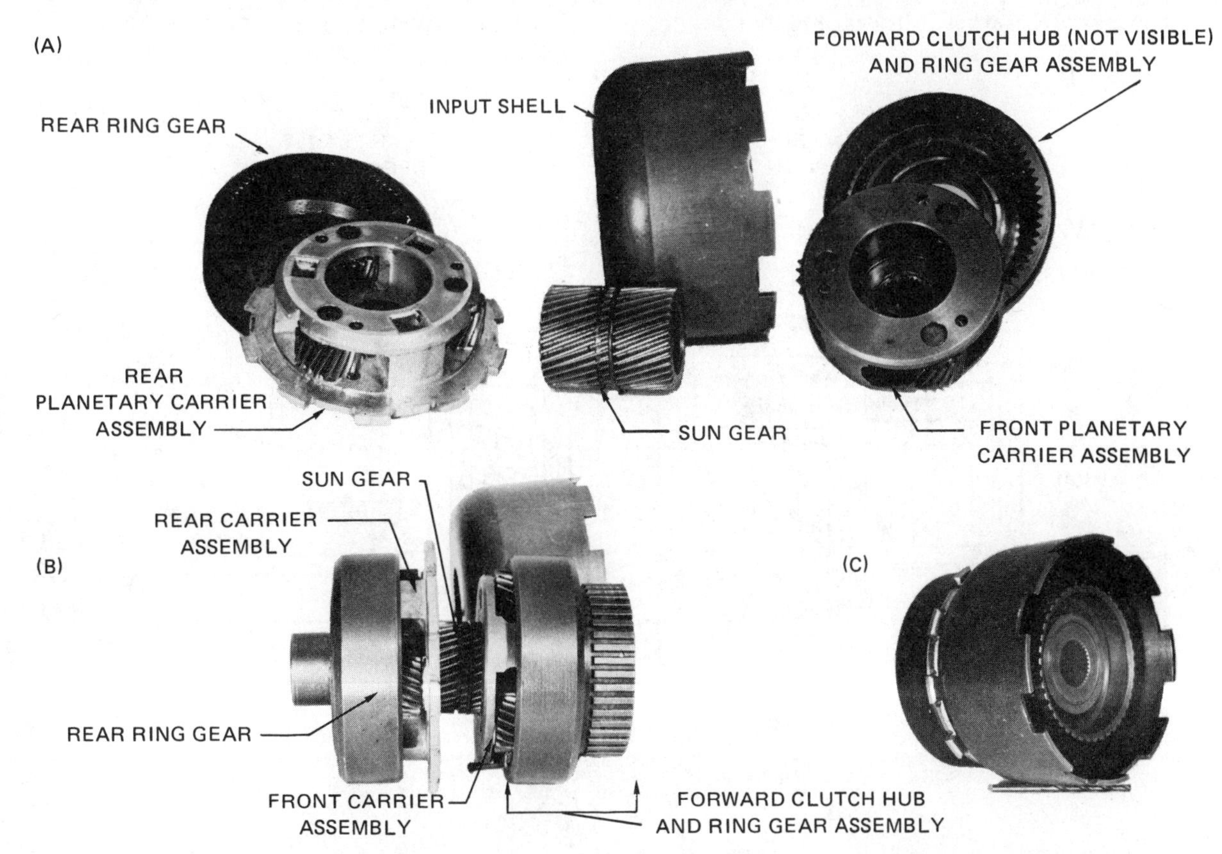

Fig. 10-1 The Simpson gear train (A) disassembled, (B) partly assembled, (C) assembled.

includes a front and rear ring gear, front and rear planetary carriers, and one sun gear that joins the front and rear sets. The sun gear is hollow and rides on the output shaft. It is not connected to the output shaft, but instead, is free to turn in either a forward or reverse direction.

POWER FLOW

The Simpson gear set is controlled by a forward clutch, a reverse and high clutch, an intermediate band, a low and reverse band, and a one-way roller clutch. A gear set of this type is shown in figure 10-1 (A–C). In this unit, the names applied to the members of the gear train are those used by Ford Motor Company. (Chrysler and General Motors use other names for some of the parts. This will be noted in the units that deal with these transmissions.) Regardless of the names of the clutches and bands, or, in some cases, the substitution of a clutch for a band, the basic parts serve the same purpose in all Simpson gear trains.

Drive Range Low

With the manual valve in D, line pressure applies the forward clutch and locks the input shaft to the front ring gear, figure 10-2. Note that the front carrier is splined to the output shaft. This has the same effect as holding the carrier and results in reverse rotation of the sun gear.

This setup is the same as that for reverse overdrive in a simple planetary gear set. It results in a reduction because the carrier is turning with the output shaft, but at a slower rate than the front ring gear. The planet pinions are still allowed to act as idlers and turn the sun gear in a reverse rotation.

The sun gear serves as an input member to the rear planetary set, and sets up a force

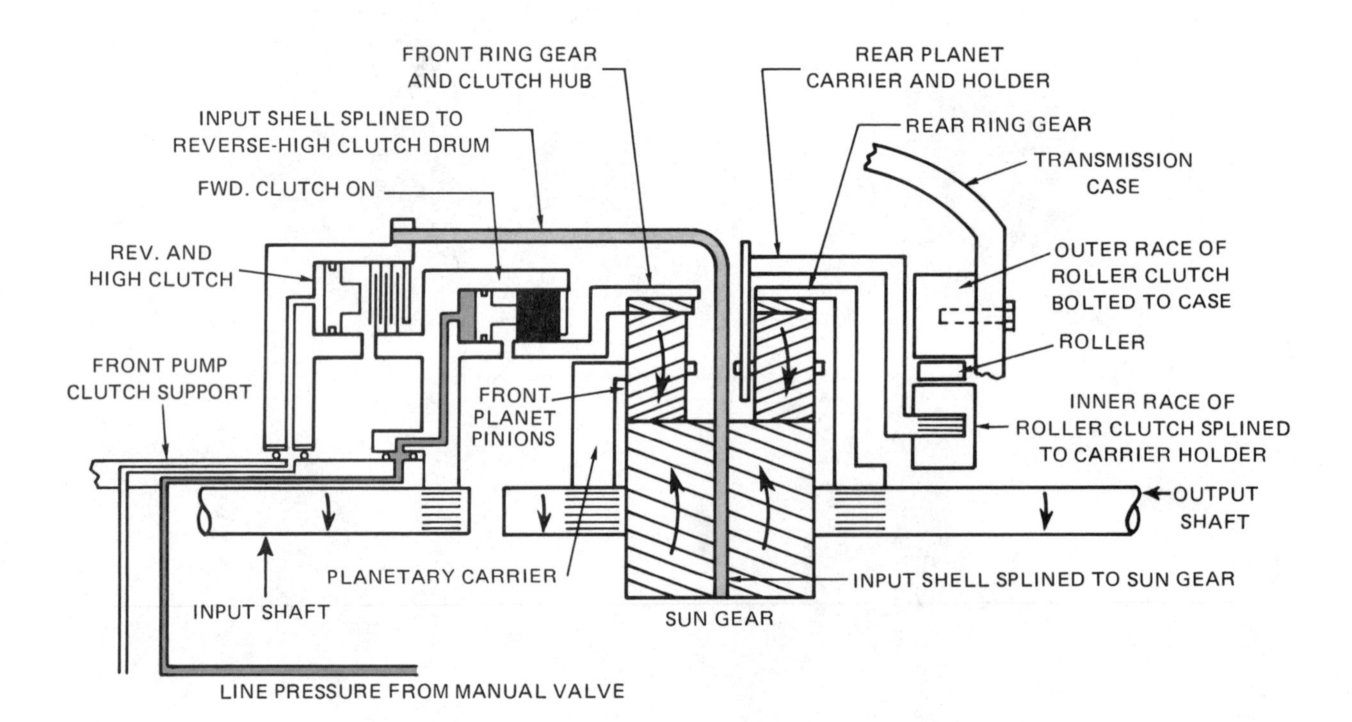

Fig. 10-2 Drive range low.

that tries to turn the rear carrier assembly in a reverse rotation. The one-way clutch, which is splined to both the carrier and the transmission case, locks up in reverse rotation and keeps the carrier from turning. This gives another reverse on the rear ring gear. Since the ring gear and output shaft are splined together, the output shaft rotates forward at a reduction of about 2.46:1.

The setup for drive range low can be thought of as two reverses. The front carrier, in effect, is held by the output shaft, and the rear carrier is held by the one-way clutch. The use of a one-way clutch helps to give a smooth upshift to intermediate.

Drive Range Intermediate

As car speed increases, governor pressure moves the 1-2 shift valve. Line pressure then passes through the 1-2 shift valve to the intermediate servo to apply the band and stop the sun gear, figure 10-3. The forward clutch is still on, and the front ring gear is driving. However, the front carrier must now walk around the stopped sun gear. Since the front carrier is splined to the output shaft, the output shaft is turned at a ratio of about 1.46:1.

Note that forward rotation unlocks the one-way clutch, and the rear planetary set turns with the unit. This has no effect on the ratio. Because only the front gear set is being used for this range, the setup is the same as for intermediate in a simple planetary gear set.

Drive Range High

When the shift to high is made, line pressure is sent to release the intermediate band while, at the same time, the reverse

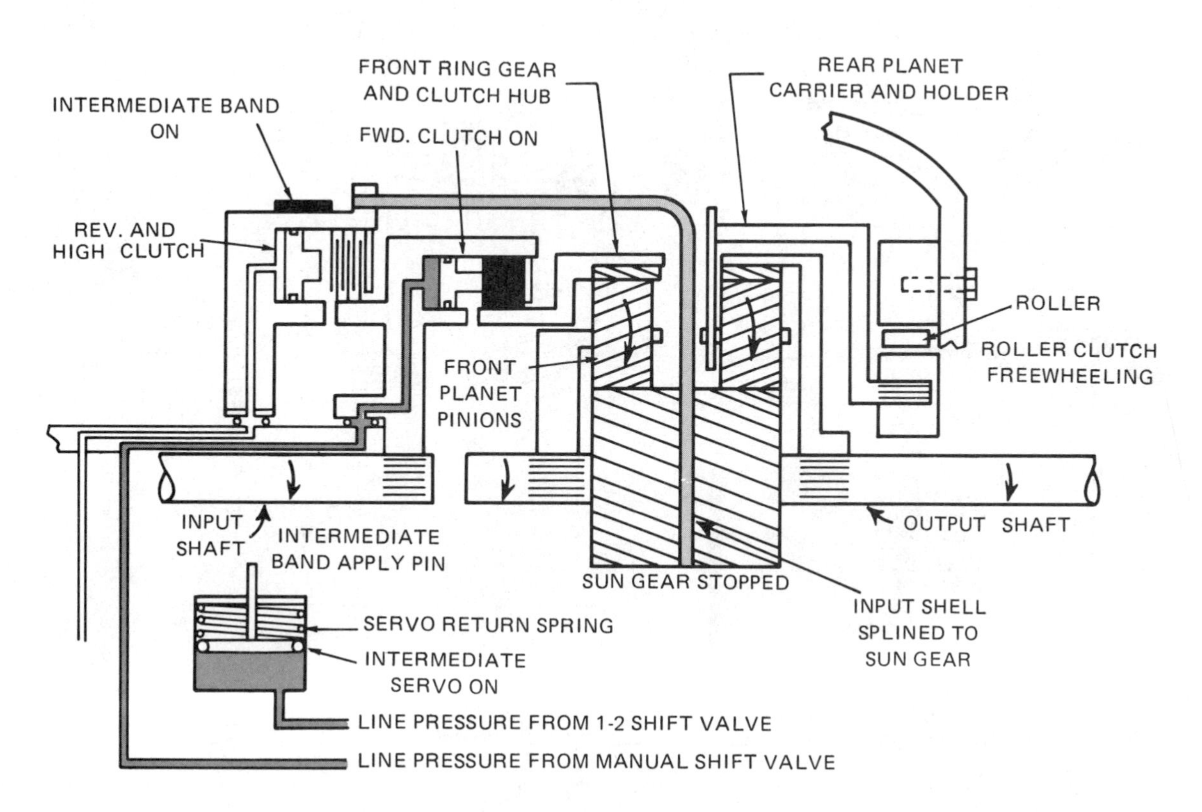

Fig. 10-3 Drive range intermediate.

and high clutch is applied, figure 10-4. The forward clutch is still being applied and turns the front ring gear. However, now the sun gear is also being driven at the same speed by the reverse and high clutch. This means that the input shaft, front gear set, and output shaft are locked in a direct or 1:1 ratio.

Note that the rear gear set is still turning, but as before, has no effect on the ratio. The front gear set is locked and in use, as if it were a simple planetary gear set.

Reverse

With the manual valve in the R position, line pressure is sent to the reverse and high clutch, and to the low and reverse servo to apply the band, figure 10-5. The reverse and high clutch turns the sun gear in a forward direction. The sun gear thus acts as the input member to the rear planetary gear set. The low and reverse band holds the rear carrier, making the pinions act as idlers. The ring gear and output shaft turn in a reverse direction at a ratio of about 2.17:1. Note that the forward clutch is off, and the front part of the gear set has no effect on the ratio.

Manual 1

The power flow and ratio in manual 1 is the same as in drive range low. That is,

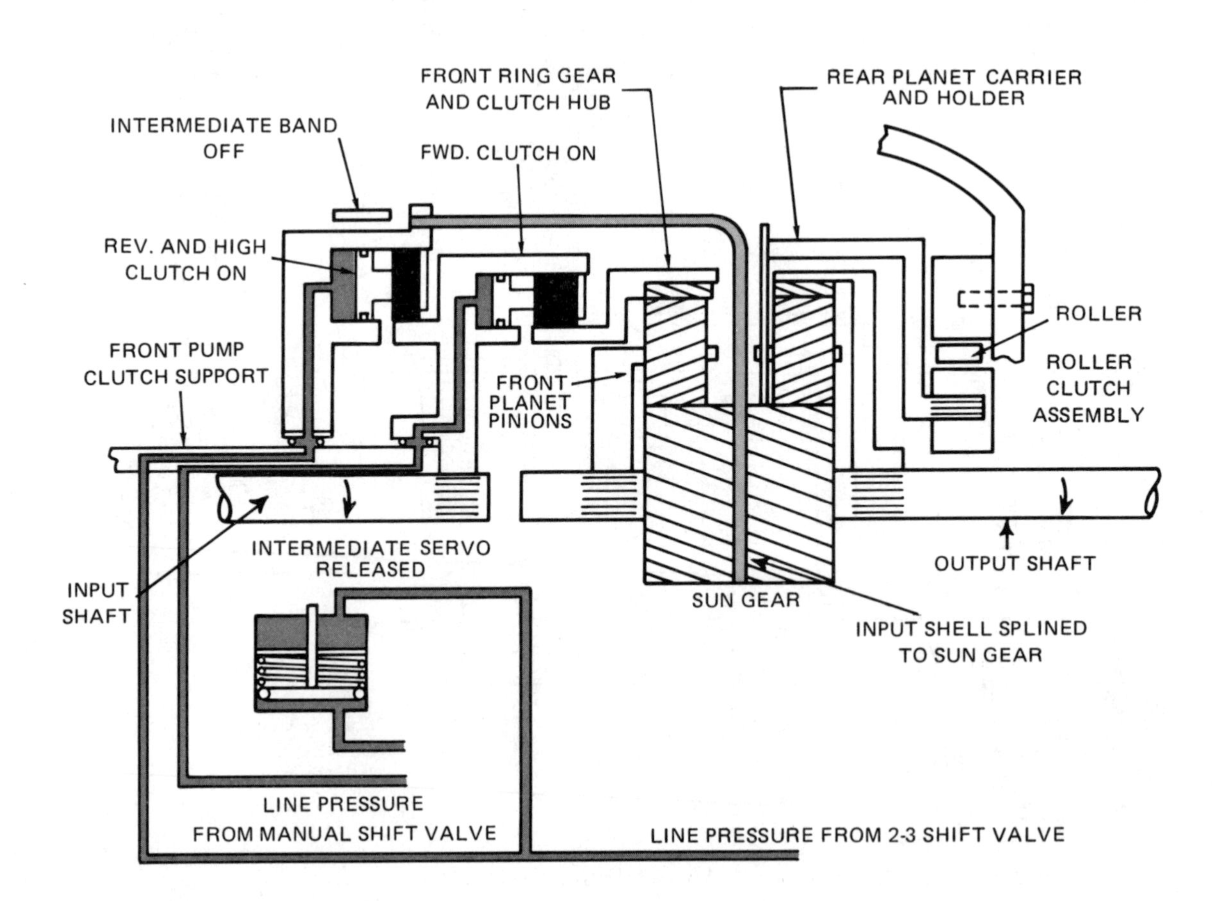

Fig. 10-4 Drive range high.

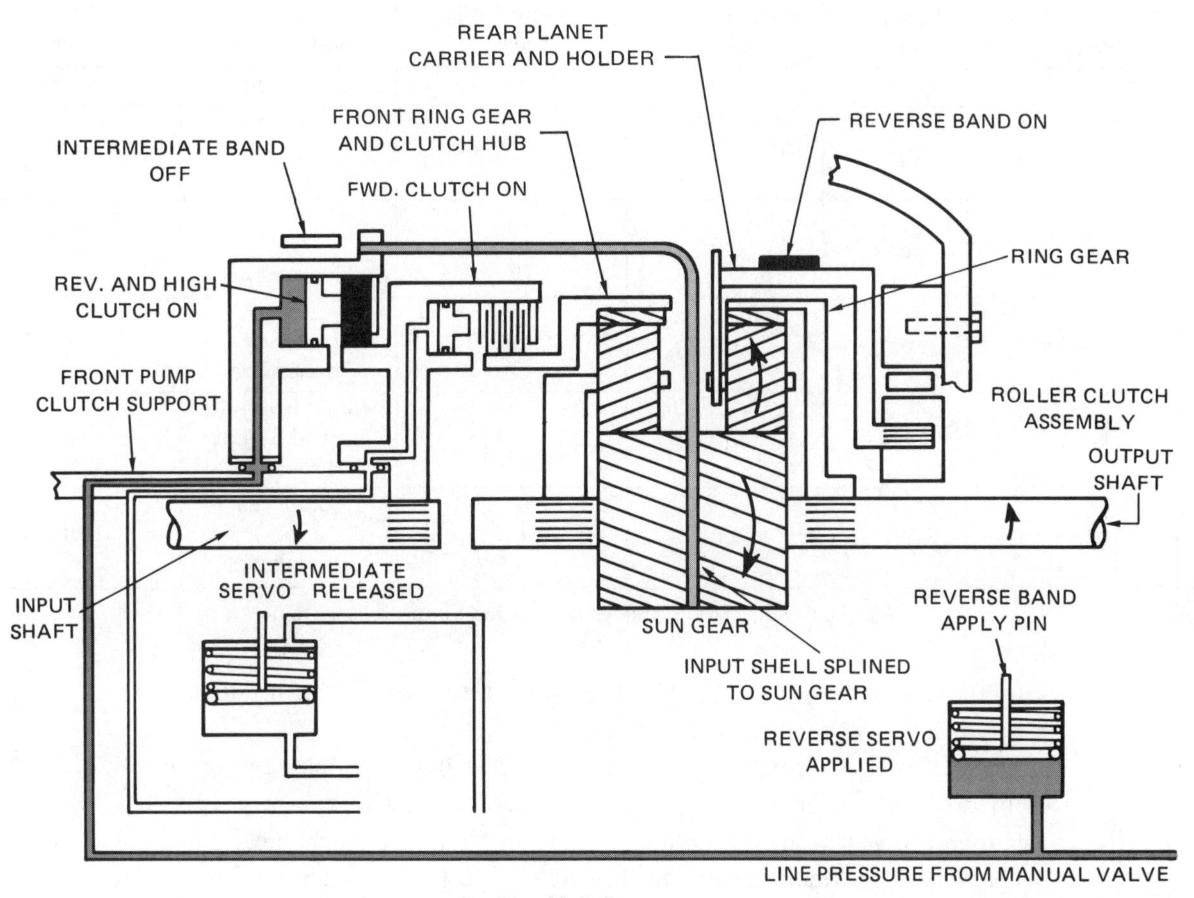

Fig. 10-5 Reverse.

power flows through the forward clutch to the front ring gear, resulting in two reverses. In manual 1, however, the rear carrier is held by the low and reverse band. This provides engine braking on hills, where the one-way clutch would freewheel.

Manual 2

In manual 2, the power flow is the same as in drive range intermediate. The difference between these two ranges is in the valve body. This is dealt with in the units covering the individual transmissions.

SUMMARY

Figure 10-6 summarizes the power flow in the Simpson gear train. A good mechanic can trace the power flow in all ranges and use this knowledge to diagnose problems in the gear train. For example, if a customer complains of no drive in D, a look at the chart shows that the trouble could be either the forward clutch or the one-way clutch. The mechanic then tries 1 range. If there is drive in 1, the trouble would have to be in the one-way

Range	Forward Clutch	Intermediate Band	Reverse & High Clutch	One-way Roller Clutch	Low & Reverse Band
D Low	on	off	off	Holding	off
D Intermediate	on	on		Unlocked or Overrunning	off
D High	on	off	on	Unlocked or Overrunning	off
Reverse	off	off	on	Locked by Low & Reverse Band	on
Manual 1	on	off	off	Would Hold, But Locked by Low & Reverse Band	on
Manual 2	on	on	off	Unlocked or Overrunning	off

Fig. 10-6 Summary of power flow in the Simpson gear train.

clutch. As the chart shows, the one-way clutch should be holding in D, whereas in 1, the band holds the rear carrier.

By thinking through the power flow and knowing which band or clutch is in use, the mechanic should be able to solve most of the problems that occur in the gear train. Of course, not all problems are mechanical; but even when solving hydraulic problems, it is helpful to know which servo or clutch is in use.

REVIEW

Select the best answer from the choices offered to complete the statement or answer the question. (Use the chart and the descriptions of power flow given in the text as necessary.)

1. Which of the following units control the gear set in drive range high?
   a. the forward clutch, the intermediate band, and the reverse and high clutch
   b. the forward clutch and the reverse and high clutch
   c. the intermediate band and the reverse and high clutch
   d. the reverse and high clutch and the one-way clutch

2. When the forward clutch is on, it drives the:

   (I) Sun gear.
   (II) Front ring gear.

   a. I only
   b. II only
   c. both I and II
   d. neither I nor II

3. The low and reverse band holds the:

   a. front ring gear.
   b. front planetary carrier.
   c. sun gear.
   d. rear planetary carrier.

4. Which of the following units control the gear set in reverse?

   a. the forward clutch and the reverse and high clutch
   b. the reverse and high clutch and the intermediate band
   c. the reverse and high clutch and the one-way clutch
   d. the reverse and high clutch and the low and reverse band

5. Which of the following units control the gear set in drive range low?

   a. the forward clutch and the low and reverse band
   b. the forward clutch, the low and reverse band, and the one-way clutch
   c. the forward clutch and the one-way clutch
   d. the forward clutch and the intermediate band

6. In drive range low, the rear planetary carrier is held by the

   a. low and reverse band.
   b. forward clutch.
   c. reverse and high clutch.
   d. one-way clutch.

7. When the forward clutch is applied it drives the:

   (I) Sun gear.
   (II) Front ring gear.

   a. I only
   b. II only
   c. both I and II
   d. neither I nor II

8. Which of the following units control the gear set in intermediate?

   a. the forward clutch and the intermediate band
   b. the forward clutch, the intermediate band, and the one-way clutch
   c. the forward clutch and the low and reverse band
   d. the forward clutch and the reverse and high clutch

9. If both the forward clutch and the reverse and high clutch are applied, the transmission is in

   a. low.
   b. intermediate.
   c. high.
   d. reverse.

10. Which of the following units control the gear set in manual 1?

    a. the forward clutch and the low and reverse band
    b. the forward clutch and the one-way clutch
    c. the forward clutch and the intermediate band
    d. the reverse and high clutch and the reverse and low band

11. When the reverse and high clutch is applied, it drives the:

    (I) Sun gear.
    (II) Front ring gear.

    a. I only
    b. II only
    c. both I and II
    d. neither I nor II

12. The intermediate band holds the

    a. front ring gear.
    b. front planetary carrier.
    c. sun gear.
    d. rear planetary carrier.

EXTENDED STUDY PROJECTS

Since the Simpson gear train is used by all American manufacturers, it is essential for the mechanic to know its operation and power flow. The following exercises provide experience in dealing with the Simpson gear train.

1. Examine a Simpson gear train from either a Ford C-4 or a Chrysler TorqueFlite. Both clutches can be removed from these gear trains, and the front ring gear can be turned by hand.
    a. By working with the gear set, demonstrate that if the front ring gear is turned in a forward direction, the sun gear and shell turn in a reverse direction, while the output shaft turns in a forward direction in the lowest reduction.
    b. While still turning the front ring gear, stop the sun gear. Note that the one-way clutch releases, and the output shaft turns at a faster speed.
    c. Turn the ring gear and sun gear at the same speed. Note that the output shaft turns in a 1:1 ratio.
    d. To demonstrate reverse, turn only the sun gear, and hold the rear carrier.
2. A customer complains of no drive in D, but there is drive in manual 1.
    a. What clutch or band would most likely cause this problem?
    b. Explain the diagnosis.
3. A common customer complaint is that the transmission will not shift to high. A road test shows that the transmission does shift to high but skips intermediate in manual 2. This is called *1-3 upshift*. A further check shows that there is no intermediate in manual 2.
    a. What clutch or band would most likely cause this problem?
    b. Explain the diagnosis.

4. A customer complains that the transmission slips in reverse. A road test shows that the transmission also slips in high, but manual 2 operates correctly. There is engine braking in manual 1.

   a. What clutch or band would most likely cause this problem?
   b. Explain the diagnosis.

5. A customer complains that the transmission slips in reverse. A road test shows that it does slip in reverse. In addition, there is no engine braking in manual 1, but it is operating correctly in all other ranges.

   a. What clutch or band would most likely cause this problem?
   b. Explain the diagnosis.

CHRYSLER TORQUEFLITE TRANSMISSIONS

OBJECTIVES

After studying this unit, the student will be able to:

- Identify the clutches and bands used to control the gear set in each speed or range of the TorqueFlite transmission.
- Trace the power flow and hydraulic circuit through D, R, manual 1, and manual 2.
- Diagnose problems that deal with the power flow and hydraulic system.

APPLICATION

The Chrysler TorqueFlite is a fully automatic three-speed transmission. It uses a torque converter and a Simpson gear train. The TorqueFlite transmission is made in two different models: the A904 and A727. The A904 is generally used on small-block V8 and six-cylinder engines, while the A727 is a heavy-duty transmission used with the larger V8s. However, for heavy-duty use, the A727 may be coupled with a 223-cubic inch six-cylinder or a 318-cubic inch V8. In some instances, the A904-LA is used with a 360-cubic inch engine.

OPERATION

Both the A904 and A727 use a Simpson gear train. The only difference between the gear trains in these models and that described in unit 10 is in the naming of the clutches and bands. The parts of the TorqueFlite are shown in figure 11-1. The

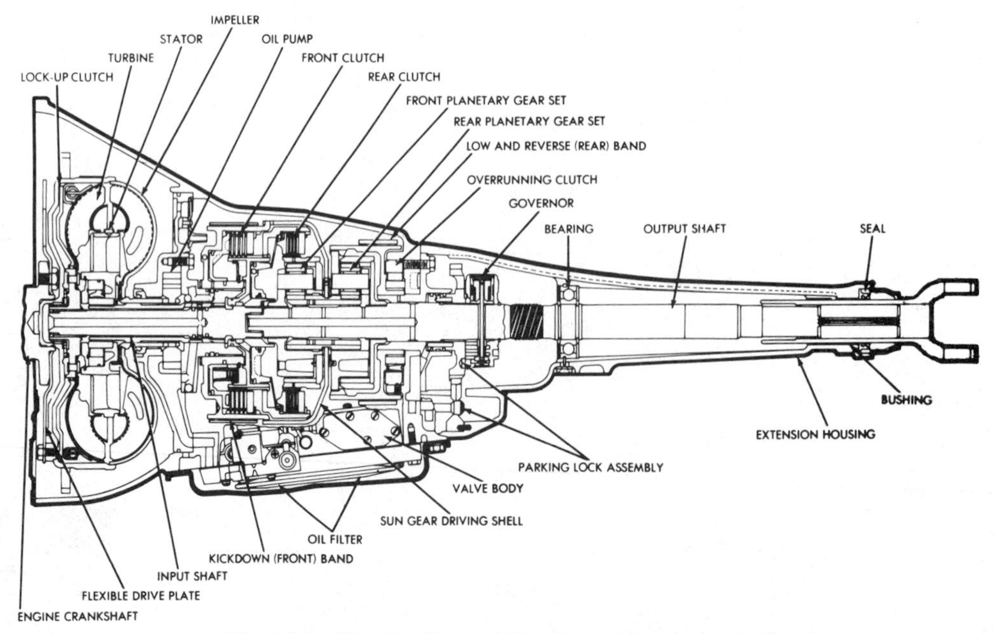

Fig. 11-1 Chrysler TorqueFlite Transmission (A-727).

Lever Position	Gear Ratio	Start Safety	Parking Sprag	Clutches			Bands	
				Front	Rear	Over-running	(Kickdown) Front	(Low-Rev.) Rear
P—PARK		X	X					
R—REVERSE	2.21			X				X
N—NEUTRAL		X						
D—DRIVE								
First	2.45				X	X		
Second	1.45				X		X	
Direct	1.00			X	X			
2—SECOND								
First	2.45				X	X		
Second	1.45				X		X	
1—LOW (First)	2.45				X			X

Fig. 11-2 Clutches and bands in use at each position of the selector lever.

rear clutch in the TorqueFlite has the same use as the forward clutch in the basic Simpson gear train described in unit 10. The front clutch serves as the reverse and high clutch, and the kickdown band has the same function as the intermediate band.

The *overrunning clutch* is a one-way roller clutch. It prevents the rear carrier from turning in a reverse direction.

For low, the power flow in the Torque-Flite is the same as that described in unit 10. When the rear clutch is applied, the front ring gear turns the planet pinions as idlers. This turns the sun gear in reverse rotation. Since the rear carrier is held by the over-running clutch, the ring gear and output shaft must turn in a forward direction at a ratio of about 2.45:1.

For second, the kickdown band is applied to hold the sun gear. This causes the front carrier to walk around the stopped sun gear and turns the output shaft in a reduction of about 1.45:1.

For direct, the kickdown band is released and the front clutch is applied. Since both the front ring gear and sun gear are now driving, the gear set is locked in direct at 1.00:1 ratio. The chart in figure 11-2 shows the clutch or band that is in use for all shift lever positions.

For reverse, the front clutch is applied to turn the sun gear. The rear band holds the rear carrier assembly. This causes reverse rotation of the rear ring gear and output shaft at a ratio of 2.21:1.

HYDRAULIC SYSTEM

The hydraulic control system consists of the various valves and hydraulic control devices described in units 5 and 6. If there is any question as to their basic operation, these units should be reviewed.

Pressure Regulator

The main pressure supply is from an internal-external rotor-type pump, figure 11-3. This pump differs from the internal-external gear type pump in that it does not have a crescent. Line pressure from the pump is regulated by a balanced, bypass pressure regulator valve to about 57 psi (393 kPa) in neutral.

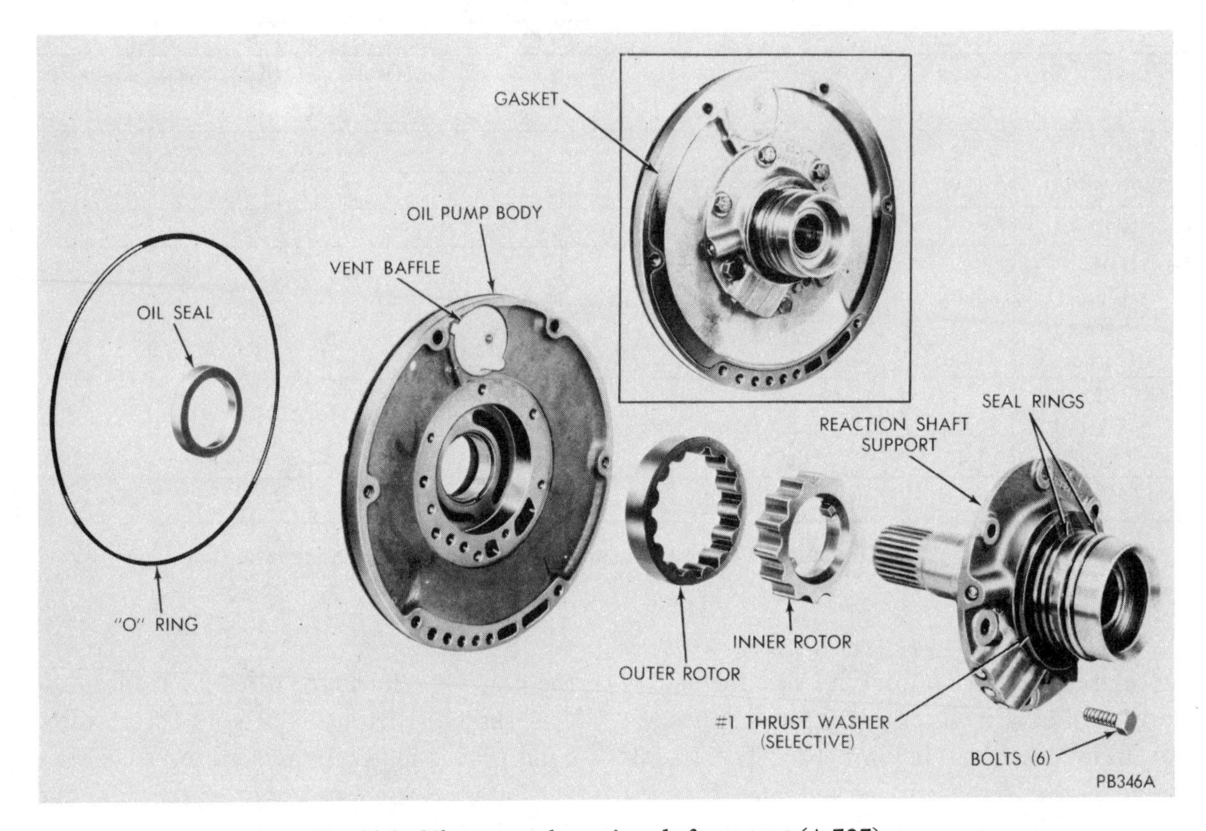

Fig. 11-3 Oil pump and reaction shaft support (A-727).

Torque Converter Control Valve

From the pressure regulator valve, line pressure is sent to the torque converter (TC) control valve, figure 11-4. This valve is a balanced restriction-type valve, and regulates converter pressure between 10 and 75 psi (69 to 517 kPa). This is done to stop *ballooning* or swelling of the converter due to high pressure caused by large throttle openings.

Due to the fluid flow caused by the controlled leakage at the bushings and cast iron seal rings, the cooling and lubrication system is held at a pressure of 5 to 30 psi (34 to 207 kPa).

DRIVE RANGE

In drive range low (or *breakaway,* as it is called in Chrysler vehicles) line pressure is sent to the rear clutch, the throttle valve, the governor, the 1-2 shift valve, and the accumulator, figure 11-5, page 138. With the rear clutch applied, the front ring gear is connected to the input shaft. Power flow in this condition is as described in unit 10 for drive range low of the Simpson gear train.

Throttle Valve

The throttle valve (TV) is a balanced, restriction-type valve and regulates TV pressure from 0 to 94 psi (0 to 648 kPa). TV pressure acts on the pressure regulator valve, the 1-2 shift valve, the 2-3 shift valve, and the shuttle valve. The shuttle valve is described in detail later in the unit.

TV pressure, at the pressure regulator valve, causes line pressure to vary with throttle opening. Depending on throttle opening, line pressure varies between 57 and 94 psi (393 to

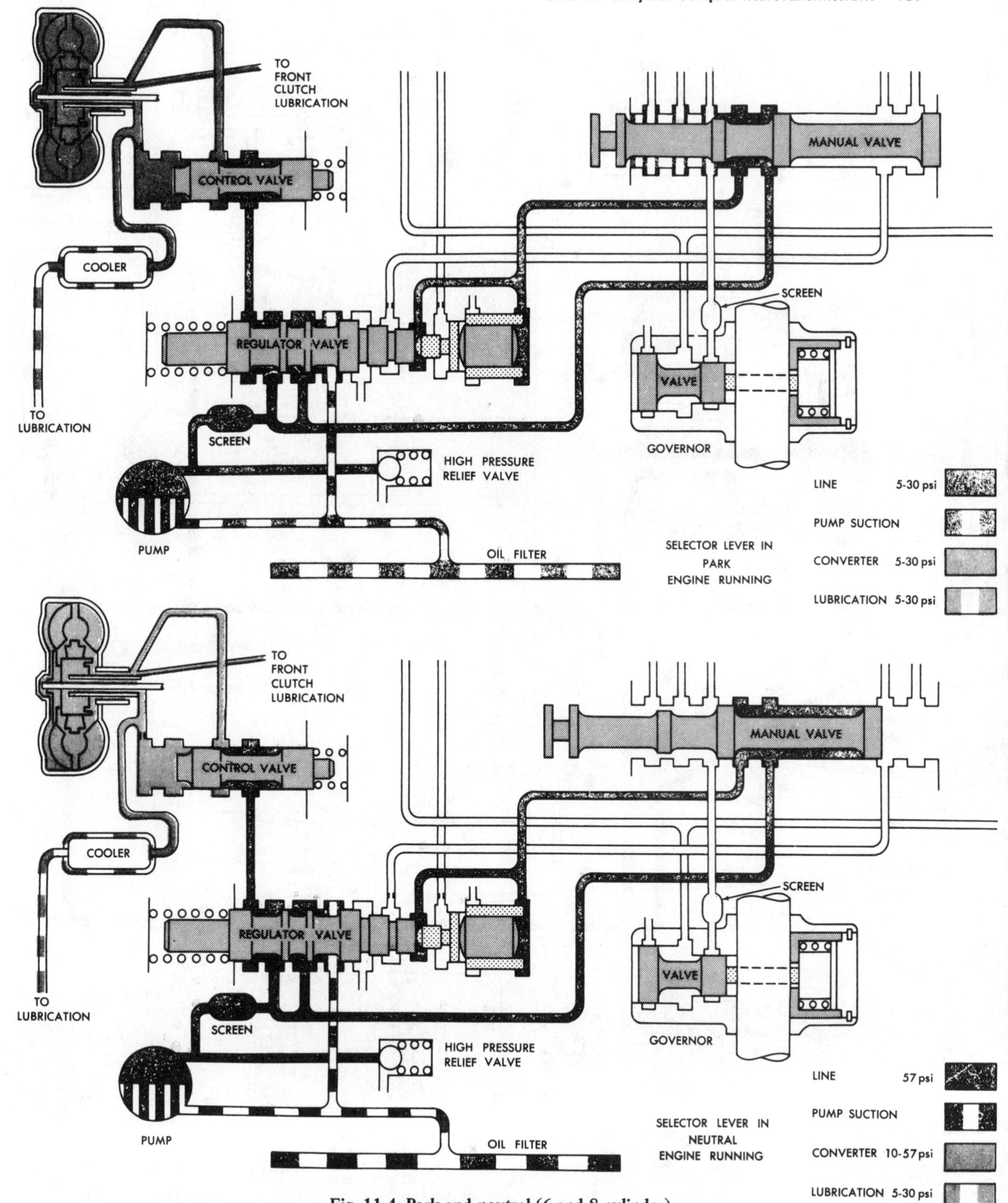

Fig. 11-4 Park and neutral (6 and 8 cylinder).

[For color diagram, see page 312.]

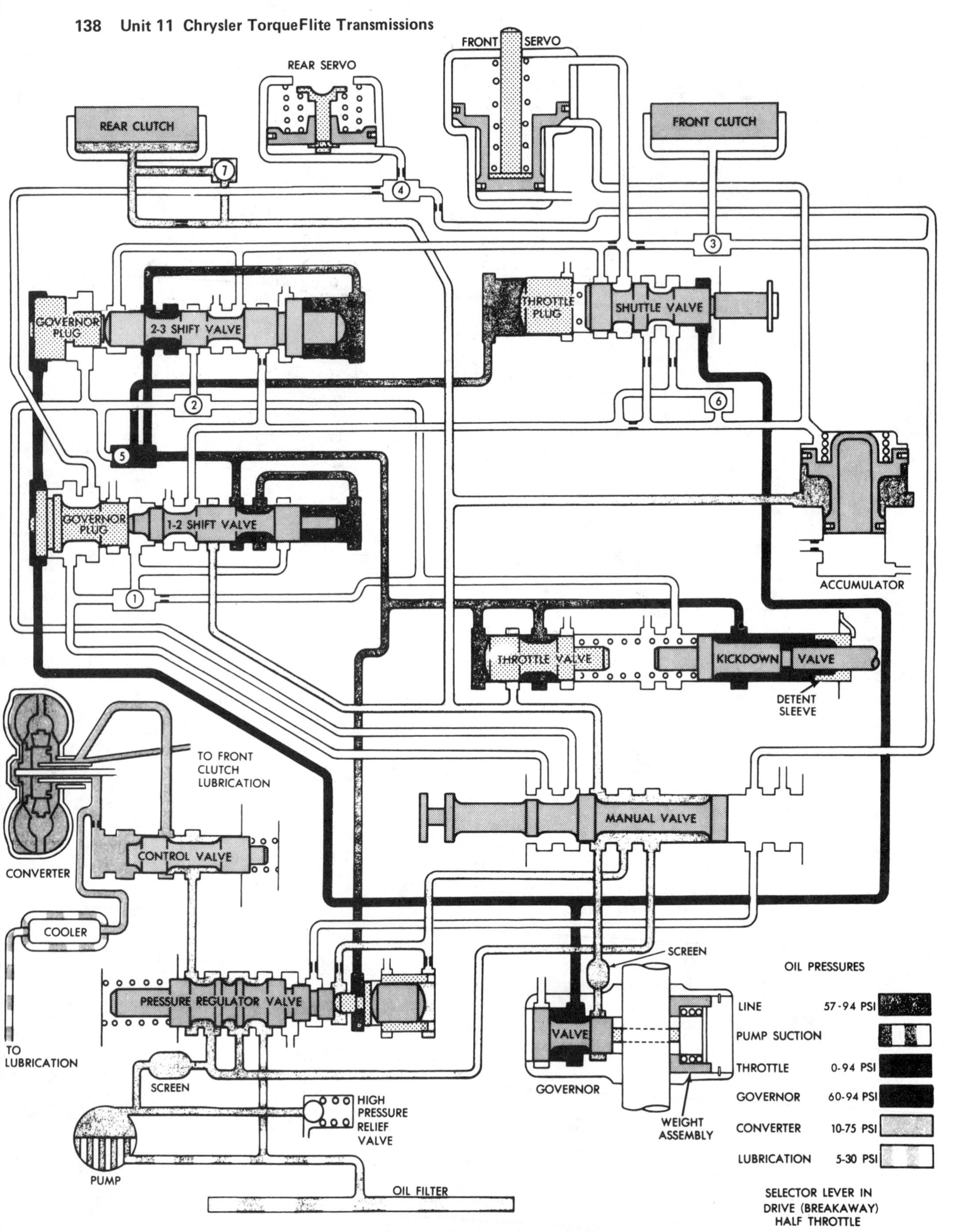

Fig. 11-5 Drive-Breakaway (6 cylinder).

[For color diagram, see page 313.]

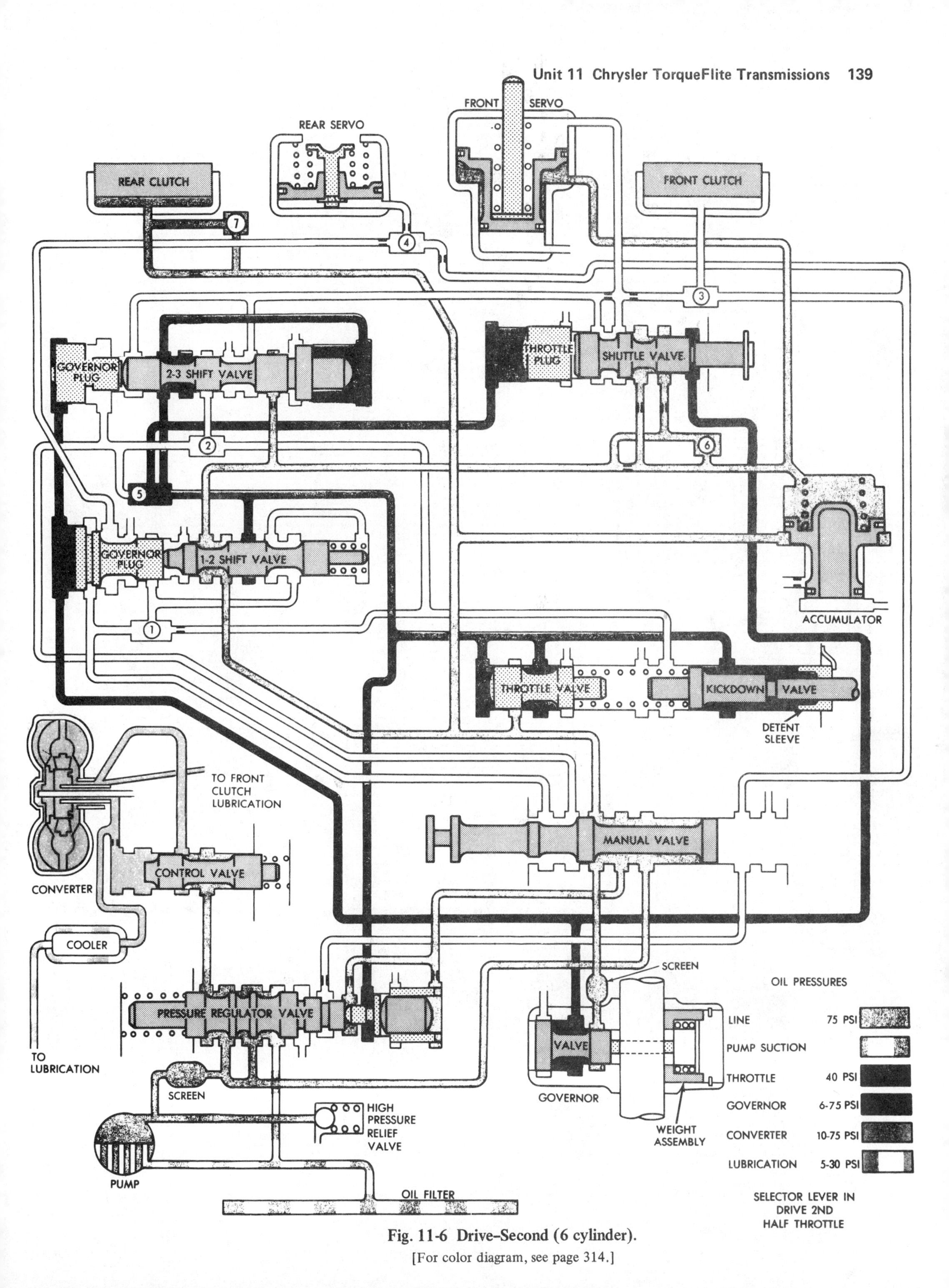

Fig. 11-6 Drive–Second (6 cylinder).

[For color diagram, see page 314.]

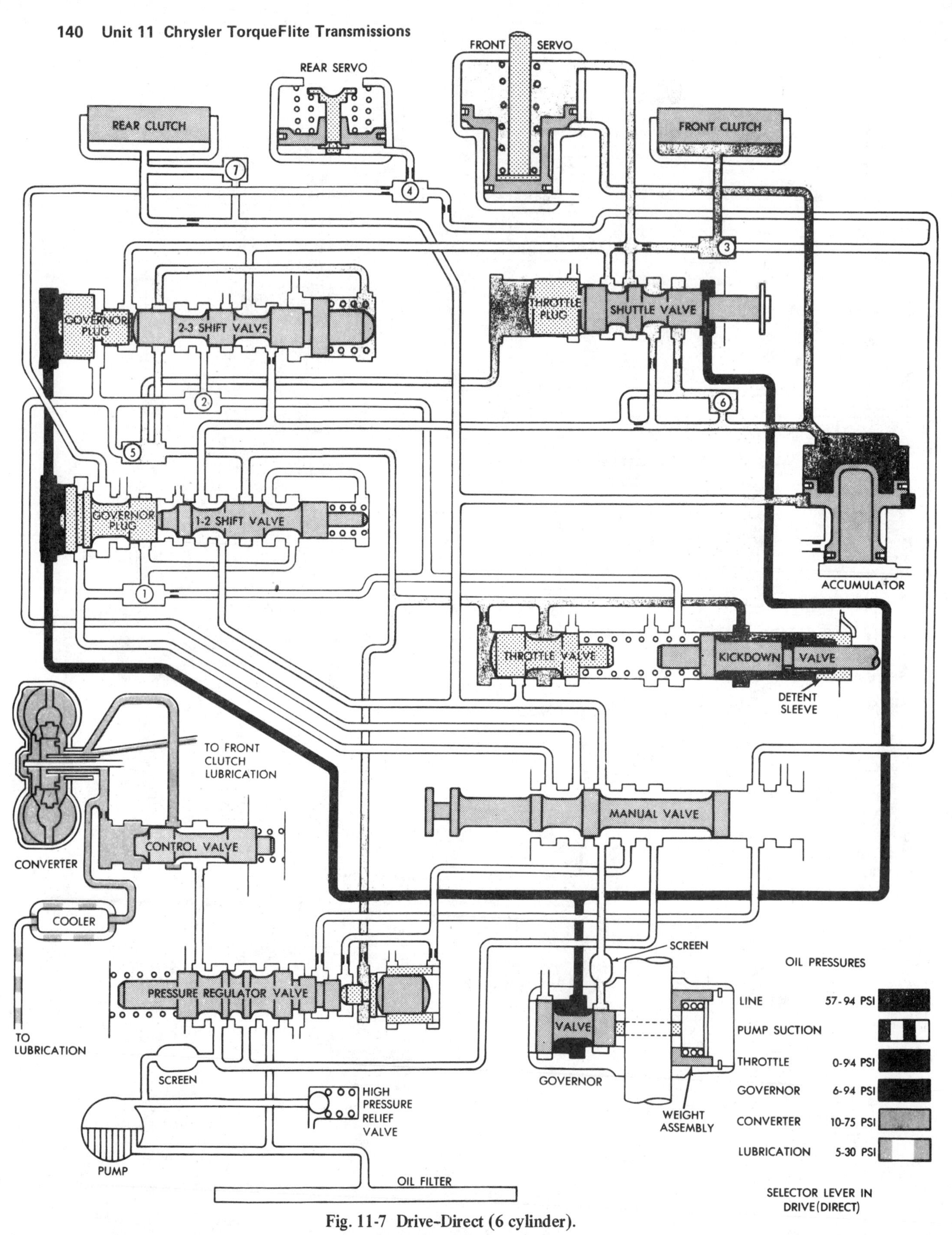

Fig. 11-7 Drive-Direct (6 cylinder).

[For color diagram, see page 315.]

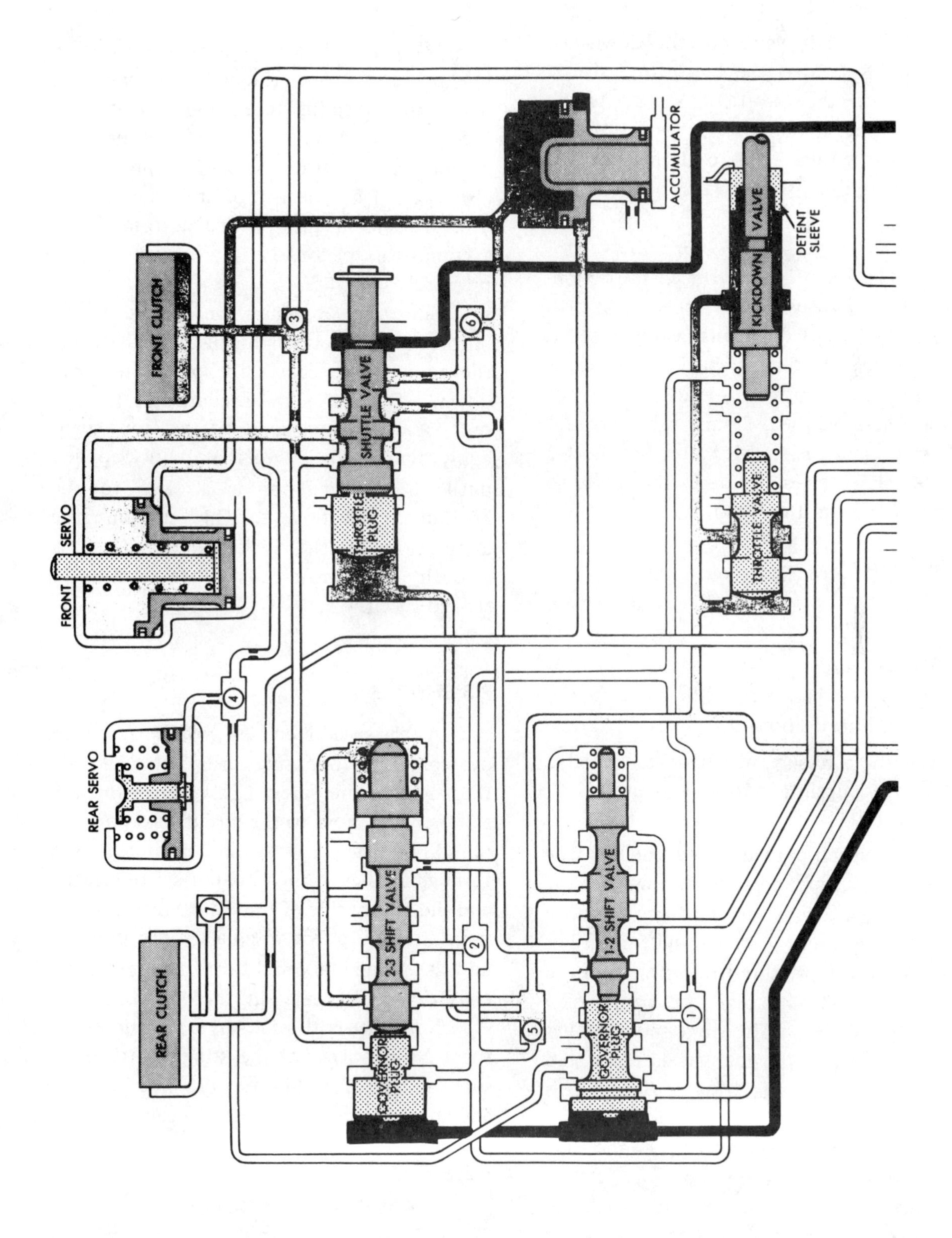

Fig. 11-8 Shuttle valve action.

648 kPa). TV pressure also acts to keep the 1-2 and 2-3 shift valves in the downshift position. Note that the kickdown valve is connected to the throttle linkage. As the throttle is opened, more spring force is placed on the throttle valve. This results in higher TV pressure with increased throttle opening.

Governor

The governor is a balanced restriction-type valve. It regulates line pressure at 0 to 1.5 psi (0 to 10 kPa) when the car is stopped, and at 94 psi (648 kPa) when the vehicle is operating at high road speeds. Governor pressure acts on the 1-2 shift valve, the 2-3 shift valve, and the shuttle valve. At the 1-2 and 2-3 shift valves, it causes the valves to move to the upshift position.

1-2 Shift Valve

The forces acting on the 1-2 shift valve are:

- spring force, plus the force of TV pressure, which keeps the valve in the downshift position, and
- governor pressure, which moves the valve to the upshift position, figure 11-6, page 139.

At a specific road speed and throttle opening, governor pressure, acting on the governor plug, exerts enough force to move the 1-2 shift valve to the upshift position. This allows line pressure to pass through the valley of the 1-2 shift valve to the apply side of the front servo. The front servo applies the kickdown band to hold the sun gear, and the transmission is in drive second. This range can be compared to drive intermediate in the basic Simpson gear train. As the shift valve moves, line pressure also moves to the 2-3 shift valve, the shuttle valve, and the accumulator. Note that line pressure is blocked by the land of the 2-3 shift valve.

Accumulator

When the shift lever is moved to drive, line pressure strokes the accumulator piston up against a spring force, figure 11-5. As the 1-2 shift valve moves and line pressure is sent to the front servo, line pressure is also exerted on the top of the accumulator. The pressure at the top and bottom of the accumulator piston is now equal to line pressure.

Since the pressure on the top and bottom of the accumulator piston is the same, the spring force causes the piston to move down. Ths allows the accumulator to take on, or accumulate, some of the front servo apply fluid. The servo is not fully applied until the accumulator is stroked to the bottom of its travel. This cushions the application of the kickdown band for a smooth shift from first (breakaway) to second speed.

2-3 Shift Valve

As car speed increases, governor pressure overcomes the opposing spring force and the throttle pressure force. This causes the 2-3 shift valve to move to the upshift position, figure 11-7, page 140. Line pressure flows through a valley in the valve to the front clutch and the release side of the front servo. At this point, release pressure equals apply pressure and the spring is allowed to stroke the servo to the OFF position, releasing the kickdown band.

Note that both valves have snap action – the 1-2 shift valve by exhausting throttle pressure, and the 2-3 shift valve by line pressure added to the governor end. Of course, as car speed decreases, automatic downshift takes place.

In a vehicle with a typical (225 in^3) six-cylinder engine, a small- or minimum-throttle 1-2 upshift will take place at 9 to 16 mph (14

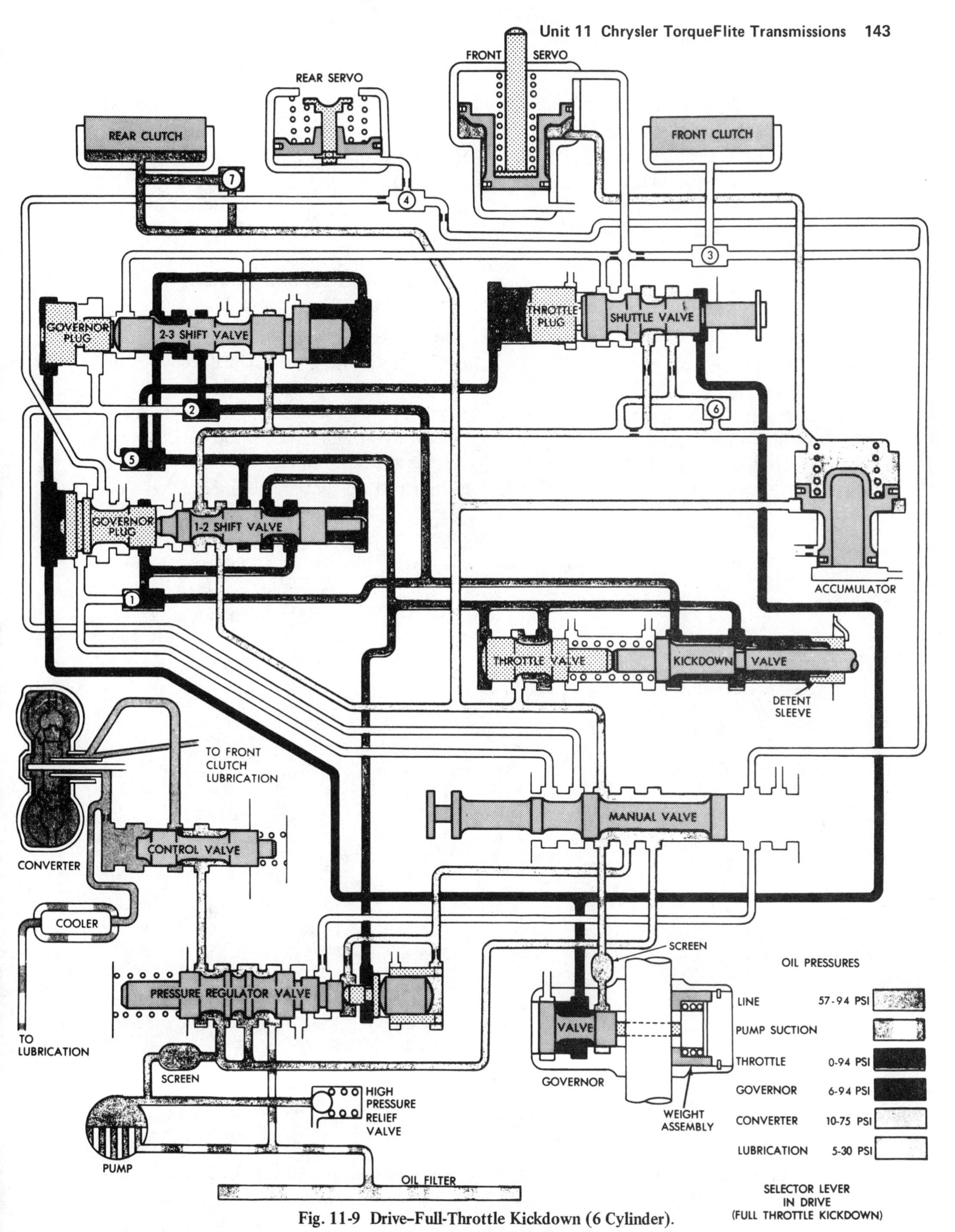

Fig. 11-9 Drive–Full-Throttle Kickdown (6 Cylinder).

[For color diagram, see page 316.]

to 26 km/h), and a 2-3 upshift will take place at 15 to 25 mph (24 to 40 km/h). A closed-throttle 3-1 downshift will take place at 8 to 13 mph (13 to 21 km/h). Wide-open throttle upshift will take place at 31 to 43 mph (49 to 69 km/h) for a 1-2 upshift, and at 63 to 76 mph (101 to 122 km/h) for a 2-3 upshift.

Lift-Foot Shift

If the driver's foot were lifted from the gas pedal as the transmission shifts from second to high, the band might not release before the front clutch was applied. This would result in a rough upshift due to clutch-band *fight*. A *shuttle valve*, figure 11-8, page 141, eliminates this problem. During a lift-foot shift, throttle pressure drops off, and the shuttle valve moves to the left. This lets servo-release oil flow through the valley of the valve to the non-orificed release line. This fast band release makes for a smooth 2-3 upshift.

Part-throttle and Kickdown Shifts

The driver can force downshift at part-throttle, or wide-open throttle (*kick-down*). By pressing the gas pedal toward the wide-open throttle position until a resistance is felt, a 3-2 part-throttle downshift can be caused at between 41 and 61 mph (66 to 99 km/h). A 3-2 kickdown shift is caused by pushing the throttle through the resistance (detent) and

Carline	VLHN	VLHN	RW	SX	PDC	PDC
Engine Cu. In.	225	318	360-4 Hi. Perf.	360-2	400-2 400-4	400-4 & 440 Hi. Perf.
Axle Ratio	2.76	2.45	3.21	2.45	2.71	2.71
Tire Size	6.95x14	E78x14	H78x14	GR78x15	HR78x15	JR78x15
Throttle Minimum						
1-2 Upshift	9-16	8-16	8-15	9-16	9-16	8-15
2-3 Upshift	15-25	15-25	15-23	17-25	15-25	15-23
3-1 Downshift	8-13	9-14	8-13	9-14	8-13	8-13
Throttle Wide Open						
1-2 Upshift	31-43	39-54	43-56	41-57	37-52	38-53
2-3 Upshift	63-76	79-95	78-93	83-100	77-92	77-93
Kickdown Limit						
3-2 WOT Downshift	60-73	76-92	75-90	79-96	73-89	74-90
3-2 Part Throttle Downshift	46-61	30-56	34-57	31-58	30-56	29-54
3-1 WOT Downshift	28-35	30-44	34-47	31-46	29-43	29-43
Governor Pressure*						
15 psi	20-22	21-23	20-22	22-24	21-23	21-23
40 psi	47-53	60-66	60-66	62-69	59-66	58-65
60 psi	66-72	83-90	81-88	87-94	82-90	81-88

*Governor pressure should be from zero to 1.5 psi at stand still or downshift may not occur.

NOTE: Figures given are typical for other models. Changes in tire size or axle ratio will cause shift points to occur at corresponding higher or lower vehicle speeds.

Fig. 11-10 Automatic shift speeds and governor pressure chart (approximate miles per hour).

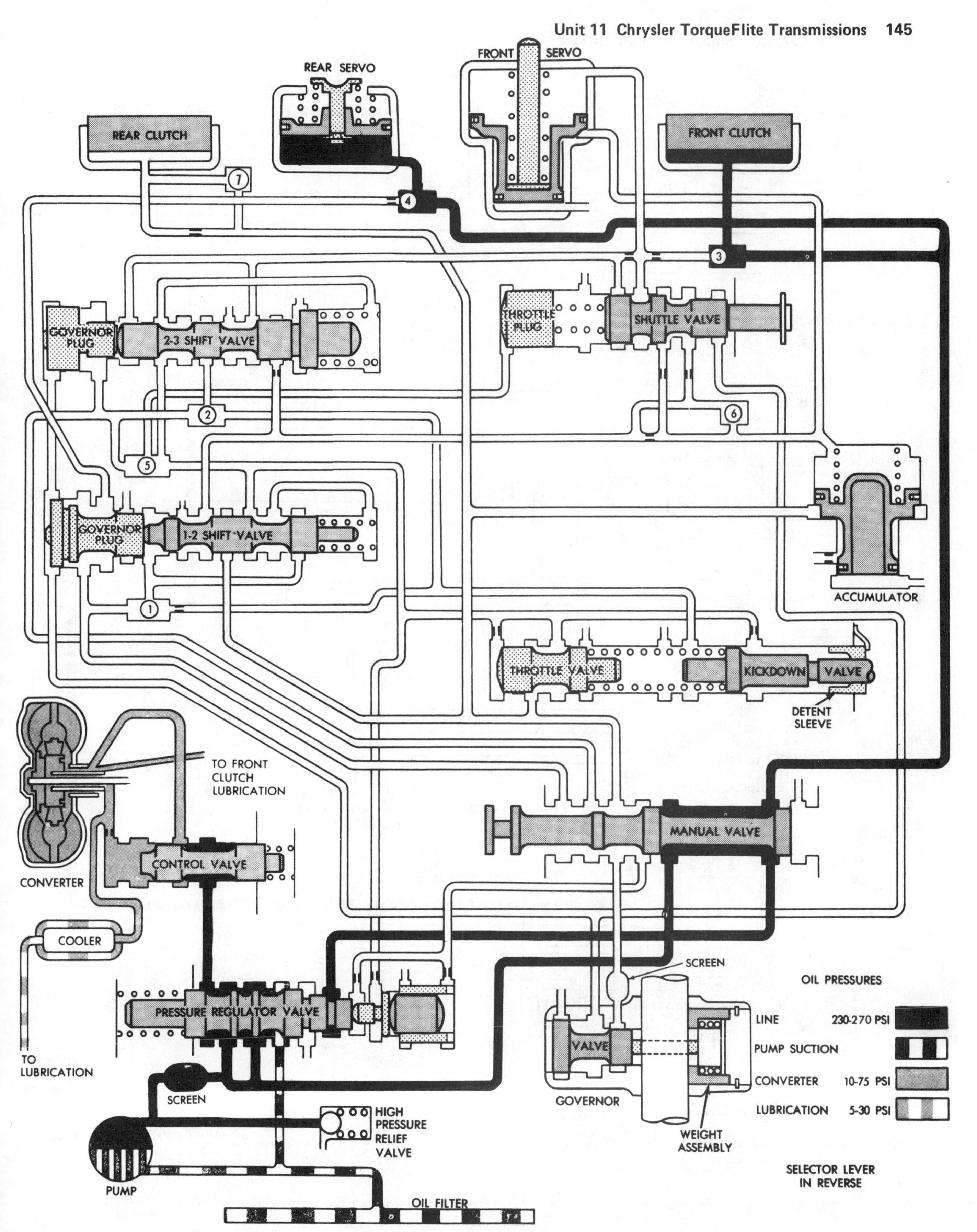

Fig. 11-11 Reverse (6 Cylinder).

[For color diagram, see page 317.]

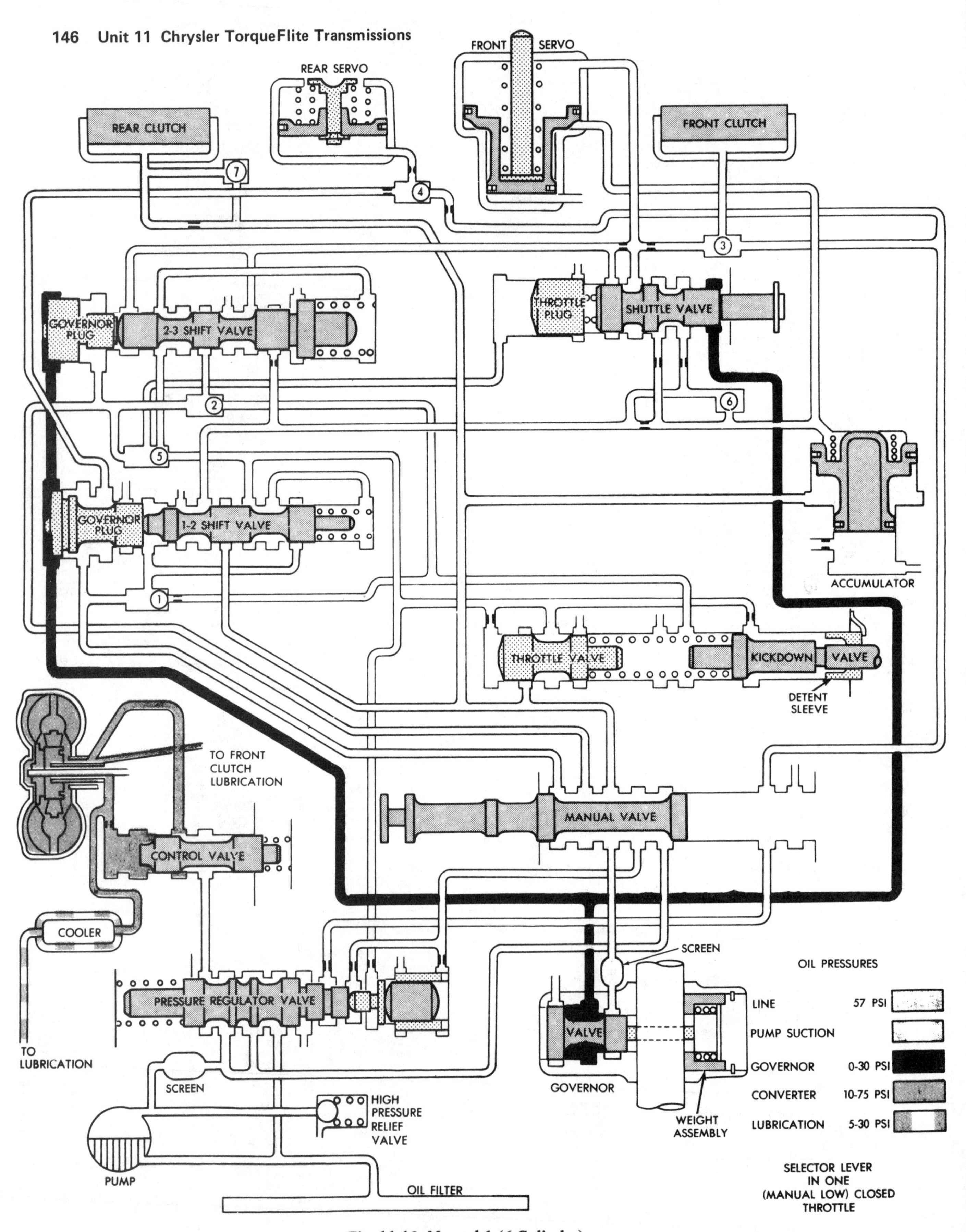

Fig. 11-12 Manual 1 (6 Cylinder).

[For color diagram, see page 318.]

will take place between 60 to 73 mph (97 to 117 km/h), figure 11-9, page 143. A 3-1 kick-down shift will take place between 28 to 36 mph (45 to 56 km/h).

Keep in mind that the shift points will vary with engine size, performance, axle ratio, and tire size as shown in figure 11-10, page 144. For example, note that a 2-3 wide-open throttle upshift can take place at from 63 to 100 mph (101 to 169 km/h) according to the combinations noted above and in the chart.

REVERSE

With the manual valve in the reverse position, line pressure seats check balls 3 and 4, and applies the front clutch and rear servo, figure 11-11, page 145. The front clutch drives the sun gear while the rear servo applies the low and reverse band to hold the rear planet carrier from clockwise rotation. This turns the ring gear and output shaft in reverse at a ratio of 2.21:1.

On figure 11-11, note that line pressure has been cut off from the large area at the right of the regulator valve. This allows the spring force to decrease the bypass of fluid and increase line pressure to between 230 to 270 psi (1 586 to 1 862 kPa). This increase is necessary to handle the high torque load that the band must hold in reverse. Note that the pressure regulator valve is still a balanced valve because of the line pressure sent to the differential force area at the right of the valve.

MANUAL 1

In manual 1, line pressure seats check ball 1 and acts on the downshift side of both areas of the governor plug, figure 11-12. This prevents upshift movement of the shift valve regardless of the road speed that may be reached. Line pressure passes through the valley of the governor plug, seats check ball 4, and applies the rear servo. This application of the rear servo and low-reverse band allows engine braking on steep hills, where the overrunning clutch would freewheel.

Note that line pressure also seats check balls 5 and 2 and acts on the downshift side of the governor plug and 2-3 shift valve. Remember, exceeding the wide-open throttle upshift speeds in manual 1 may cause serious damage to the engine or transmission.

Manual 1 can be selected at any road speed. At high road speeds (within the kickdown limits) the transmission will shift to 2 but not to 1. The reason for this can be seen by comparing figure 11-7 to figure 11-12. Note that the line pressure that seats check balls numbers 1, 2, and 5, works on only the small differential force area of the governor plug. This keeps the valves in the upshift position until governor pressure drops to a safe road speed value. By referring to figure 11-12, it can be seen that once the transmission is in low, line pressure will stop upshift because line pressure is acting on force areas that exert more force than the force of governor pressure.

MANUAL 2

In manual 2, line pressure seats check balls 2 and 5 and acts on the throttle valve side of the 2-3 shift valve and the downshift side of the governor plug. In this range, the transmission will start in 1 and shift to 2. However, when the shift is made to 2, the transmission will not upshift to direct or high.

In drive range, a manual 3-2 shift can take place, within kickdown limits, if the driver selects manual 2. This is a useful range for providing engine braking when going down steep hills.

SUMMARY

The Chrysler TorqueFlite is a fully automatic three-speed unit using a torque converter and Simpson gear train. Figure 11-2 and the oil circuit diagrams in the unit are a helpful reference for use in diagnosing problems in the TorqueFlite transmission.

By knowing the power flow and being able to trace the oil circuits, the mechanic can quickly pinpoint the cause of transmission trouble. For example, if the problem is no drive in reverse, the mechanic knows that the problem must be in the front clutch or low reverse band. If, on a road test, the transmission slips in drive range high, the mechanic can deduce that the front clutch or front clutch circuit is at fault. If the transmission does not slip in high, but there is no engine braking in manual 1, the mechanic will know that the problem is somewhere in the low-reverse band, servo, or circuit.

REVIEW

Select the best answer from the choices offered to complete the statement or answer the question.

1. In drive range second gear, the sun gear in a TorqueFlite is held by

   a. the kickdown band. c. the front clutch.
   b. the rear band. d. the rear clutch.

2. In the Chrysler TorqueFlite transmission, when the rear clutch and the kickdown band are applied, the transmission is in

   a. first speed. c. direct speed.
   b. second speed. d. reverse.

3. The effective elements for low or 1 range in the TorqueFlite transmission are

   a. the front clutch and the low-reverse band.
   b. the front clutch and the roller clutch.
   c. the front clutch and the kickdown band.
   d. the rear clutch and the low-reverse band.

4. In the Chrysler TorqueFlite transmission, the driving shell connects the front clutch to the

   a. ring gear c. sun gear
   b. carrier d. input shaft

5. The effective elements for drive range first speed, or breakaway, in the TorqueFlite transmission are

   a. the rear clutch and the low-reverse band.
   b. the rear clutch, the low-reverse band, and the overrunning clutch.
   c. the rear clutch and the overrunning clutch.
   d. the front clutch and the overrunning clutch.

6. The effective elements for reverse in the TorqueFlite transmission are

   a. the rear clutch and the reverse high clutch.
   b. the rear clutch and the low-reverse band.
   c. the front clutch and the overrunning clutch.
   d. the front clutch and the low-reverse band.

7. A customer complains of no reverse. This could be caused by:

   (I) Rear clutch slippage.
   (II) Front clutch slippage.

   a. I only
   b. II only
   c. both I and II
   d. neither I nor II

8. The effective elements for drive range third speed in the TorqueFlite transmission are

   a. the rear clutch, the kickdown band, and the front clutch.
   b. the rear clutch and the front clutch.
   c. the kickdown band and the front clutch.
   d. the front clutch and the overrunning clutch.

9. In manual low in the TorqueFlite, the low-reverse band is applied to:

   (I) Hold the rear planetary carrier.
   (II) Provide engine braking.

   a. I only
   b. II only
   c. both I and II
   d. neither I nor II

10. If an owner of a TorqueFlite-equipped car complains of harsh shifting, the first thing to check is

    a. throttle linkage adjustment.
    b. line pressure and throttle pressure.
    c. governor pressure.
    d. kickdown servo release pressure.

11. The effective elements for drive range second speed in the TorqueFlite transmission are

    a. the rear clutch and the kickdown band.
    b. the rear clutch, the kickdown band, and the overrunning clutch.
    c. the front clutch and the kickdown band.
    d. the front clutch and the rear clutch.

12. With the TorqueFlite transmission in reverse, line pressure would be

    a. higher than in drive.
    b. the same as in drive.
    c. lower than in drive.
    d. sent to the rear clutch.

13. A condition of no drive in drive range could be caused by:

    (I) A faulty overrunning clutch.
    (II) A faulty rear clutch.

    a. I only
    b. II only
    c. both I and II
    d. neither I nor II

14. In the TorqueFlite transmission:

    (I) Governor pressure works on the downshift side of the shift valves.
    (II) A spring force and the force of TV pressure oppose governor pressure.

    a. I only
    b. II only
    c. both I and II
    d. neither I nor II

15. The driving shell of the TorqueFlite is splined to:

    (I) The front ring gear.
    (II) The rear clutch.

    a. I only
    b. II only
    c. both I and II
    d. neither I nor II

EXTENDED STUDY PROJECTS

I. On blank oil circuit diagrams, complete each of the following exercises:

1. Indicate the oil circuit for D range low.
2. Indicate the oil circuit for D range high.
3. Indicate the oil circuit for full-throttle kickdown.
4. Indicate the oil circuit for manual 2.
5. Indicate the oil circuit for manual 1.
6. Indicate the oil circuit for reverse.

II. Using the oil circuit diagrams and figure 11-2, answer the following questions.

7. State two reasons for harsh engagement from neutral to drive or reverse.
8. List four reasons for delayed engagement from neutral to drive or reverse.
9. Assuming that the oil level is correct, give three reasons for no upshift.

UNIT 12 FORD JATCO, C-3, C-4, AND C-6 TRANSMISSIONS

OBJECTIVES

After studying this unit, the student will be able to:

- Identify the clutches and bands used to control the gear set in each speed or range.
- Trace the power flow and hydraulic circuit for D, R, manual 1, and manual 2.
- Identify the operating characteristics in the various ranges.
- Diagnose problems that deal with the power flow and hydraulic system.

APPLICATION

At this time, the Jatco transmission is in use mainly in the Ford Courier. The C-3 transmission is used on the Bobcat, Mustang, and Pinto. In some cases, the C-4 is also used in these cars, and with engines up to 351 cubic inches (5.75 L). The C-6 transmission is used with engines having a displacement of from 351 to 460 cubic inches (5.75 L to 7.54 L).

The Jatco, C-3, C-4, and C-6 transmissions all use the Simpson gear train. The Jatco, C-3 and C-4 power flow is the same as that in the TorqueFlite, except for the names of the clutches and bands.

As shown in figure 12-1, page 152, the C-4 uses a forward clutch, intermediate band, reverse-high clutch, low-reverse band, and a one-way roller clutch. The Jatco and C-3 transmissions are not shown, but their clutch and band setup is essentially the same as the C-4. Except for the names of the clutches and bands, the power flow is the same as that in the Chrysler TorqueFlite transmission described in unit 11. Therefore, the rest of this unit deals with the C-6 transmission.

OPERATION

The C-6 transmission differs from the C-3 and C-4 in that it has a low and reverse clutch instead of a low and reverse band. The low and reverse clutch holds the rear planetary carrier in both manual 1 and reverse.

In drive range low, the forward clutch is applied. This drives the front ring gear, and the power flow is as described in unit 10. That is, the gear train goes through two reverses that result in a forward reduction of 2.46:1.

As road speed increases, governor pressure moves the 1-2 shift valve to the upshift position. Control (line) pressure is then sent to the intermediate servo to apply the band. This stops the sun gear and makes the front planetary gear set walk around the stopped sun gear. The condition of the gears is now the same as a simple planetary gear set in intermediate. The ratio is 1.46:1.

As road speed continues to increase, governor pressure moves the 2-3 shift valve to the upshift position. Control pressure releases the intermediate band, while at the same time, the reverse and high clutch is applied. This locks the front ring gear to the sun gear,

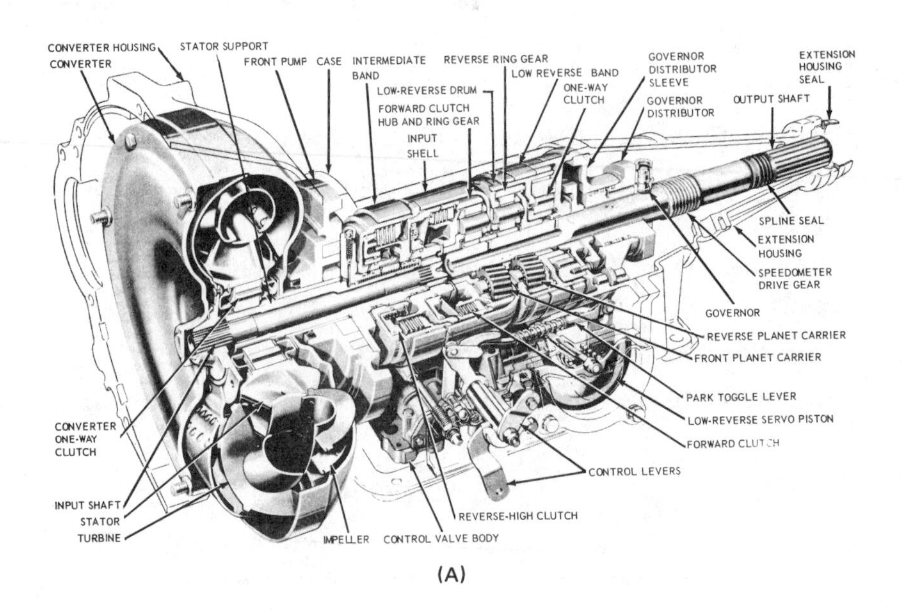

(A)

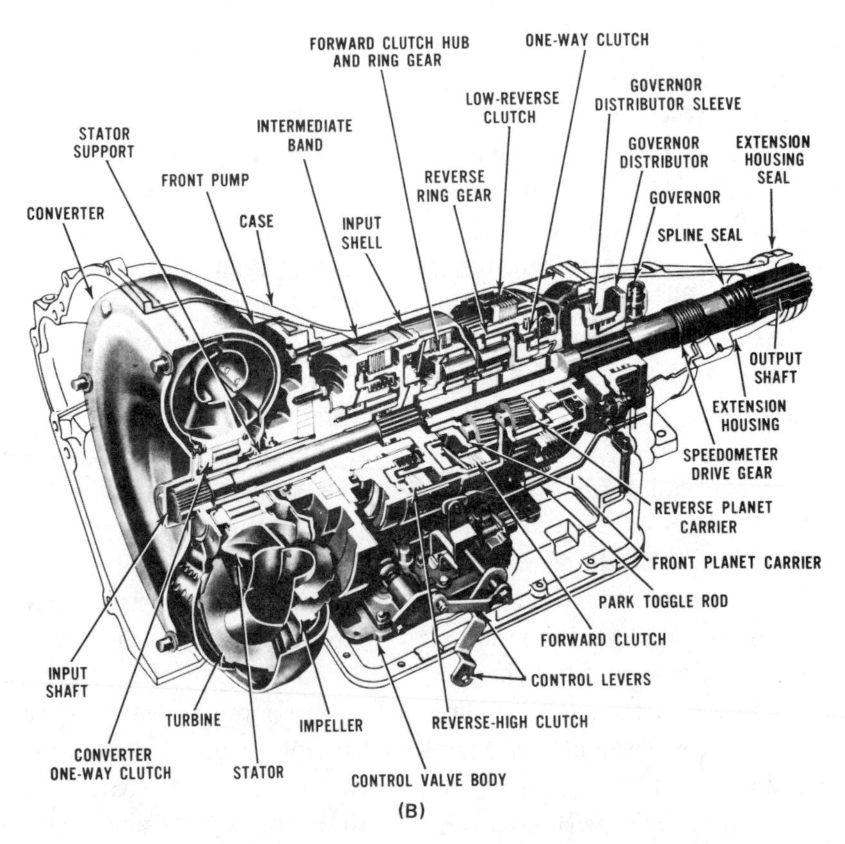

(B)

Fig. 12-1 Cutaway of (A) Ford C-4 and (B) Ford C-6 transmission.

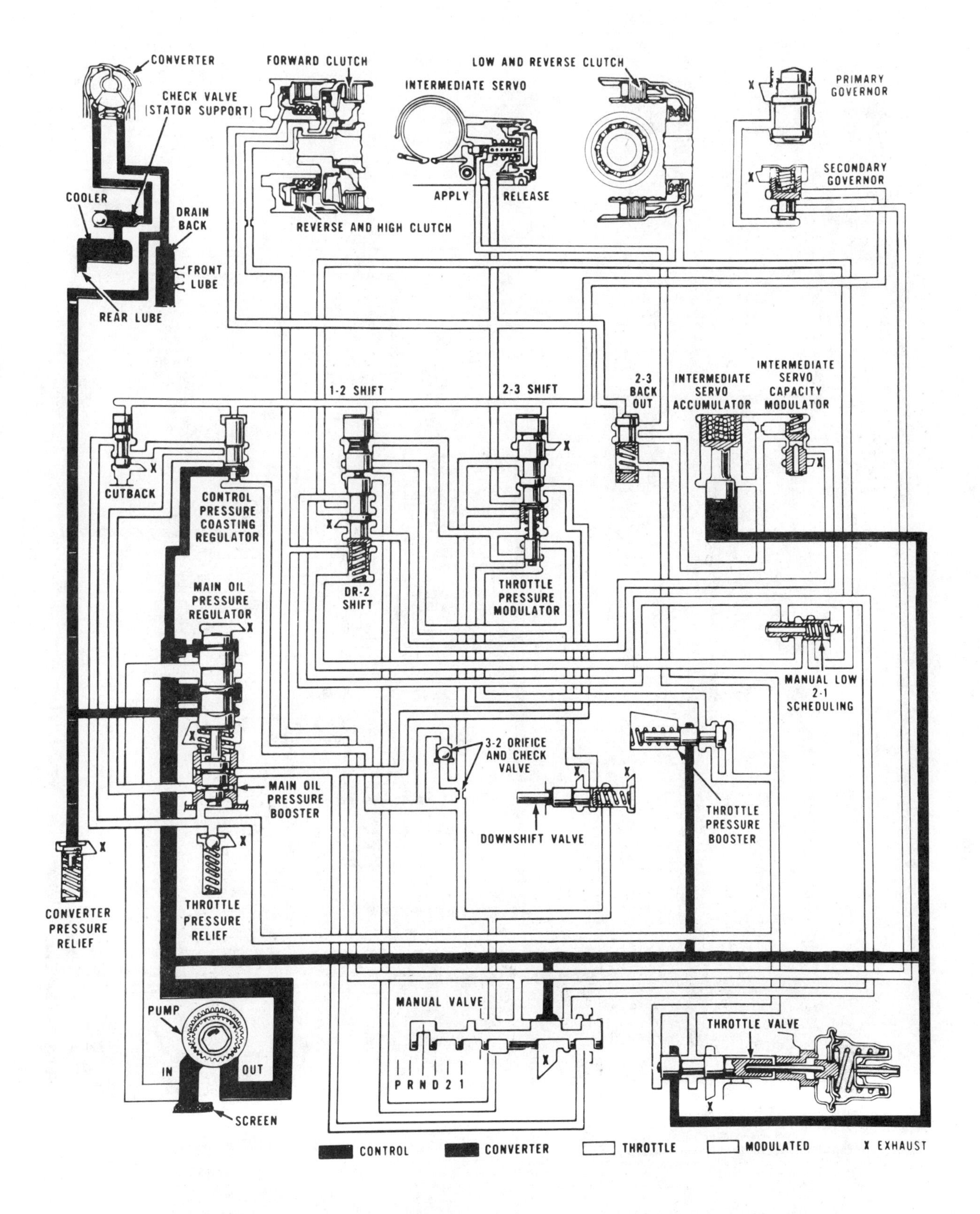

Fig. 12-2 Control pressure flow and regulation (engine idling and car stationary).

[For color diagram, see page 319.]

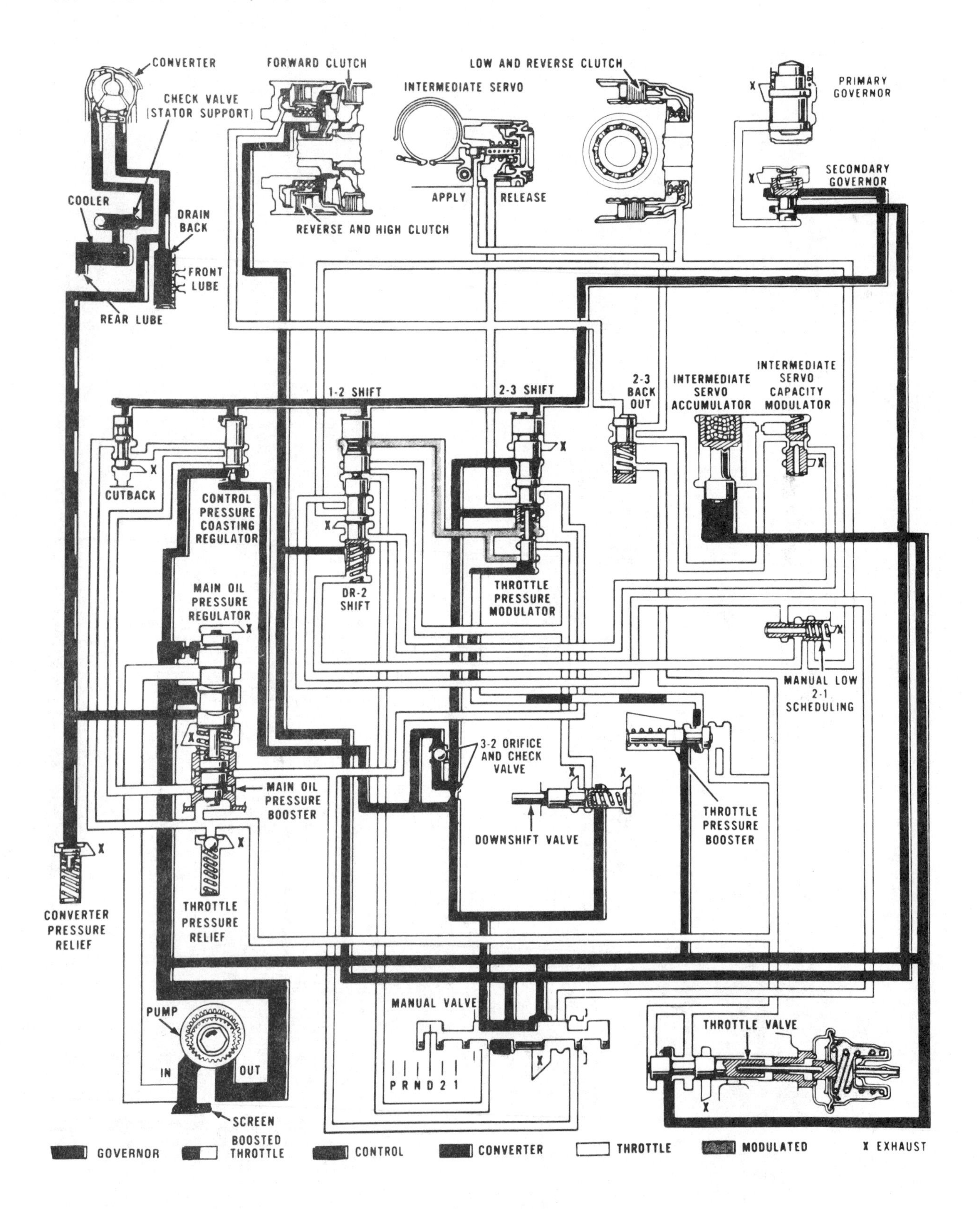

Fig. 12-3 First gear-D.

[For color diagram, see page 320.]

causing them to turn as one unit. In other words, all tooth-by-tooth gear rotation stops, and the gear set turns as one unit in a ratio of 1:1. (If this power flow is not clear, review unit 10.)

HYDRAULIC SYSTEM

An internal-external gear-type pump that is driven by the torque converter provides the hydraulic pressure necessary to run the transmission. As noted earlier, this type of pump provides more pressure and volume than is actually needed by the transmission. Therefore, it must be regulated to be useful.

Pressure Regulator

The *main oil pressure regulator* regulates to what Ford calls *control pressure.* The oil circuit for the C-6 transmission is shown in figure 12-2, page 153. Note that in this system the bypass is directed to the converter, cooler, and lubrication circuits. If too much pressure builds up in the converter, it will be exhausted to the sump by the converter pressure-relief valve.

The drain-back valve opens under low pressure to supply the front lubrication system. With the engine off, the drain-back valve is closed. In this way, the converter oil is prevented from draining back to the sump.

In figure 12-2, it can be seen that throttle pressure works on the main oil pressure regulator to increase control pressure. Throttle pressure is controlled by engine vacuum. This means that throttle pressure increases with increased throttle opening. Thus, any increase in throttle pressure will increase control pressure.

Throttle pressure starts at about 20 in Hg (138 kPa) manifold vacuum, and at 10 in Hg (69 kPa) it is about 55 psi (379 kPa). This results in control pressure being regulated between 60 and 90 psi (414 and 621 kPa). At wide-open throttle, manifold vacuum will be less than 1 in Hg (3.37 kPa). This causes a rise in throttle pressure to about 80 psi (552 kPa). This rise in throttle pressure works on the boost side of the main pressure regulator. The resulting line pressure is about 170 psi (1 172 kPa).

DRIVE RANGE

When the manual valve is moved from neutral to drive, control pressure flows to the forward clutch, the shift valves, the control pressure coasting regulator, the downshift valve, and the governor, figure 12-3.

Shift Valves

The shift valves have four forces acting of them:

- spring force
- throttle pressure
- downshift valve pressure
- governor pressure

The force of governor pressure tends to move the shift valves to the upshift position, while the three other forces tend to downshift or keep the valves in the downshift position.

Governor

Governor pressure is regulated by two valves that rotate with the output shaft, figure 12-4, page 156.

Note that, at speeds below 10 mph (16 km/h), control pressure passes by flats in the secondary valve. This moves the valve in until it is balanced by a spring force and centrifugal force. Control pressure is blocked by the valve land and cannot regulate in this position. Control pressure also flows to the primary valve and is trapped in the valley of the valve. A spring force is acting on the primary valve to keep the valve moved in.

At speeds of about 10 mph (16 km/h) centrifugal force causes the primary valve to

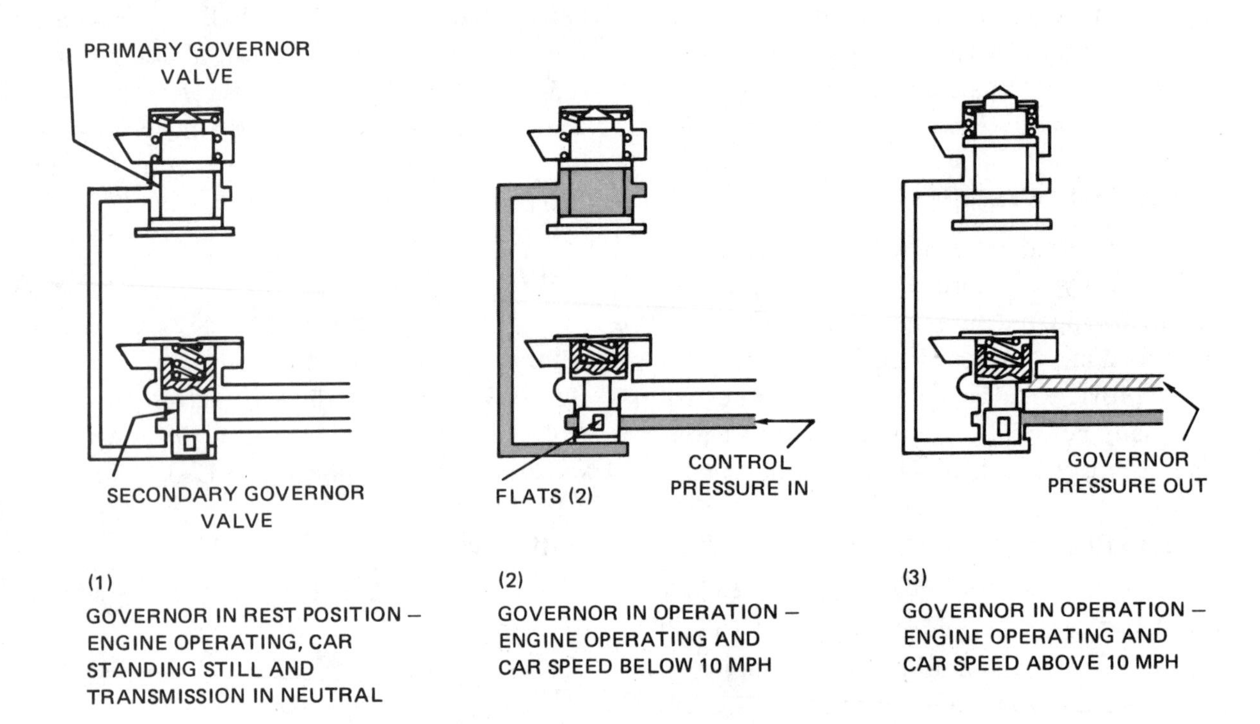

Fig. 12-4 Governor Operation.

move out against the spring force. When this happens, the primary to secondary control pressure is exhausted. This allows the secondary valve to move out and regulate control pressure to governor pressure. The forces acting on the secondary valve are:

- spring force and centrifugal force, which tend to move the valve out, and
- governor pressure working on the differential force area, which tends to move the valve in.

This creates a balanced valve, and therefore, governor pressure will vary directly with road speed.

Throttle Valve

The throttle valve is a balanced, restriction-type valve that is acted on by a vacuum unit. This vacuum unit may be of the noncompensated or the compensated type. Review unit 6 for a detailed explanation of this type of system.

Throttle Pressure Booster

At throttle openings of more than 50 degrees, intake manifold vacuum changes only a small amount. To keep the shift delay system working correctly, throttle pressure is boosted by the throttle pressure booster valve, figure 12-5. The throttle pressure booster valve operates to delay shifting at large throttle openings until a higher road speed is reached. The throttle pressure booster system also controls torque demand downshifts at large throttle openings and under heavy load conditions.

Downshift Valve

The downshift valve is connected to the throttle linkage so that when the throttle is wide open, the 2-3 upshift can take place at as

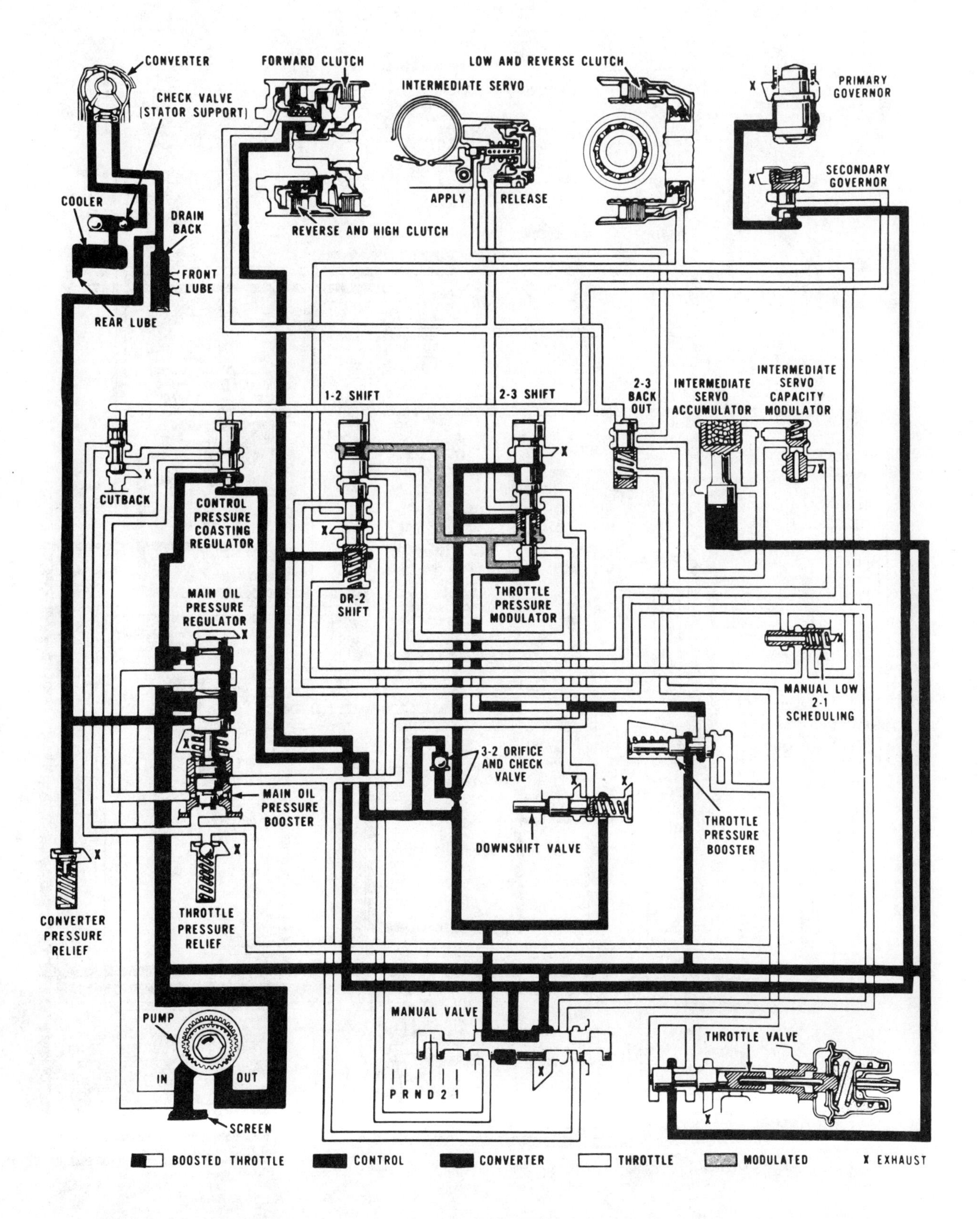

Fig. 12-5 Transmission pressures at less than one inch of vacuum, throttle pressure booster in operation.

[For color diagram, see page 321.]

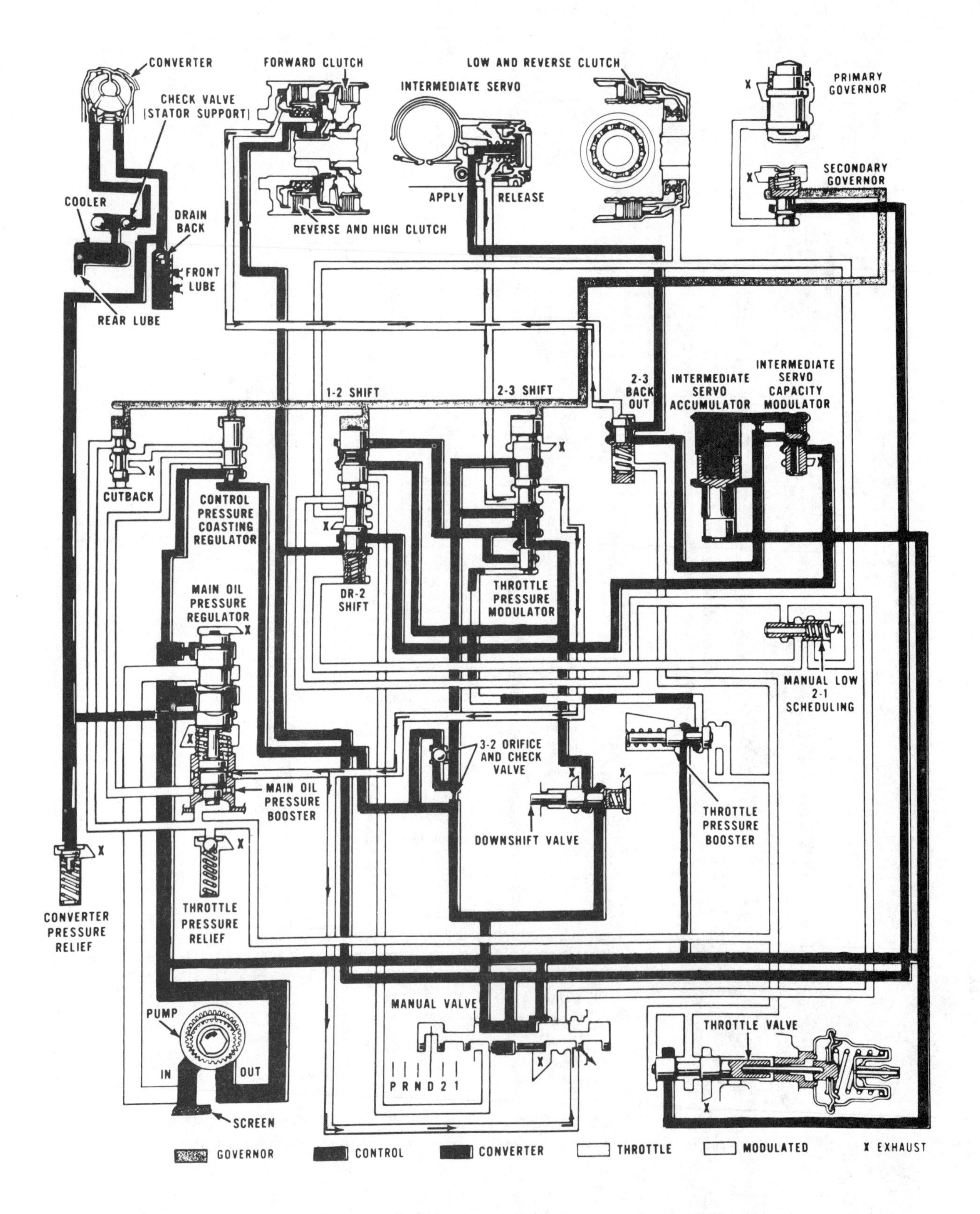

Fig. 12-6 3-2 downshift.

[For color diagram, see page 322.]

high as 100 mph (161 km/h). A 3-2 downshift, figure 12-6, can take place at road speeds as high as 90 mph (145 km/h). This difference in upshift and downshift speeds is related to control pressure acting on a differential force area of the shift valve. It is this difference in force that gives the valve snap action.

Downshift pressure also works on the 1-2 shift valve and will give a 2-1 or 3-1 downshift at about 30 mph (48 km/h). Note that when the downshift valve is open, full control pressure works to overcome governor pressure.

In summary of shifts in the drive range, both automatic upshifts and downshifts depend on road speed (governor pressure) and throttle opening. For example, a typical closed- or minimum-throttle 1-2 upshift takes place when the governor starts to regulate – about 10 mph (16 km/h). At larger throttle openings the shifts are delayed until a higher road speed is reached.

SHIFT QUALITY CONTROLS

Automatic shifts at different road speeds and throttle openings require different shift quality control. At low road speeds and small throttle openings, the shift should be soft and just barely noticeable. At large throttle openings and high speeds, the shift should be fast and firm. As explained in units 5 and 6, orificing and accumulators are used to control shift quality. The following sections deal with the shift quality controls of the C-6 transmission.

Intermediate Servo Accumulator and Capacity Modulator

During a 1-2 upshift, an accumulator is used to soften the shift. The intermediate servo capacity modulator controls the pressure to the accumulator according to control (line) pressure. Due to spring force, the capacity modulator senses a difference of 5 psi (34 kPa) between the top and bottom areas of the valve. This means that the accumulator will vary intermediate servo apply according to control pressure. Since control pressure is dependent on throttle opening, the shift will be firmer at large throttle openings, and softer at small throttle openings.

2-3 Backout Valve

At steady throttle openings, control pressure is equal at both the reverse and high clutch apply, and the servo release. This gives smooth clutch application and band release. (The release of the servo acts as an accumulator.)

If the driver releases the throttle as the 2-3 shift is about to happen, a clutch-band fight could take place. The reason this might occur is because there might not be enough pressure to release the band before the clutch was applied. To avoid this problem, a 2-3 backout valve is used, figure 12-7, page 160. When the driver releases the accelerator pedal, throttle pressure is 0 psi. When this happens, the 2-3 backout valve moves down and clutch apply, servo release, and servo apply pressure are the same. Since the pressures are the same, the clutch cannot apply before band release takes place. This gives a smooth lift-foot or backout shift.

Cutback Valve

A cutback control valve, figure 12-8, page 161, is used to give a soft shift at small throttle openings and a firmer shift at large throttle openings. At small throttle openings, governor pressure moves the cutback control valve down. Because part of the throttle pressure is exhausted from the main pressure regulator valve, the control pressure decreases. This results in smooth upshifts at small throttle openings. In addition, control pressure is reduced at cruising speeds. Depending on throttle opening, cutback can take place between 10 and 30 mph (16 and 48 km/h).

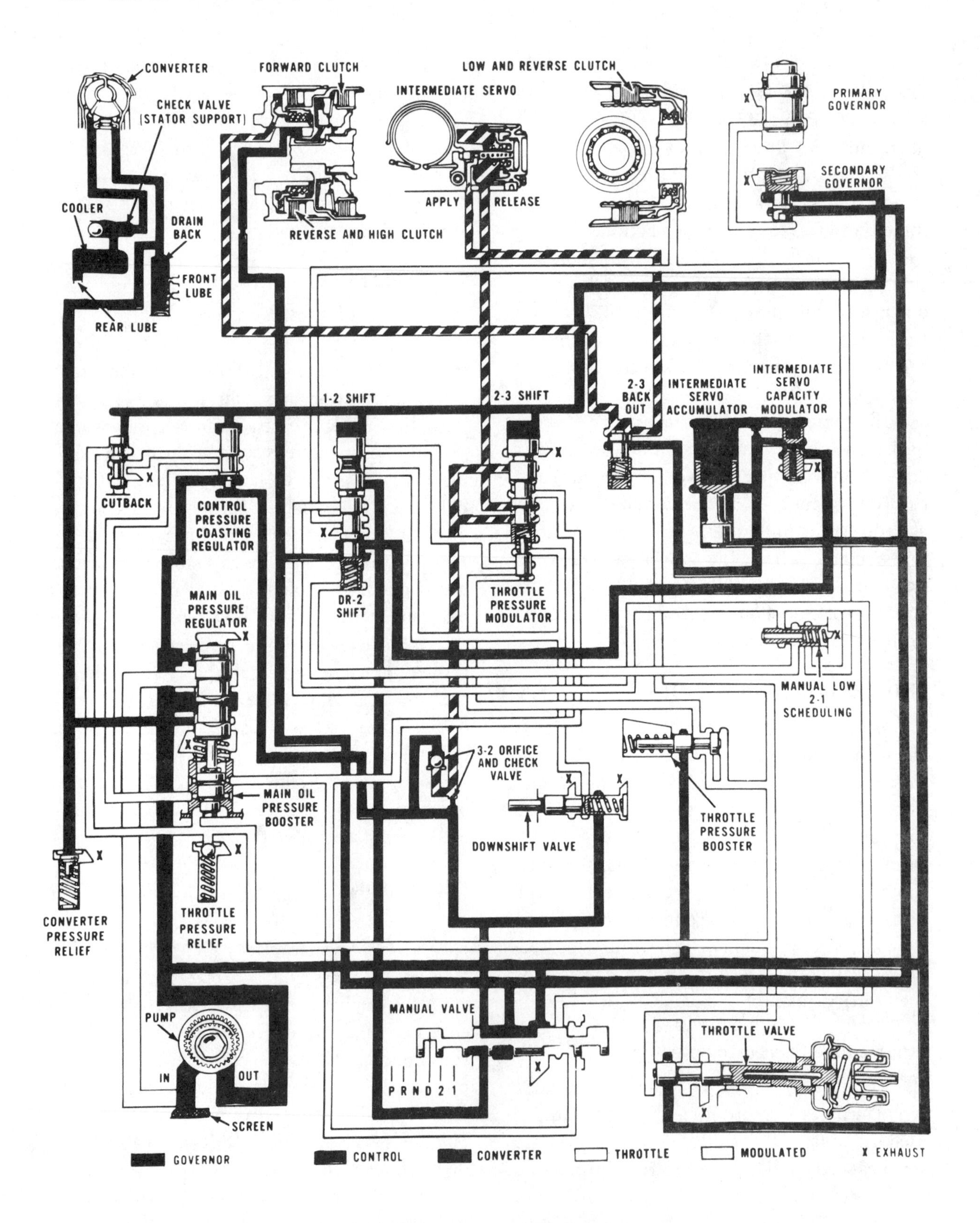

Fig. 12-7 2-3 backout shift.

[For color diagram, see page 323.]

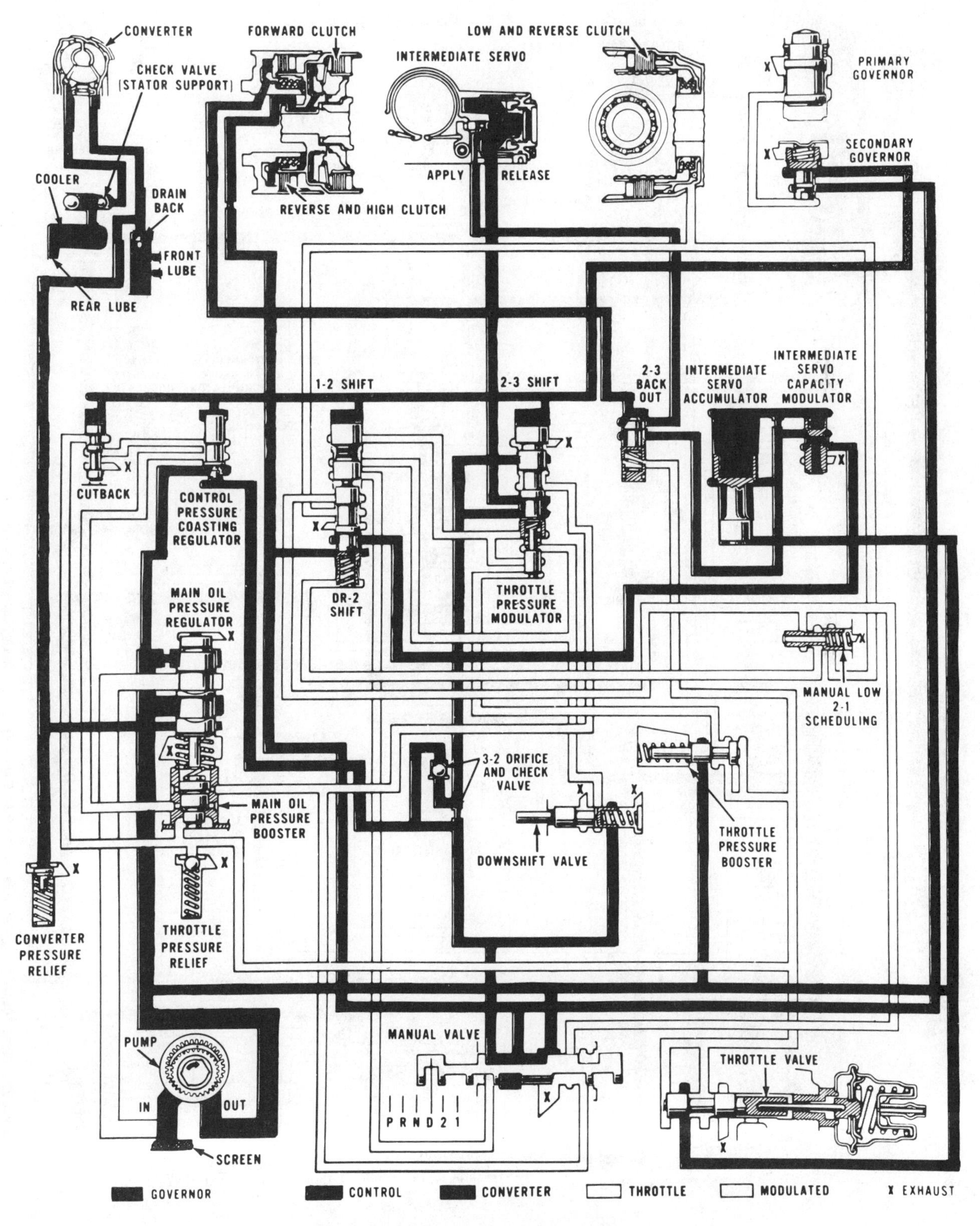

Fig. 12-8 High gear–D, cutback in use.

[For color diagram, see page 324.]

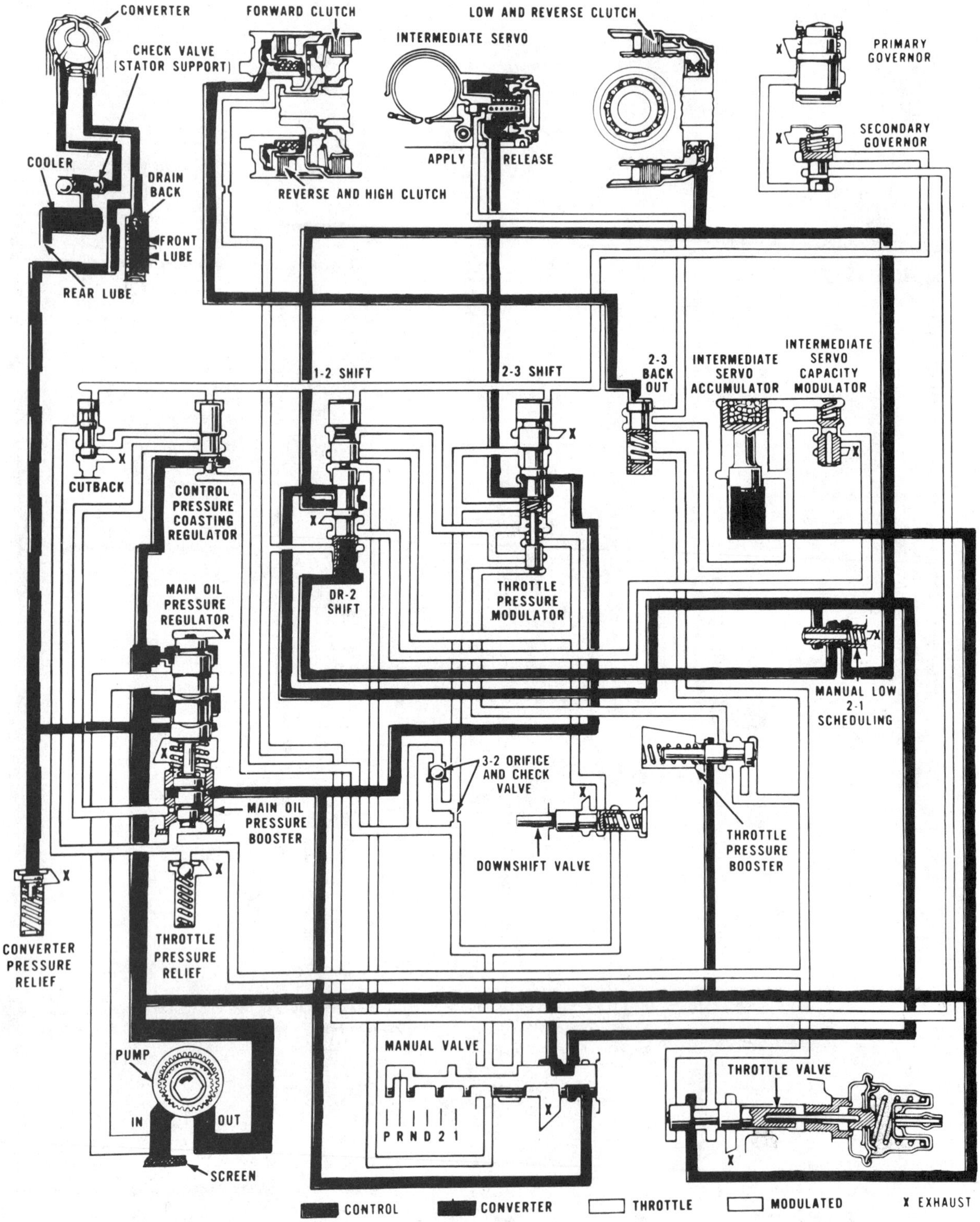

Fig. 12-9 Reverse.

[For color diagram, see page 325.]

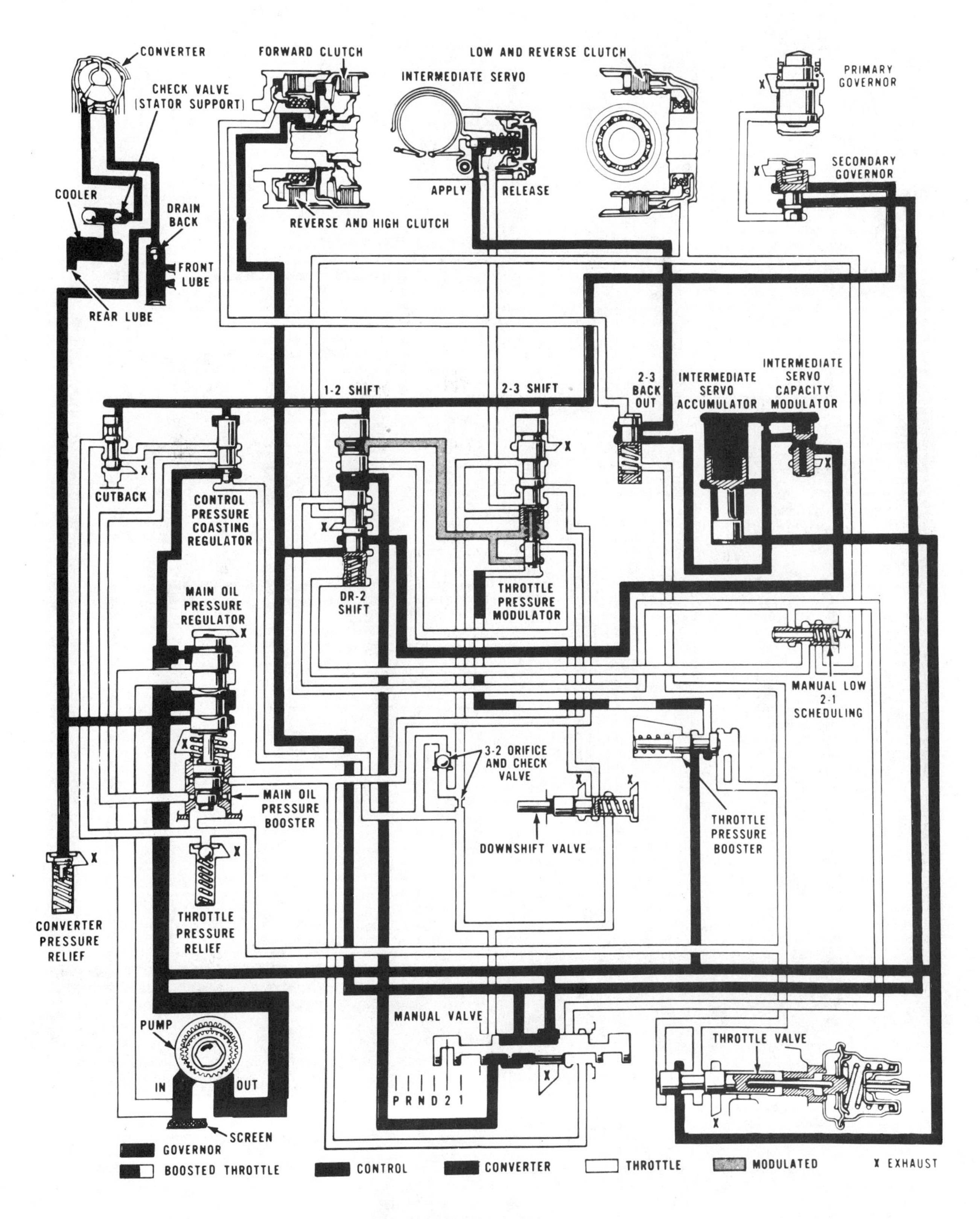

Fig. 12-10 Second gear – 2.
[For color diagram, see page 326.]

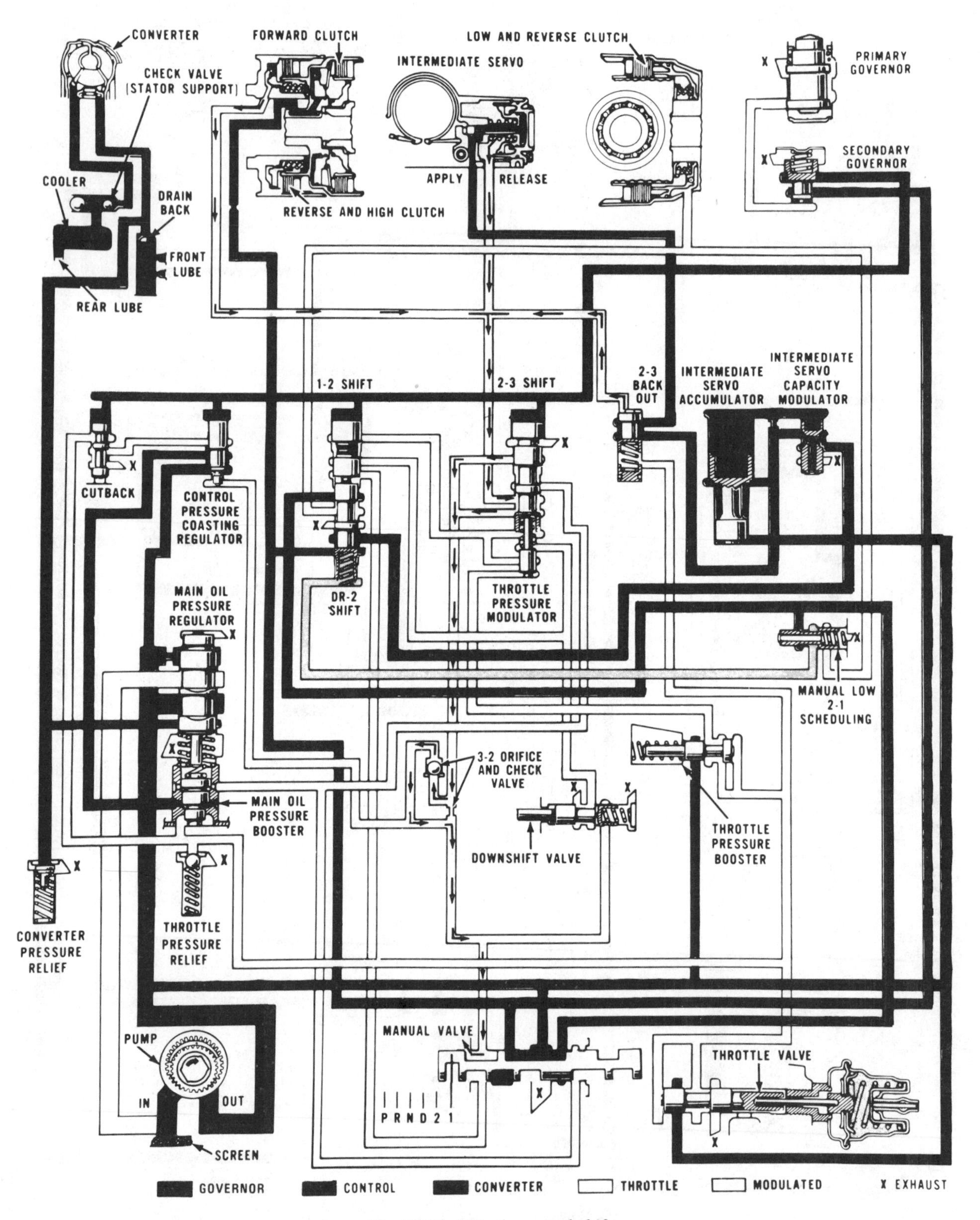

Fig. 12-11 D to 1 manual shift.

[For color diagram, see page 327.]

REVERSE

When the driver moves the shift lever to reverse, the manual valve sends control pressure to the reverse and high clutch, the low and reverse clutch, and to a boost side of the main pressure regulator, figure 12-9, page 162. At 1 in Hg (3.57 kPa), this will increase control pressure to 242 to 300 psi (1 655 to 2 069 kPa). An increase in pressure is necessary because there is more torque placed on the gear train in reverse. The roller clutch is not holding. Thus, all of the torque must be held by the low and reverse clutch.

MANUAL 2

In manual 2, the transmission cannot shift to high or 1. That is, if the transmission is placed in 2, it will stay in 2 until the manual valve is moved to another position. For engine braking, the manual valve can be moved to second, and a 3-2 downshift will take place.

With the manual valve in the 2 position, control pressure is exhausted at the manual valve from the reverse and high clutch, and from the release side of the intermediate servo, figure 12-10, page 163. As shown, control pressure from the manual valve is sent to an area between the 1-2 shift and DR 2 shift valves. This moves the DR 2 shift valve down and allows control pressure to apply the intermediate servo. Since control pressure at the high clutch and low passage is open to exhaust at the manual valve, the transmission is locked in second gear.

MANUAL 1

In manual 1, the manual valve sends control pressure to the forward clutch and the low and reverse clutch. Full line pressure at the DR 2 shift valve will prevent governor pressure from upshifting the 1-2 valve no matter what road speed is reached. Also, a 2-3 shift cannot take place because control pressure is cut off from the 2-3 valve by the manual valve.

A D to 1 manual shift can take place when the manual valve is moved to the 1 position, figure 12-11. As shown, servo release and reverse and high clutch oil are exhausted at the manual valve. Due to governor pressure, the 2-3 shift valve stays down, and the 3-2 check valve unseats to allow quick exhaust because the restriction orifice is bypassed. Figure 12-11 shows a D to 1 shift at about 60 mph (97 km/h). Note that the transmission has shifted to second but not to first. Since the DR 2 valve will not move up at 60 mph (97 km/h) (because of governor pressure), control pressure passes through it and applies the intermediate band.

Manual Low 2-1 Scheduling Valve

With the manual valve in 1, control pressure flows to the left end of the manual low 2-1 scheduling valve. This prevents the valve from moving to the right. As the car slows, governor pressure allows the 1-2 and DR 2 valves to move to the downshift position. Control pressure from the DR 2 valve moves the 2-1 scheduling valve to the right. Control pressure passes through the valley of the valve, and applies the low and reverse clutch. At the same time, servo apply oil exhausts at the 1-2 shift valve. In 1 position this shift will take place at speeds between 33 and 19 mph (53 and 31 km/h).

Control Pressure Coasting Regulator

Under engine braking or coast conditions, throttle pressure is 0 psi, and there is no boost to the main oil pressure regulator. To avoid this problem, control pressure is cut off from the bottom of the control pressure coasting regulator. This allows governor pressure to move the valve down. Line pressure can then pass through the valley of the valve to the main oil pressure booster as shown in figure 12-11. Under these conditions, control pressure is being regulated to about 100 (690 kPa). It should be noted that cutback occurs before the coasting regulator operates.

SUMMARY

The Ford C-3, C-4, and C-6 transmissions are used in all Ford Motor Company products. All of these transmissions use the Simpson gear train. The only difference is in the C-6, which uses a low and reverse clutch instead of a low and reverse band. This does not change the power flow in any way.

Several new valves have been introduced in the hydraulic system of the C-4 and C-6 transmissions. What should be remembered is that all hydraulic systems have much in common. That is, they all have main pressure regulators, a governor that senses road speed for upshift, a shift delay and downshift system that senses throttle opening, and a manual valve to choose the operating range.

In addition to the gear set and hydraulic system, the C-6 transmission has other parts to refine its performance. These devices vary from transmission to transmission; but serve the same purpose. For example, in the C-6, the 2-3 backout valve can be compared to the shuttle in valve in the TorqueFlite transmission.

With familiarity with these shift control devices, a knowledge of the power flow, and an understanding of basic hydraulics, the mechanic can read oil circuit diagrams and correctly diagnose the cause of the problem.

REVIEW*

Select the best answer from the choices offered to complete the statement or answer the question.

1. Upshifting and downshifting of the C-4 transmission in manual 2 is prevented by:

   (I) Cutting off control pressure from the 2-3 valve.
   (II) Cutting off control pressure from the DR 2 valve.

   a. I only
   b. II only
   c. both I and II
   d. neither I nor II

*The explanations and illustrations in the text may be used to answer these questions.

2. On the Ford C-6 transmission, the units in use for reverse are the

   a. reverse and high clutch and low and reverse clutch.
   b. forward clutch and reverse and high clutch.
   c. reverse and high clutch and one-way roller clutch.
   d. reverse and high clutch and reverse and low band.

3. In the Ford transmissions described in this unit, the members in use for drive range third speed are the

   a. forward clutch, intermediate band, and the reverse and high clutch.
   b. intermediate band and the reverse and high clutch.
   c. reverse and high clutch and the one-way roller clutch.
   d. forward clutch and the reverse and high clutch.

4. In the Ford transmission studied, the units in use for drive range intermediate are the

   a. forward clutch and reverse and high clutch.
   b. forward clutch and one-way roller clutch.
   c. forward clutch and intermediate band.
   d. forward clutch and intermediate clutch.

5. In drive range high gear, the forward and high clutches drive the

   a. front ring gear and front planetary carrier.
   b. front ring gear and sun gear.
   c. rear ring gear and sun gear.
   d. rear ring gear and front ring gear.

6. In second gear the intermediate band holds the

   a. sun gear.
   b. front ring gear.
   c. rear planetary carrier.
   d. front ring gear and sun gear.

7. In drive range low, power from the input shaft flows to the

   a. front planetary carrier.
   b. sun gear.
   c. rear ring gear.
   d. front ring gear.

8. To prevent upshift of the C-6 transmission in manual 1:

   (I) Governor pressure is cut off from the shift valve.
   (II) Full control pressure flows to the DR 2 valve.

   a. I only
   b. II only
   c. both I and II
   d. neither I nor II

9. In the Ford C-6 transmission, the units in use for manual 1 are the

   a. forward clutch and intermediate band.
   b. forward clutch and low and reverse clutch.
   c. forward clutch and low and reverse band.
   d. forward clutch and reverse and high clutch.

10. In the Ford C-3 or C-4 transmission, the units in use for drive range low are the

    a. forward clutch and the one-way roller clutch.
    b. forward clutch and the low and reverse band.
    c. reverse and high clutch and the one-way roller clutch.
    d. forward clutch and the low and reverse clutch.

11. A cause of late upshift could be:

    (I) A faulty governor.
    (II) A faulty vacuum unit.

    a. I only　　c. both I and II
    b. II only　　d. neither I nor II

12. A *to the detent* or *torque demand* 1-2 shift should take place at about 26 to 39 mph, and a 2-3 shift at 48 to 67 mph. If the shift points take place at the correct time, the mechanic knows that:

    (I) The downshift system is operating correctly.
    (II) The shift delay system is operating correctly.

    a. I only　　c. both I and II
    b. II only　　d. neither I nor II

13. A cause of too soft a 1-2 upshift could be:

    (I) A leaking vacuum unit.
    (II) High control pressure.

    a. I only　　c. both I and II
    b. II only　　d. neither I nor II

14. A shift that takes place late, and is also harsh, could be caused by a:

    (I) Leaking vacuum unit.
    (II) Stuck throttle booster valve.

    a. I only　　c. both I and II
    b. II only　　d. neither I nor II

15. Closed- or minimum-throttle upshift will take place at different road speeds. A typical closed-throttle 1-2 shift (above 17 in Hg) should take place at about 6-12 mph, and a 2-3 shift should take place at 11 to 20 mph. If, on a road test, these shift points are occurring correctly, the mechanic has checked the governor circuit and the:

    (I) 1-2 and 2-3 shift valves.
    (II) Shift delay system.

    a. I only　　c. both I and II
    b. II only　　d. neither I nor II

EXTENDED STUDY PROJECTS

1. Complete an oil circuit diagram for each range in the C-4 and C-6 transmission as assigned by the instructor.
2. In your own words, describe the operation of the C-6 transmission in the following ranges:
   A. *Drive range low.* The description should include the friction and holding units that are in use, how they are applied hydraulically, and what effect they have on the planetary gear set.
   B. *Drive range second.* This should include the description outlined in A, and what part the one-way clutch plays in this gear.
   C. *Drive range third.* This should include the description as outlined in parts A and B.
3. What valve controls the pressure to the intermediate servo accumulator so that the shift will be soft or firm as needed?
4. What valve controls pressures during a 2-3 lift-foot shift?
5. How does the valve named in question 4 control pressure to the friction units?
6. At approximately what speeds does cutback take place, and what effect does this have on control pressure?
7. How is control pressure affected during a manual 3-1 shift, and what valve is used to give this effect?
8. What will control pressure be in reverse, and how is this accomplished?
9. Why is there a difference between upshift speeds for a to detent (torque demand), and a through detent (wide-open throttle) shift?
10. With the manual valve in the 2 position, why will the gear train be locked in second?

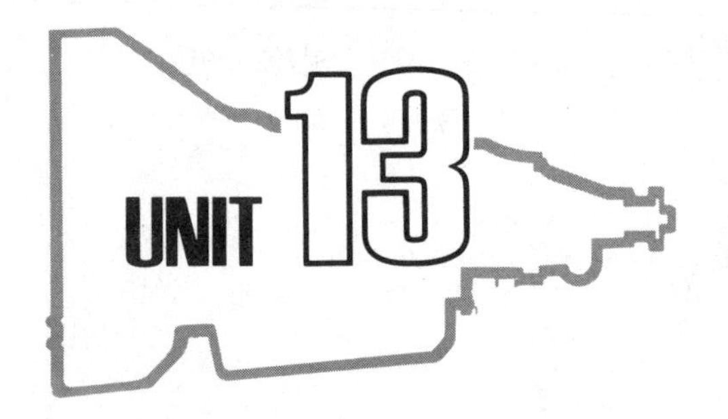

GENERAL MOTORS 200, 250, AND 350 TURBO HYDRA-MATIC TRANSMISSIONS

OBJECTIVES

After studying this unit, the student will be able to:

- Compare the gear set control of the Turbo Hydra-Matic 200 and 250 to that of the Ford C-6.
- Trace the power flow and hydraulic circuit of the Turbo Hydra-Matic 350 through all ranges.
- Identify the operating characteristics of the Turbo Hydra-Matic 350 in the various ranges.
- Diagnose power flow and hydraulic system problems in the Turbo Hydra-Matic 350.

APPLICATION

The Turbo Hydra-Matic 200 is built according to metric specifications. It is used in the Buick Skyhawk and Skylark; the Chevrolet Monza, Nova, and Vega; and the Oldsmobile Omega and Starfire.

The Turbo Hydra-Matic 250 is used in the Chevrolet Camaro, Chevelle, six-cylinder Nova, four-cylinder and V-8 Monza 212, and the four-cylinder Pontiac Astre. The Turbo Hydra-Matic 350 is used in the larger Chevrolet, Buick, and Oldsmobile automobiles. A variation on this transmission, called the 375B, is used in some models of the Buick. A variation used by Pontiac is the M38. All Turbo Hydra-Matic transmissions use a Simpson gear train.

OPERATION: TURBO HYDRA-MATIC 200 AND 250

The friction elements in the Turbo Hydra-Matic 200 and 250 are a forward clutch, a direct clutch, an intermediate band, a low and reverse clutch, and a low roller clutch, figure 13-1. The setup is very much like that of the Ford C-6. The power flow is as described in unit 12.

The forward clutch is applied for all forward gears. The intermediate band is applied for second, and the direct clutch for third or direct speed. For reverse, both the direct and the low and reverse clutch are applied. The forward clutch and the low and reverse clutch are also applied for manual low. This provides engine braking in low range. When placed in low, the transmission will not upshift to second or direct.

A shift to low range can be made at any road speed. At high road speeds, the transmission will shift to second speed, and remain there until road speed drops to about 30 mph (48 km/h). At this point, the transmission will shift to low and remain in low regardless of the road speed reached.

In manual second or intermediate, the transmission will start out in low, shift to second, but will not shift to high regardless of the road speed that is reached. A shift to second can be made at any road speed. The transmission will shift to second speed and stay there until road speed drops

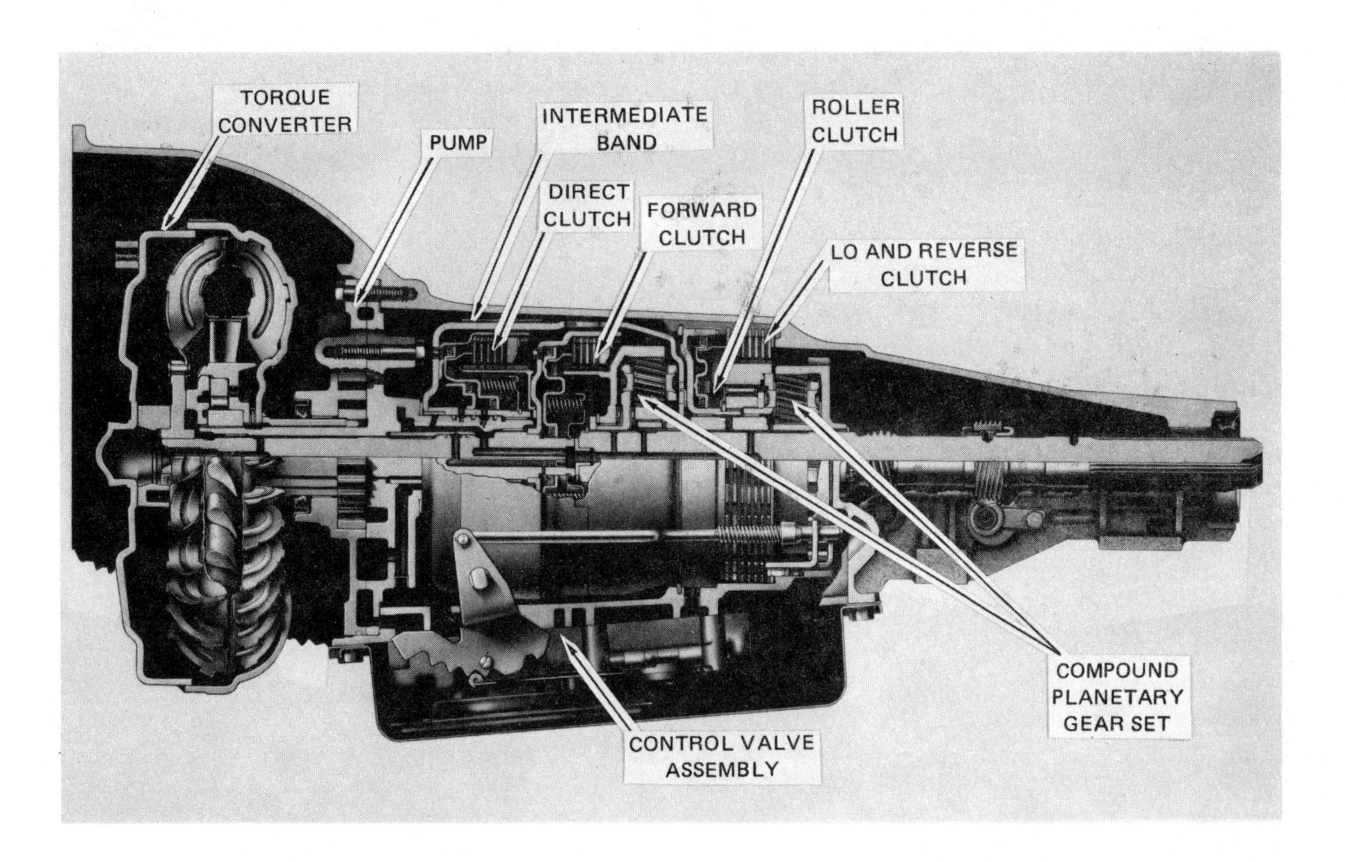

RANGE	GEAR	DIRECT CLUTCH	INTERMEDIATE BAND	FORWARD CLUTCH	ROLLER CLUTCH	LO-REVERSE CLUTCH
Park-Neutral						
Drive	First			Applied	Holding	
	Second		Applied	Applied		
	Third	Applied		Applied		
Intermediate	First			Applied	Holding	
	Second		Applied	Applied		
Lo	First			Applied	Holding	Applied
	Second		Applied	Applied		
Reverse		Applied				Applied

Fig. 13-1 Cutaway view and summary of power flow for the Turbo Hydra-Matic 200.

or the throttle is opened to force a shift to low.

The valve bodies of the Turbo Hydra-Matic 200 and 250 are very similar. The main difference is that the Turbo Hydra-Matic 200 uses a throttle valve that is connected to the throttle linkage, while the 250 uses a vacuum modulator and valve to help control shift points and line pressure.

OPERATION: TURBO HYDRA-MATIC 350

To control the gear set, the Turbo Hydra-Matic 350 uses a forward clutch, an intermediate clutch, a direct clutch, a low and reverse clutch, a low roller clutch, an intermediate roller clutch and an intermediate band, figure 13-2. Although the 350 has a Simpson gear train, the control of the gear train is quite a bit different from that in the transmissions discussed thus far. As in other Simpson gear trains, the forward clutch is applied in all forward gears. However, for drive intermediate, this transmission does not need to time the release of a band to the apply of a clutch. Refer to figure 13-3 while studying the following description of the power flow.

In drive range, the forward clutch is applied and power flow is the same as in the other Simpson gear trains studied thus far. That is, the front ring gear acts on the pinions to give a reverse or counterclockwise rotation to the sun gear. With the low roller clutch holding the rear carrier, the rear ring gear and output shaft rotate in a forward direction.

The 350 varies from other transmissions with Simpson gear trains in the control of the shift to drive-range second, or intermediate. In place of an intermediate band stopping the sun gear, an intermediate clutch is applied. This holds the outer race of the intermediate roller clutch which, in turn, stops the sun gear.

As road speed increases, the transmission shifts to high. Then, as in other Simpson gear trains, the direct clutch is applied. With the direct clutch on, both the sun and ring gear are driven. This locks the front gear set, and the transmission is in high. The intermediate roller clutch freewheels, allowing the sun gear to move in a forward or clockwise direction.

The result of this setup is that the sun gear and ring gear are locked to the input

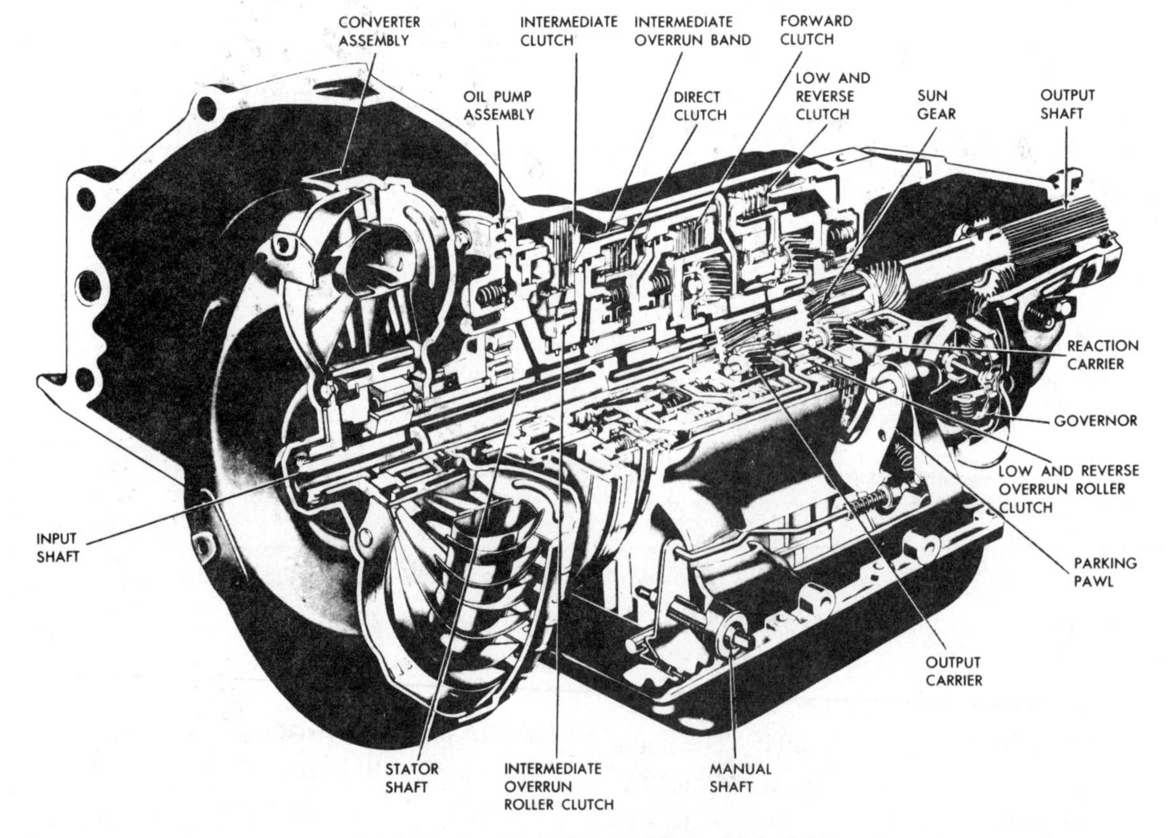

Fig. 13-2 Cutaway of Turbo Hydra-Matic 350.

shaft, and the transmission is in direct. However, the intermediate roller clutch is freewheeling. Thus, there is no need to release the intermediate clutch. Note that the intermediate band is not used for drive range intermediate.

For reverse, the power flow is the same as that in the Ford C-6. That is, the direct clutch is applied to drive the sun gear, and the low and reverse clutch is applied to hold the rear planetary carrier. This, as in other Simpson gear trains, causes the output shaft to move in a reverse direction.

For low range, the forward clutch is applied. To provide engine braking, the low and reverse clutch is also applied. In manual second, the power flow is the same as intermediate in drive except that the intermediate overrun band is on. Again, the band is applied to stop the intermediate roller clutch from freewheeling, and thus provide engine braking in manual second.

HYDRAULIC SYSTEM

The hydraulic system of the Turbo Hydra-Matic 350 is similar to other hydraulic systems. Therefore, only its distinctive features are discussed here.

Pressure Regulation

Main pressure regulation is accomplished in much the same way as in other transmissions studied thus far. An internal-external gear-type pump supplies pressure for the hydraulic system. This pressure is regulated to line pressure by a pressure regulator of the bypass type, figure 13-4. The converter, cooling, and lubricating systems, as shown, are also similar to other transmissions studied.

DRIVE RANGE

In drive range, all shifts are automatic and take place according to road speed and throttle opening. As in other automatic

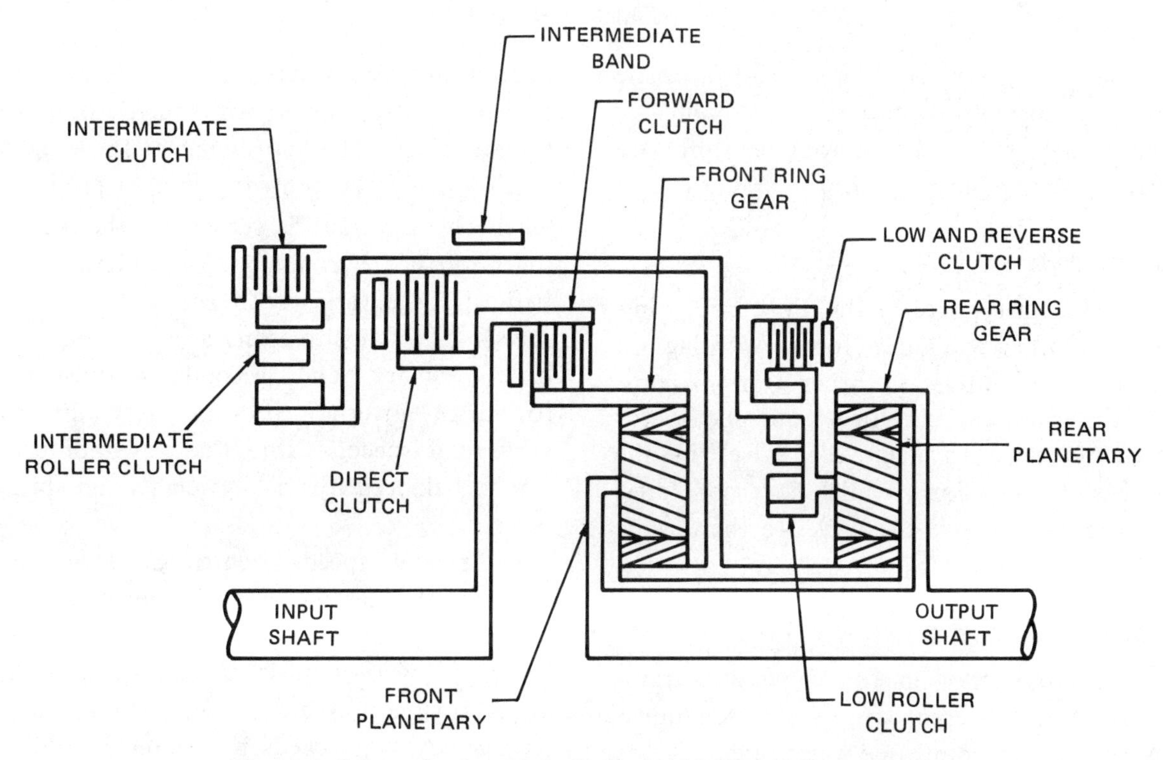

Fig. 13-3 Schematic view of the Turbo Hydra-Matic 350 power train.

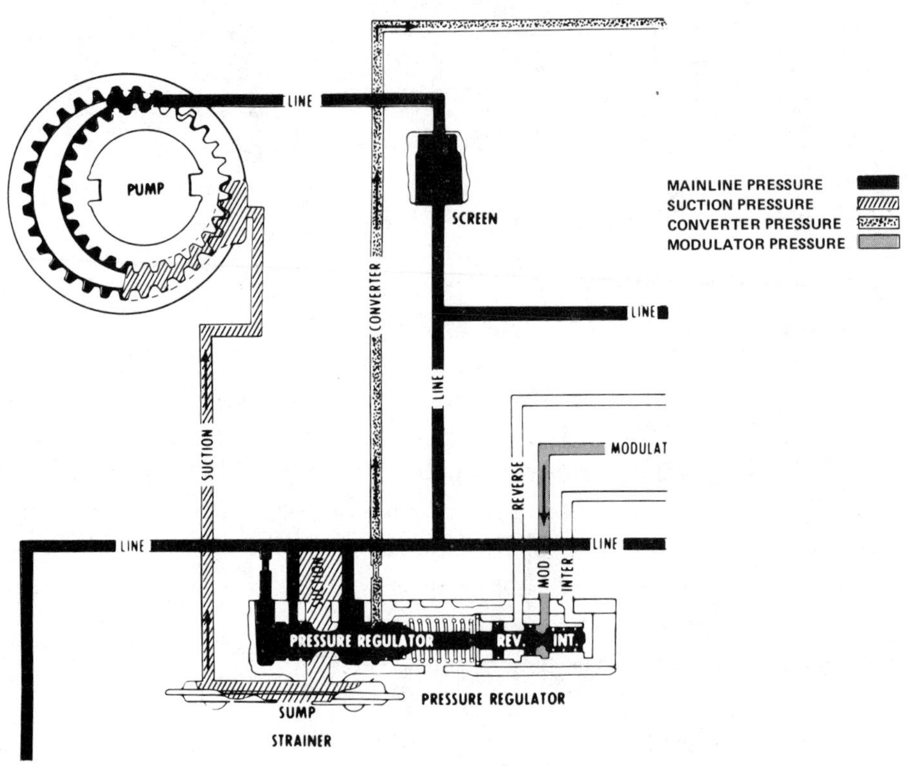

Fig. 13-4 Pump pressure regulation.
[For color diagram, see page 328.]

transmissions, the road speed and throttle opening signals are fed to the "brain" – the valve body – and automatic shifts are effected at the proper speeds.

Shift Valves

The shift valves in the Type 350 transmission have the same forces working on them as the valves in other transmissions. In this case, spring force and modulator pressure tend to keep the valves in the downshift position, while governor pressure tends to move the valves to the upshift position.

Governor

The governor in the Type 350 transmission is different from others studied in that it has only one valve, figure 13-5. As shown, the governor has a balanced, restriction-type valve, and primary and secondary flyweights that tend to move the valve up. This movement of the weights is a result of the centrifugal force produced when the governor is turned by the output shaft of the transmission. Recall that centrifugal force varies with the square of the speed. For this reason, it is necessary to have primary and secondary flyweights to make governor pressure vary directly with road speed. Note that governor pressure is balanced by the flyweights and spring force.

At slow speeds, centrifugal force acts only on the heavier, primary weights. This moves the valve up against the balancing force of governor pressure, and allows more drive oil (line pressure) to enter the governor passage. At high speeds, the lighter, secondary weights move out and allow additional drive oil to enter the governor circuit and increase the pressure. In this way, governor pressure is adjusted to match road speed.

Modulator

The modulator is made up of a vacuum unit and valve. General Motors refers to the vacuum unit as a *modulator*, and the pressure it regulates as *modulator pressure*, figure 13-6. Modulator pressure is the same as TV pressure and acts on the downshift side of the shift valves. Modulator pressure also acts to boost line pressure, depending on throttle (vacuum signal) opening.

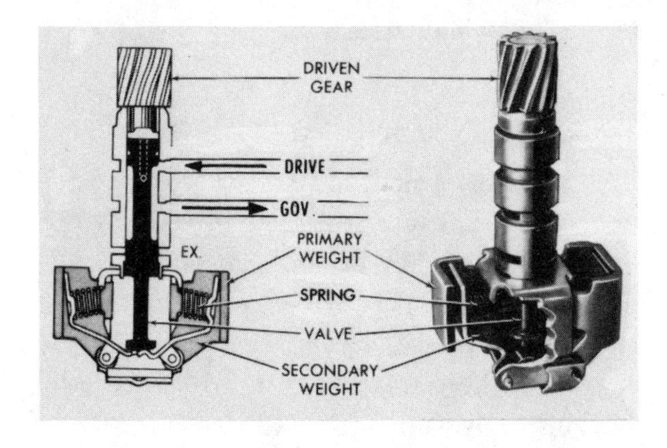

Fig. 13-5 Governor assembly.

Detent Valve and Detent Pressure Regulator

To force a downshift, a detent valve is used, figure 13-7. The detent valve is linked to the throttle by a cable and may come into use at part-throttle or wide-open throttle position. Line pressure is regulated by the detent pressure regulator valve to 80 psi (55 kPa).

At small throttle openings, modulator oil passes through the detent valve to the modulator shift valve (MOD SV) passage to delay upshift. At large throttle openings (detent touch) and road speeds below about 50 mph (80 km/h), a 3-2 downshift will take place. This is due to modulator pressure filling the 3-2 part-throttle (PT) passages which act on another area of the shift valve to cause downshift. This same action also delays 2-3 upshift during hard acceleration.

At speeds below about 75 mph (121 km/h) a 3-2 downshift will take place when the throttle is moved to the wide-open position (through detent). This allows full detent pressure to work on the downshift side of the shift valves, and the transmission shifts to second gear. Also, at speeds below about 40 mph (64 km/h) a 2-1 or 3-1 downshift will take place at wide-open throttle.

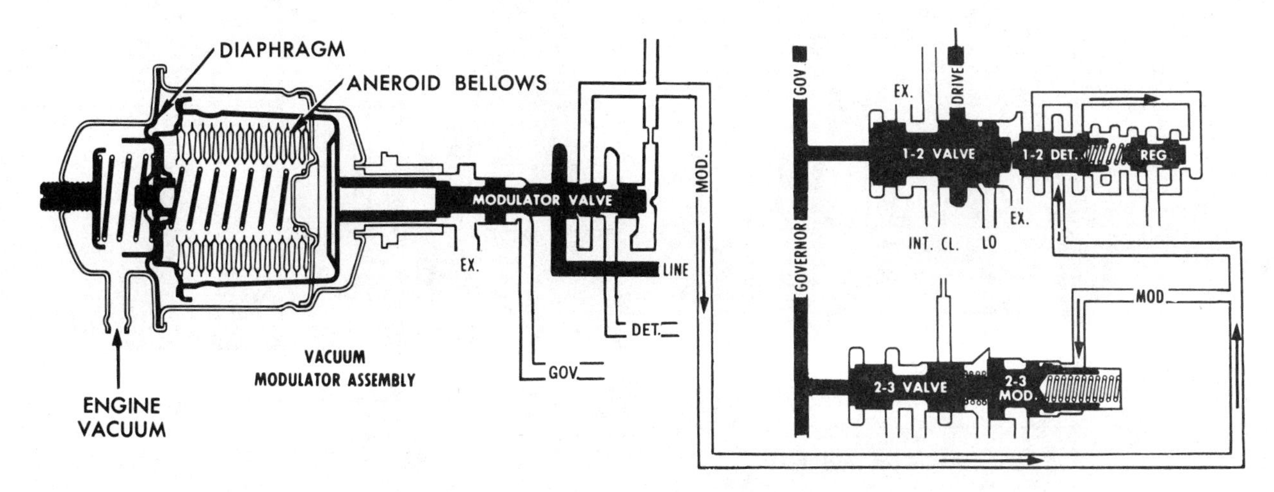

Fig. 13-6 Typical General Motors modulator circuit.

[For color diagram, see page 329.]

Keep in mind that due to governor action on the modulator valve, modulator pressure decreases as road speed increases. The job of the detent valve and detent regulator is to cause downshifts, or to delay upshifts at large, detent touch, or wide-open throttle positions.

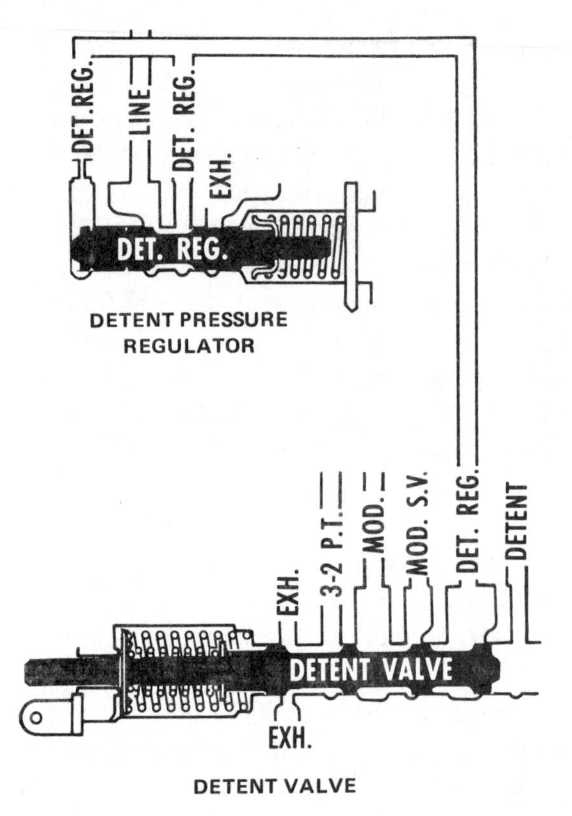

Fig. 13-7 Detent system.

SHIFT QUALITY CONTROL

Shift quality control in the Turbo Hydra-Matic depends on accumulators, main pressure regulation, and orificing.

From the manual valve, line pressure is sent to the 1-2 accumulator piston, figure 13-8. This strokes the piston against a spring force, and then readies the 1-2 accumulator for a 1-2 upshift. Since the pressure exerted on the 1-2 accumulator flows directly from the main pressure regulator, it is present in all ranges. When drive range is selected, line pressure, called *drive oil*, is sent to the forward clutch, the 1-2, and 2-3 shift valves. The operation of the pressure control devices for the 1-2 and 2-3 shifts are discussed in detail in the following paragraphs.

1-2 and 2-3 Accumulators

In reverse, neutral, and drive, line pressure (known as *reverse neutral drive*, or *RND, oil*) is sent to the release side of the intermediate servo. This prevents the servo from applying in reverse, neutral, and drive.

With the transmission in drive range, the 1-2 shift valve moves to cause upshift at the correct road speed according to throttle opening, figure 13-9.

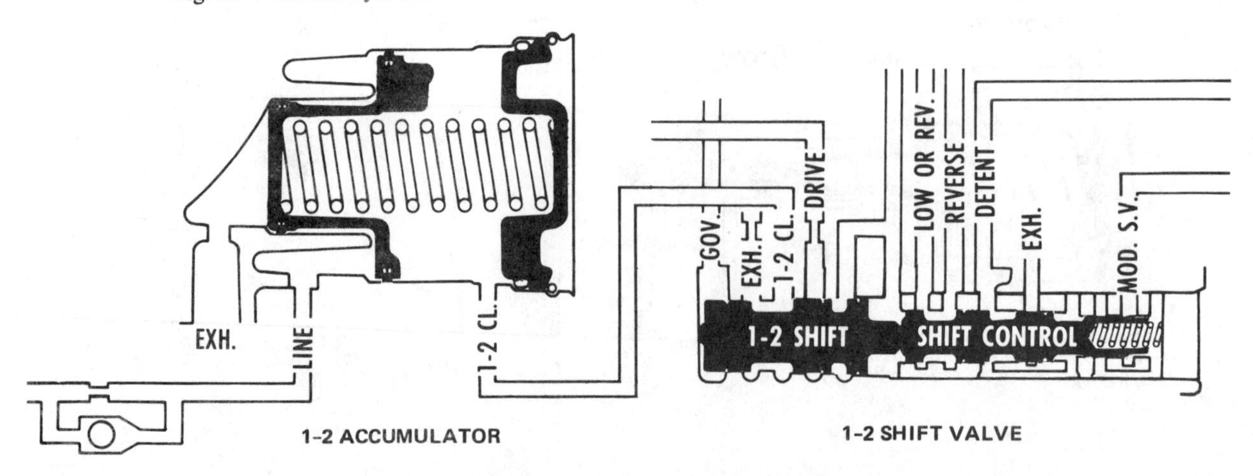

Fig. 13-8 1-2 shift valve and accumulator at rest.

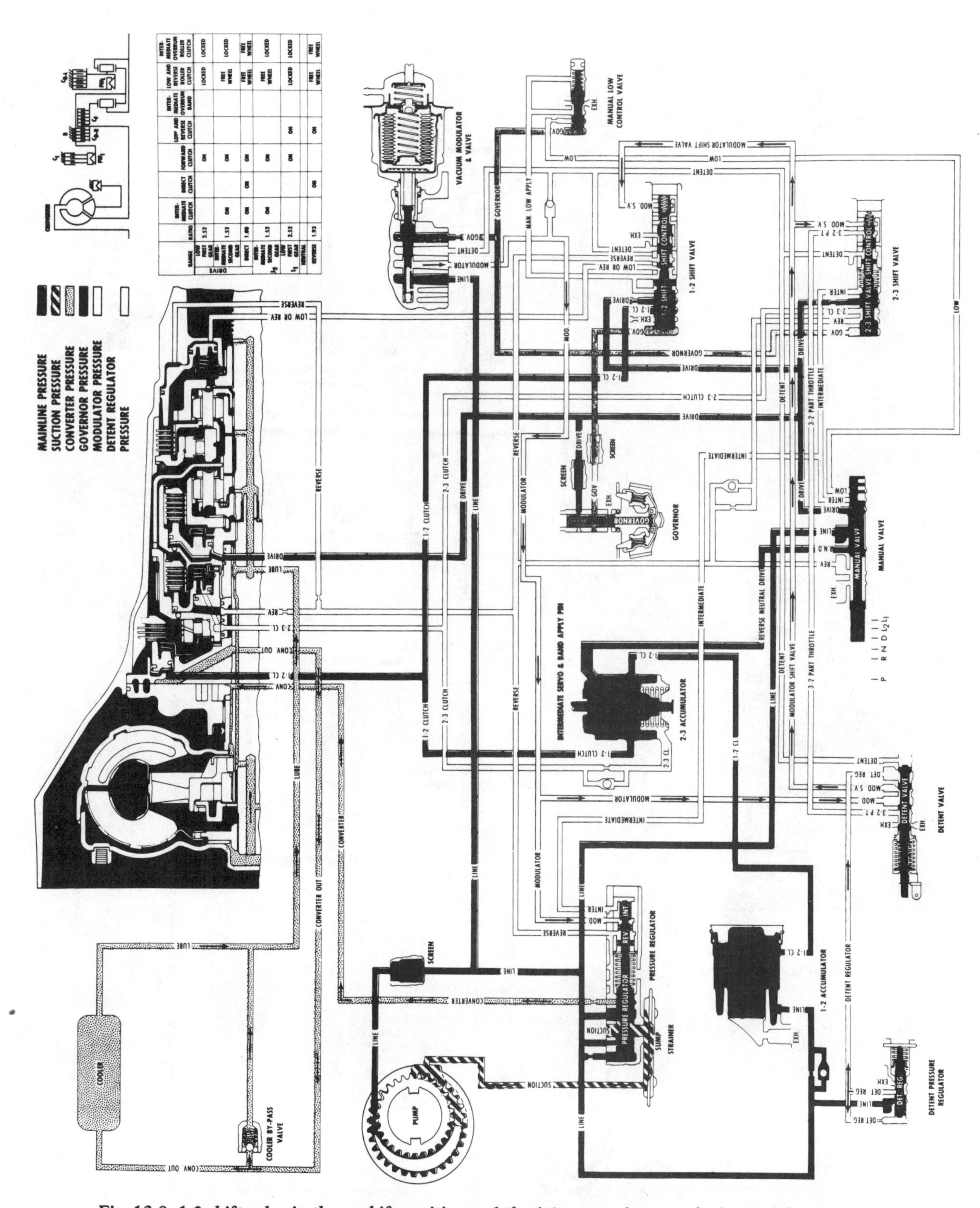

Fig. 13-9 1-2 shift valve in the upshift position and the 1-2 accumulator stroked against line pressure.

[For color diagram, see page 330.]

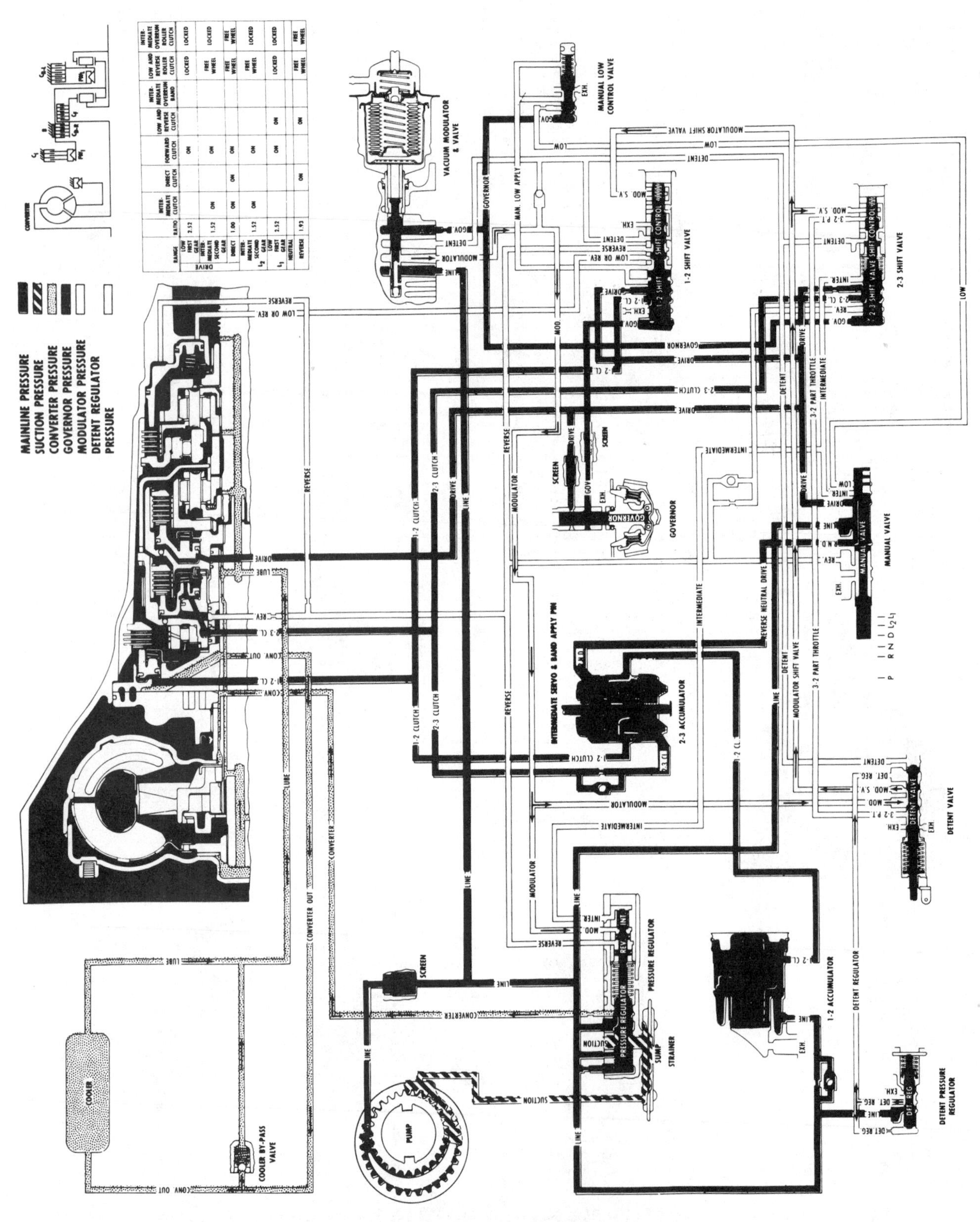

Fig. 13-10 Valves and accumulators in the upshift position.

[For color diagram, see page 331.]

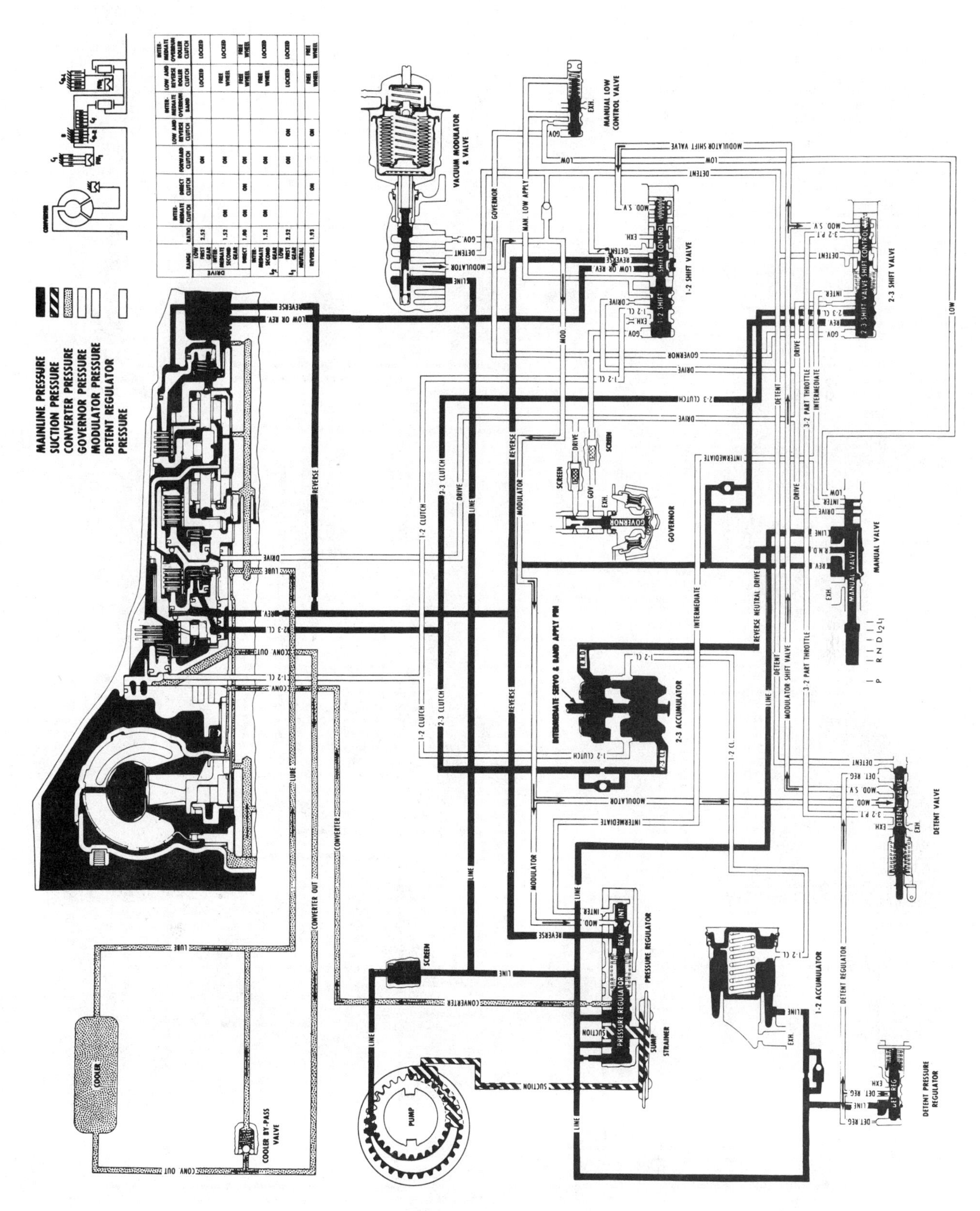

Fig. 13-11 Reverse range.

[For color diagram, see page 332.]

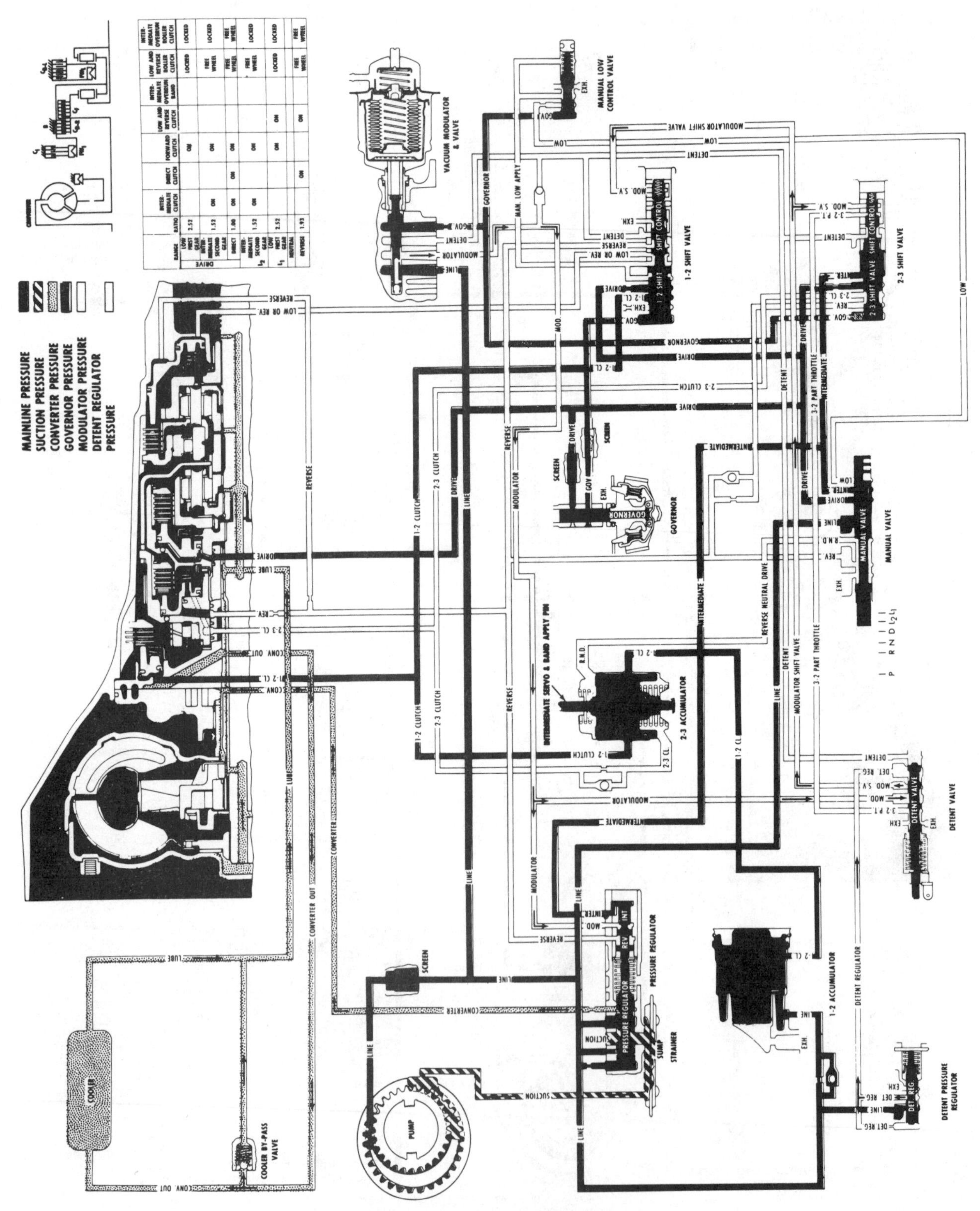

Fig. 13-12 L2 range: manual second gear.

[For color diagram, see page 333.]

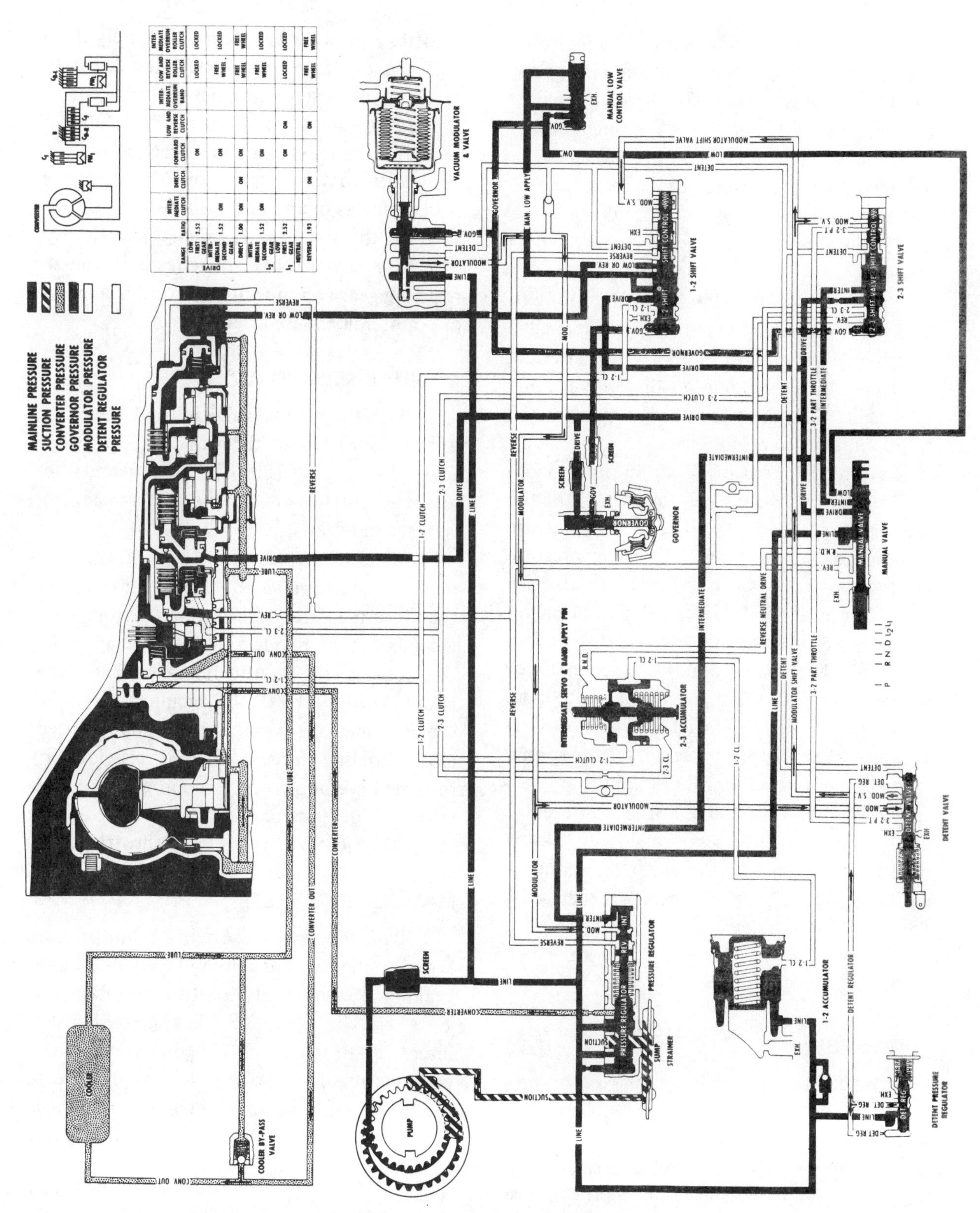

RANGE		RATIO	INTERMEDIATE CLUTCH	DIRECT CLUTCH	FORWARD CLUTCH	LOW AND REVERSE CLUTCH	INTERMEDIATE OVERRUN BAND	LOW AND REVERSE ROLLER CLUTCH	INTERMEDIATE OVERRUN ROLLER CLUTCH
DRIVE	LOW FIRST GEAR	2.52			ON			LOCKED	LOCKED
DRIVE	INTERMEDIATE SECOND GEAR	1.52	ON		ON			FREE WHEEL	LOCKED
DRIVE	DIRECT	1.00	ON	ON	ON			FREE WHEEL	FREE WHEEL
L2	INTERMEDIATE SECOND GEAR	1.52	ON		ON			FREE WHEEL	LOCKED
L1	LOW FIRST GEAR	2.52			ON	ON		LOCKED	LOCKED
	NEUTRAL								
	REVERSE	1.93		ON		ON		FREE WHEEL	FREE WHEEL

Fig. 13-13 L1 range: manual first gear.

[For color diagram, see page 334.]

When this happens, line pressure, now called *1-2 CL*, (for 1-2 or intermediate clutch) passes through an orifice, the 1-2 shift valve, and to the intermediate clutch. A branch circuit passes through the intermediate servo, between the servo and 2-3 accumulator piston, and to the 1-2 accumulator. (Note that this will not apply the servo due to the force of RND oil.) The pressure is equal on both sides of the 1-2 accumulator piston. This allows the spring to stroke the accumulator piston and take on 1-2 clutch oil for smooth apply of the intermediate clutch. Depending on throttle opening, a 1-2 shift will take place at road speeds of 12 to 50 mph (19 to 80 km/h).

At road speeds of 22 to 85 mph (35 to 137 km/h), a 2-3 shift will take place. As the 2-3 valve snaps to the right, line pressure passes through the valve. This line pressure is now called *2-3 CL oil*, figure 13-10, page 178. From the 2-3 valve, 2-3 CL oil moves to the small area of the direct clutch piston. A branch line leads to the 2-3 accumulator where some 2-3 CL oil is taken on to help smooth out the 2-3 shift. Due to RND pressure plus the heavy servo return spring, the intermediate servo will not apply.

The check balls and orifices on the 1-2 and 2-3 accumulator serve to slow up clutch apply and give smooth shifting. During downshift, the check balls are forced off their seats for fast clutch release and a quick, smooth downshift.

REVERSE

With the manual valve in the reverse position, figure 13-11, page 179, line pressure is sent to both the large and small areas of the direct clutch, the large and small areas of the low and reverse clutch, and a branch line that leads to a boost area of the main pressure regulator. This increases line pressure to about 250 psi (1 724 kPa) at full throttle. In order to cushion the apply of the clutches, the large areas of both the direct and the low and reverse clutches are orificed, and the 2-3 accumulator takes on 2-3 clutch oil sent to the small area of the direct clutch.

Note that the operation of the gear set is the same as in any other Simpson gear train. That is, the sun gear is driving, and the rear planetary carrier is held. The result is that the rear ring gear and output shaft turn in a reverse or counterclockwise direction.

MANUAL SECOND (L2)

With the manual valve in the L2 position, line pressure is sent through the 1-2 CL line, figure 13-12, page 180. The forward and intermediate clutches are on and the transmission is in second gear.

Since RND pressure is cut off, the intermediate servo applies the band. This prevents the intermediate roller clutch from freewheeling and provides engine braking in L2. The apply of the intermediate clutch and band is smoothed out by the 1-2 accumulator. Since the intermediate band is quite small it needs extra holding force. This is provided by intermediate boost on the main pressure regulator. Line pressure is thus increased to at least 80 psi (552 kPa) at closed throttle.

MANUAL FIRST (L1)

With the manual valve in L1, line pressure is sent to the forward clutch. To provide engine braking, pressure is also sent to the low and reverse clutch, figure 13-13, page 181. The intermediate passage is open to the main pressure regulator to boost line pressure to the same value as in manual second range.

Manual first gear can be selected at any road speed, but will not shift to low until speeds below about 50 mph (80 km/h). This shift delay is controlled by the manual low control valve, which is controlled by governor pressure, line pressure, and a spring force.

SUMMARY

The Turbo Hydra-Matic 200, 250, and 350 transmission are three-speed, fully automatic transmissions. They are used in the full line of General Motors automobiles. All three of these transmissions use the Simpson gear train.

The power flow and the clutch and band setup of the Types 200 and 250 are very much like that of the Ford C-6. The main differences between these transmissions and the Ford C-6 are in the valve bodies and the method of shift quality control.

The Turbo Hydra-Matic 350, however, uses a different setup; roller clutches are used: one for drive range low, and one for drive range intermediate. To make the intermediate roller clutch effective, an intermediate clutch is used. The clutch discs are splined to the outer race of the roller clutch and the transmission case. When the shift is made to drive range third, the intermediate clutch does not have to be released because the roller clutch will freewheel and allow the sun gear to turn.

Knowledge of the operation of the Simpson gear train and its control in these transmissions is important when diagnosing transmission problems. The oil circuit diagrams for these models should also be used to locate trouble spots.

REVIEW

Select the best answer from the choices offered to complete the statement or answer the question.

1. In the GM Turbo Hydra-Matic 200 and 250, the control of the gear set may be compared to that in the

   a. Ford C-6.
   b. Ford C-4.
   c. Chrysler TorqueFlite.
   d. GM Type 300.

2. In a GM Turbo Hydra-Matic 350 transmission, a late 1-2 upshift could be caused by:

   (I) Governor pressure being too high.
   (II) A vacuum leak.

   a. I only
   b. II only
   c. both I and II
   d. neither I nor II

3. In the Type 350 transmission, when the selector lever is placed in L2, the transmission will

   a. start in intermediate.
   b. start in first and shift to intermediate.
   c. not leave first.
   d. shift through all the gears.

4. When the Type 350 transmission is in drive range low, the effective members are the
   a. forward clutch and the low roller clutch.
   b. forward clutch and the low and reverse clutch.
   c. direct clutch and the low roller clutch.
   d. forward clutch and the low band.
5. When the Type 350 transmission is in drive range second, the effective members are the
   a. forward clutch, intermediate clutch, and the low and reverse clutch.
   b. forward clutch, intermediate clutch, and low roller clutch.
   c. forward clutch, intermediate clutch, and intermediate band.
   d. forward clutch, intermediate clutch, and intermediate roller clutch.
6. When the forward clutch of the Type 350 is applied, it locks the input shaft to the
   a. sun gear.
   b. front planetary carrier.
   c. front ring gear.
   d. rear planetary carrier.
7. In the Type 350 transmission, no full-throttle downshift could be caused by:
   (I) Throttle linkage adjustment.
   (II) A vacuum leak.
   a. I only
   b. II only
   c. both I and II
   d. neither I nor II
8. When the Type 350 transmission is in reverse, the effective members are the
   a. forward clutch and the low and reverse clutch.
   b. intermediate clutch, and the low and reverse clutch.
   c. direct clutch, and the low and reverse band.
   d. direct clutch, and the low and reverse clutch.
9. In the Type 350 transmission, the governor is driven by the
   a. input shaft.
   b. sun gear.
   c. output shaft.
   d. ring gear.
10. When the intermediate band of the Type 350 is applied it
   a. holds the sun gear.
   b. holds the ring gear.
   c. holds the pinion carrier.
   d. locks the ring gear and pinion carrier.

11. When the Type 350 transmission is in super or L2 range and second gear, the effective members are the

    a. forward, intermediate, and direct clutches.
    b. forward clutch, intermediate clutch, and intermediate-roller clutch.
    c. forward clutch, intermediate clutch, and intermediate band.
    d. intermediate clutch, direct clutch and intermediate band.

12. The detent regulator valve of the Type 350 changes

    a. governor pressure to main line pressure.
    b. modulator pressure to main line pressure.
    c. line pressure to detent pressure.
    d. governor pressure to detent pressure.

13. The Type 350 transmission includes a torque converter, a compound planetary gear set, and

    a. two multiple-disc clutches, and one roller clutch.
    b. three multiple-disc clutches, and two roller clutches.
    c. four multiple-disc clutches, and two roller clutches.
    d. three multiple-disc clutches, and two bands.

14. During engine braking, when the Type 350 transmission is in low gear in L range, the effective members are the

    a. forward clutch and the low and reverse clutch.
    b. forward clutch and the low roller clutch.
    c. direct clutch and the low roller clutch.
    d. forward clutch and the low and reverse band.

15. When the Type 350 transmission is in direct drive, the members that are applied are the

    a. direct clutch and the intermediate and low roller clutches.
    b. forward clutch, the intermediate clutch, and the direct clutch.
    c. direct clutch, the intermediate clutch, and the intermediate roller clutch.
    d. forward clutch, the intermediate clutch, and the low and reverse band.

EXTENDED STUDY PROJECTS

I. On blank oil circuit diagrams, trace the oil circuits for each of the ranges of the Turbo Hydra-Matic 350 transmission.

II. Select the best answer for each of the following problems.

1. A customer complains that the Type 350 in his car slips in direct. The fluid level and linkage are checked and are correct. A road test shows that the transmission also slips in reverse, but not in L1. The most likely cause of this problem would be

   a. leaking low and reverse clutch seals.
   b. a slipping direct clutch.
   c. a defective intermediate clutch.
   d. Not enough information is given.

2. A customer complains that the Turbo Hydra-Matic 350 transmission in his car will not upshift to second or third until very high road speeds are reached. This could be caused by:

   (I) A faulty modulator unit.
   (II) A stuck detent valve.

   a. I only
   b. II only
   c. both I and II
   d. neither I nor II

3. In the Type 350 transmission a cause of no drive in D could be:

   (I) A faulty roller clutch.
   (II) A faulty direct clutch.

   a. I only
   b. II only
   c. both I and II
   d. neither I nor II

4. In the Turbo Hydra-Matic transmissions, a rough or harsh shift could be caused by:

   (I) A leaking clutch seal.
   (II) A leaking accumulator seal.

   a. I only
   b. II only
   c. both I and II
   d. neither I nor II

5. A customer complains that the Turbo Hydra-Matic in his car slips in D range low. The fluid level is correct and the linkage is adjusted properly. The most probable cause of this problem would be

   a. a faulty low roller clutch.
   b. a slipping forward clutch.
   c. leaking forward clutch seals.
   d. Not enough information is given.

GENERAL MOTORS TURBO HYDRA-MATIC 400

OBJECTIVES

After studying this unit, the student will be able to:

- Identify the clutches and bands used to control the gear set in each range of operation.
- Identify the operating characteristics in each range.
- Trace the power flow through each range of operation.
- Trace the hydraulic circuit through each range.
- Diagnose problems that stem from the power flow and hydraulic system of the Turbo Hydra-Matic 400.

APPLICATION

The General Motors Turbo Hydra-Matic 400 is used in Buick, Cadillac, Chevrolet, and Oldsmobile passenger cars, and in certain truck models. Variations on this transmission are the Type 375, used in Buick, Chevrolet and Oldsmobile passenger cars; the Type M-40, used in Pontiac cars; and the Type 425,

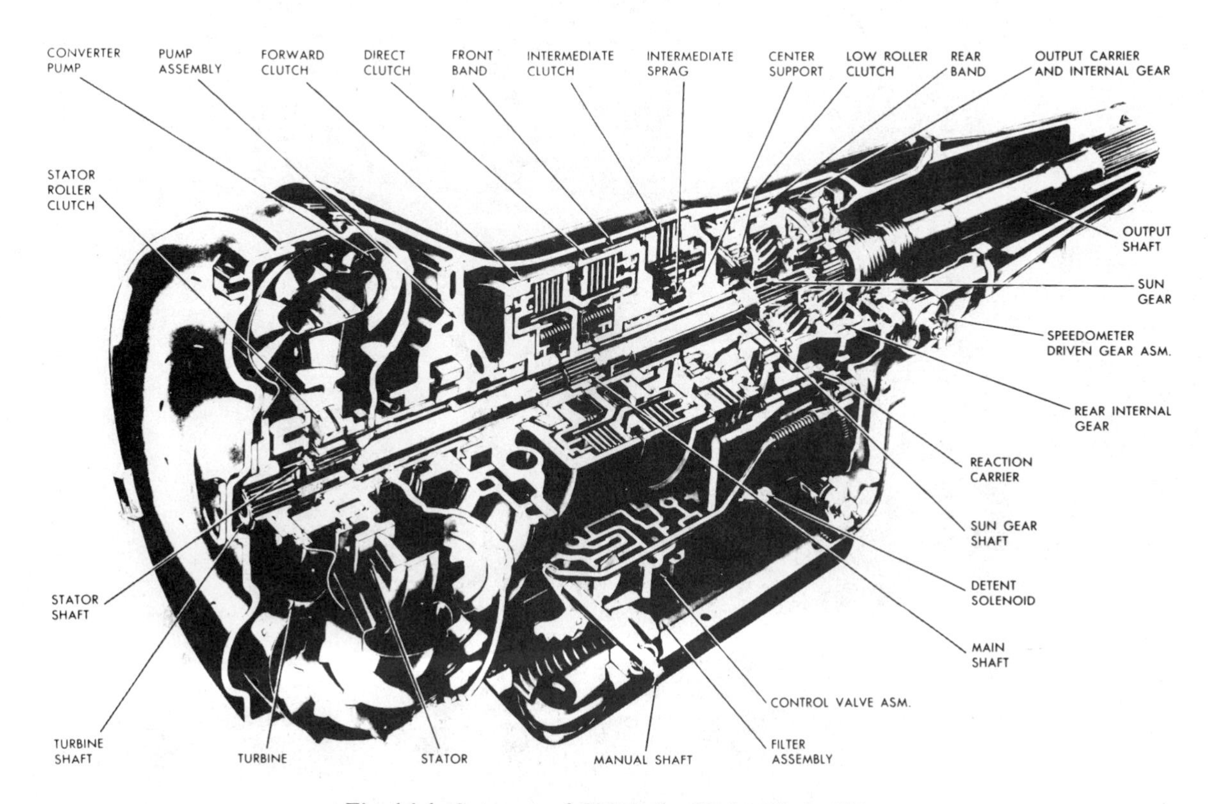

Fig. 14-1 Cutaway of GM Turbo Hydra-Matic 400.

figure 14-2, modified for use in front-wheel drive passenger cars.

OPERATION

Although the gear setup on the Type 400 is different from any of the transmissions studied thus far, it is still a Simpson gear train. As usual, the turbine or input shaft connects to a forward clutch, figure 14-3. In this case, however, the forward clutch connects the turbine shaft to a main shaft. The main shaft is splined to, and drives, the rear ring gear. This means that in forward speeds, input power must go through the rear gear set first.

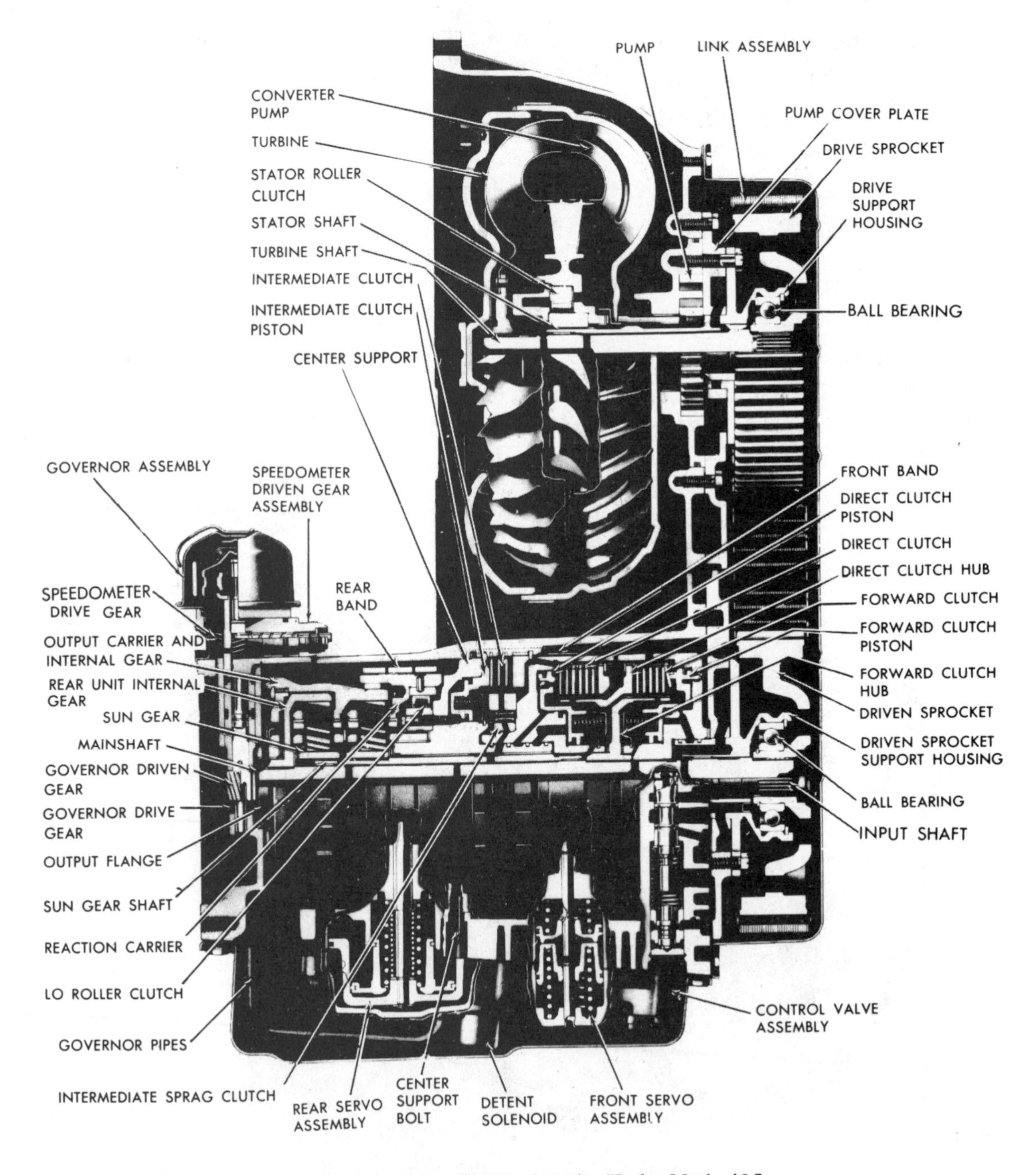

Fig. 14-2 Cutaway of typical Turbo Hydra-Matic 425.

This does not change the basic use of the Simpson gear train, but it should be kept in mind when studying the power flow of this model.

To control the gear set, the Type 400 uses a forward clutch, an intermediate clutch, a direct clutch, front and rear bands, a low roller clutch, and an intermediate roller clutch. (An intermediate sprag is used on earlier model transmissions.)

Drive Range Low

When the shift lever is moved to the D position, the forward clutch is applied, figure 14-4, page 190. With the forward clutch on, power flows from the forward clutch to the main shaft. The main shaft passes through the hollow sun gear shaft and drives the rear ring gear. Because the rear carrier is part of the output shaft, the planet pinions act as idlers and turn the sun gear in a reverse rotation. The sun gear then acts as an input for the front gear set. Since the front carrier (reaction carrier) is held by the low roller clutch, there must be another reversal of rotation on the front ring gear. The ring gear, rear carrier, and output shaft assembly turn in a forward rotation at a ratio of about 2.5:1.

Note that although the power flow is from the rear gear set to the front gear set, the result, for drive range low, is the same as in other Simpson gear trains. That is, there is a ring gear driving in a forward direction, a reverse rotation on the sun gear, and another reversal of direction on the other ring gear. For drive range low, this provides the two reversals of rotation of the Simpson gear train necessary to obtain a forward rotation of the output shaft.

Drive Range Intermediate

Depending on the make of car, tire size, axle ratio, and throttle opening, a shift to intermediate or second speed will take place between 8 mph (13 km/h) for a closed throttle shift, to 60 mph (97 km/h) for a wide-open throttle shift. As the 1-2 shift valve moves, line pressure passes to the intermediate clutch, figure 14-5, page 191. With the intermediate clutch applied, the intermediate roller clutch locks up and stops the reverse rotation of the sun gear.

Power flow is still from the forward clutch to the rear ring gear. However, the rear carrier and planet gears are now forced to walk around the stopped sun gear, and the carrier and output shaft turn at a ratio of about 1.5:1. Note that the power flow is the same as in other Simpson gear trains, but in this case the rear gear set is the effective

FORWARD CLUTCH RELEASED
DIRECT CLUTCH RELEASED
REAR BAND RELEASED
INTERMEDIATE SPRAG INEFFECTIVE
LO ROLLER CLUTCH INEFFECTIVE
FRONT BAND RELEASED
INTERMEDIATE CLUTCH RELEASED

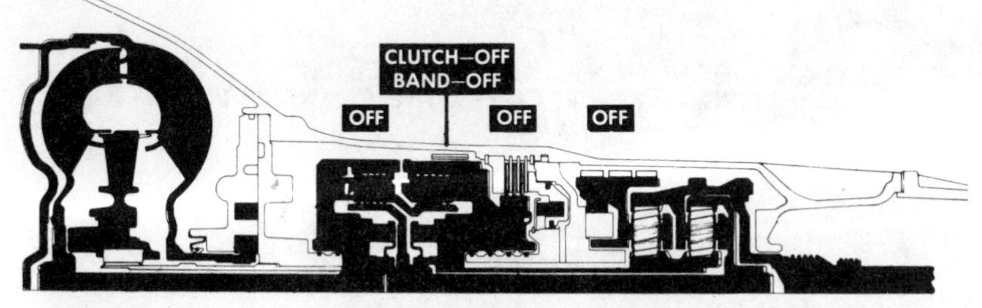

Fig. 14-3 Members of the gear set in the Turbo Hydra-Matic 400 (neutral).

[For color diagram, see page 335.]

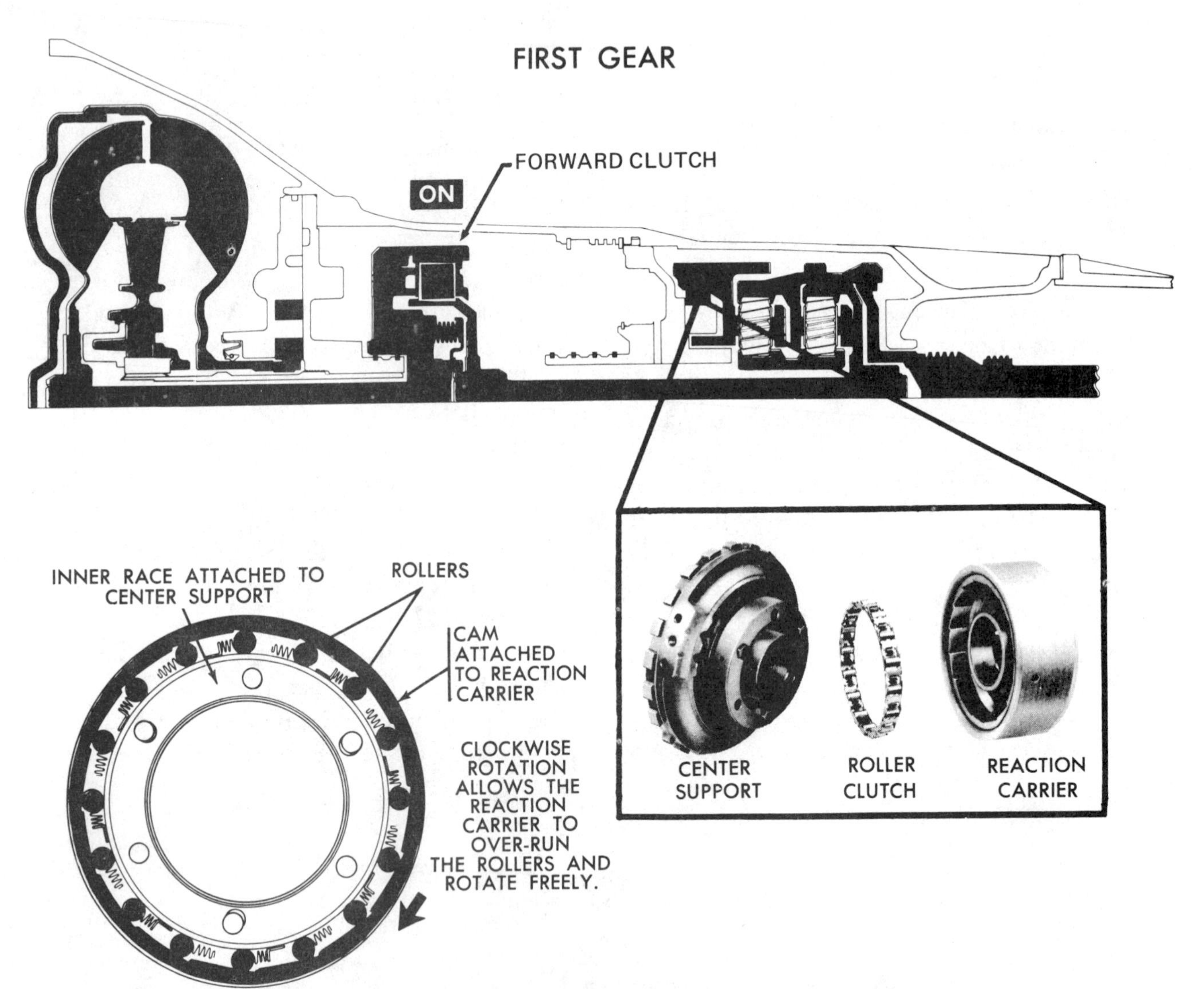

Fig. 14-4 Drive range first gear.

[For color diagram, see page 336.]

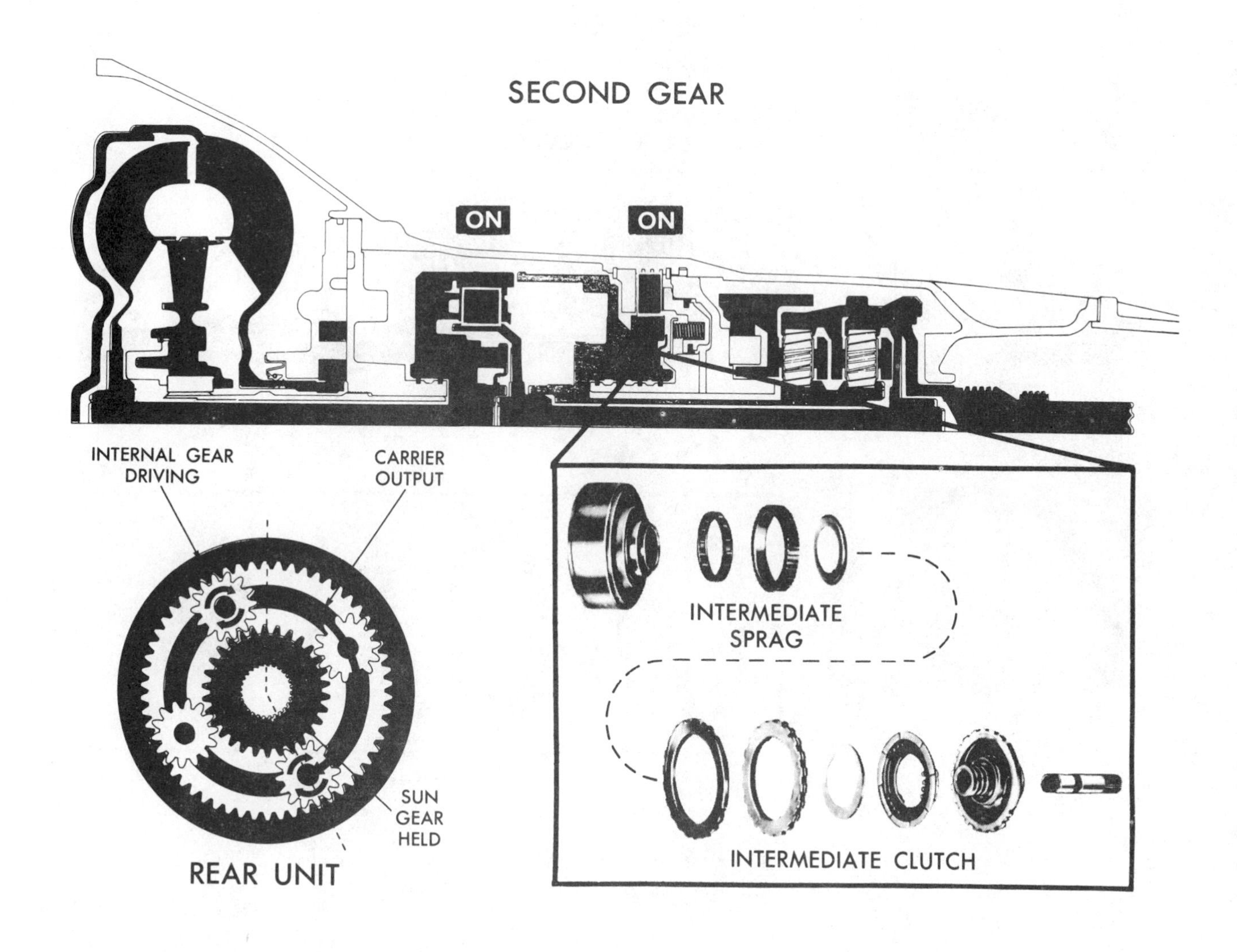

Fig. 14-5 Drive range second gear.
[For color diagram, see page 337.]

unit. The low roller clutch overruns and allows the front gear set to freewheel.

Drive Range High (Direct)

A shift to high will occur at road speeds between 20 and 90 mph (32 and 145 km/h). The road speed at which this takes place depends on make of car, tire size, axle ratio, and throttle opening. The movement of the 2-3 shift valve allows line pressure to move to the small, inner area of the direct clutch piston and to apply the direct clutch, figure 14-6, page 192.

With the transmission now in high or direct, power flow is through the forward clutch to the rear ring gear, and through the direct clutch to the hollow sun gear shaft and sun gear. With both the rear ring gear and sun gear driving, the rear set locks up for direct drive. As is the case with the Type 350, it is not necessary to release the intermediate clutch. The intermediate roller clutch will

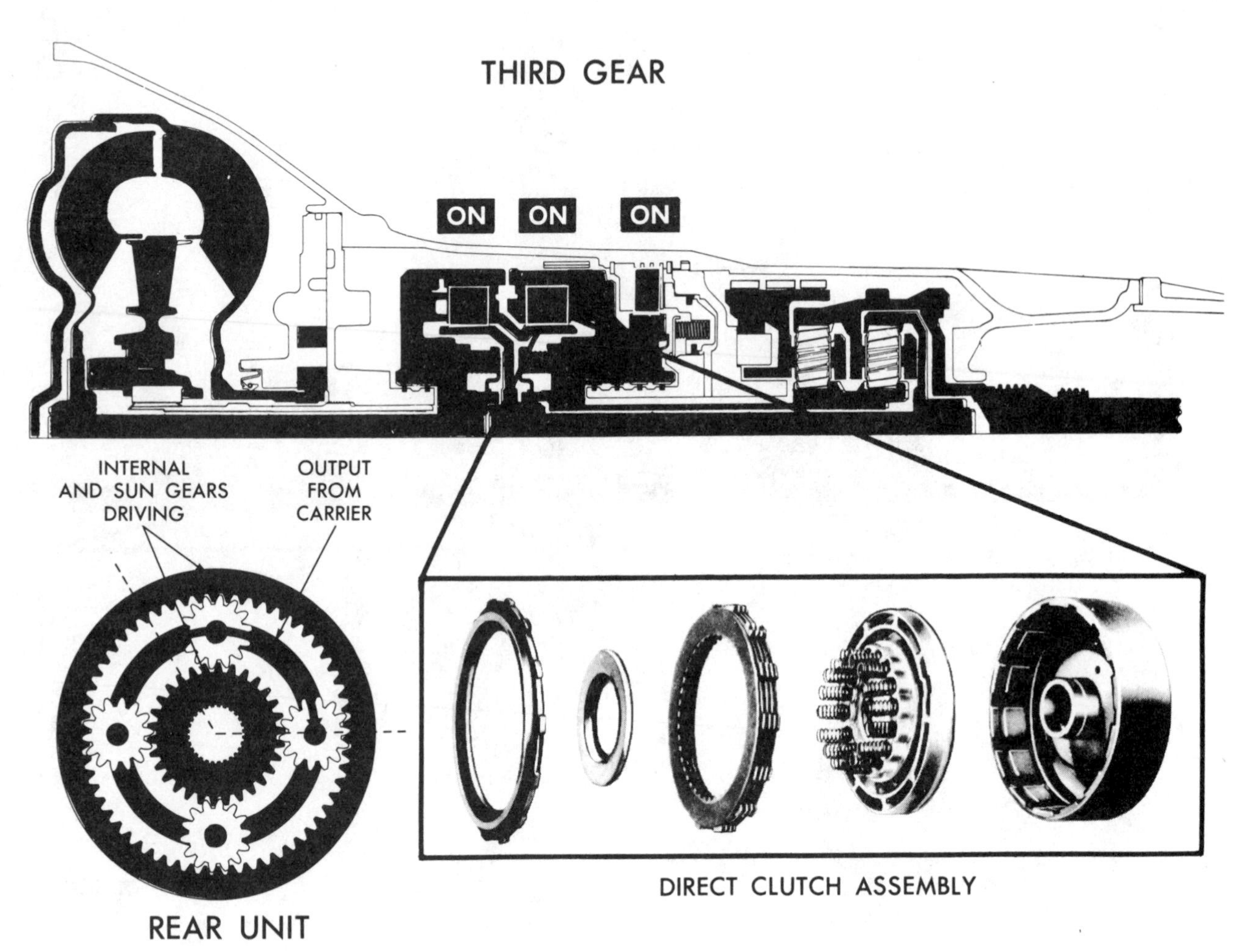

Fig. 14-6 Drive range third gear.
[For color diagram, see page 338.]

freewheel and allow the sun gear to turn in a forward rotation.

Manual 2 (Second)

In this transmission, manual second may be marked on the shift quadrants as I, D▪, S, or L2, depending on the make and model of car. Regardless of its name, manual second has the same functions in the Type 400 as in the Type 350. That is, in manual second the transmission will start in low, shift to, and stay in second, regardless of the road speed that is reached. For engine braking, the driver may choose manual second, and the transmission will shift to second speed. For the purpose of engine braking, the front band is on in manual second.

Manual 1, L or L1 (First)

Manual first in the Type 400 also operates as in other transmissions. When manual first is selected at higher road speeds, the transmission will downshift to second. At speeds below about 40 mph (64 km/h) the transmission will shift to low.

To provide engine braking in this range, the rear band is applied to hold the front carrier. It should be noted that due to the design of the hydraulic system, the trans-

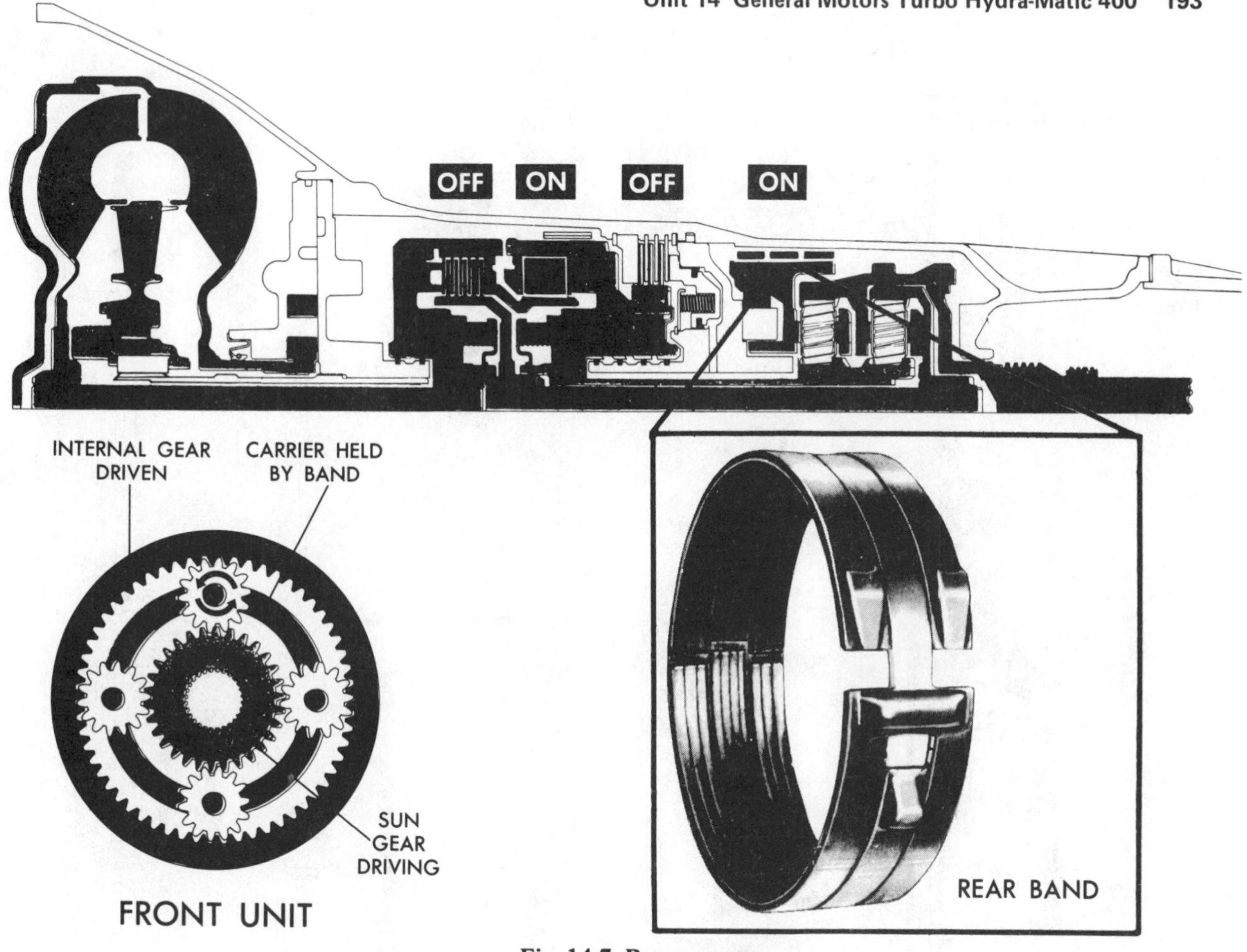

Fig. 14-7 Reverse.
[For color diagram, see page 339.]

mission will not upshift out of manual low regardless of the road speed that is reached. (Vehicles with a very low-ratio rear axle are an exception to this rule.)

Reverse

For reverse, the rear servo applies the rear band. Line pressure, directed to the large area of the direct clutch piston, applies the direct clutch, figure 14-7. The direct clutch drives the sun gear, and the front carrier is held by the rear band. This turns the front ring gear and output shaft in a reverse rotation at a ratio of about 4:1. Note that the Type 400 differs from other transmissions with Simpson gear trains in the operation of reverse. In the Type 400, the front gear set is used for reverse.

HYDRAULIC SYSTEM

The hydraulic system of the Type 400 is very much like other systems studied thus far. Hence, only its fine points are discussed here in detail. As in other automatic transmissions, the Type 400 hydraulic system has a main pressure regulation system, shift valves, and governor, modulator, and detent systems to control shift points, figure 14-8, page 194.

Note that the pressure regulator for this system has differential force areas for reverse, modulator, and manual intermediate boost. A boost in manual low is also provided by the use of the intermediate boost passage. Depending on throttle opening, line pressure is regulated to between 70 and 150 psi (483 and 1 034 kPa). This pressure is a result

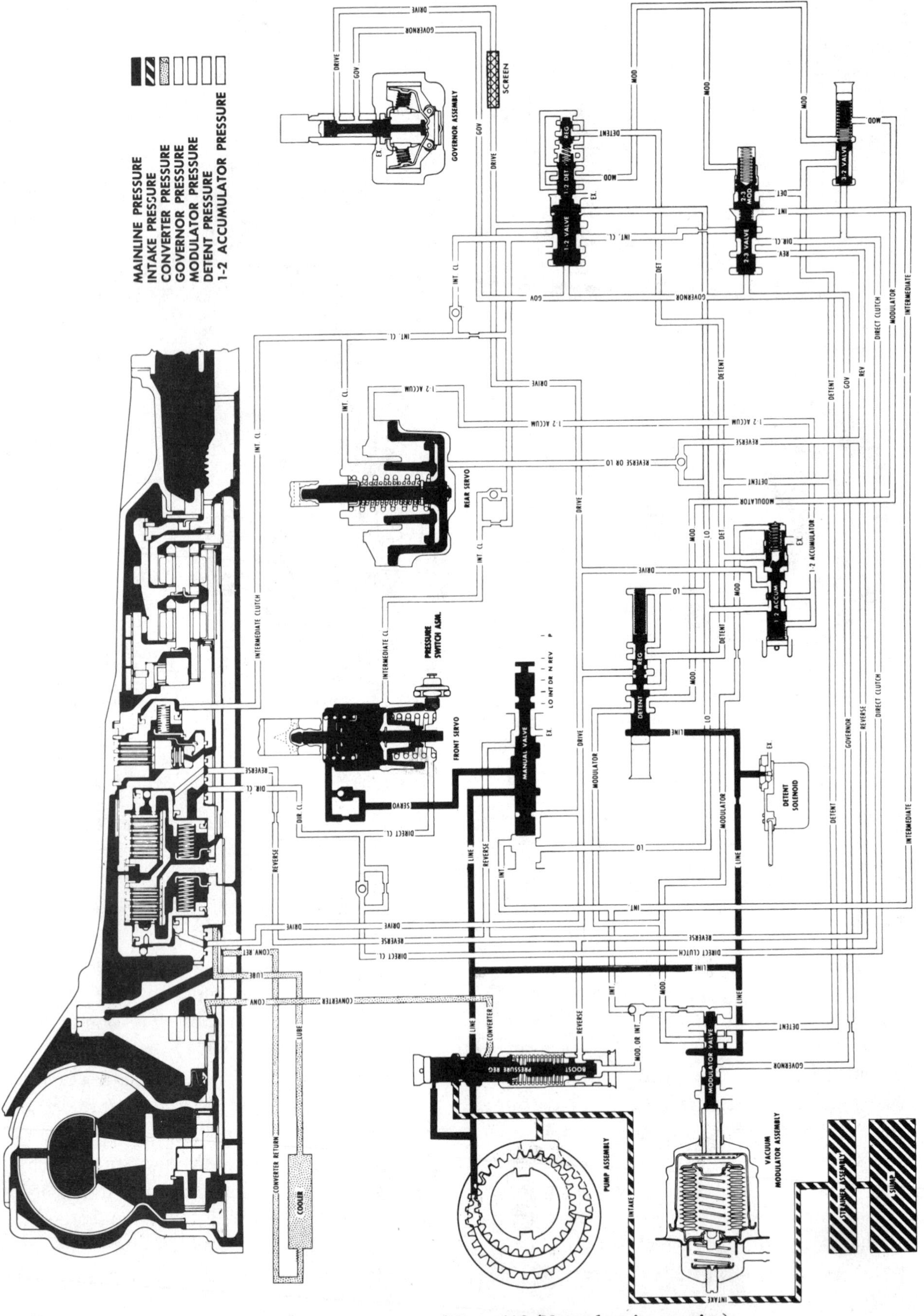

Fig. 14-8 Hydraulic system of Type 400 (Neutral engine running).

[For color diagram, see page 340.]

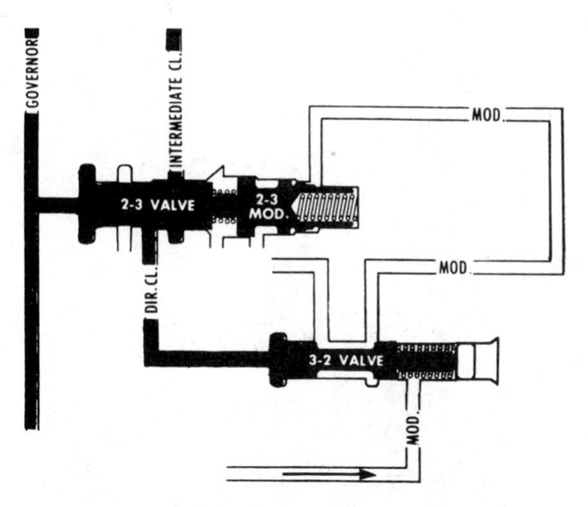

VALVES IN 3RD. GEAR POSITION MODULATOR PRESSURE UNDER APPROX. 90 P.S.I.

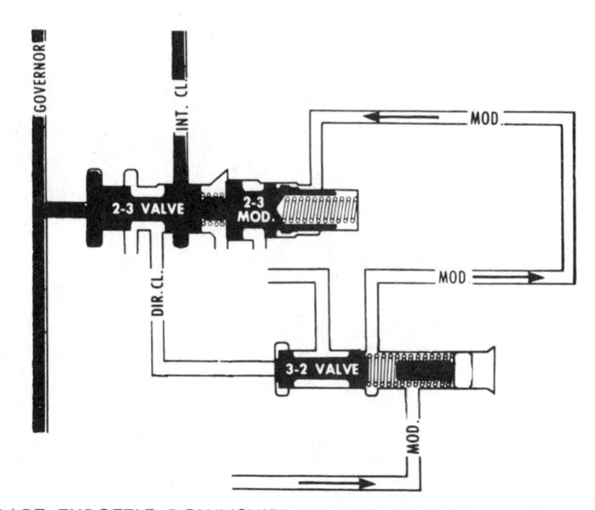

PART THROTTLE DOWNSHIFT VALVES IN 2ND. GEAR POSITION MODULATOR PRESSURE OVER 90 P.S.I.

Fig. 14-9 Part-throttle downshift 3-2.
[For color diagram, see page 341.]

of modulator pressure on the main pressure regulator valve.

In manual low or intermediate, line pressure from the manual valve moves a check ball to seal off the modulator passage. Line pressure then passes through the line marked *MOD* or *INT*, giving a boost in line pressure to about 150 psi (1 034 kPa). For reverse, line pressure in the reverse passage boosts main line pressure to about 260 (1 793 kPa). For even tighter lockup in reverse, the larger, outer area of the direct clutch piston is also used.

3-2 Valve

To allow light to medium throttle use in third gear, without causing a downshift, a 3-2 valve is used, figure 14-9. With the transmission in third, the 3-2 valve exhausts modulator pressure at the 2-3 valve through the detent passage and valve, and to the Lo passage at the manual valve. (See figure 14-8.) The transmission will remain in third as long as modulator pressure remains below about 90 psi (621 kPa), and road speed remains above 33 mph (53 km/h). If the throttle is opened wide enough to raise modulator pressure above 90 psi (621 kPa), a 3-2 downshift will take place below speeds of about 33 mph (53 km/h).

As shown in figure 14-9, the force of modulator pressure and the spring force move the 3-2 valve to the left. This allows modulator pressure to work on the 2-3 valve again and downshift the transmission to second.

Detent Valve and Solenoid

Forced, or detent, downshifts can be made by moving the throttle to the wide-open position. A 3-2 downshift can take place at road speeds below about 70 mph (113 km/h). A 2-1 or 3-1 downshift will take place at speeds below about 20 mph (32 km/h). A detent shift in the Type 400 is accomplished in much the same way as in other transmissions. However, in this case, there is no mechanical hookup to the throttle linkage. Instead, the Type 400 makes use of a detent valve, detent regulator valve, and a detent solenoid, figure 14-10. The detent valve is brought into

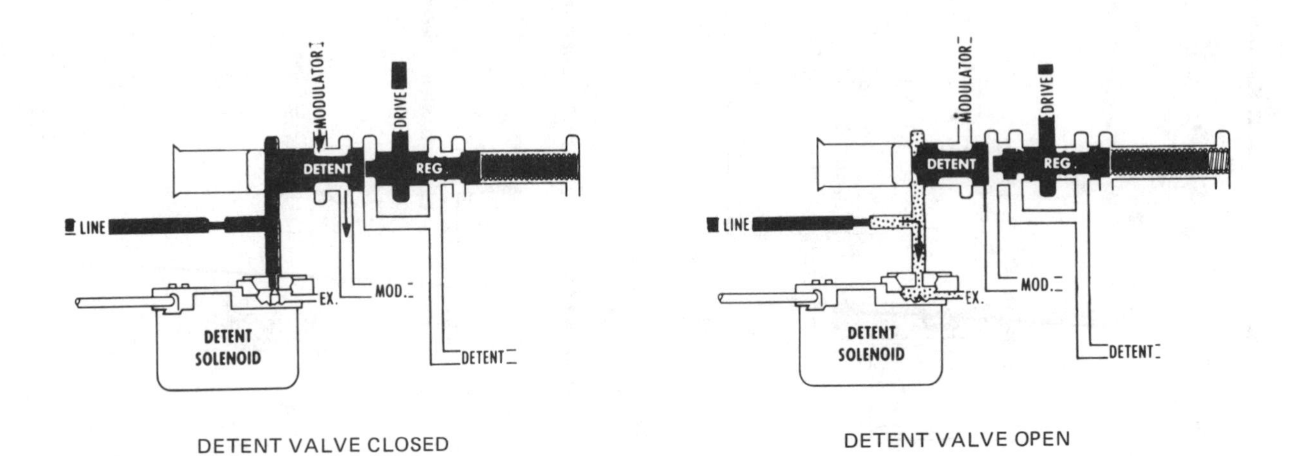

DETENT DOWNSHIFT – VALVES IN SECOND GEAR POSITION

Fig. 14-10 Detent downshift.

[For color diagrams, see page 342.]

use by an electric switch connected to the throttle linkage. At wide-open throttle, the switch is closed, and current passes through a wire to the detent solenoid in the valve body.

At all throttle positions below wide open, the detent valve is closed. As shown in the illustration, line pressure holds the detent valve and detent regulator valve to the right against a spring force. When the detent solenoid is energized, an exhaust is opened to the line pressure acting on the detent valve. Since line pressure must pass through an orifice before reaching the detent valve and solenoid, a pressure drop takes place on the detent side of the orifice. The spring force causes the detent and detent regulator valves to move to the left. This regulates drive oil (line pressure) to about 70 psi (483 kPa) and fills the modulator and detent passages. Detent pressure then acts on the shift valves to downshift the transmission at the road speeds noted above.

Detent oil is also sent to the modulator valve to prevent modulator pressure from dropping below 70 psi (433 kPa). This maintains line pressure at about 105 psi (724 kPa) at all road speeds and altitudes. This gives firm, slip-free detent shifts under all operating conditions.

SHIFT QUALITY CONTROLS

To control shift quality, the Turbo Hydra-Matic 400 uses controlled pressure regulation and other familiar hydraulic devices. Those devices that have not been discussed previously, or which are used in a different way in the Type 400, are described here.

Forward Clutch Apply

For a smooth application of the forward clutch, the Turbo Hydra-Matic 400 uses a two-diameter clutch piston, figure 14-11.

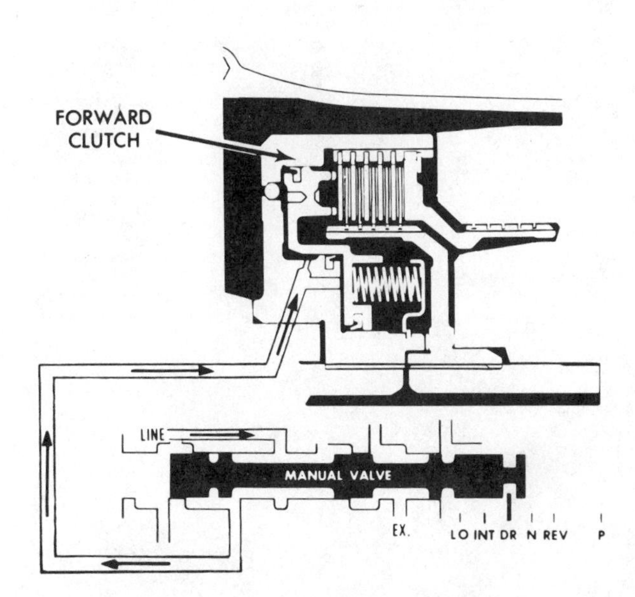

Fig. 14-11 Forward clutch applied.

When line pressure moves to the forward clutch, the small piston area fills first. This is due to the orifice at the large piston area. The clutch plates are thus brought together with less force, giving a smooth shift from neutral to drive. When the large area is filled, full holding force is used to apply the clutch.

Rear Servo and Accumulator

The rear servo is used to apply the rear band for reverse and manual low. A separate piston inside the rear servo is used as an accumulator for the intermediate clutch. One difference in this system is that a separate accumulator pressure, also known as *trim pressure,* is used to control intermediate clutch application, figure 14-12.

The 1-2 accumulator valve is a balanced, restriction-type regulator controlled by modulator pressure and a spring force. An increase in modulator pressure results in an increase in 1-2 trim pressure. Hence, at large throttle openings, intermediate clutch apply oil must work against high 1-2 trim pressure in the accumulator. This gives a firm 1-2 shift for high-torque, heavy-throttle operation. At small

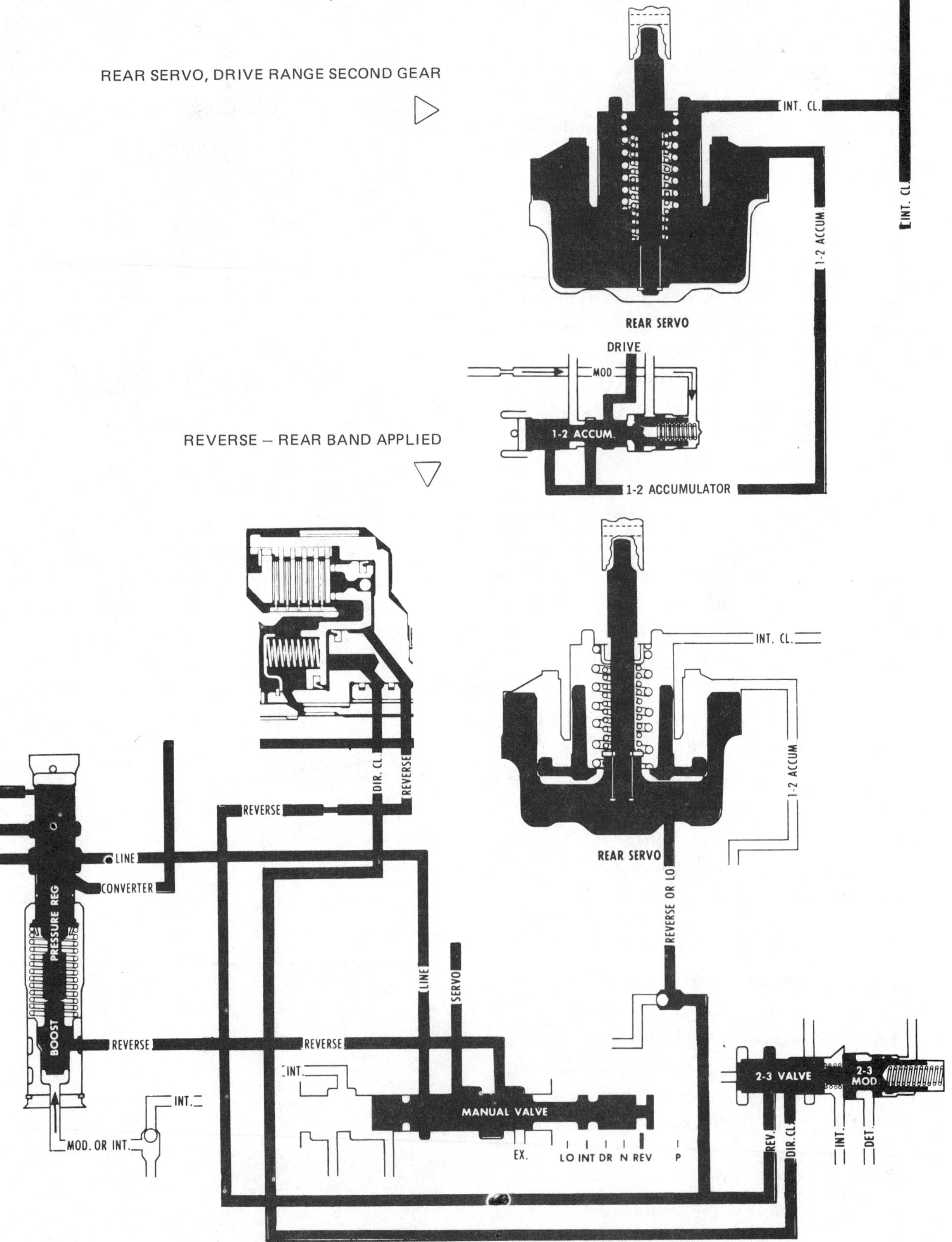

Fig. 14-12 Rear servo and accumulator.
[For color diagrams, see page 343.]

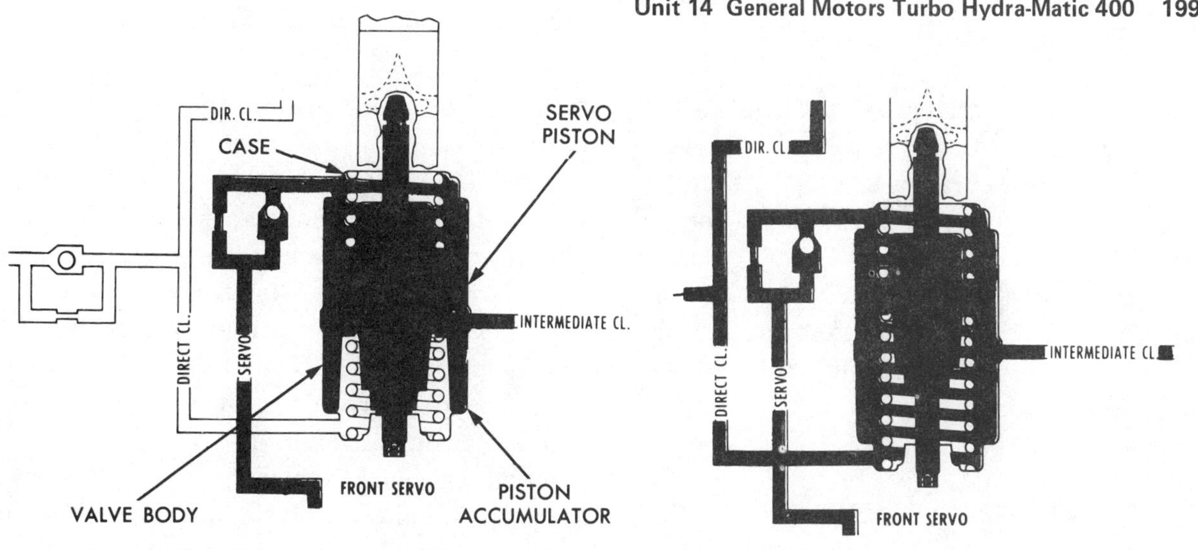

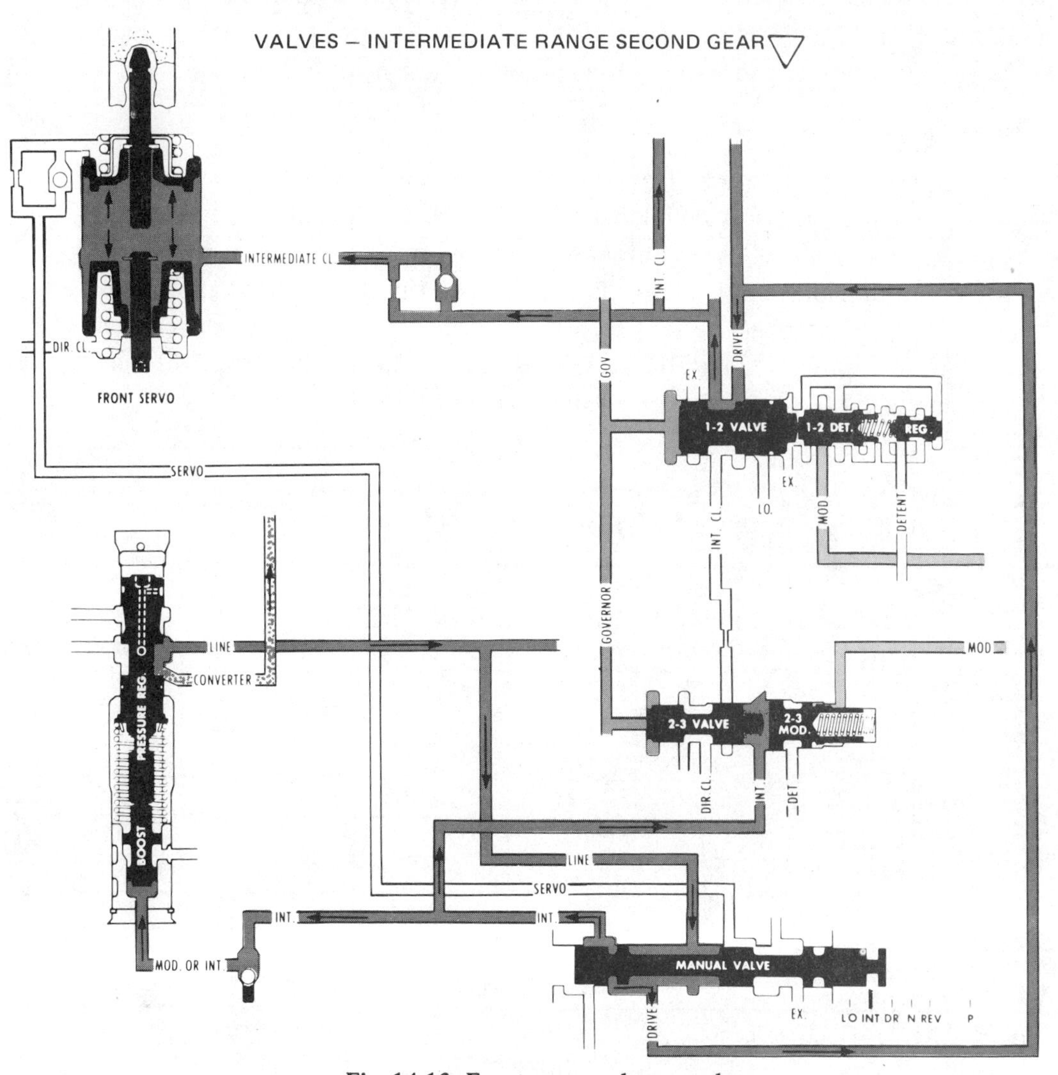

Fig. 14-13 Front servo and accumulator.

[For color diagrams, see page 344.]

throttle openings, trim pressure will be low and shift feel will be soft and smooth. In this way, the trim pressure varies to supply the right shift feel for all throttle openings and torque conditions.

Note that the accumulator piston is spring loaded and is not connected to the band apply pin. This allows the accumulator piston to move without applying the band.

Front Servo and Accumulator

The front servo is used to apply the front band for manual second, figure 14-13. In drive range second, intermediate clutch oil enters the servo, but cannot apply the band. This is because of the force of the servo oil at the top of the piston. Since servo oil and intermediate clutch oil are both at line pressure, the forces on the top and bottom of the piston are the same. This prevents band application. In manual second, servo oil is exhausted at the manual valve, and the servo can then apply the band.

The front servo also serves as an accumulator for the direct clutch. In this case, servo oil prevents the band from applying and acts as an accumulator pressure for the apply of the direct clutch. Direct clutch oil enters the accumulator, and the accumulator spring strokes both pistons up until contact is made with the servo return spring.

SUMMARY

The General Motors Turbo Hydra-Matic 400 is a fully automatic, three-speed transmission coupled to the engine by a torque converter. A Simpson gear train is used, but power flow is not exactly the same as in other transmissions studied. In this case, the rear gear set is used as an input for low and as the effective set for forward speeds.

Aside from this variation, the Simpson gear train in the Type 400 functions in much the same way as that in other transmissions.

- Two planetary gear sets are served by a common sun gear.
- For low, two reverses are used, with one carrier splined to the output shaft and the other held by a one-way clutch.

RANGE	GEAR	FORWARD CLUTCH	DIRECT CLUTCH	FRONT BAND	INT. CLUTCH	INT. SPRAG	LO ROLLER CLUTCH	REAR BAND
PARK - NEUTRAL		O	O	O	O	O	O	O
DRIVE	FIRST	X	O	O	O	O	X	O
	SECOND	X	O	O	X	X	⊗	O
	THIRD	X	X	O	X	⊗	⊗	O
INTERMEDIATE	FIRST	X	O	O	O	O	X	O
	SECOND	X	O	X	X	/	⊗	O
LO	FIRST	X	O	O	O	O	/	X
	SECOND	X	O	X	X	/	⊗	O
REVERSE		O	X	O	O	O	O	X

Key: O not effective X in use ⊗ overrunning / partial hold

Fig. 14-14 Summary of power flow in the Turbo Hydra-Matic 400.

- For intermediate, direct, and reverse, one gear set is used as a simple planetary gear set.

These basic points, and the summary of power flow for the type 400 shown in figure 14-14 should be kept in mind when diagnosing power flow problems in this transmission.

The hydraulic system in the Type 400 works in much the same way as those studied previously. Its distinctive features are an electrically operated detent system, the use of the 3-2 valve, and variable 1-2 accumulator pressure.

REVIEW*

I. Select the members that are effective in the Type 400 Turbo Hydra-Matic transmission for the range named in each question.

1. Drive range – first speed
   a. roller clutch, direct clutch, front band
   b. forward clutch, front band
   c. forward clutch, roller clutch
   d. forward clutch, roller clutch, front band
2. Drive range – second speed
   a. forward clutch, intermediate clutch, intermediate sprag
   b. forward clutch, intermediate clutch, rear band
   c. forward clutch, intermediate clutch
   d. forward clutch, intermediate clutch, direct clutch
3. Drive range – third speed
   a. forward clutch, roller clutch, direct clutch
   b. direct clutch, roller clutch, intermediate sprag
   c. forward clutch, direct clutch, rear band
   d. forward clutch, intermediate clutch, direct clutch
4. Reverse
   a. forward clutch, rear band
   b. direct clutch, front band
   c. direct clutch, rear band
   d. forward clutch, front band
5. To permit heavy-throttle operation in direct, the 3-2 valve
   a. regulates modulator pressure to 105 psi to the 2-3 valve.
   b. cuts off modulator pressure to the 2-3 valve.
   c. directs detent pressure to the 2-3 valve.
   d. directs governor pressure to the 2-3 valve.

**Note:* For some questions, it may be necessary to refer to the units dealing with the basic principles of automatic transmissions.

II. Select the best answer from the choices offered to complete the statement or answer the question.

6. Which of the following statements apply (applies) to the detent valve of the Type 400 transmission?

   (I) It directs detent oil to the governor passage for a forced downshift.
   (II) It directs detent oil to the direct clutch for a forced downshift.

   a. I only
   b. II only
   c. both I and II
   d. neither I nor II

7. A condition of delayed upshift, or full-throttle upshift only, could be caused by

   a. governor pressure that is too high.
   b. an electrical short in the detent system.
   c. disconnected detent wire.
   d. none of the above

8. To control direct clutch apply in drive range third, direct clutch oil:

   (I) Is directed to the large area of the direct clutch piston.
   (II) Is directed to the front servo which acts as a 2-3 accumulator.

   a. I only
   b. II only
   c. both I and II
   d. neither I nor II

9. 1-2 accumulator pressure is controlled by the 1-2 accumulator valve. The forces affecting this valve are spring forces and:

   (I) Modulator pressure.
   (II) Governor pressure.

   a. I only
   b. II only
   c. both I and II
   d. neither I nor II

10. During a 1-2 shift, intermediate clutch pressure is controlled by

   a. the modulator valve
   b. the governor
   c. the 1-2 shift valve
   d. the 1-2 accumulator valve

11. To obtain smooth initial application of the forward clutch when the manual valve is moved to D, drive oil (line pressure) is:

   (I) Metered through an orifice to the large area of the clutch apply piston.
   (II) Metered through an orifice to the small area of the clutch apply piston.

   a. I only
   b. II only
   c. both I and II
   d. neither I nor II

12. The purpose of the orifice in the intermediate clutch passage that runs from the 1-2 to the 2-3 valve is to

    a. help control intermediate clutch application.
    b. help control direct clutch application.
    c. reduce the pressure on the 2-3 valve.
    d. help time the release of the direct clutch.

EXTENDED STUDY PROJECTS

I. Complete hydraulic circuit diagrams for the Turbo Hydra-Matic 400 as assigned by the instructor.

II. Answer the following questions, making full use of the text, illustrations, and hydraulic circuit diagrams. Give reasons for the answer to each question.

1. A customer complains of no drive in D. Fluid level and condition are normal. This condition is probably caused by:

   (I) Burned direct clutch facings.
   (II) A leaking rear servo.

   a. I only c. both I and II
   b. II only d. neither I nor II

2. A customer complains that the transmission has no reverse. Fluid level and condition are normal, but a road test shows no engine braking in L. This condition is probably caused by:

   (I) A broken rear band apply pin.
   (II) A leaky rear servo piston.

   a. I only c. both I and II
   b. II only d. neither I nor II

3. A customer complains of a soft or slipping 2-3 shift. Fluid level and condition are normal. This condition is probably caused by:

   (I) A stuck rear accumulator piston.
   (II) Damaged or leaky passages in the valve body.

   a. I only c. both I and II
   b. II only d. neither I nor II

4. To control intermediate clutch application in drive range second, intermediate clutch oil:

   (I) Passes through an orifice before reaching the clutch.
   (II) Is directed to that part of the rear servo which acts as a 1-2 accumulator.

   a. I only c. both I and II
   b. II only d. neither I nor II

5. A customer complains of a harsh 2-3 shift in the transmission. Fluid level and condition are normal. This condition is probably caused by:

   (I) A leak to the outer area of the direct clutch piston.
   (II) Leaking modulator bellows.

   a. I only
   b. II only
   c. both I and II
   d. neither I nor II

6. Governor pressure is directed to the modulator valve to:

   (I) Increase modulator pressure.
   (II) Increase main line pressure.

   a. I only
   b. II only
   c. both I and II
   d. neither I nor II

7. A customer complains of no detent downshift. This condition is probably caused by:

   (I) Improper accelerator linkage adjustment.
   (II) No current to the detent switch.

   a. I only
   b. II only
   c. both I and II
   d. neither I nor II

8. A customer complains of slipping in drive range second, but super is functioning properly. The fluid level is normal but has a burnt odor and color. This condition is probably caused by:

   (I) A burned front or intermediate band.
   (II) A burned intermediate clutch.

   a. I only
   b. II only
   c. both I and II
   d. neither I nor II

9. A customer complains of a soft or slipping 1-2 shift. Fluid level and condition are normal. This condition is probably caused by:

   (I) A leaking modulator diaphragm.
   (II) A stuck detent valve.

   a. I only
   b. II only
   c. both I and II
   d. neither I nor II

10. A customer complains of a delayed 2-3 upshift. Fluid level and condition are normal. A closed-throttle shift test gives a 1-2 shift at 11 mph and a 2-3 shift at 40 mph. This condition is probably caused by:

    (I) A sticking 2-3 shift valve.
    (II) A short circuit in the detent switch.

    a. I only
    b. II only
    c. both I and II
    d. neither I nor II

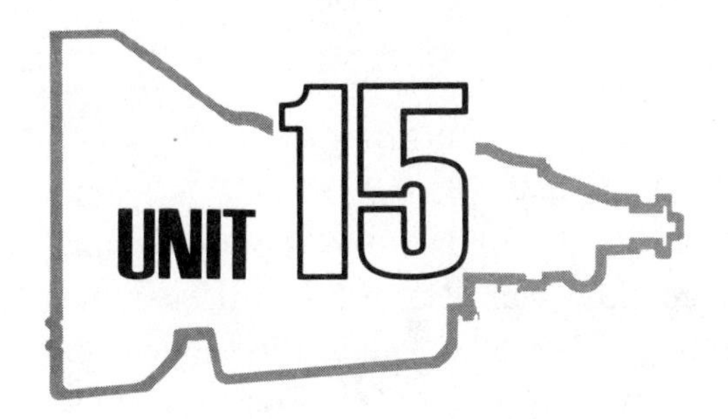

AUTOMATIC TRANSMISSION SERVICE

OBJECTIVES

After studying this unit, the student will be able to:

- Explain the relationship between transmission performance and periodic service.
- Perform in-car service and the necessary adjustments for proper transmission performance.
- Describe the operation of a vehicle with an automatic transmission under various driving and road conditions.

DEPENDABILITY AND SERVICE

Although most car owners realize the value of having the car's engine serviced regularly, few are aware that periodic service is also important to transmission performance.

Transmission, engine, and overall car performance can be improved, and the life of the transmission extended, by proper service at regular intervals. For example, proper adjustment of a transmission that shifts sluggishly due to improper band or linkage adjustment can help to improve engine performance. Transmission life can be extended by changing the oil and filter regularly and providing proper service when needed.

This unit deals with common transmission problems and the correct service for these problems.

CHECKING FLUID LEVEL AND CONDITION

Unit 7 describes the necessary characteristics of transmission fluid. The mechanic must know these traits and be able to detect not only correct fluid level but contaminated fluid, or fluid containing foreign materials.

Checking the fluid level in a transmission may appear to be a simple job. However, shop manuals show that there are different procedures for various types of cars. For example, some manufacturers recommend that the oil level be checked with the transmission in park and the engine running. In some transmissions, leakage at the manual valve in park will give a false reading of oil level. These transmissions must be checked in neutral. At least one model of transmission must be checked with the engine off. These different ways of checking the fluid are due to the differences in construction and engineering of the automobiles. Because of the differences, the mechanic should check the shop manual for the correct procedure before testing the oil level.

Preparation

In general, when checking the fluid level, the transmission should be at normal operating temperature. Normal operating temperature is produced by about 10 to 15 miles (16 to 24 kilometers) of highway driving and is equal to a temperature of about 180°F (82°C). Some cars can be checked with the fluid cold, but most manufacturers recommend a warm-up procedure.

Checking the Level

Before checking the fluid level, be sure all dirt is cleaned from around the dipstick tube. With the fluid warmed, the transmission should be shifted through all manual ranges and returned to park or neutral, whichever is called for in the manual. This assures that the servos, clutches, and fluid passages are filled. If all passages are not filled to operating capacity, a false reading will be given. (The level will appear to be higher than it actually is.)

The dipstick should be removed, wiped clean with a lint-free cloth, and returned to the tube. The stop on the dipstick must make contact with the top of the tube. The dipstick is then removed a second time, and the level noted, figure 15-1. In general, the level should be between the ADD and FULL as shown. The level should never be below the ADD, or above the FULL mark.

Too little fluid, indicated by a level below the ADD mark on the dipstick, can result in slipping of bands and clutches. What is often not understood though, is that **too much** fluid can also result in these same problems. The foam inhibitors added to transmission fluid can only handle a certain amount of churning or mixing. If the fluid level is too high, the fluid will be churned and aerated by the gear train. Since air is compressible, this will lead to soft sluggish shifts and slippage. This will, in time, lead to burned clutches and/or bands.

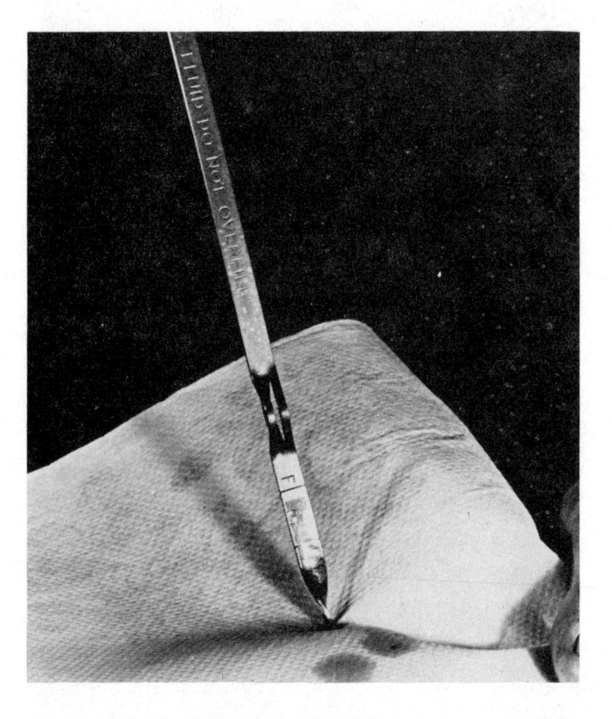

Fig. 15-1 Oil level should be between the ADD and FULL marks.

A fluid level that is too high can be corrected by removing the pressure test plug or *cracking* a cooler line, and starting the engine. This pumps fluid out of the transmission. By draining the fluid into a container, the mechanic can judge when enough fluid has been removed. At this point, the engine should be stopped, the plug replaced or the cooler line tightened, and the fluid level rechecked. If too much fluid has been drained, only clean fluid of the correct type should be added.

Checking Appearance and Smell

It is important to note the appearance and smell of the fluid when checking the level. Fluid that has a dirty, black color and a burned smell (much like that of burning brakes or a slipping clutch) usually indicates that a major overhaul is needed. If the fluid is in this condition, the mechanic can assume that the problem is internal. The black color and burned odor tell the experienced mechanic that a clutch or band has been slipping and major repairs are needed.

If the fluid has a thick, brownish, milky or varnished look, moisture is present and sludge is forming. In this case, the fluid should be changed and the car road tested for proper operation.

Fluid that has a brown color and a sour, stale smell, shows that the fluid is oxidized. This condition results when the fluid has not been changed for some time, or when it has been overheated. Overheating can be caused by hard use such as towing heavy loads or trailers. If the fluid shows signs of oxidation, it should be changed and the car road tested for proper transmission performance. If the overheating has been caused by hard use, the customer should be advised to have an auxiliary cooler installed.

If a white, lint-free cloth is used to wipe and hold the end of the dipstick, any material that should not be in the fluid can be easily seen. Brass, aluminum, steel, or plastic particles, that may look like filings, should not be present in the fluid. Small amounts of particles may be considered normal. However, the presence of large amounts of foreign materials indicates internal wear, and a major overhaul may be necessary.

When to Change the Fluid

Some manufacturers suggest regular fluid changes based on mileage. Others do not recommend periodic fluid changes. The recommendations made in this text are intended only as guidelines. Sludged, varnished, or oxidized fluid should be changed at once, and further checks made to find and correct the cause of these abnormal conditions.

Changing the Fluid and Filter

When it is decided either because of mileage recommendations or fluid condition, that the fluid should be changed, it is wise to replace the filter as well. Even though the screen that serves as the filter may look clean, a varnish *web* may be built up which will block fluid flow. This may be especially true in the case of oxidized fluid. Filter screens should be examined carefully for varnish buildup since this could cause slipping and burning of clutch discs or bands. Since the varnish is difficult to remove, the filter should be changed.

The procedure for changing the fluid and filter is simple and straightforward. The shop manual for a particular model will outline the steps and state the amount and type of fluid to be used. In general, the following steps are recommended.

1. Remove the oil pan plug or filler tube and drain the transmission. In some cases, the oil pan must be removed and tilted to one side to allow the fluid to drain.
2. Remove the converter cover and locate the converter drain plug by turning the converter and flywheel until the plug is straight down in the six o'clock position. Remove the plug, and drain the fluid.

   Note: Some converters do not have a drain plug, and the converter fluid is not normally changed. If the fluid condition in a converter of this type is such that it must be changed, drill and tap the converter for a pipe plug. The factory manual should first be consulted for the proper steps and location of the hole.
3. The filter is changed by removing and replacing the retainer screws or clip(s). If necessary, O-rings on the filter may be held in place with petroleum jelly.
4. Replace the oil pan and/or plug, and the converter plug and cover.
5. Pour five or six quarts of the correct fluid through the filler pipe.

   Note: The amount of fluid needed for the initial filling will vary; consult the shop manual.
6. Start the engine and shift the transmission through all ranges. Check the fluid level, as outlined earlier. Add additional fluid if necessary to bring the level to, or slightly above, the ADD mark.
7. Thoroughly warm up the transmission, and if necessary, add additional fluid to bring the fluid to the correct level.
8. Road test and check for leaks.

These steps do not replace the instructions in the shop manual. They are given as guidelines only. Remember that only the correct type of fluid, as specified by the manufacturer, should be used.

Caution: Under certain conditions, hydraulic fluid can cling to the dipstick tube and give a false reading. This may show up as a normal or high level when, in fact, the actual level is low.

A false reading may occur when the transmission is being filled after a fluid change or when the car has just been driven. A false level can usually be detected by an uneven level or by dry spots on the dipstick. It is due to fluid clinging to the inside of the dipstick tube. To guard against this problem, the transmission should be left in neutral or park with the engine idling, for five or ten minutes before checking.

Another cause of false fluid level is an internal leak. Worn bushings, case cross-leaks, or valve body leaks are some of the causes of internal leaks. An exceptionally high fluid level after warm-up, with no dry spots on the dipstick, could be an indication of an internal leak. The subject of internal leaks is covered in greater detail in unit 16.

Band Adjustment

Since most transmissions have at least one band adjustment that requires pan removal, it is usually best to schedule band adjustment at the time of fluid and filter change. The steps and tools necessary for band adjustment should be found in the shop manual for the make and model of transmission being worked on. Adjustment specifications (specs) may vary between models of the same make of transmission. Although some manufacturers call for special band adjustment tools, a good job can be done with hand tools, an accurate inch-pound torque wrench, and adapters as needed, figure 15-2.

As an example of band adjustment, the specifications for the intermediate band of one model of the Ford C-4 transmission calls for the band adjustment screw to be tightened to a torque of 10 lb-ft (13.6 N•m), and then backed off 1 3/4 turns. Since, for greater accuracy, most band adjusting torque wrenches are calibrated to pound-inches, the torque required on the adjusting screw would be 120 lb-in (10 lb-ft × 12 = 120 lb-in).

Note: Metric torque wrenches for band adjustment use should be calibrated in increments of 0.5 N•m.

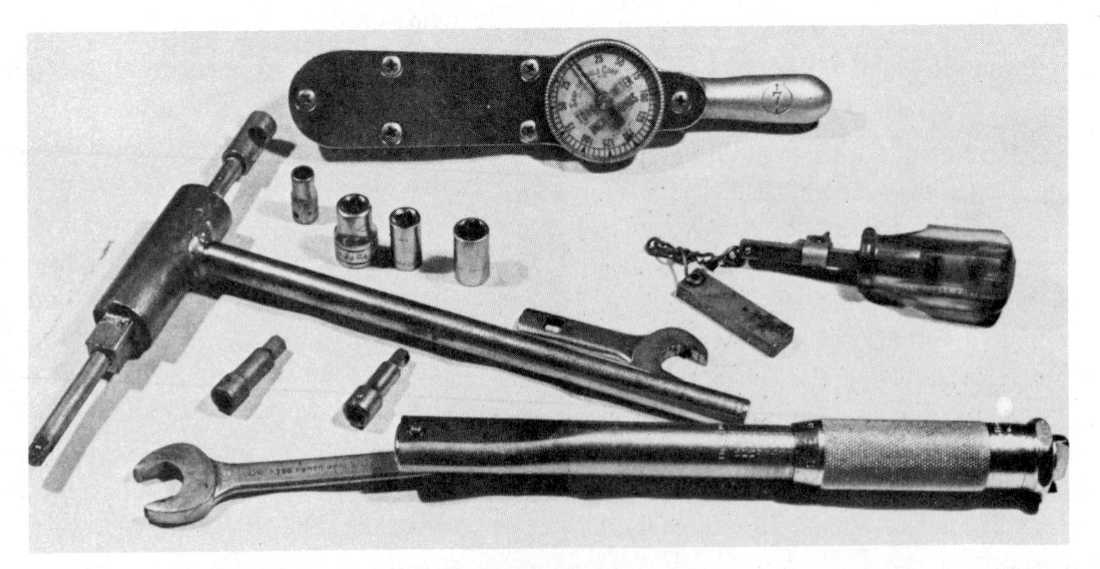

Fig. 15-2 Band adjusting tools.

Using the above specifications, the following steps should be used for band adjustment:

1. Clean all dirt and grease from the area of the band adjusting screw.
2. Loosen the locknut, and while holding the band adjusting screw, back the locknut off four or five turns, figure 15-3. This keeps the locknut from bottoming and giving a false torque reading.

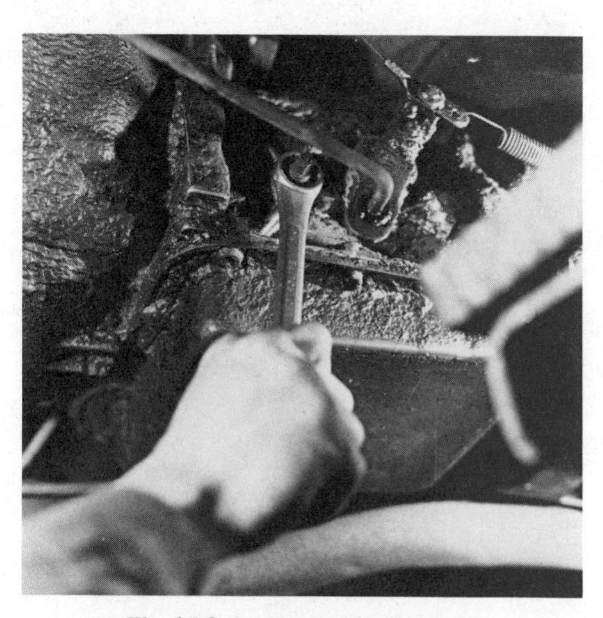

Fig. 15-3 Loosening the locknut.

Caution: Do not back off band adjusting screw. This could cause a band apply strut to fall out of place, and the pan and valve body would have to be removed in order to replace the strut.

3. Turn band adjusting screw in and torque to 120 lb-in (13.6 N•m), figure 15-4. It is best to back off and re-torque the screw several times to be sure the band is seated properly. Also, be sure that the locknut does not bottom and give a false torque reading.
4. Back off the adjusting screw exactly 1 3/4 turns. Since band adjustment screws are sometimes hard to get at, it is best to mark a flat on the screw with chalk or paint as an aid to counting the correct number of turns. A small open-end wrench can then be used to back off the screw, figure 15-5.
5. While holding the band adjusting screw, tighten the locknut securely.

The low and reverse band is adjusted in the same way, except that the adjustment screw is backed off 3 turns. (Check the shop manual). It should be noted that Ford recommends that the locknuts be replaced with

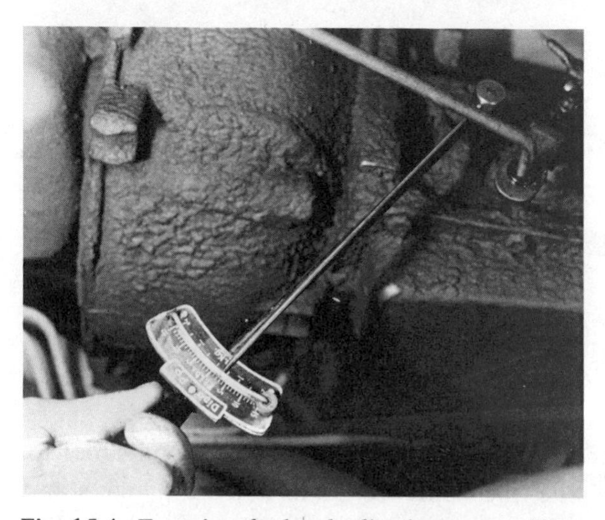

Fig. 15-4 Torquing the band adjusting screw. Note locknut-to-case clearance.

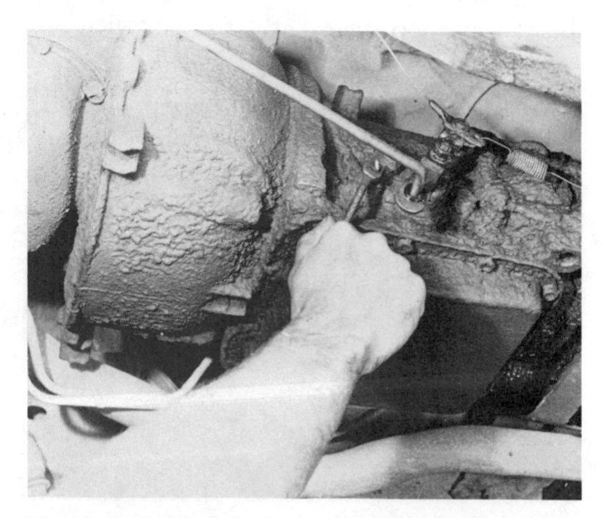

Fig. 15-5 Backing off the band adjusting screw. Note the chalk mark used to help count the number of turns.

new nuts when the bands are adjusted to avoid oil leaking past the adjusting screw.

The steps given here can be used as a guide for adjusting the bands on other makes of transmissions. Always refer to the specifications in the proper shop manual. In some cases it may be necessary to remove the pan or use other special procedures for band adjustment.

Torque Wrench Formula

In some cases, it may be necessary to use an adapter on the end of a torque wrench in order to turn a hard-to-reach band adjusting screw, figure 15-6. Since this increase in leverage results in more torque on the screw, a correction must be made. For example, most Chrysler bands call for a torque of 72 lb-in (8.1 N•m). This would be the true torque required on the screw. With the special Chrysler adapter, a torque of 47 lb-in. (5.3 N•m) is applied.

If special adapters are not available, an open-end box wrench can sometimes be used or special adapters made up, figure 15-7. If any type of adapter is used, a correction must be made to arrive at the true torque value.

Since the torque on a bolt or screw is directly proportional to the force applied and the length of the lever arm (T = F X D), the problem can be set up as a simple proportion. For example, if the true torque required is 50 lb-in, the effective length of the torque wrench is 7 in, and the effective length of the adapter is 4 in, the torque reading necessary to obtain a true torque of 50 lb-in can be found as follows:

Fig. 15-6 Open-end wrench being used as an adapter on the end of a torque wrench.

7″ + 4″ = 11″ total length of the lever arm

$$11:7 = 50:X$$

$$\frac{7 \times 50}{11} = 31.8$$

Thus, the torque reading on the wrench must be 31.8 to obtain a true torque of 50 lb-in on the screw. The mechanic would round this to 32 lb-in when making the reading.

Another formula that can be used is:

$$\frac{A + B}{A} = F \qquad \text{and } T \div F = C$$

or

$$\frac{A}{A + B} = F \qquad \text{and } T \times F = C$$

Where: A = effective length of torque wrench

B = effective length of adapter

F = factor

T = true torque

C = scale reading on torque wrench

Using the same wrench and adapter and a specification of 50 lb-in, the scale reading can be found as follows. (See figure 15-8.)

$$\frac{7 + 4}{7} = 1.57 \text{ and } 50 \div 1.57 = 31.8$$

or

$$\frac{7}{7 + 4} = .636 \text{ and } 50 \times .636 = 31.8$$

With this method it is possible for the mechanic to make adapters and stamp or mark them with the proper factor. Provided they are used on the same length

Fig. 15-7 Torque wrench and adapters.

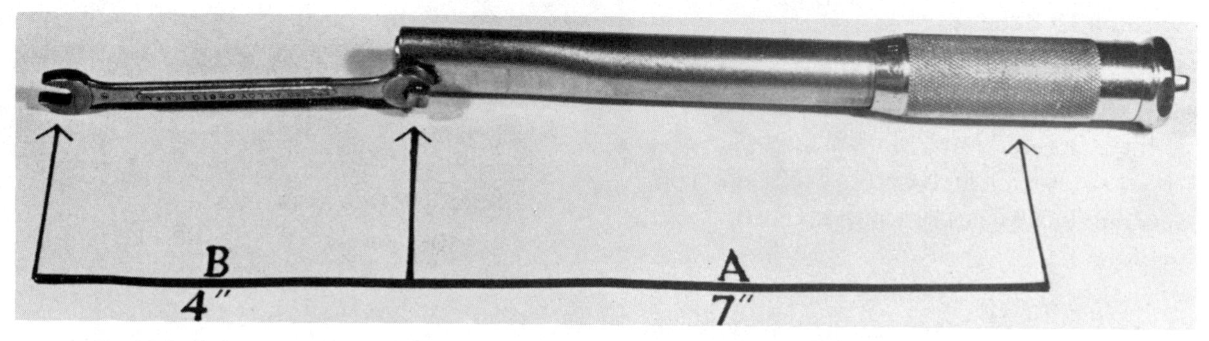

Fig. 15-8 Unless marked on the wrenches, the measurements for effective length are taken as shown.

torque wrench they can be used easily to calculate the scale reading for any required true torque.

Linkage Adjustments

Due to wear, both the manual valve linkage, and throttle linkage require periodic adjustment. The exact linkage adjustment for the various makes and models of cars can be found in the proper shop manuals. As a general rule, the manual linkage should be adjusted so that when the shift quadrant indicates the drive range, the correct manual valve detent is engaged.

If the neutral safety switch is in good condition and properly adjusted, a quick check of manual valve linkage and adjustment can be made by turning the ignition switch to the start position in neutral and park. The starter should engage without having to move the shift lever up or down. Any movement of the shift lever necessary to engage the starter could mean improper adjustment, linkage wear, or both.

Throttle linkage adjustment is of two kinds:

- throttle valve linkage (found in Chrysler TorqueFlite transmissions), and
- downshift valves (found in Ford and General Motors transmissions).

Before any throttle linkage adjustments are carried out, it is necessary to adjust engine idle speed to specifications.

Adjustments to TorqueFlite linkage are usually made by moving the transmission TV (throttle valve) rod to the forward, or minimum TV pressure position at the correct engine idle speed. The throttle is then moved to the wide-open position to be sure that there is no binding, and that the TV rod has moved the kickdown valve to the wide-open throttle position.

The linkage-type downshift valve should be adjusted to the wide-open position with the throttle pressed firmly to the floor of the car. To adjust solenoid-operated downshift valves, a test light is connected to the switch on the accelerator linkage. The test light should come on just as the accelerator pedal makes firm contact with the floor.

Vacuum Modulators

Vacuum units or modulators should be checked for proper vacuum with the engine idling. At sea level, the vacuum should be about 18 to 20 in Hg (61 to 67 kPa). The vacuum is checked by using a T fitting at the modulator.

Low vacuum readings can mean a faulty modulator, a vacuum leak, or poor engine performance. This can be checked by removing the T fitting, and attaching the vacuum gauge directly to the vacuum line. If vacuum is still low, poor tune-up or a vacuum leak in the line is indicated. A normal reading indicates a leaking modulator diaphragm. A leaking diaphragm can also be detected by the presence of transmission fluid in the vacuum hose or connection on the modulator. The presence of fluid in the vacuum hose or modulator can also be checked with a pipe cleaner.

A further check of the modulator can be made by using an outside vacuum source, figure 15-9. The modulator is removed, and a vacuum pump is connected to the vacuum port. The push rod or valve should move in at about 15 to 20 in Hg (51 to 67 kPa), and the vacuum gauge should hold steady. Failure to hold vacuum indicates a defective diaphragm, and the unit should be replaced. The vacuum advance tester of a distributor scope may also be used for modulator testing, figure 15-10. The tester is set to about 18 in Hg (61 kPa) with the thumb closing off the vacuum hose. The vacuum hose is then attached to the modulator, and the same vacuum reading should be obtained.

Modulators should also be checked for correct spring and bellows force. General Motors modulator spring and bellows force is checked by the use of a gauge pin and a new or good modulator of the correct type, figure 15-11. Gauge pins are made of round or flat stock measuring 11/32 or 3/8 inch (8.7 or 9.5 mm) in diameter and 1 inch (25.4 mm) in length and have a line scribed in the exact center. The gauge pin is installed in the sleeve end of each modulator as shown in the illustration. Pressure is then placed on each modulator to compress the springs. The scribed line should stay within 1/16 of an inch (1.6 mm) of one modulator sleeve, when it is just touching the other sleeve end.

Ford vacuum units may be checked by using an accurate platform scale and the throttle valve push rod, figure 15-12. The vacuum unit and push rod are placed on

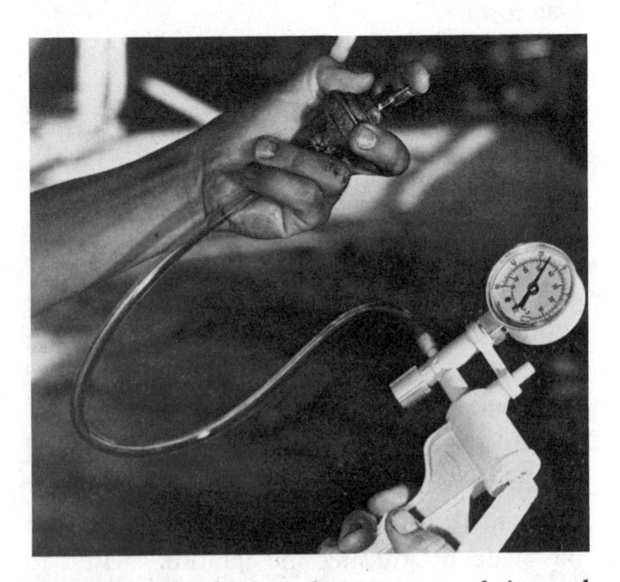

Fig. 15-9 A hand-operated vacuum pump being used to check a modulator.

(A) Setting the vacuum tester.

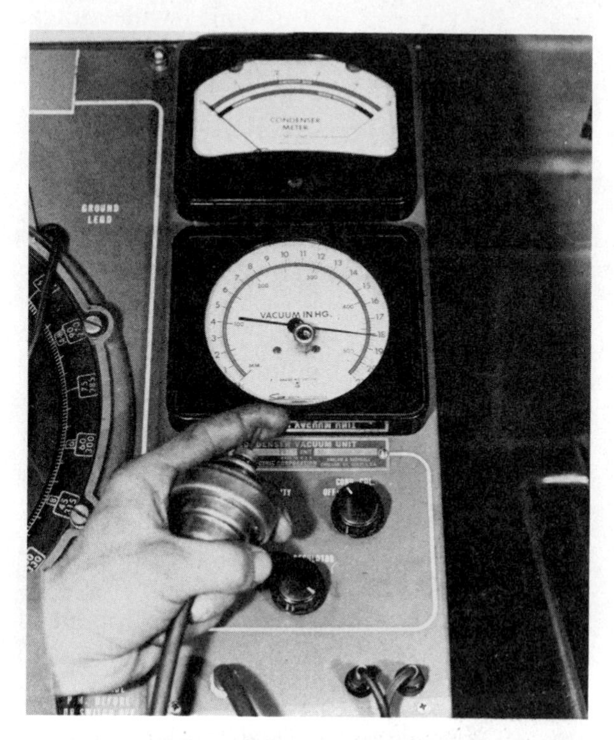

(B) Testing the modulator.

Fig. 15-10 Use of a vacuum advance tester of a distributor scope.

the scale and pressure is applied until the push rod just begins to move. (A pencil mark scribed on the push rod will show this.) The scale reading is noted at this point. Depending on the transmission model, the scale reading should be between 8 and 12 pounds (3.6 and 5.4 kg).

Original units that do not meet specifications should be replaced. Replacement units

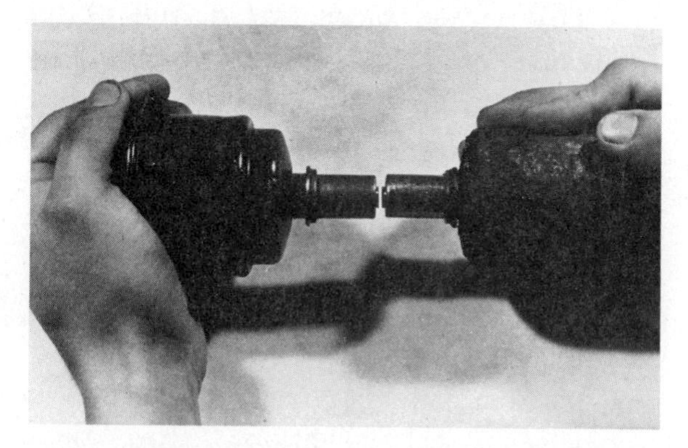

Fig. 15-11 Checking General Motors modulator spring and bellows force.

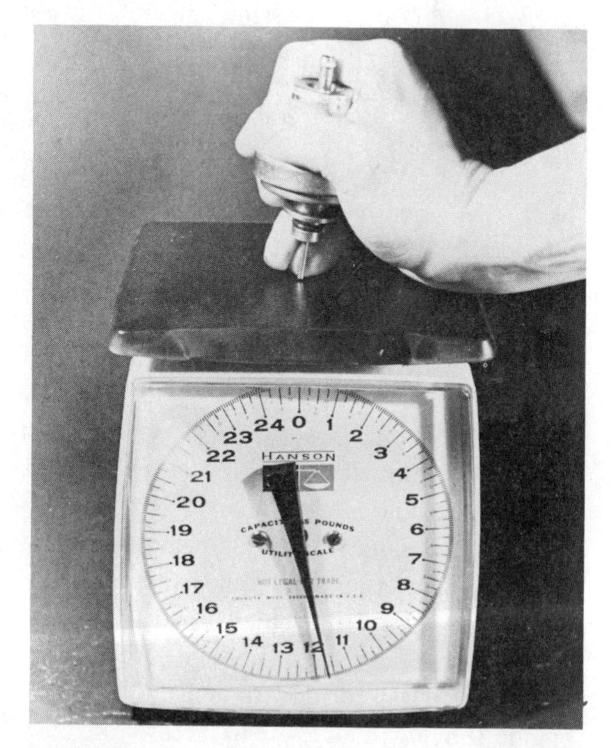

Fig. 15-12 Checking Ford vacuum unit (modulator) spring force.

are usually adjustable, and should be checked and adjusted before installation. Remember, more spring force means later shift points; less spring force means earlier shift points. Line pressure should always be checked after each adjustment of the vacuum unit to be sure it is not below the minimum or above maximum pounds per square inch specified.

Low engine vacuum at the modulator can cause late, harsh shifting. The cause of low vacuum must be detected and corrected for proper transmission performance. Vacuum leaks can occur at the modulator hose, vacuum fittings at the intake manifold, and at vacuum operated accessories such as heater controls, headlight doors, or power brake units. Other sources of low vacuum may be traced to engines in poor mechanical condition (valves, rings, and so forth), or an engine in need of a tune-up. The mechanic should not overlook any of these possible trouble spots when searching for causes of poor transmission performance.

TRANSMISSION TUNE-UP

When good mechanics consider an engine tune-up, they think in terms of making the engine run better or solving a performance problem for the customer. Changing plugs and points is of little value if the engine won't start or perform properly due to bad valves or piston rings. The alert mechanic checks everything that could affect engine performance and reliability.

By the same token, it does little good to change transmission fluid and filter or adjust bands and linkage if a sticking shift valve or a leaking clutch seal is overlooked. The goal of transmission tune-up, like that of engine tune-up, should be to restore like new, or better than new, performance. With this goal in mind, the following points should be considered when undertaking a transmission tune-up:

- Verify customer's complaints (if any).
- Check fluid level and condition.
- Stall test the vehicle.
- Give the vehicle a short road test.
- Change the fluid and filter.
- Check for external fluid leaks.
- Make air pressure test for internal leaks.
- Check band adjustment.
- Check linkage adjustment (including modulator and vacuum test).
- Make fluid pressure test.
- Make final road test.

From this list, it can be seen that a tune-up is actually a series of diagnostic or trouble-shooting tests and adjustments. Some of these tests are covered in this unit, while others are described in following units. At this point, keep in mind that the goal of any automatic transmission service is to maintain, restore, or improve performance.

TOWING CARS WITH AUTOMATIC TRANSMISSIONS

It is sometimes necessary to tow automatic transmission equipped cars to the shop for service. In this case, certain precautions should be taken to prevent damage due to lack of lubrication.

Since the modern transmission uses only a front pump, no lubrication is provided unless the engine is running. When towing is necessary, most manufacturers recommend that speeds be kept below 30 mph (48 km/h) for a distance of not over 25 to 50 miles (40 to 80 kilometers). However, transmission damage may occur when cars are towed as short a distance as 15 to 20 miles (24 to 32 kilometers).

The safest way to tow an automatic transmission equipped car is to remove the drive shaft or use a tow truck and rear-end pickup to raise the rear wheels from the pavement. When either of these methods are used, no damage will result to the transmission, and the car can be towed any distance.

DRIVING A VEHICLE WITH AN AUTOMATIC TRANSMISSION

When driving automatic transmission equipped cars, most drivers use the D position for all forward speeds. The D range actually does give the best all-around performance and economy for normal driving conditions. However, it is sometimes necessary or desirable to use other selector positions.

Of course, to back up, the driver moves the shift lever to R. However, a shift to R should not be made while still moving forward, nor should a shift be made to D, 2 or 1 while still moving in reverse. To do so places undue strain on the transmission and drive train, and could result in early failure of some drive train parts.

Engine Braking

When slowing down on steep hills in a car equipped with a standard transmission, many drivers use a lower gear, figure 15-13. With an automatic transmission, however, most drivers do not use this feature and may be unaware that this feature can be used. Failing to use a lower gear may result in the brakes wearing faster and being subject to fading. This could lead to loss of control and an accident.

Caution: With any transmission, whether standard or automatic, shifting to a lower ratio at too high a road speed could cause damage to the engine from overspeeding or place too much strain on the drive train. For these reasons, lower manual ranges should not be used at road speeds above wide-open throttle shift points.

Snow and Ice

Driving on snow and ice in a vehicle equipped with an automatic transmission is much the same as driving a vehicle with a

Fig. 15-13 Under these conditions, the wise driver will shift to a lower transmission ratio.

standard transmission under the same conditions. That is, wheel-spinning starts should be avoided, and a lower gear or range should be used when going down hills. For some drivers, avoiding wheel spin is not as difficult with an automatic as with a standard. The torque converter provides some slippage and all that is needed is a slow, gentle push on the gas pedal. The second gear hold on some transmissions may also help to prevent wheel spin on snow and ice.

At times, it may be necessary to *rock* the car when stuck in deep snow. The best way to do this is to shift between reverse and manual low, since manual low gives the extra holding power of a band or clutch. This avoids strain on the one-way clutch which would occur if drive range were used.

Caution: To avoid too much strain on the transmission and drive train, the driver should release the gas pedal and make sure that the wheels and car have stopped before shifting back and forth between reverse and low.

TOWING HEAVY LOADS OR TRAILERS

Before using an automatic transmission equipped car or truck to haul heavy loads or pull trailers, the driver should consult the owners manual, or check with a reputable dealer or mechanic as to maximum loads and the availability of heavy-duty or trailer-towing accessories. Most automatic transmissions are strong enough to take the added strain of hauling light loads or small utility trailers. However, when hauling heavy loads, heat buildup can lead to engine or transmission failure.

Heavy-duty clutches and bands, high-volume fans, shift or programming kits, torque converters, and auxiliary coolers are available in towing packages or for off-the-road recreational vehicle (RV) use. (See unit 18.) Whether or not a towing package is installed, the manual ranges should be used on steep hills or when getting under way to lessen strain on the one-way clutch or clutches.

SUMMARY

For the most part, the modern automatic transmission is a trouble-free unit. Hence, many owners and mechanics tend to neglect regular servicing. However, regular service is an important factor in transmission life and performance. Under hard use, the transmission may need even more frequent service such as linkage adjustments, band adjustments, fluid and filter changes, or complete transmission tune-up.

From time to time, customers need advice on service and use, or intended use, of their automatic transmission equipped vehicles. Owner's manuals and service intervals are intended as guidelines only and may not apply in situations of hard use. In these cases, the customer must turn to a reliable mechanic for advice.

REVIEW

Select the best answer from the choices offered to complete the statement or answer the question.

1. A car with an automatic transmission must be towed a long distance to the shop.

Mechanic A wants to remove the drive shaft before towing.
Mechanic B wants to use a tow truck and tow with the rear wheels clear of the road.
Which mechanic is right?

a. A only
b. B only
c. both A and B
d. neither A nor B

2. When going down a steep hill with an automatic transmission equipped car, it is best to

a. shift to N to avoid wear on the bands.
b. shift to R for added engine braking.
c. shift to P in order to slow down.
d. shift to 1 or 2 to avoid wear on the brakes.

3. The manual linkage of an automatic transmission is usually adjusted

a. in D position.
b. at the shift quadrant.
c. every 10,000 miles.
d. to give 18 inches of vacuum at idle.

4. Mechanic A says that screen-type filters should be rinsed and reused. Mechanic B says filters should be replaced every other fluid change. Which mechanic is right?

a. A only
b. B only
c. both A and B
d. neither A nor B

5. For dependability and long life, servicing of automatic transmissions

a. should be carried out every 12,000 miles.
b. should be timed to service conditions and use.
c. should be carried out every 25,000 miles.
d. is not necessary during the life of the transmission.

6. Mechanic A says a fluid level that is too high could cause a clutch or a band to slip.
Mechanic B says that a fluid level that is too low could cause a clutch or a band to slip.
Which mechanic is right?

a. A only
b. B only
c. both A and B
d. neither A nor B

7. Before using a car equipped with an automatic transmission to tow a trailer, the owner should first

a. install a high-volume fan.
b. install an auxiliary cooler.
c. install a shift kit.
d. check the owner's manual for recommended loads and intended use.

8. Mechanic A says that, before band adjustment, the adjusting screw should be backed off 5 turns.
   Mechanic B says that for band adjustment, you must have special tools.
   Which mechanic is right?

   a. A only
   b. B only
   c. both A and B
   d. neither A nor B

9. If stuck in a snow bank, the driver of an automatic transmission equipped car should

   a. remove the drive shaft before towing the car out of the snow bank.
   b. call a tow truck.
   c. try rocking the car between D and R.
   d. try rocking the car between L and R.

10. In general, the fluid level in an automatic transmission should be

    a. slightly above the FULL mark when cold.
    b. between the ADD and FULL mark when hot.
    c. slightly below the ADD mark when hot.
    d. checked with the transmission in D.

11. A mechanic finds that he must use an adapter on the end of a torque wrench in order to reach a band adjusting screw. The true torque required is 50 lb-in, the torque wrench is 8 inches long, and the adapter is 2 inches long, what scale reading should the mechanic use on the torque wrench?

    a. 40 lb-in
    b. 12.5 lb-in
    c. 10 lb-in
    d. 62.5 lb-in

12. A car with an automatic transmission has 12,000 miles on the odometer. The fluid has a clear, red color and no odor, but the level is slightly low. The customer says that the transmission seems to be operating properly.
    Mechanic A says the fluid and filter should be changed.
    Mechanic B says that fluid should be added to the proper level and the transmission checked for leaks.
    Which mechanic is right?

    a. A only
    b. B only
    c. both A and B
    d. neither A nor B

13. In general, downshift linkage should be adjusted

    a. at wide-open throttle.
    b. every 5,000 miles.
    c. at closed-throttle position.
    d. to give 18 inches of vacuum at wide-open throttle.

14. The fluid in an automatic transmission is at the proper level but has a dirty black color and a burned odor.

Mechanic A says that a burned band could be the cause.
Mechanic B says that a burned disc clutch could be the cause.
Which mechanic is right?

a. A only
b. B only
c. both A and B
d. neither A nor B

15. The fluid in an automatic transmission is at the proper level but has a clear, brown color and a sour odor.
    Mechanic A says the fluid and filter should be changed and the transmission should be road tested for proper performance.
    Mechanic B says the transmission should be overhauled.
    Which mechanic is right?

    a. A only
    b. B only
    c. both A and B
    d. neither A nor B

16. Mechanic A says that the fluid level in an automatic transmission should be at the ADD mark when hot.
    Mechanic B says the fluid level should be checked in drive when the engine is cold.
    Which mechanic is right?

    a. A only
    b. B only
    c. both A and B
    d. neither A nor B

17. When adjusting a vacuum modulator, the mechanic must keep in mind that

    a. more spring force means earlier shift points.
    b. line pressure should be checked after each adjustment.
    c. less spring force means later shift points.
    d. shift points will not change.

18. Mechanic A says vacuum modulators can be tested by comparing spring and bellows force with a new modulator.
    Mechanic B says vacuum modulators can be tested with a vacuum pump and gauge.
    Which mechanic is right?

    a. A only
    b. B only
    c. both A and B
    d. neither A nor B

EXTENDED STUDY PROJECTS

1. In the proper shop manuals, find the fluid-checking steps for some of the different makes of domestic and foreign transmissions. Note the differences.
2. Perform fluid level and condition checks on as many different makes of transmissions as possible.

3. Using the steps outlined in this unit and a shop manual, perform at least one band adjustment.
4. Using a shop manual, perform a manual linkage adjustment on at least one make of transmission.
5. Make at least one automatic transmission throttle linkage adjustment on a Chrysler-built car and road test for proper operation.
6. Make downshift linkage adjustments on General Motors and Ford-built cars and road test for proper operation.
7. Test several vacuum modulators for spring and bellows force and compare results.
8. Test several vacuum modulator diaphragms by using a vacuum gauge and T fitting. Repeat these tests using an outside vacuum source.
9. Road test several cars with automatic transmissions noting the following:
   a. Closed or minimum throttle upshift and downshift points.
   b. Medium throttle upshift points.
   c. Heavy throttle (through detent) 1-2 upshift point.
   d. Wide-open throttle (through detent) 1-2 upshift point.
   e. Torque demand (to detent) downshift points.
   f. 3-2, 2-1, and 3-1 detent downshift points.
   g. Downshift points from D to manual 2 and manual 1.
   h. Quality of manual upshifts from 1 to 2 to D at various throttle openings.

Caution: To avoid exceeding the speed limit, make only the 1-2 upshift test at heavy and wide-open throttle positions.

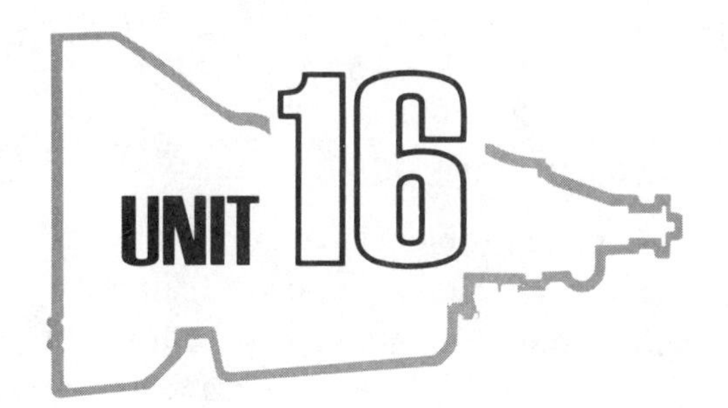

AUTOMATIC TRANSMISSION DIAGNOSIS

OBJECTIVES

After studying this unit, the student should be able to:

- Interpret manufacturers' diagnosis guides.
- Apply automatic transmission theory to transmission diagnosis.
- Use the senses of sight, smell, touch, and hearing in automatic transmission diagnosis.
- Perform diagnosis procedures in logical order.

Diagnosis is a logical, step-by-step plan designed to help the mechanic close in on the cause of a specific problem. When a doctor diagnoses an illness, he or she notes the patient's symptoms. This may include inquiring about past and present illnesses, and, among other steps, testing the patient's heart, lungs, blood pressure, temperature, and pulse.

In the same way, the trained mechanic diagnoses automotive problems by using a series of tests and checks. The diagnostic steps may include the use of diagnostic guides, information from the customer, fluid and linkage checks, stall testing, road testing, and pressure testing.

DIAGNOSIS GUIDES

The diagnosis guides supplied by the automobile manufacturers provide a logical, step-by-step plan to solve transmission problems. A typical guide is shown in figure 16-1.

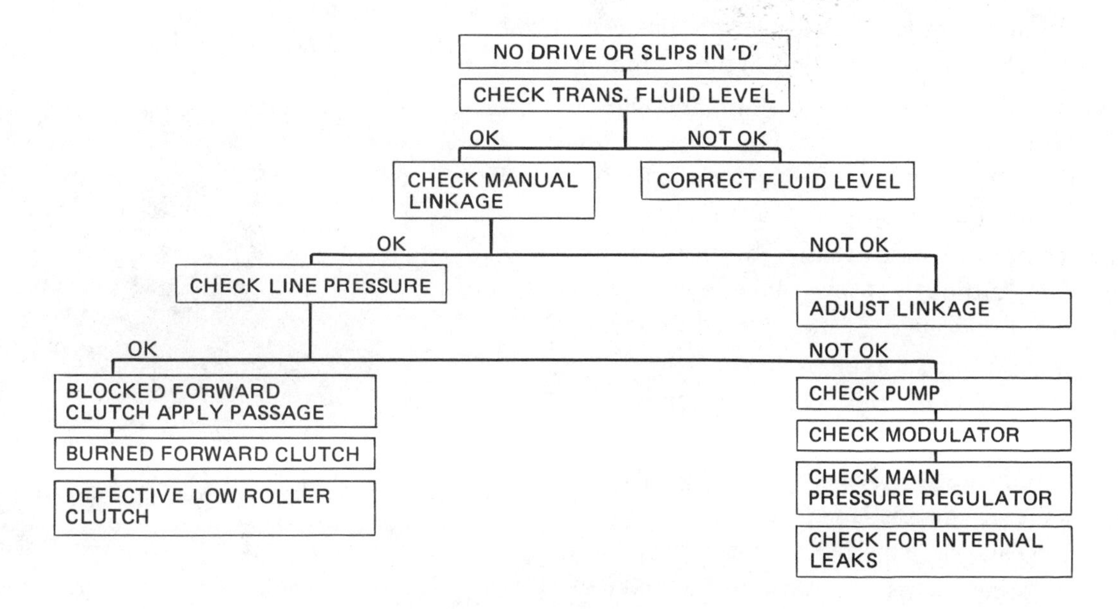

Fig. 16-1 A typical diagnosis guide.

The manufacturers' diagnosis guides are based on solid theory and can be a worthwhile aid to the beginning mechanic, as well as to the experienced mechanic who may not be familiar with a specific transmission. In using these guides, the mechanic should start at the beginning and go through each step until the problem is found. Very often, this takes a great deal of time. The experienced mechanic who is well-trained in transmission operation, may omit some or all of the steps and still come up with the correct answer to the problem. By reviewing the theory of operation of the specific transmission, the mechanic can visualize the power flow and the hydraulic units involved in each range.

For example, in transmissions using the Simpson gear train, either a faulty one-way clutch or forward clutch could be the cause of no drive in D. By trying manual low, the mechanic narrows the problem to either the forward clutch or the one-way clutch. If the problem turns out to be the forward clutch, more checks can be made to find out whether the problem is mechanical or hydraulic.

The use of this reasoning method does not lessen the importance of factory diagnosis charts. They are very valuable to the beginning mechanic, as well as to the experienced mechanic who may not be familiar with the specific transmission being worked on. Following the step-by-step plan shown in the chart will also result in locating the problem, although it may take longer than using a knowledge of the power flow and hydraulics to deduce the problem.

The rest of this unit deals with the application of transmission theory to automatic transmission diagnosis. The steps and checks that are explained are the result of many years of experience in the field and should bring success regardless of the make of transmission being tested.

THREE MAJOR PROBLEMS

Three major problem areas are common in automatic transmissions: no drive (slippage), shift timing, and shift quality. No drive, or slippage, may take place at a standstill or under way in drive, second, first, or reverse. Shift timing problems include early or late shifts, or failure to shift. Shift quality problems include mushy or too soft shifts, or harsh, jerky shifts.

Using the Four Senses to Diagnose Transmission Problems

Four of the human senses – sight, touch, hearing, and smell – are used in automatic transmission troubleshooting. The fluid level and linkage adjustments can be seen. A harsh or too soft shift can be felt. Odd or unusual noises can be heard. A burned clutch or band can be smelled. If there is a "sixth" sense, it is a thorough knowledge of transmission operation. These senses are sharpened by experience and training in the field.

Talking with the Customer

Many times, a problem can be located by talking with the customer, figure 16-2. The mechanic should direct the interview by asking

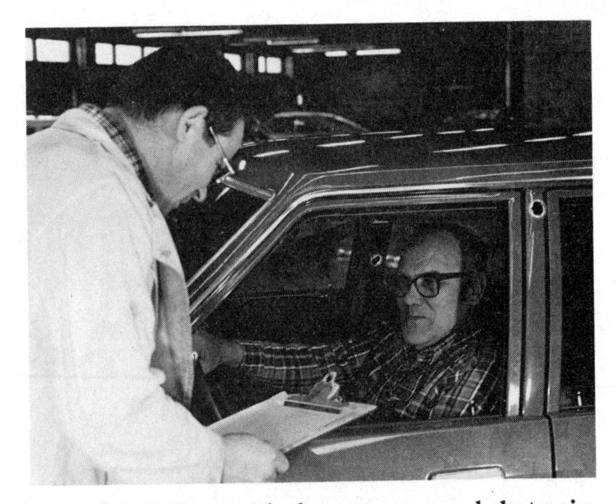

Fig. 16-2 Talking with the customer can help to pinpoint problems if the right questions are asked.

the right questions. For example, if the problem is slippage, the mechanic should ask if the transmission slips in all ranges or just one. By talking with the customer, it may be discovered that what the customer thinks is a problem is actually a characteristic of normal operation. By careful questioning, the mechanic may be able to pinpoint the problem, or know which tests to use next.

Checking the Fluid

After talking with the customer, the trained mechanic normally performs the fluid checks presented in unit 15. The mechanic must remember that an incorrect fluid level can cause slippage, as well as shift timing and quality problems. The appearance of the fluid also tells a great deal about the mechanical condition of the transmission. Further checks or tests are not necessary if the condition of the fluid shows that the transmission needs rebuilding.

Checking Linkage and Engine Tune-up

Remember that linkage problems can cause slippage as well as shifting complaints. Poor engine tuning and the condition of the vacuum lines and modulator can also cause shifting problems. The trained mechanic does not overlook these items when performing linkage checks.

STALL TESTING

Probably the best check for slippage, converter operation, and loss of power is the stall test.

Caution: Stall testing can be dangerous to bystanders and to the mechanic doing the test. Also, if not performed properly, can seriously damage the transmission. For these reasons, stall testing should be done only under the direct supervision of an instructor or experienced mechanic.

Before stall testing, a fluid check should be performed and the fluid brought to the correct level. If the fluid shows signs of a burned clutch or band, a stall test should not be carried out.

To conduct a stall test, a shop manual, a marking pen, and a tachometer are needed. First, the stall speed specifications are found for the engine, transmission, and converter combination being tested, figure 16-3. The correct stall speed is marked on the tachometer, figure 16-4.

ENGINE SIZE (CID)	CONVERTER SIZE (DIAMETER)	STALL SPEED (R/MIN)
170	10-1/4	1400-1600
200	10-1/4	1500-1720
225	10-3/4	1800-2100
250	11-1/4	1620-1820
302	11-1/4	1780-1980
351	12	1520-1720
360	10-3/4	2200-2500
360	11-3/4	1775-2075
440	10-3/4	2600-2900
440	11-3/4	2100-2400

Fig. 16-3 Some typical converter-to-engine size stall speeds. Stall speeds vary with engine horsepower output. (Check shop manual for exact speeds.)

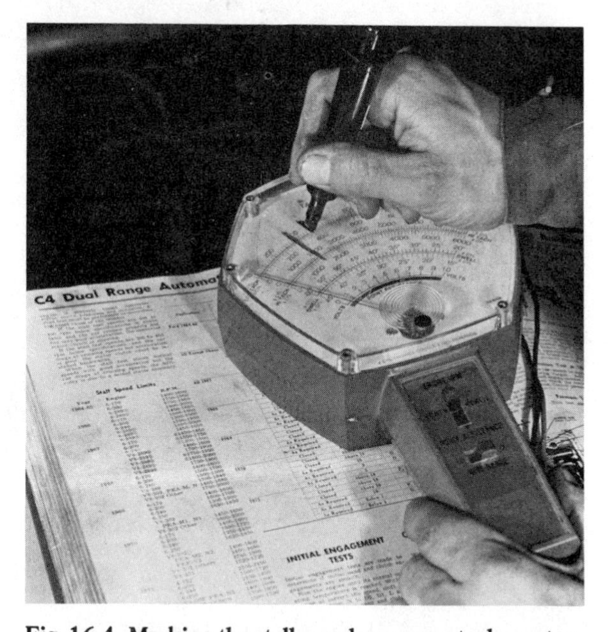

Fig. 16-4 Marking the stall speed range on a tachometer.

The tachometer is connected, and the engine is run to bring the transmission to normal operating temperature (160° to 180°F or 71° to 82°C). Next, the front wheels are blocked, and the parking and foot brake firmly applied. The transmission is then shifted into the range to be tested.

Caution: Be sure that no one is standing at the front or rear of the car while a stall test is being performed.

With the transmission in the range to be tested, the accelerator pedal is pushed firmly to the floor and the highest tachometer reading reached is noted.

Caution: Do not hold the pedal to the floor any longer than it takes for the tachometer needle to come to a steady reading. Under no conditions should it be held more than five seconds. Holding the throttle wide open any longer than this could cause serious transmission damage due to overheating.

In some cars, with large displacement or high performance engines, the rear brakes may not hold under a stall test. In this case, note the highest tachometer reading reached as the brakes slip, and let off on the accelerator pedal at this point, figure 16-5. Next, the transmission is shifted to neutral and run at a fast idle for one minute. This serves to cool the transmission fluid. These steps should be repeated for each range to be tested, remembering to allow the transmission to cool down between each test.

Fig. 16-5 Stall speed within range.

If the engine tends to race above stall speed in any range, let off on the throttle at once. This indicates slippage, and could cause further damage to the transmission. The results of each test should be noted as normal, high, or low stall speeds.

Stall Speed Normal

Normal stall speeds indicate that the clutch and/or band in the range tested is functioning properly; that the engine is well tuned; and that the one-way stator clutch in the torque converter is not slipping. This test does not rule out a seized stator clutch however, and a road test should be performed to check for this problem. (See the road test section in this unit, page 225.)

Stall Speed High

High stall speeds indicate that a clutch or band is slipping. Just which clutch or band is determined by checking the ranges tested with the power flow. In the Simpson gear train, this would be as follows:

- Slips in D, all right in L means low roller clutch problems.
- Slips in D, slips in L means forward clutch problems.
- Slippage in reverse could mean either direct clutch or low reverse band or clutch problems. A road test is needed to find the exact cause (See road test section in this unit.)

The cause of slippage could be either mechanical or hydraulic, and further tests may be needed to close in on the exact cause.

Stall Speeds Low

Low stall speeds indicate one of two things: the engine is in need of a tune-up, or the one-way stator clutch in the torque converter is slipping. Further tests are needed to locate the exact cause of the problem.

ROAD TESTING

The road test is probably the most used but least understood diagnostic test for transmissions. Little is gained by taking a short ride with the transmission in drive. In some cases, all ranges should be tested as well as light, high-torque, and wide-open throttle shift points. Slippage, shift timing, or shift quality problems can be caused by a number of things in the transmission, and the proper type of road test can help pinpoint the cause.

Converter Operation

A slipping converter stator clutch may be identified by a stall test, as outlined earlier. On the road, a slipping stator clutch acts in much the same way. The engine seems to lack power when getting under way because a freewheeling stator will not redirect fluid to the impeller, and no torque multiplication will take place. At cruising speed however, the car appears to run normally and cruise with no effort. This is because during the coupling stage of the converter, the stator is freewheeling anyway, and has no effect on performance.

A seized stator, which cannot be detected during a stall test, would redirect fluid. Torque multiplication and engine performance would be normal when getting under way, or during heavy throttle use. At the coupling stage, however, the seized stator would be in the way of rotary flow. More throttle than normal would be needed to maintain a steady speed, and the engine would seem to lack power when cruising.

When testing for shift problems, always keep in mind that harsh shifts usually mean high pressure, and soft, mushy shifts usually mean low pressure. Also, a harsh shift is usually a late shift, and a soft, mushy shift is usually an early shift. This is due to the effects of modulator or TV pressure on main line pressure.

Light Throttle Shifts

With shift complaint problems, the first tests made are usually those at closed or light throttle shift points. During a shift test, keep in mind that governor pressure tends to upshift, while spring force and TV or modulator pressure tends to downshift or keep the shift valves in the downshift position. At closed or light throttle, there is little TV or modulator pressure, and shift points depend mainly on governor pressure and spring force working on the shift valves. Therefore, light throttle shift tests can be used to check governor action and the fit of the shift valves in their bores.

A shop manual is necessary to check the shift points for the particular make and model of car, transmission used, engine size, axle ratio, and tire size, figure 16-6, page 226. Before performing any tests, the engine and transmission should be brought to normal operating temperature (unless the problem only occurs when the transmission is cold). Sluggish shifting when the transmission is cold should not be mistaken for a problem. Most automatics will shift a bit sluggishly until normal or near normal operating temperature is reached.

ENGINE SIZE 200 CID			AXLE RATIO / TIRE SIZE			
THROTTLE	RANGE	SHIFT	3.25:1 FR 70X14	3.25:1 F70X14	3.00:1 C78X14	3.00:1 F70X14
			mph	mph	mph	mph
LIGHT	D	1-2	8-10	9-10	9-10	9-11
	D	2-3	11-20	12-21	12-21	13-22
	D	3-1	10	10	10	11
	1	2-1	16-25	17-26	17-26	19-29
HIGH TORQUE	D	1-2	17-29	18-30	18-30	20-33
	D	2-3	31-49	33-51	33-51	36-56
	D	3-2	32	34	34	36
	D	2-1 or 3-1	18	19	19	20
WIDE OPEN THROTTLE	D	1-2	26-35	28-37	28-37	30-40
	D	2-3	50-65	53-68	53-68	57-74
	D	3-2	69	72	72	77
	D	2-1 or 3-1	27	28	28	30

Fig. 16-6 Some typical shift points.

To perform a closed or light throttle upshift test, the car is started from a standstill in drive, using just enough throttle to cause a 1-2 and 2-3 upshift. Note the speed at which the upshifts take place. Usually the 1-2 shift occurs between 7 to 12 (11 to 19 km/h) and the 2-3 shift between 8 to 21 mph (13 to 34 km/h). These shift points depend on the conditions noted earlier. In high gear drive, the throttle is backed off to the closed position, and the speed at which a 3-1 downshift takes place is noted.

If both the closed throttle upshift and downshift points are correct, the governor is probably working properly, and the shift valves are not sticking in their bores. Early upshifting at closed throttle could mean that the governor valve is sticking open. In this case, downshifts at closed throttle may not take place at all, or may take place when the car is stopped or almost stopped.

Late shifting at closed throttle can be caused by a governor valve that is stuck closed. Other causes of this problem could be: a vacuum leak or faulty modulator; modulator or TV valve stuck; improper TV linkage adjustment; detent or downshift valve sticking, or linkage out of adjustment; or sticky shift valves. Finally, if either, but not both, the 1-2 and 2-3 closed throttle upshift points are late, sticky shift valves are the most probable cause. A torque demand or high-torque shift point test is used to check the shift delay system (modulator or TV systems).

High-Torque Shifts

With the throttle moved almost to the wide-open position, full modulator or TV pressure should act on the shift valves, and hold off upshifts until very high road speeds are reached.

Caution: At large throttle openings, a 2-3 upshift takes place at very high road speeds –

RANGE	RATIO	INT. CLUTCH (HOLDS SUN)	DIRECT CLUTCH (DRIVES SUN)	FORWARD CLUTCH (DRIVES FT. RING)	LOW AND REVERSE CLUTCH (HOLDS REAR CARRIER)	INT. ROLLER CLUTCH (HOLDS SUN)	LOW ROLLER CLUTCH (HOLDS REAR CARRIER)	INT. BAND (HOLDS SUN)
PARK-NEUT	O	O	O	O	O	O	O	O
DRIVE	FIRST	O	O	X	O	O	X	O
	SECOND	X	O	X	O	X	⊗	O
	THIRD	X	X	X	O	⊗	⊗	O
INT (L^2)	FIRST	O	O	X	O	O	X	O
	SECOND	X	O	X	O	/	⊗	X
LOW (L^1)	FIRST	O	O	X	X	O	/	O
	SECOND	X	O	X	O	/	⊗	X
REVERSE		O	X	O	X	O	O	O

Key: O - not effective X - in use ⊗ - overrunning / - partial hold

Fig. 16-7 Clutch and band use chart – Turbo Hydra-Matic Type 350.

in some cases as high as 75 to 80 mph (121 to 129 km/h). Therefore, only the 1-2 shift point should be checked. Since the shift delay system acts the same on both shift valves, a 1-2 shift test is sufficient. This avoids testing the vehicle at high speeds. Sticky shift valves can be checked with a closed-throttle shift test.

In high gear, high-torque 3-2 or 3-1 downshift should take place at speeds of about 20 mph (32 km/h) less than upshift points. If high-torque shift points take place at the correct speeds, the shift delay (modulator or TV) system is functioning properly.

If high-torque shifts take place early, and light-throttle tests show no problem, the trained mechanic looks for a cause of low modulator or TV pressure. If, on the other hand, high-torque shifts are late, the cause could also be in the downshift (detent) system. A wide-open throttle shift test is used to check this system.

Wide-Open Throttle Shifts

At wide-open throttle, a 1-2 shift should take place at a higher road speed than a 1-2 high-torque shift. From high gear, a 3-2 or 3-1 wide-open throttle downshift should also take place at a higher road speed than high-torque downshifts. If wide-open throttle shifts take place at the correct time, the downshift or detent system is working properly.

If high-torque shift points are high, but wide-open throttle shifts are correctly timed, check for a stuck downshift valve, or downshift linkage or switch out of adjustment.

If high-torque shift points are correct, and wide-open throttle shifts take place at the same speed, check the downshift linkage and valve; or the detent switch, wiring, solenoid, and the detent regulator or detent regulator valve.

If the problem is slippage, a road test can be used to close in on the clutch or band causing the problem. This requires a knowledge of the power flow in each range. A chart such as the one shown in figure 16-7 is helpful in finding a clutch or band problem.

Manual Ranges

After the drive range shift tests have been made, the manual ranges are checked. The chart in figure 16-7 shows the band and clutch use for the General Motors Turbo Hydra-Matic 350 transmission. This chart is used as an example in the explanation of the rest of the road test results. The trained mechanic uses the chart to compare one range to another. For example, it has been pointed out that a cause of no drive in D, but drive in manual low, would be a faulty low roller clutch. A look at the chart shows that a faulty low roller clutch will not affect manual low operation. This is so because in manual low, the low and reverse clutch is holding the rear planetary carrier.

Slippage in Reverse

The problem of slippage in reverse can also be checked using the chart. The chart shows that the cause could be either the direct clutch or the low-reverse clutch. The mechanic checks this by testing for slippage in high gear. Slippage, of course, indicates direct clutch problems. If the direct clutch is functioning properly, the mechanic should try manual low on a steep downgrade. If there is no engine braking, or slippage takes place during engine braking, the problem is in the low and reverse clutch.

1-3 Upshift

A slipping 1-2 shift or 1-3 upshift could be caused by a faulty intermediate clutch or a faulty intermediate roller clutch. This can be checked by selecting manual 2. No engine braking in manual 2 points to intermediate band problems.

In the examples just described, the problems could be mechanical (band or clutch failure) or hydraulic (low pressure). A pressure test is needed to pinpoint the problem.

PRESSURE TESTS

Line pressure for each range must stay within certain limits, and it should increase with increasing throttle opening. Also, in most cases, line pressure will decrease at cruising speed due to governor pressure acting on the modulator or cutback system, figure 16-8.

The details of pressure testing for each individual make and model of transmission cannot be covered here. However, the following methods can be used with most transmissions.

THROTTLE OPENING	MANIFOLD VACUUM (in Hg)	RANGE	PRESSURE (psi)
CLOSED	ABOVE 18	P,N,D	55 - 61
CRUISING	18 to 14	D	55 - 81
MODERATE Line pressure cutback should take place at 10 to 20 mph.	10	D	93 - 101
HEAVY TO WIDE OPEN Line pressure cutback should take place at 30 to 50 mph.	0 to 2	D	142 - 150
CLOSED	ABOVE 18	R	55 - 182
AS REQUIRED, BRAKES SET (FULL STALL)	BELOW 1	R	254 - 268

Fig. 16-8 Typical transmission hydraulic pressures. (Check manuals for exact pressures.)

In any case, pressures and pressure testing steps should always be checked in the correct shop manual.

Equipment for pressure testing includes: a 0 to 400 psi (0 to 2 758 kPa) pressure gauge with the correct fittings for connecting to the transmission; a tachometer; a vacuum gauge; and T fitting or an outside vacuum pump and gauge, such as a distributor tester or a hand pump.

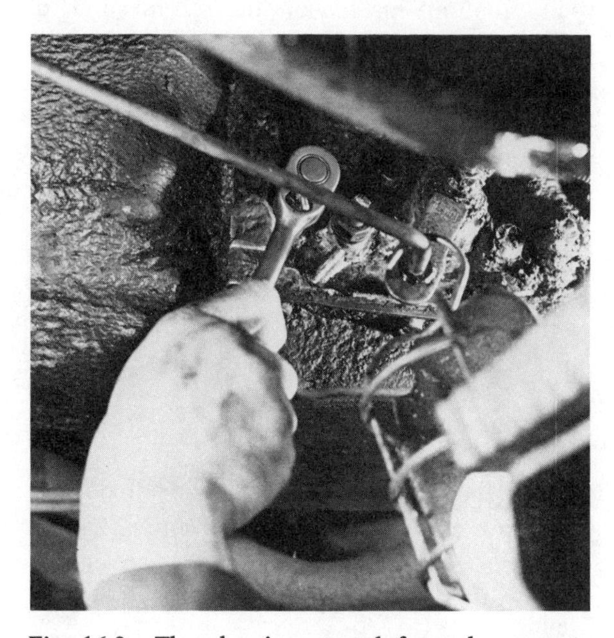

Fig. 16-9 The plug is removed from the pressure takeoff port and the pressure gauge is attached to the transmission.

Two different methods are used for pressure testing:

- the stall method for varying engine manifold vacuum, or
- use of an outside vacuum source connected directly to the modulator.

Test results should be the same whichever method is used. The stall method is the most common, and is presented here. *Note:* In a variation of the stall method, used with some transmissions, engine speeds rather than vacuum readings are used for pressure tests in the various ranges. Check the proper shop manual for this procedure.

Steps in the Stall Test Method

First, the pressure gauge is connected to the pressure takeoff port on the transmission, figure 16-9. Next, the vacuum gauge is connected in parallel with the modulator. This setup allows engine vacuum to act on the modulator and also register on the gauge, figure 16-10. The engine and transmission

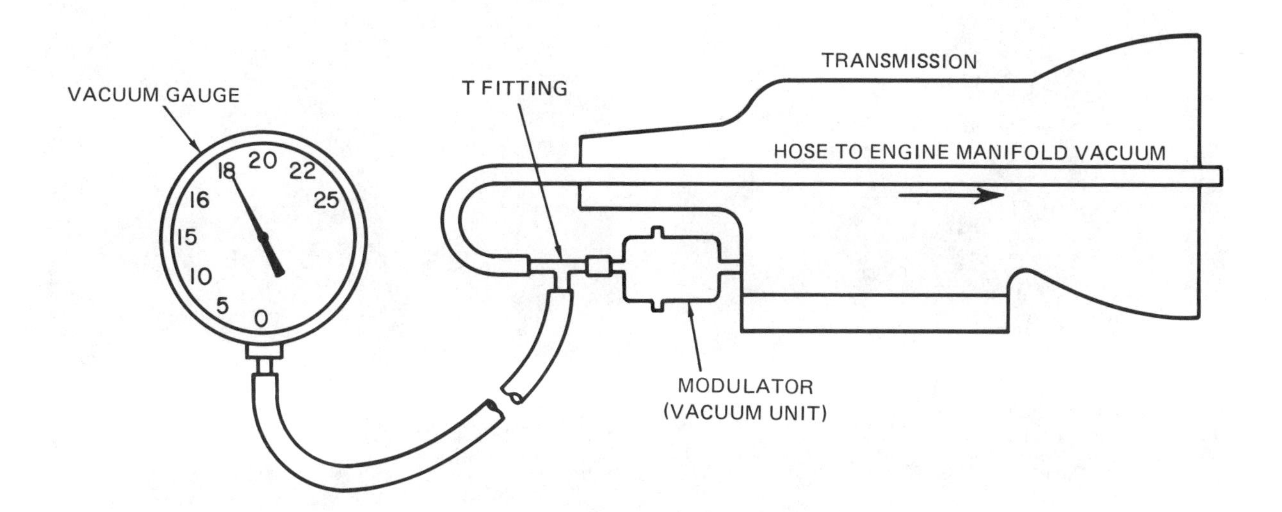

Fig. 16-10 Vacuum hookup for pressure test.

should be warmed up to normal operating temperature, and engine idle adjusted to the correct speed, figure 16-11.

With the engine warmed up to operating temperature and engine idle adjusted to the correct speed, engine vacuum should hold steady at about 18 in Hg (61 kPa) at sea level, and pressure should be within specifications. Low vacuum may be due to a vacuum leak, leaking modulator diaphragm, or poor engine tune-up. This should be corrected before going on with the test. Also, at high altitudes, engine vacuum may register low. For each 1 in Hg (3.368 kPa) drop in vacuum, pressure can be expected to increase about 4 to 9 psi (28 to 62 kPa) depending on make and model of transmission and whether a compensated or non-compensated modulator is used. Check with the correct shop manual.

Line pressure at idle should be checked for each manual range and compared with the correct specifications. Next, line pressure is checked in all ranges at 10 in Hg (34 kPa), figure 16-12, and then at 1 in Hg or less (3.368 kPa.) To do this, the mechanic first selects the range to be tested. (Note: Neutral and park cannot be checked using the stall method.) The foot and parking brakes are set and the throttle is opened just long enough to get the correct vacuum reading and note the pressure – never over five seconds. *Note:* When using a vacuum pump instead of the stall method to check pressure, it may be necessary to set the throttle to give an engine speed of 1,000 to 1,500 r/min during the tests. Some transmission pumps will not produce enough pressure at idle speeds.

Caution: Be sure that the transmission is allowed to cool down between tests by operating at 1,000 r/min in neutral for one minute.

Pressure Test Results

If line pressure is correct in all ranges, the pump and hydraulic system and the seals in the forward clutch, intermediate clutch and/or servo, direct clutch, the low and reverse

Fig. 16-11 Adjusting the idle speed.

Fig. 16-12 Pressure test at 10 in Hg using stall method.

clutch or servo and the accumulators are all right. If pressure is low in any range, compare the clutch and band application to the range being tested. This is much the same procedure as that used on the road test, but now the hydraulic system is being tested. Hydraulic circuit diagrams are very helpful in locating problems when this testing system is used.

For example, low pressure in neutral or park could be caused by the pump, main pressure regulator, modulator or valve body, or case leaks. Using this same method of reasoning, low pressure in drive range low would mean forward clutch problems. Low pressure in manual low, but correct pressure in drive range low, would mean low and reverse servo or clutch problems. Low pressure in reverse, but correct pressure in manual low, would mean direct clutch problems. By comparing the hydraulic system to the power flow, causes of trouble can be further pinpointed.

High pressures could be caused by a sticking main pressure regulator valve. Other causes of high pressure may be found in the modulator or TV circuit, or the downshift circuit. In this case, linkage adjustment should not be overlooked as a possible cause.

Governor Checks

Chrysler TorqueFlite transmissions have a separate takeoff port for use in checking governor pressure. On other transmissions, it is still possible to find out if the governor is operating. To do this, the mechanic must apply basic transmission theory. It is known that governor pressure will not be exerted until the output shaft of the transmission is moving, and governor pressure tends to reduce line pressure. Using this knowledge, the following check can be made.

To make this check, the rear wheels of the car must be raised clear of the floor.

Caution: Be sure the car is securely supported on safety stands or a lift, and never operate at speeds faster than 60 mph (97 km/h).

With the rear wheels clear of the floor, and the foot brakes on, select drive range. Slowly release the foot brakes while watching the speedometer and pressure gauge, figure 16-13.

Fig. 16-13 Governor pressure check.

With moderate throttle use, a drop in line pressure should take place when the governor starts to operate – usually at not over 10 to 20 mph (16–32 km/h). *Note:* Under some conditions, pressure drop may be more noticeable in manual second or first. Also, some transmissions call for a pressure drop of 7 to 10 psi (48–60 kPa) between 1,000 and 3,000 r/min in drive. Check the proper shop manual for the specifications for a particular model.

AIR PRESSURE TESTS

At times, it is necessary to further pinpoint the cause of a clutch or band problem. In most transmissions, this can be done with an air pressure test. Shop manuals show the points at which to apply air pressure to test the clutches and servos, figure 16-14. A dull thud can be heard or felt if the clutch is applying. The servos can be seen as they apply, figure 16-15. Using this method, any leaks will also show up in the clutch seals, servo seals, or case passages.

By applying air pressure to the governor feed port, a governor check can be made. If a sharp click can be heard, the governor is free in its bore.

Caution: Apply air pressure to the clutch, servo, and governor ports only. A shower of transmission fluid could result if air pressure is applied to some of the other ports.

COOLER FLOW TEST

Overheating of the transmission can cause burnouts or shortened transmission life. Overheating can be caused by a plugged cooler or cooler lines, which can be discovered by use of a cooler flow test. Before beginning a cooler flow test, the mechanic must be sure that the fluid level and linkage adjustment are correct, and that line pressure is up to specifications.

Checking Flow from the Return Line

The dipstick is removed, and a clean funnel placed in the dipstick tube. The cooler return line is then disconnected. A neoprene hose of the proper size for a snug fit is slipped over the return line and run to the funnel,

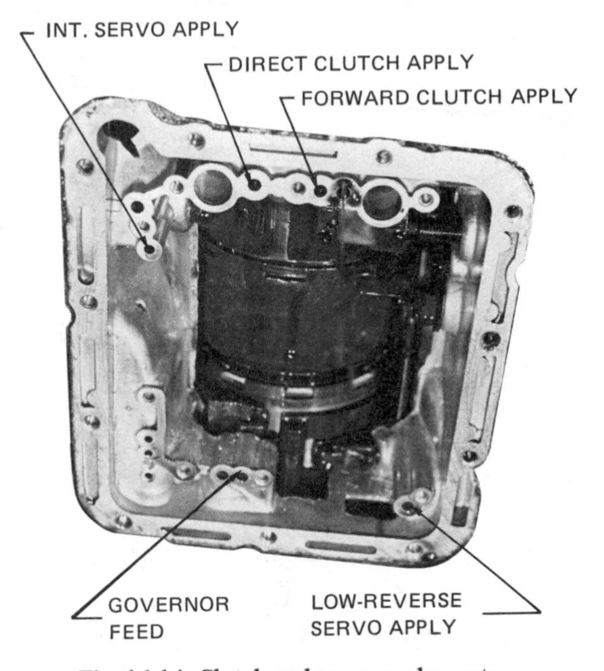

Fig. 16-14 Clutch and servo apply ports.

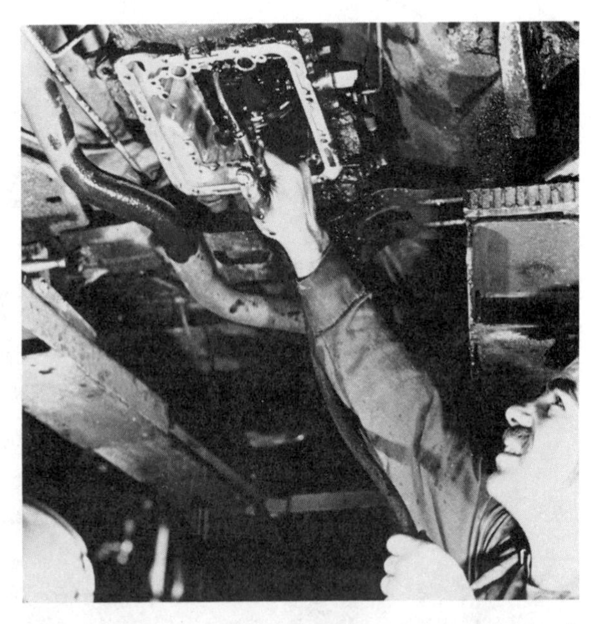

Fig. 16-15 Air pressure testing a forward clutch with a neoprene-tipped blowgun.

Fig. 16-16 Hose attached to cooler for flow test.

Fig. 16-17 Cooler flow test.

figure 16-16. The engine is started and run about 1,000 r/min in neutral. At first, some air will flow through with the fluid. After the air has been bled from the system, the fluid should be free of air and flow in a large, solid stream, figure 16-17.

Caution: Do not run this test for too long a time. Remember, cooler return flow is used for lubrication.

After bleeding, air bubbles in the stream could mean an internal transmission leak. If the flow seems small, the cooler may be partially blocked with sludge or the pump capacity may be low due to wear. *Note:* Even though pump pressure seems all right, wear could cause a loss of volume.

Checking Flow from the Supply Line. A further check for cooler blockage can be made by checking the flow from the cooler supply line. The cooler return line is reconnected to the transmission, and the same test is repeated with the hose connected to the cooler supply line. A good flow in this setup means that the cooler is partially blocked. If the flow is still poor, internal transmission leaks or a worn pump may be the cause.

FLUID LEAKS

Transmission fluid leaks are sometimes hard to detect and correct. This is especially true when leaks are large or when engine oil leaks are also present. In these cases, the entire transmission and underside of the car tend to become soaked with fluid. This makes it difficult to pinpoint the source of the leak. Steam cleaning the engine, transmission, and underside of the car help in finding the leak. The mechanic should also keep in mind that serviceable transmission fluid is red or amber-orange, while fresh motor oil is a golden color.

Transmission leaks can be caused by a bad seal at the converter, at the output shaft, or where the manual linkage shafts and speedometer adapter enter the transmission. Other sources of leaks are the gaskets at the pump, extension housing, servo covers, or pan. Leaks can also occur at the cooler connections and the pressure takeoff plugs.

Leaks from the dipstick tube can be caused by air pressure buildup due to a plugged vent, or foaming due to an internal leak or an overfull transmission. A customer may complain that the transmission is using fluid, but no visible source of a leak can be found. Here, a good possibility for the source of the problem is the vacuum modulator. A bad diaphragm will allow manifold vacuum to draw transmission fluid into the engine where it can be burned during combustion.

A transmission cooler leak can cause serious trouble. Indication of this problem can usually be found at the radiator filler neck and/or the transmission dipstick. Any leakage of permanent-type antifreeze into the transmission means that at least a complete fluid change and converter and cooler flushing are necessary. A bad cooler leak, or one that has existed for some time, may mean that a complete transmission overhaul will be needed.

SUMMARY

All of the tests presented in this unit may sometimes be necessary to locate a transmission problem. At other times, some steps can be omitted, as long as the cause of the problem is found. Thus, no hard and fast rules for testing can be established. However, it is best to first look for the problems that can most easily be checked and repaired. This includes any problems that do not require the removal of the transmission or any major parts in order to find and repair the trouble.

It has been shown that the wrong fluid level or contaminated fluid can cause many problems, and the first check is made here. In most cases, this is followed by linkage checks and checks of the modulator and vacuum lines if the transmission is equipped with these parts. In checking and testing these simple points, the cause of the problem may be found and repaired. If not, other tests are made with the knowledge that the basic and most common trouble spots resulting in shifting and slippage problems have been checked.

Through experience, the mechanic learns to choose the next diagnosis step. It could be a road test for shift problems, a stall test for slippage, or pressure tests made to close in on hydraulic problems. The test will vary according to the particular problem being handled.

Automobile dealers and garage owners want to hire mechanics with diagnostic skills. Diagnosis is probably the most important skill a mechanic can acquire. The skills needed to repair or replace parts are certainly necessary, but they are of little value if the mechanic cannot locate the part that needs repair. Diagnosis skills can prevent the mistake of rebuilding an entire transmission when only a linkage adjustment is needed. These skills are gained by careful study of transmission theory and by practice of its application to actual transmission problems.

REVIEW*

Select the best answer from the choices offered to complete the statement or answer the question.

1. A mechanic road tests a car having a TorqueFlite transmission by comparing the operation of one range with another. This procedure is based on the theory of

   a. governor operation.
   b. hydraulic flow and pressures.
   c. the power flow of the Ravingeau gear train.
   d. the power flow of the Simpson gear train.

2. During a road test at light throttle, a 1-2 shift takes place at 12 mph and a 2-3 shift at 18 mph.
   Mechanic A says that this test shows that the governor is operating correctly.
   Mechanic B says that this test shows the modulator is operating correctly.
   Which mechanic is right?

   a. A only c. both A and B
   b. B only d. neither A nor B

3. During a stall test, a transmission slips in drive range.
   Mechanic A says this could be caused by a faulty roller clutch.
   Mechanic B says this could be caused by a slipping forward clutch.
   Which mechanic is right?

   a. A only c. both A and B
   b. B only d. neither A nor B

4. A mechanic is pressure testing a transmission in D range and the rear wheels are raised so that they are free to turn. At about 12 mph at light throttle, a drop in pressure takes place.
   Mechanic A says that this means that the governor is working.
   Mechanic B says that this means that the shift delay system is working.
   Which mechanic is right?

   a. A only c. both A and B
   b. B only d. neither A nor B

5. The results of a stall test show 200 r/min under specifications in all ranges.
   Mechanic A says that this could be caused by a slipping clutch or band.

**Note:* To answer some of the following questions, it may be necessary to refer to the units dealing with basic hydraulics and mechanics, and the theory of operation of the various transmissions.

Mechanic B says that this could be caused by a slipping one-way stator clutch.
Which mechanic is right?

a. A only
b. B only
c. both A and B
d. neither A nor B

6. A stall test shows slippage in reverse only, and a road test shows that there is no engine braking in manual low.
   Mechanic A says this could be caused by a slipping forward clutch.
   Mechanic B says this could be caused by a faulty low roller clutch.
   Which mechanic is right?

   a. A only
   b. B only
   c. both A and B
   d. neither A nor B

7. Using the steps outlined in figure 16-1, mechanic A says it is possible to detect a faulty forward clutch.
   Mechanic B says that, in using these steps, it is possible to turn up a faulty low roller clutch.
   Which mechanic is right?

   a. A only
   b. B only
   c. both A and B
   d. neither A nor B

8. A customer complains of late shifting. The mechanic finds the fluid level and condition to be correct. The next step would be a

   a. stall test.
   b. linkage and vacuum check.
   c. pressure test.
   d. air pressure test.

9. In figure 16-1, the purpose of a pressure test is to rule out or detect a faulty

   a. low band.
   b. forward clutch seal.
   c. shift valve.
   d. governor.

10. In solving the problem in figure 16-1, a stall test in manual low could rule out or detect a faulty

    a. low roller clutch.
    b. low band.
    c. direct clutch.
    d. governor.

EXTENDED STUDY PROJECTS

1. Perform a stall test on a car with an automatic transmission and write a report covering the following points:

   a. State the stall speed specifications from the shop manual for the vehicle being tested, engine size (CID), horsepower, and converter size.

b. List the actual stall speeds obtained for each range.
c. Using the stall speeds obtained on the test, write an opinion on the mechanical condition of the transmission and torque converter.

Caution: Perform stall tests only under the supervision of an instructor or a trained mechanic.

2. Perform a road test on a car with an automatic transmission and write a report covering the following points:

a. Shift point specifications from the shop manual, and normal operating characteristics of the transmission being tested.
b. Actual closed-throttle shift points obtained on the test. Include both upshift and downshift points and note shift quality.
c. Actual high-torque 1-2 and 3-1 shift points and shift quality.
d. Actual wide-open throttle 1-2 and 3-1 shift points and shift quality.
e. Actual downshift points in manual 2 and 1.
f. Check for proper park and reverse operation and creep in neutral.
g. Using the road test results, write an opinion on the overall performance of this transmission.

3. Perform a pressure test on a car with an automatic transmission using both the stall and outside vacuum source methods. Write a report covering the following points:

a. Correct pressure specifications for the make and model of transmission being tested.
b. Actual pressures obtained at 18 in Hg for each range. (Engine at idle speed.)
c. Actual pressures obtained at 10 in Hg for each range. (Engine speed as required.)
d. Actual pressures obtained at 1 in Hg or less for each range. (Engine speed as required.)
e. Using the test results, write your opinion of the overall hydraulic system of this transmission.

4. Remove the valve body from an automatic transmission and air pressure test the clutches, servos, and governor. Then answer the following questions:

a. How were you able to tell whether or not the clutches and servos were applying?
b. How were you able to tell whether or not the governor was free?
c. How would it be possible to check for internal leaks using air pressure?

VALVE BODY OVERHAUL

OBJECTIVES

After studying this unit, the student will be able to:

- Differentiate between serviceable and unserviceable valve bodies and valve body parts.
- Disassemble, repair, and reassemble a valve body to factory specifications.
- Make in-car repairs to other automatic transmission subassemblies.

Unit 16 described a variety of tests and checks that help in pinpointing the cause of a transmission problem. When the results of these tests point definitely toward valve body problems, then, and only then, should the valve body be overhauled. Valve bodies may need service because of burnouts, oxidized fluid, or mechanical failures that contaminate the fluid.

Valve bodies are usually serviced at the time of transmission overhaul. However, the valve body can be removed and serviced with the transmission in the car. The fluid is drained and the pan removed. After removing the manual linkage, the valve body can then be unbolted and removed. Keep in mind that contamination means that a complete fluid change and cooler flushing should be done.

To those who are untrained in automatic transmission repair, valve body overhaul may appear to be too difficult to undertake. The possibility of losing or mixing up some of the many valves, springs, check balls, washers, clips, pins, and bolts discourages some mechanics from attempting an overhaul. However, by using the tools and methods that are described in this unit and by practicing this skill, the mechanic can approach valve body overhaul with confidence.

THE NEED FOR VALVE BODY OVERHAUL

Sludge and varnish buildup are probably the main causes of valve body failure. Although they may be badly sludged and varnished from overheating or a burnout, most valve bodies can be made to perform like new with careful reconditioning. For example, the valve body shown in figure 17-1 was overhauled and returned to service. All that was needed was proper cleaning and care in reassembly.

A valve body in need of service must be completely disassembled in order to do a good job of cleaning. In most cases, removing

Fig. 17-1 Although badly sludged, this valve body was reconditioned and returned to use.

one valve at a time will not do the job properly. Sludge and dirt lodged in the passageways is almost impossible to clean out unless every part is removed from the valve body, figure 17-2.

Fig. 17-2 To remove this amount of sludge, the valve body must be completely disassembled.

EQUIPMENT

The following equipment and tools are needed for a valve body overhaul:

- a shop manual showing a clear exploded view of the valve body to be overhauled
- a numbered spring holder
- several trays for holding valve body parts
- a pail of carburetor cleaner
- a strainer basket
- a spray can of carburetor and choke cleaner (varnish and gum solvent)
- hard Arkansas bench stone and/or #600 crocus cloth
- mineral spirits
- hand tools for disassembly and assembly of valve body, figure 17-3.

Fig. 17-3 Only simple tools and equipment are needed for valve body service.

Making a Spring Holder

Although it is possible to sort springs with the use of a spring chart, most mechanics find it faster and easier to separate them as the valve body is being disassembled. This can be done by using a numbered spring holder, figure 17-4. This tool cannot be purchased, but can be easily constructed by the mechanic. A pattern for a spring holder is shown in figure 17-5.

Fig. 17-4 A homemade spring holder in use.

The plans for, and use of the spring holder shown in figure 17-4 and 17-5 are from a service publication developed by Ford Motor Company. The holder is simple to make, inexpensive, and does the job. Most mechanics who have used a spring holder find that the time invested in its construction was well spent.

Materials:
- (2) 4″ round electrical box covers
- (2) 1/4–20x2″ screws, threaded full length
- (2) 1/4–20 nuts
- (1) 1/4–20 wing nut
- (15) 6–32 x 1 1/2″ machine screws, threaded full length

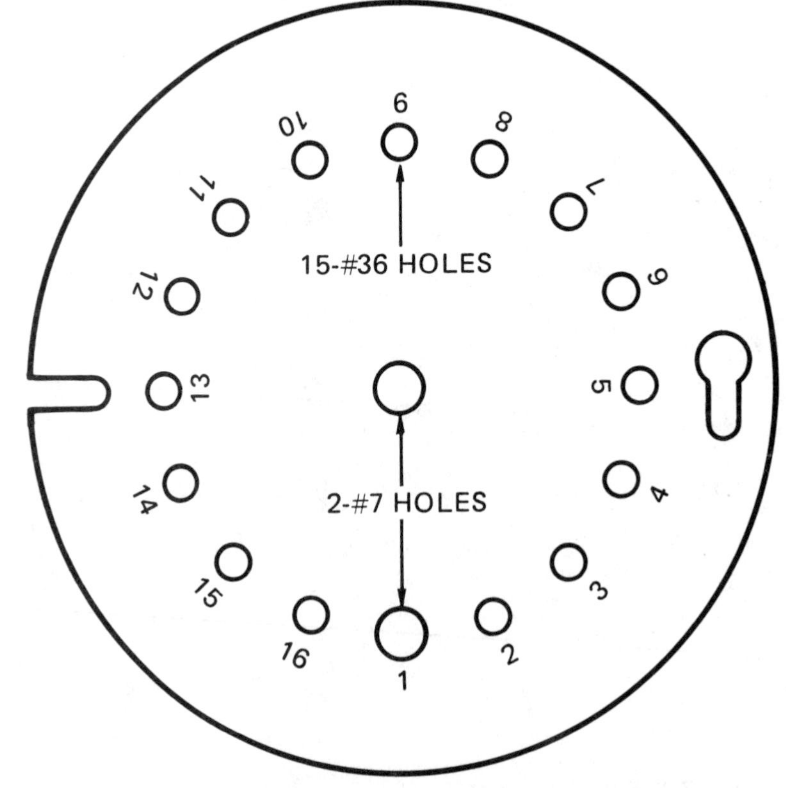

Fig. 17-5 Full-size pattern for spring holder plate (see text for instructions).

Assembly:

1. Trace the pattern shown in figure 17-5 on a piece of paper or cardboard. Cut out the pattern and paste or tape it to one of the electrical box covers.
2. Center punch all holes.
3. Bolt the two plates together, and drill fifteen #36 holes and two #7 holes.
4. Take the plates apart and thread the holes in one plate with a 6-32 tap and a 1/4-20 tap. (Self-tapping screws may also be used.) This will become the bottom plate.
5. Number the holes as shown with a stamp or tool marker.
6. Drill out the holes in the other (top) plate to 5/32" (15) and 1/4" (2).
7. Install the screws in the bottom plate. Use the two 1/4-20 nuts on the center screw to space the top plate so that the 6/32 screws just come through the top plate. Jam the nuts together to hold this adjustment. (In use, the top plate is held in place with the 1/4-20 wing nut. The actual use of the spring holder will be dealt with under disassembly and repair.)

DISASSEMBLY AND REPAIR

Guidelines

When working on the valve body, the following guidelines should be observed:

- After cleaning and during reassembly, parts, tools, equipment, work area, and the mechanic's hands must be clean and free of dirt and grit. The valves are machined to a very close fit in their bores, and small particles of dirt or grit could cause the valves to stick. They must fit free.
- Read the step-by-step instructions in the shop manual. Do not try to work solely by looking at the illustrations. Important assembly information, specifications, and special instructions are found in the text.
- Completely disassemble the valve body.
- During reassembly, torque all screws and bolts.
- Do not clamp the valve body, or any parts of the valve body in a vise. Use a special holding fixture, or place the body on a clean bench while working on it.
- Do not use shop towels to dry valve body parts. Lint trapped in the body could cause sticking valves. Dry with compressed air only.

Preparation

Before disassembly, open the shop manual and locate and correct exploded view for the valve body being worked on. Next, write a number for each of the springs on the exploded view, figure 17-6. Be sure that all of the springs are numbered. Note that in some manuals, the valve body halves may be shown on two different pages.

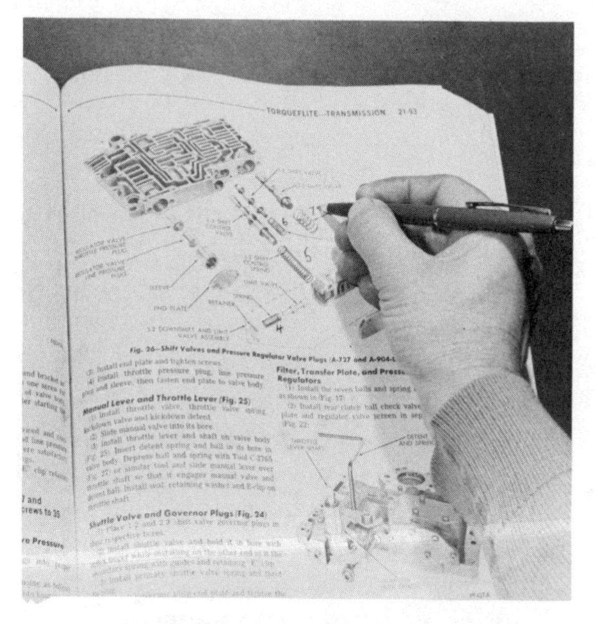

Fig. 17-6 Numbering the valve springs in the shop manual. Each spring must be numbered so that it can be identified for reassembly.

Fig. 17-7 Most valve bodies have two or more check valves or balls.

Removing the Valves

As the valves and springs are removed from the valve body, place the valves in the cleaning basket. Place each spring on the bolt in the spring holder that has the same number as that assigned to the spring on the exploded view.

In some cases it is necessary to separate the valve body halves before all of the valves can be removed. Some valves are held in by pins or clips that can only be removed when the valve body halves have been separated. Follow the steps in the shop manual for this operation. Also be careful not to lose any of the check valves or balls located between the valve body halves, figure 17-7.

Avoid nicking the machined surfaces of the valve body halves or the separator plate. Also, save the old valve body gasket. It will not be reused, but can be compared to the gaskets provided in the overhaul kit so that the correct gasket can be chosen for replacement.

Most valves are under spring tension in the valve body. When removing retaining pins, clips or plates, care must be taken that the springs or valves are not propelled into the

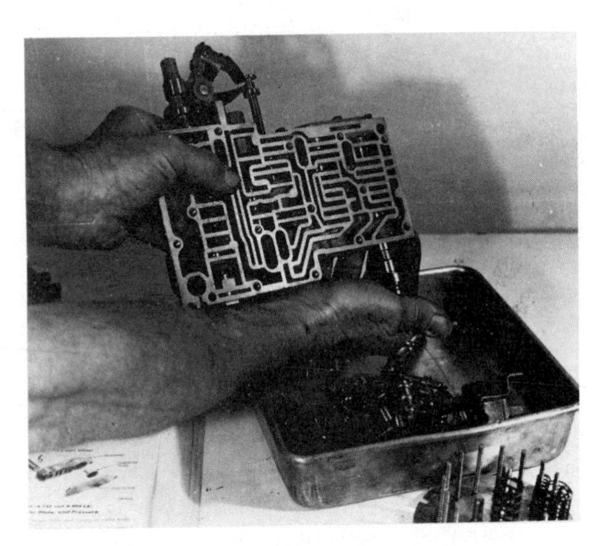

Fig. 17-8 Correct way of removing valves from the valve body.

shop. Most valves, if not too badly stuck, can be removed by giving the valve body a sharp rap with the heel of the hand, figure 17-8. If this does not work, spraying the valve and passages with gum and varnish solvent is helpful. It may be necessary to spray, and work the valve back and forth several times before it is freed. A screwdriver or other tool should not be used to pry the valve from the bore. This could ruin the valve body by causing nicks or cracks in the passages.

Removing Burrs and Nicks

As each valve is removed, it is checked for nicks and burrs. Any nicks and burrs should be removed on a hard Arkansas bench stone or on #600 crocus cloth spread flat on a piece of plate glass, figure 17-9.

Caution: Be sure that the land of the valve is kept flat on the stone or crocus cloth – the sharp edges must not be rounded off.

On some valves, the sharp edge works very close to a port or passage in the valve body and rounding the edges could cause leakage past the valve. The sharp edges also give the valves a self-cleaning action by scraping the bores clean. If the edges are rounded, the valve will tend to ride over the dirt and, in time, stick. Valves with deep scores or nicks that cannot be removed should be replaced.

Small nicks or scratches in the mating surfaces of the valve body can be removed with an Arkansas stone stroked flat across the mating surface. Stoning with a figure eight motion helps to keep the surfaces flat. Nicks can also be removed by rubbing the mating surface on #600 crocus cloth spread flat on a piece of plate glass.

Fig. 17-9 A hard Arkansas bench stone removes nicks without rounding the edges of the valve.

After deburring, each valve is checked for a free fit in its own bore. The valve should move back and forth in its bore of its own weight, or with a very slight tap on the valve body with the heel of the hand. Working the valve back and forth in its bore will sometimes free it up. If the valve does not move freely, there may be varnish, sludge, or imbedded dirt in the valve body. In this case, further cleaning should be done. (Other ways of freeing up the valve are discussed later in this unit.)

When all of the springs have been placed on the spring holder, the spring holder cover should be installed and secured in place with the wing nut.

Cleaning

All valve body parts, with the exception of gaskets and rubber check valves or balls, should be placed in the cleaning basket and soaked in carburetor cleaner for fifteen to twenty minutes.

After soaking, the dissolved sludge and varnish should be washed off with hot water. Compressed air is used to remove most of the water. The rest of the water is removed by rinsing the parts in mineral spirits, figure 17-10.

Fig. 17-10 Water is removed from parts by soaking them in mineral spirits. Note the beads of water in the bottom of the pan.

As shown, the water beads up and settles to the bottom of the pan. As each part is removed from the mineral spirits, it is dried further by blowing with compressed air. It is then placed in a clean tray for reassembly.

REASSEMBLY

In preparation for reassembly, check to see that the following clean parts are at hand: the valve body halves and separator plate; and a tray of valves, clips, pins, and screws. The clean springs are still in place on the spring holder. Every effort must be made to keep all parts, tools, and equipment clean during the reassembly process. The mechanic must have clean hands and a work area that can be kept free of airborne dirt or grit, such as paint overspray or grinding dust.

Before beginning reassembly, read the instructions in the shop manual carefully. On some valve bodies, the valves must be replaced before bolting the valve body halves together. As a general rule, it is usually better to replace the valves after the valve body halves have been bolted together. A sticking valve, resulting from valve body warpage as the screws are tightened, may show up after the halves have been bolted together. The manual will tell how to handle this problem.

When ready to bolt the valve body halves together, select a new gasket that matches the old gasket, figure 17-11. All holes that are in the old gasket must also be in the new gasket and they must line up properly. It is also a good idea to check the new gasket with the valve body separator plate. A final check should be made to see that all varnish has been removed. If there is still varnish present, a few squirts of the gum and varnish solvent, followed by a blast of dry compressed air should remove it. If all valves now fit freely, the valve body halves can be bolted together.

Torquing

The valve body halves should be assembled with the screws finger tight, and then tightened in the proper sequence, and to the correct torque, figure 17-12. (*Note:* Most valve bodies have different length screws, and it is important not to mix them up.) It is advised that the screws be torqued in three stages. If, for example, the correct torque is 30 lb-in (3.39 N·m), the first torque for all screws is to 10 lb-in, the next to 20 lb-in, and the last to 30 lb-in. If no torque sequence is given in the manual, the center screws should be torqued first, working towards each end in turn. A typical torque sequence is shown in figure 17-13. *Note:* The filter or screen is not replaced at this time. All screws are torqued first, including the filter screws. Then the filter screws are removed, a new filter installed, and the screws retorqued.

Installing Valves

Dirt, grit, varnish, and scratches or nicks can prevent the free movement of valves in

Fig. 17-11 Note that the holes in this gasket do not match up with those in the old gasket.

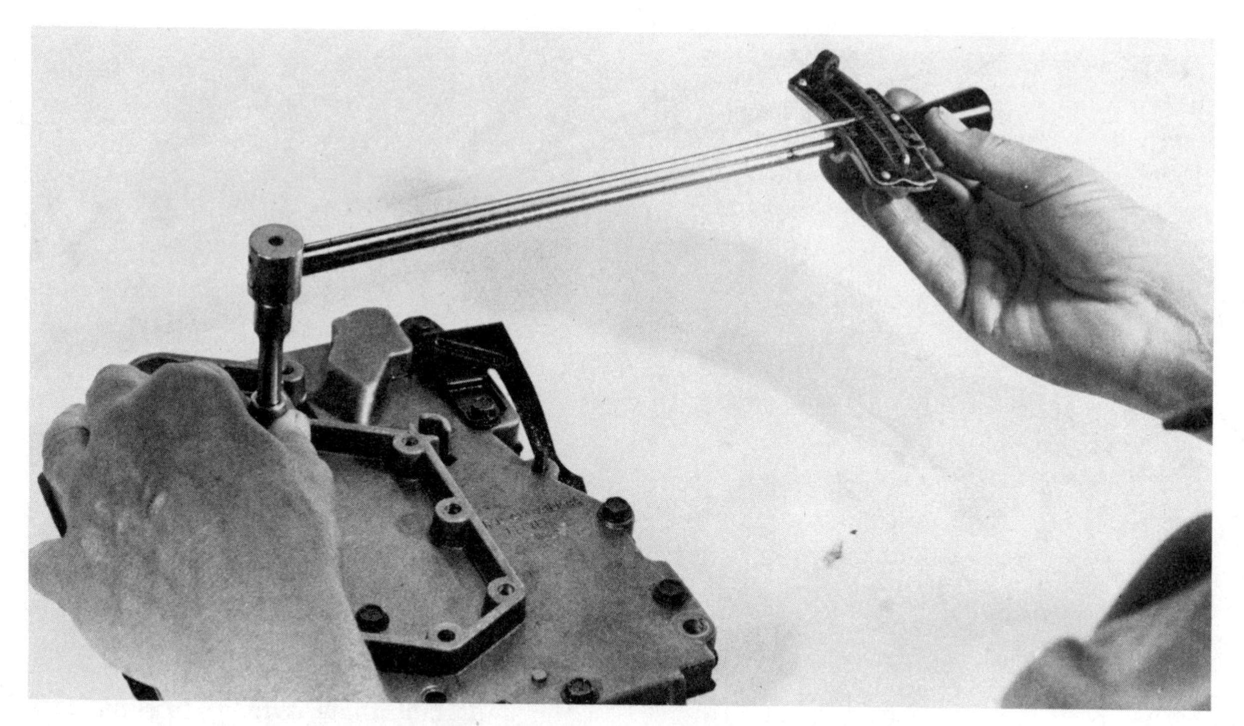

Fig. 17-12 All valve body screws must be torqued properly.

their bores. If, on disassembly, it was found that the valves were sticking, they should be checked and freed up before the valve body halves are bolted together. The valve body should be placed on the bench so that it is facing the same way as in the exploded view in the manual. Next, match up a valve from the tray with the exploded view, figure 17-14. Be sure that the valve is free from nicks and burrs. If it is, place it in its proper bore. It should drop freely of its own weight, or with a slight shake, to the bottom of its bore. When the valve body is turned over, the valve should drop into the mechanic's hand of its own weight. If the valve must be pushed into its bore, no matter how lightly, it is too tight. Check the valve and bore again for signs of varnish, and, if present, remove with gum and varnish solvent.

If the valve sticks and has a dry, gritty feel, small particles of dirt may be partly sunk into the bore. Sometimes these particles can be removed with a piece of rolled up #600 crocus cloth, figure 17-15. Allow the crocus cloth to unroll in the bore – do not use any more pressure than this. Give the cloth a few twists, blow out the bore with compressed air, and try the valve again. Working the valve back and forth a few times may help free it up.

If these methods do not work, dirt is probably imbedded in the valve bore. This cuts down on the clearance and causes the

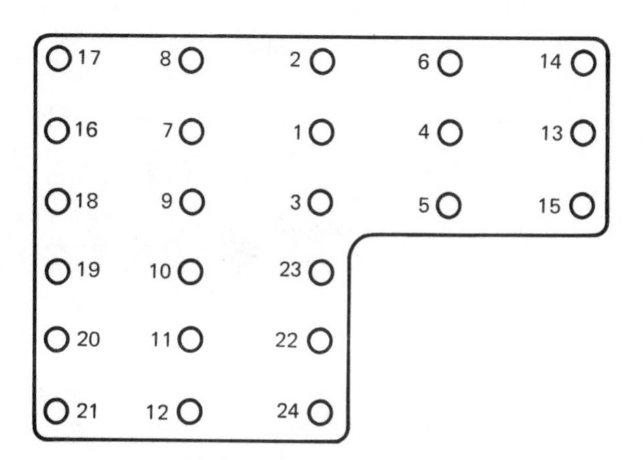

Fig. 17-13 Typical torque sequence.

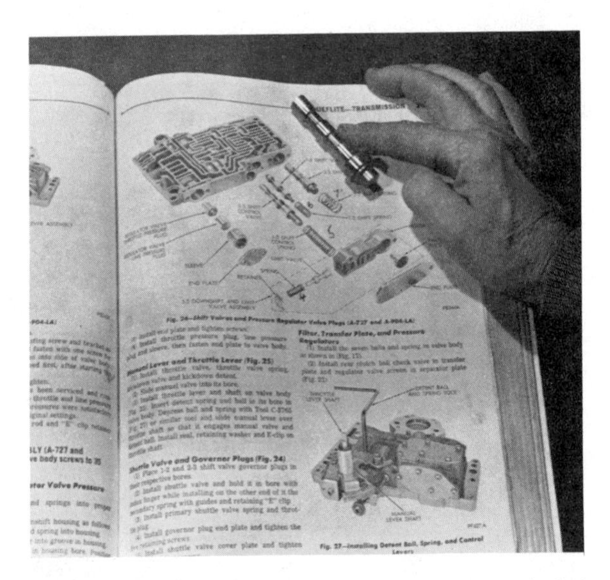

Fig. 17-14 Compare each valve with the exploded view in the manual. A 2-3 shift valve is shown here.

Fig. 17-15 A piece of rolled up crocus cloth can be used to clean the valve bores.

valves to stick, figure 17-16(A). To correct this problem, the dirt must be imbedded deeper so that it is flush with the bore, figure 17-16(B). To do this, first cover a wooden work bench, or a piece of wood with a surface area larger than that of the valve body, with a few layers of shop towels. Next, push the valve into the bore until it sticks. Then, grasping the valve body so that the machined surface is down, raise the valve body about 1 foot (30 cm) and, keeping it level and square, slam it flat on the cloth-covered surface, figure 17-17. Although this may seem like rough treatment of the valve body, it does the job. If done properly, this procedure will not harm the valve body. The important thing is to be sure the valve body is flat when it strikes the padded surface.

After bringing the body down hard on the cloth, turn the valve body over. Insert a thin-bladed screwdriver through a passage so that the blade rests in a valley of the valve, and lightly tap the screwdriver handle with a plastic hammer, figure 17-18. Use light taps, and work around the circumference of the valve as shown. If necessary, move the valve to a new spot in its bore and repeat the above steps until the valve is free.

This imbedding treatment usually works. If it does not, the valve body should probably be discarded. As a last resort, one more procedure can be tried. Mix up a thin paste

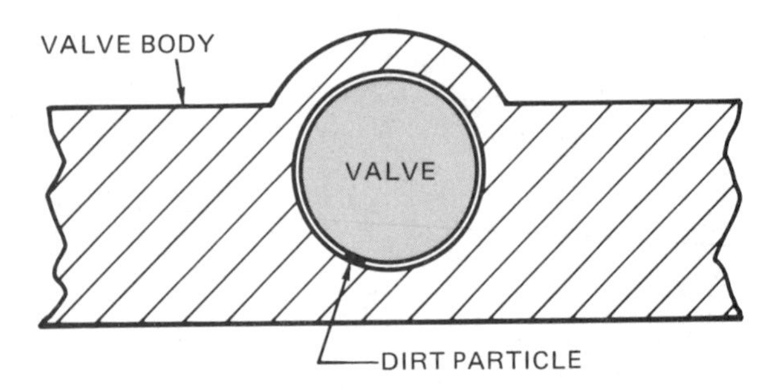

Fig. 17-16(A) Valve sticks due to partially imbedded dirt.

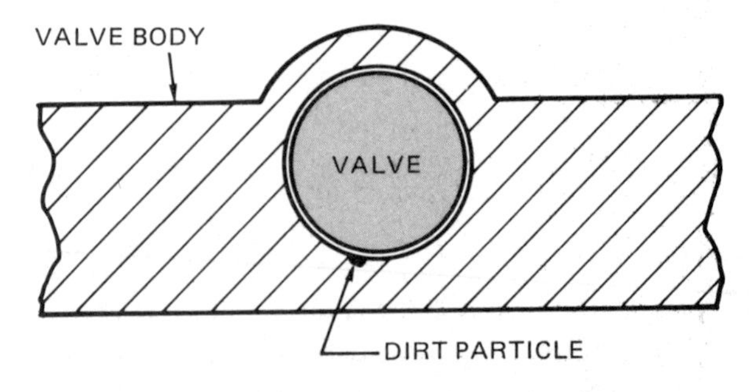

Fig. 17-16(B) Valve free – dirt fully imbedded.

of transmission fluid and household scouring cleanser. Coat the valve with this mixture and work it back and forth in its bore. Clean the valve and valve body with mineral spirits, and blow dry. Then try the fit of the valve again. If this does not work, the valve body must be replaced.

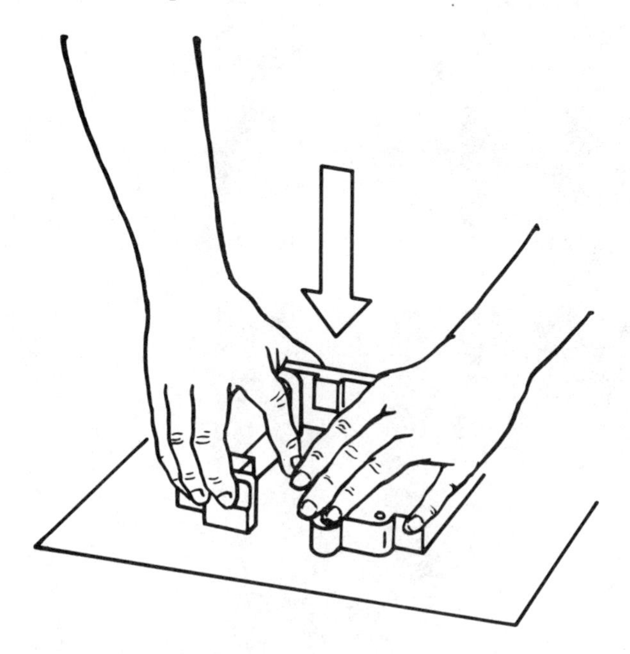

Fig. 17-17 Slamming the valve body on a padded surface imbeds the dirt in the bore so that the valve can move freely.

After all the valves have been checked and freed up, dip each valve in clean transmission fluid. Then select the proper spring from the spring holder and install the valve in its bore. Be sure all washers, clips, pins, and retainers are in place. Torque all end plate screws. The valve body is now serviceable and ready to be installed. If the valve body is not to be put back in service immediately, store it in a clean plastic bag.

Installing the Valve Body

To replace the valve body, simply reverse the steps used to remove it. (*Note:* Loose check balls can be held in place with petroleum jelly.) Before replacing the valve body, a good procedure is to air pressure test the clutches and servos. New seals should be used where the manual linkage shafts come through the transmission case. Torque the valve body to the case in three stages, and install the pan with a new gasket. Refill the transmission with fluid, and road test for proper performance.

OTHER IN-CAR SERVICE

In many cases, repairs can be made to the transmission without removing it from the car. Repairs of this kind include extension

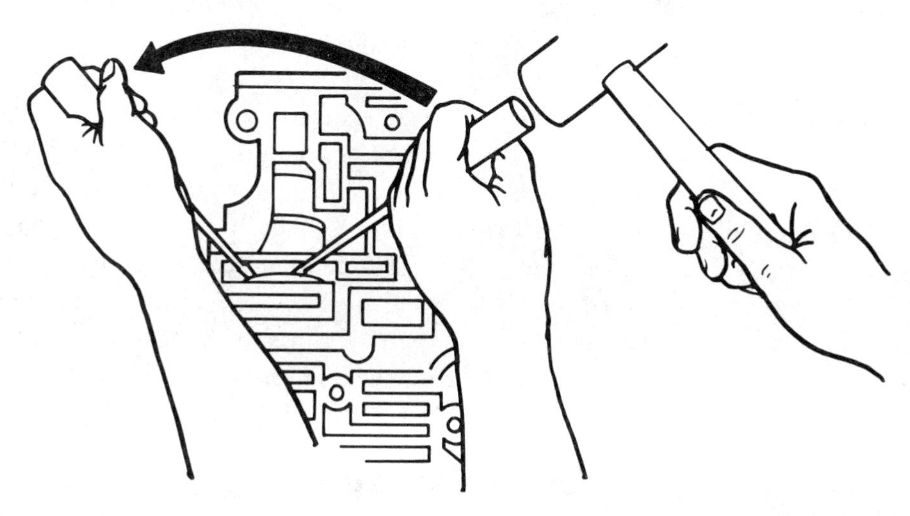

Fig. 17-18 Dirt under the valve in the upper part of the valve body is imbedded by light taps with a screwdriver in the valley of the valve, around its circumference.

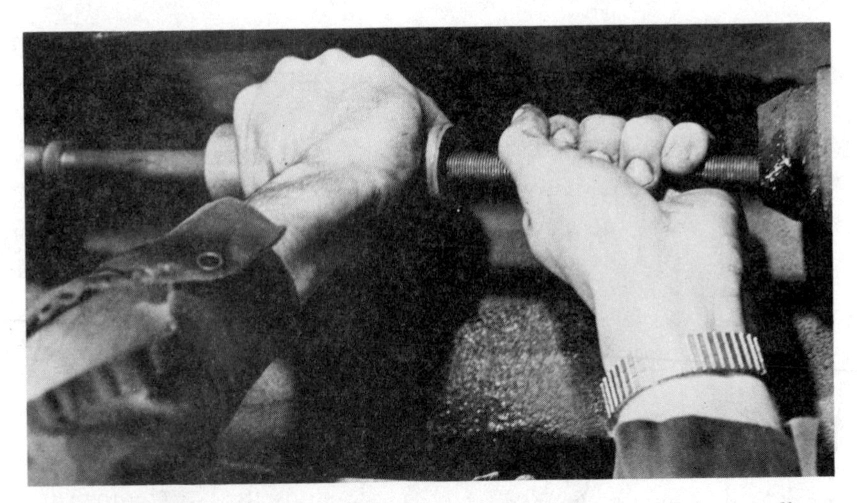

Fig. 17-19 Removing an extension housing seal with a slide hammer puller.

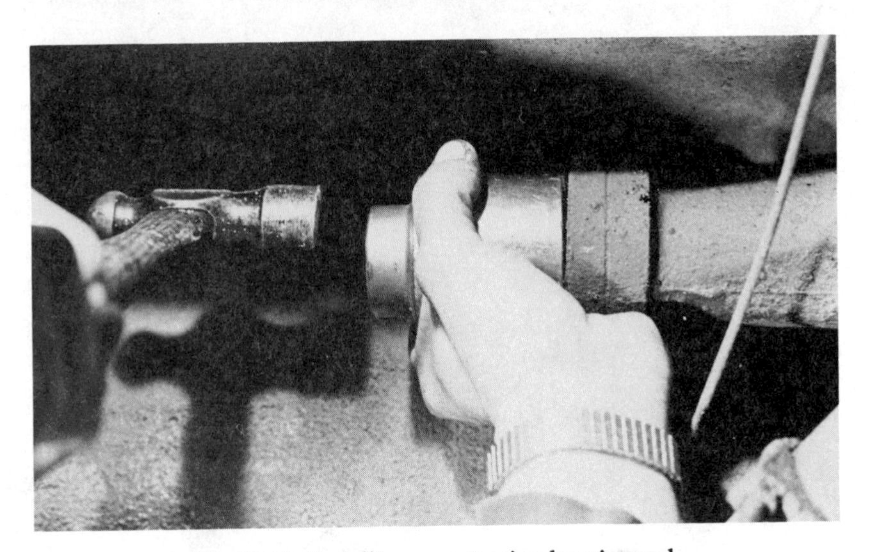

Fig. 17-20 Installing an extension housing seal.

housing seal and gasket replacement; and repairs to the servos, modulator and valve, governor assembly, speedometer drive, and parking components. For a complete list of in-car repairs and the correct procedures to use, check a shop manual. The following paragraphs present some general points to keep in mind when performing in-car repairs.

Extension Housing Gasket Seal and Bushing

Manufacturers recommend that the extension housing bushing and seal be replaced at the same time. After removing the drive shaft, a slide hammer or hook-type seal puller can be used to remove the old seal, figure 17-19. The bushing should be removed and replaced by following the steps in the shop manual. When installing the seal, be sure to use the proper installing tool, or a section of pipe that bears on the outer diameter of the seal. Lubricate the rubber part of the seal with petroleum jelly, and drive the seal in square and flush with the housing, figure 17-20.

When removing the extension housing, support the transmission with a jack. Remove the drive shaft, rear mount, and cross member. Then lower the transmission slightly and remove the extension housing for gasket replacement; or governor, speedometer gear, or parking mechanism service.

Caution: On cars with the ignition distributor located at the rear of the engine, the distributor cap should be removed before lowering the transmission. This prevents hitting and possibly breaking the cap on the firewall.

Servos and Accumulators

Manufacturers usually supply special tools for the removal of servos and accumulators. However, it may be possible to use the standard tools that are on hand. Chrysler Corporation, for instance, recommends the use of a valve spring compressor for removing servos. Check with the shop manual for this information. Always be sure that new seals are installed correctly and are lubricated with transmission fluid or petroleum jelly.

Modulator and Valve

Modulator or vacuum unit service is described in unit 15. To service the valve, remove the vacuum unit and, with a strong magnet, remove the valve. The valve can be stoned to remove nicks and burrs as described in the section on valve body overhaul. However, for in-car service, do not use crocus cloth or any other material that could leave grit in the bore and transmission. Use gum and varnish solvent and work the valve back and forth in the bore until it is free.

Governor

The governor is easily removed from most transmissions. On the General Motors Turbo Hydra-Matic transmissions, the governor is simply slipped out after removing a cover. On other makes, the extension housing must first be removed. Remember, the governor is a pressure regulating valve and should be serviced in the same way as other valves in the hydraulic system.

Speedometer Drive and Parking Components

The speedometer pinion can be removed from outside the transmission. To service the drive gear, however, the extension housing must be removed.

After removing the extension housing, the parking components on most transmissions can be serviced. The proper shop manual should be checked regarding these and other in-car services.

SUMMARY

If possible, except in cases of burnout or severe contamination, transmissions should be serviced without removing them from the vehicle. This saves time and money, and insures repeat business from the customer.

When servicing valve bodies, keep the following points in mind:

- Valve bodies have many small passages where dirt and sludge can lodge. To do a good cleaning job, the valve body must be completely disassembled.
- The valves must fit freely in their bores. They should drop in and out of the valve body of their own weight or with a very slight shake of the valve body.

- All valve body screws must be tightened to the specified torque and in the correct sequence.

REVIEW

Select the best answer from the choices offered to complete the statement or answer the question.

1. Which of the following units can be replaced or repaired with the transmission in place?

   a. torque converter
   b. servo pistons
   c. clutch pistons
   d. planetary gears

2. Before lowering the back end of a transmission in order to replace the extension housing gasket, which of the following precautions should be taken?

   a. Remove the bell housing bolts.
   b. Remove the front motor mounts.
   c. Jack up the engine as high as possible.
   d. Remove the distributor cap.

3. Mechanic A says that, when possible, valves should be installed in the valve body after the valve body halves are bolted together.
   Mechanic B says that valve bodies should be completely disassembled for cleaning and servicing.
   Which mechanic is right?

   a. A only
   b. B only
   c. both A and B
   d. neither A nor B

4. Mechanic A says that valve body screws should be torqued in three stages.
   Mechanic B says that no special torquing sequence is needed for valve bodies.
   Which mechanic is right?

   a. A only
   b. B only
   c. both A and B
   d. neither A nor B

5. Scratches or nicks in the mating surface of a valve body should be removed with

   a. A single cut mill file.
   b. A flat Arkansas bench stone.
   c. A surface grinder.
   d. A carbon scraper.

6. Mechanic A says that a valve body assembly that has been badly sludged or varnished can, in most cases, be overhauled.
   Mechanic B says that a valve body from a transmission that has been burned out should be discarded.
   Which mechanic is right?

   a. A only
   b. B only
   c. both A and B
   d. neither A nor B

7. For in-car modulator service, which of the following would be acceptable for cleaning the bore?

   a. gum and varnish solvent
   b. crocus cloth
   c. scouring cleanser and oil
   d. engine valve lapping compound

8. Mechanic A says that valves should be stoned so that the edges of the lands are slightly rounded to ease their passage through the bore.
   Mechanic B says that valves may be cleaned with #600 crocus cloth.
   Which mechanic is right?

   a. A only
   b. B only
   c. both A and B
   d. neither A nor B

9. Mechanic A says that valve bores should be honed (smoothed) with #60 emery cloth.
   Mechanic B says that new valves should be lapped to fit their bores with engine valve lapping compound.
   Which mechanic is right?

   a. A only
   b. B only
   c. both A and B
   d. neither A nor B

10. Mechanic A says the old valve body gasket should be saved and matched up with a new one.
    Mechanic B says a new valve body gasket should match up with the separator plate.

    Which mechanic is right?

    a. A only
    b. B only
    c. both A and B
    d. neither A nor B

11. In figure 17-14, a mechanic is shown checking the 2-3 shift valve with the exploded view in the manual. On which screw of the spring holder would the spring for this valve be found?

    a. 4
    b. 5
    c. 6
    d. 7

12. In disassembling a valve body, the mechanic finds a stuck valve. Which of the following would be the best way to free up the valve?

    a. Use a screwdriver and pry between the passages and land.
    b. Hold the valve body securely in a vise and drive the valve out with a punch and hammer.
    c. Work the valve back and forth after spraying with gum and varnish solvent.
    d. Strike the valve body sharply against the anvil of a vise.

EXTENDED STUDY PROJECTS

1. Construct a spring holder using the directions given in the unit.
2. Obtain several scrapped valve bodies from different makes of transmissions, and try to rebuild them to serviceable condition.
3. Using service manuals, make a list of all repairs that can be done with the transmission in place, for at least three different makes of transmissions.
4. Perform various in-car transmission services as assigned by the instructor.

TRANSMISSION OVERHAUL AND MODIFICATIONS

OBJECTIVES

After studying this unit, the student will be able to:

- Recognize the conditions that make major transmission repairs necessary.
- List the general repair and rebuilding techniques used for each phase of transmission service.
- Explain the function of shop manuals in transmission service.
- List three types of transmission modifications that can be made for special use and driving conditions.

As stated in unit 17, many automatic transmission problems can be solved without removing the transmission from the vehicle. By the proficient use of diagnostic skills, the mechanic must decide whether or not the transmission actually needs to be removed to solve a particular problem. An overhaul should not be undertaken if the problem requires only a simple adjustment or replacement of a part that can be made without removing the entire transmission. If all signs indicate a pump, clutch, band, or gear train problem, the transmission must be removed, disassembled, and repaired. In this case, a complete job should always be done – in other words, all rebuilding steps should be performed and all damaged or worn parts replaced.

TRANSMISSION REMOVAL

When major transmission overhaul is required, the transmission must be removed from the automobile. Most shop manuals provide a step-by-step plan for transmission removal. The following information is of a general nature and can be applied to all makes and models.

Preparation

The car must be properly positioned on a lift. Before raising the car, the battery ground cable should be removed. Also, if the distributor is mounted at the rear on the engine, the distributor cap should be removed to prevent breakage.

Removal

As a general rule, the following steps are used for transmission removal.

1. Remove exhaust pipes if necessary.
2. Remove cooler lines, linkage, speedometer cable, and detent, neutral safety switch, and back-up light wires.
3. Remove starter if necessary.
4. Drain the converter and transmission.
5. Mark the converter and drive plate. On some makes, the converter bolts are offset and will line up with the drive plate only one way. Marking before removal saves time during reassembly.
6. Remove converter-to-drive plate bolts.

Caution: Never attempt to remove an automatic transmission while leaving the converter bolted to the drive plate. This could cause serious damage to the drive plate, converter, or the transmission pump.

7. Remove the drive shaft.

At this point, a transmission jack should be placed under the transmission pan, and the transmission raised slightly, figure 18-1. The transmission jack is specially made so that the transmission can be chained or attached securely to the jack. This is essential for safe removal of the transmission. The jack also allows the transmission to be moved from front to rear and side to side, and also tilts both front to back and side to side. These features are important for safe removal, and allow proper positioning when the transmission must be lined up for installation.

At this point, the rear transmission mount and/or cross member is removed. An engine support fixture is installed, or a block of wood can be used between the engine oil pan and cross member to support the engine. The jack is then lowered so that the weight of the engine rests on the engine support or wood block, and the bell housing bolts are removed. The bell housing bolts can best be removed with a long (at least 24-inch or 61 cm) extension, universal socket and ratchet, or impact wrench, figure 18-2.

Fig. 18-1 A good, adjustable transmission jack eases installation and removal of the transmission.

With the bell housing bolts removed, the transmission can be worked off of the engine dowel pins and moved to the rear.

Caution: Be sure that the converter comes out of the drive plate and moves rearward with the transmission. The transmission and converter must be removed as a unit.

With the transmission moved back from the engine, fasten a small C clamp to the edge of the bell housing to hold the converter in place. The transmission can now be lowered and removed from under the car.

TRANSMISSION DISASSEMBLY

Before continuing with disassembly, the outside of the transmission should be cleaned of all grease, oil, and dirt. Absolute cleanliness during disassembly and assembly is a must. All parts, except friction materials, should be washed in cleaning solvent and dried with compressed air. Friction materials (bands and lined clutch discs) should be wiped

Fig. 18-2 A mechanic using a long extension, socket and ratchet to remove the bell housing bolts.

Fig. 18-3 A stand allows the transmission to be positioned for ease of disassembly or assembly.

or blotted with clean shop towels. Be sure not to leave any lint sticking to the ports. Band and clutch discs should not be washed since certain solvents may be harmful to the friction materials.

For ease of disassembly, the transmission is attached to a transmission stand, figure 18-3. A stand makes the job much easier, but satisfactory work can be done on a sturdy work bench. The converter is removed by pulling it straight off from the front of the transmission. The converter drive hub opening should be taped closed to keep out dirt. Set the converter aside for later checking and cleaning. Next, remove the pan and valve body. An end-play check is made before going on with the disassembly.

End-play Check. End-play in automatic transmissions is controlled by selective-fit thrust washers. The pump must be bolted in place to check end-play. By checking end-play before disassembly, the mechanic can choose the selective-fit thrust washers for reassembly that will correct end-play. This eliminates the need to partially disassemble the transmission after reassembly in order to correct end-play.

To check end-play, a dial indicator is clamped to the bell housing and the indicator zeroed on the input shaft, figure 18-4, page 256. By moving the input shaft in and out, the end-play can be read on the dial indicator. Shop manuals explain the proper way to move the input shaft to obtain this reading.

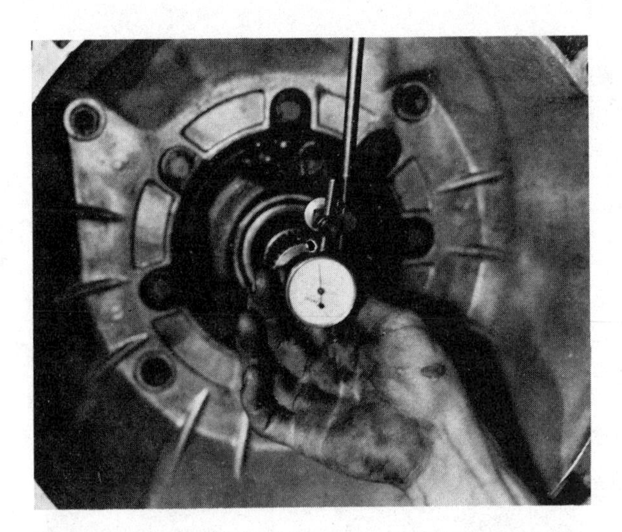

Fig. 18-4 An end-play check should be made before pump removal.

Note: Some transmissions also call for an output shaft end-play check. The shop manual will indicate if this is necessary and the correct way to take the measurement. Write down the end-play figures so that the correct size thrust washer can be chosen for reassembly.

Front Pump Removal

After the attaching bolts are removed, the front pumps on some models can be pulled out by prying forward on the gear train. Other models may require the use of a slide hammer puller. Two of the front pump attaching bolt holes are tapped so that special puller shafts can be installed. If the special puller shafts are not available, bolts can be threaded into these holes and a single slide hammer puller used, figure 18-5. When using a slide hammer puller in this way, pull the pump evenly by working from side to side between the two bolts.

Gear Train Removal

After removing the front pump, follow the steps in the shop manual for gear train removal. In general, the gear train, clutches, and bands are removed from the front of the transmission.

Fig. 18-5 A slide hammer puller being used to remove a front pump.

Caution: Do not pry, hammer, or force the parts. Some parts are held by snap rings, clips or set screws. Follow the step-by-step procedure in the shop manual and work carefully.

Next, remove the servos, linkage shafts, extension housing, parking mechanism, and one-way clutch. At this point, the transmission should be completely disassembled and ready for cleanup and reconditioning. Except in a few cases (such as a defective warranty part, for example) automatic transmissions in need of major repairs should be completely disassembled. This allows for a complete checkup of all parts and a good general cleaning.

PUMP SERVICE

The pump can be thought of as the heart of the automatic transmission hydraulic system. Because of its importance, it should receive a good, close inspection. Remember, the pump must be able to supply adequate pressure and volume of fluid for the entire transmission hydraulic system. A worn pump will be unable to do this, and transmission failure will result.

Inspection and Clearance Checks

The pump is disassembled by removing the stator support-to-pump housing bolts.

Fig. 18-6 Pump gear side clearance check (upper) and gear face-to-body clearance check.

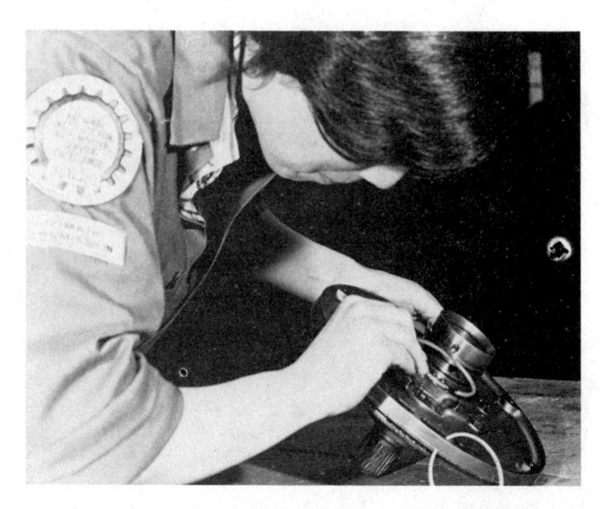

Fig. 18-7 Cast iron seal rings should fit their grooves with very little side clearance.

Before removing the pump gears, the top of the gears should be marked so that they can be reinstalled in the same way. After operating for a period of time, the pump gears tend to become *mated* to each other and to the housing. Installing them upside down could cause leak paths that would reduce pressure and capacity.

A few small score marks in the gear or housing surfaces will not harm pump performance. One way to determine if score marks are a problem is to scrape a fingernail across the score marks. If the fingernail does not stick or hang up in the grooves, the surfaces are acceptable. Students should ask the advice of an instructor or trained mechanic on this until they have more experience.

If there is no scoring or the scoring is acceptable, the next step is to check the gear-to-housing clearance. One method for this check is shown in figure 18-6. In the upper part of the photograph, the feeler gauge is placed to check gear-side clearance. At the center, a feeler gauge and straightedge is placed to show gear face-to-housing clearance. A shop manual should be checked for the proper tolerance. In general, side clearance should be no more than 0.004 to 0.008 inch (0.10 to 0.20 mm) and face clearance should be within 0.002 to 0.0025 inch (0.05 to 0.06 mm).

Some manufacturers recommend placing a 0.003-inch (0.08 mm) test shim between the gear faces and housing. With the pump reassembled to the proper bolt torque, the pump gears should bind up. To test this, the pump is placed on the converter drive hub so that the drive gear is meshed with the drive hub. If the pump body can be turned in this set up, there is too much clearance between the gear faces and pump body.

Caution: Be sure to remove the shim before returning the pump to service.

If the pump is of the type that has a clutch support hub, the seal-ring grooves should be checked for wear. A check should also be made to see that the new cast iron rings fit properly, figure 18-7. *Note:* Since this part of the front pump forms a one-piece unit with the stator support, it is called a *stator support* or *reaction support* by some manufacturers. In this text, it is called a *clutch support* because of its relationship to supporting and applying the transmission clutches.

Fig. 18-8 Front pump seals can be removed with an old screwdriver or pry bar.

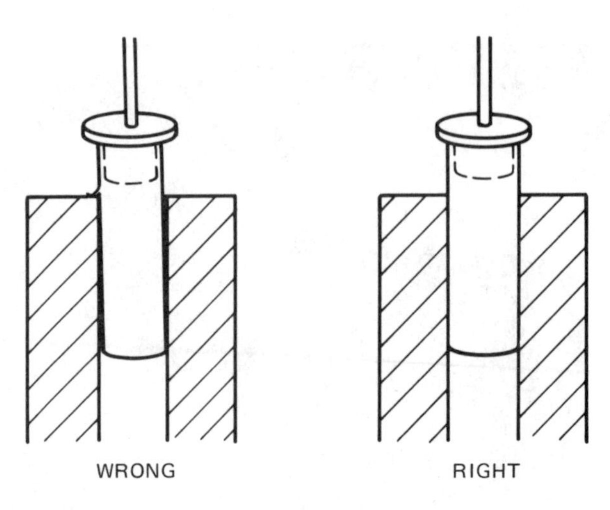

Fig. 18-9 Bushings must be installed square to the bore.

Bushing and Seal Replacement

Pry out the old seal, figure 18-8, and check the converter hub bushing and input shaft bushings for wear and scoring. If bushing replacement is necessary, work carefully, following the steps in the shop manual, and either cut or drive out the old bushings. Bushings for automatic transmissions are precision made – much like engine camshaft bushings, and do not need to be reamed or honed after installation. Care must be taken, however, that the bushings are installed square and straight in their bores, figure 18-9. Use a bushing driver that fits the bore of the bushing exactly. Line up the lubrication holes or grooves and drive the bushing in square with the bore and to the proper depth. *Note:* Some bushings must be staked in place. Follow the instructions in the shop manual for this procedure.

Assembly

The pump should now be ready for cleaning and reassembly. (*Note:* Some pumps may have rubber or neoprene check valves in the housing. These should not be cleaned in solvents that are harmful to these materials.) After cleaning, lubricate the pump parts with automatic transmission fluid, assemble the pump, and install the bolts finger tight.

Fig. 18-10 A converter is being used to align the pump body halves for torquing.

At this point, some means must be used to make sure that the housing and gears are in alignment. Some shop manuals show the use of a special strap clamp that fits the outer diameter of the pump like a large hose clamp. In some cases the pump can be placed backwards into the transmission housing or installed on the torque converter for final bolt torquing, figure 18-10. Whatever method is used, correct pump alignment is essential. Serious damage could occur if there is any binding or interference between the pump parts and converter drive hub.

SERVICING MECHANICAL AND HYDRAULIC PARTS

Other parts of the automatic transmission that will need close and careful service include the case or housing, bushings, bearings, servos, seals, clutches (including one-way clutches), gears, shafts, the hydraulic system, linkage, and the parking mechanism. When performing these services, careful workmanship, which includes cleanliness and attention to detail for each part serviced, help to insure a rebuilt transmission that will have a long and dependable service life.

Case, Bushing, and Bearing Service

Case leaks due to porosity or cracks may cause a loss of hydraulic pressure and lead to early transmission failure if not corrected. Locating case leaks can be difficult. However, if the cause of transmission failure is low pressure, and a careful examination of the hydraulic system shows no reason for this loss of pressure, a case leak could be the cause. One source of internal leaks are the bushings in the transmission case. Many case bushings, like engine camshaft bushings, are pressure lubricated. If worn too much, they can cause a loss of pressure, just as worn camshaft or connecting rod bearings can cause a loss of engine oil pressure.

The bushing should be checked for wear and replaced if necessary as described under pump service. When installing new bushings, the oil holes or grooves must be lined up and the bushing must be pressed or driven in square and true.

The case should also be checked very carefully for porosity and cracks. Larger cracks can be found by using a good strong flashlight. Examine the case carefully, changing the angle at which the light strikes the surface of the case.

Case porosity or hairline cracks are more difficult to spot. One method that can be used is to block off one end of a passage. Compressed air is then applied at the other end with a rubber-tipped blow gun. This pressurizes the passage. A soap solution is then brushed on the case and bubbles can be seen if there is a leak. If pressure testing is not possible, a cast iron case can be checked by the magnetic crack detection process. A dye penetrant test can be used on aluminum cases. These testing products are available commercially. The manufacturer's directions should be followed carefully.

The following is a second method of detecting hairline cracks.

1. Soak the area of the suspected crack with kerosene for at least one-half hour. The kerosene can be brushed on, but the case must be kept wet. Kerosene is a very good penetrant. The half hour allows adequate time for it to penetrate any fine, hairline cracks that may be present.
2. Wash the soaked area thoroughly with a non-oily solvent so that the surface is clean and dry.
3. Coat the cleaned area with a light layer of talc. (A spray can of foot powder works well for this.) After a few minutes, any cracks will show up as dark lines on the white talc. This occurs because the kerosene that cannot be cleaned from the cracks bleeds out and wets the powder.

Check all roller or ball bearings for galling or pitting and replace as necessary. A good way to clean ball or roller bearings is to hold them by the inner race and submerge them in a can of solvent. When the outer race is turned, dirt particles float out and settle to the bottom of the solvent can.

Clutch Service

Clutch discs are removed by prying out the snap ring, figure 18-11, turning the assembly over, and giving it a sharp rap on the bench top. Check the old clutch discs for flaking and wear. If the condition of the old clutch discs is questionable, compare them with new ones, figure 18-12.

Most clutch pistons can be removed by compressing the clutch piston return spring and removing a snap ring, figure 18-13. After removing the spring compressor and return spring, the clutch piston can be blown out with compressed air. A rubber-tipped blowgun is placed over the apply hole in the clutch support bore, figure 18-14, and air is applied. Remove the old seals from the piston and bore. The old seals should never be reused, but should be saved to help choose the correct new seal from the overhaul kit. Once the new seals have been selected, discard the old seals.

Fig. 18-11 Clutch-pack snap rings can be removed with a screwdriver.

Check the clutch piston and bore for grooves or damage that could interfere with proper sealing. The clutch support bore should also be checked for grooving where the cast iron rings ride. Also, check the cast iron seal rings that fit on the clutch support of the front pump for roundness and proper end-gap in the clutch support bore, figure 18-15. There must be some clearance, but usually not over about 0.012 inch (0.30 mm).

If new clutch discs are to be used, the lined discs should be soaked in transmission fluid for at least one-half hour before being installed. It is very important that the new discs

Fig. 18-12 The worn clutch disc (left) lacks grooves and has been blackened by heat.

be given this presoak. Clutch and band material is porous and must be given enough time to become completely saturated with fluid. If this is not done, the clutch could overheat and burn out with the first few applications.

While the clutch discs are soaking, select new seals, and lubricate and install them on the piston and in the clutch bore. Be sure that the seals are not twisted and are installed straight in their grooves, figure 18-16. Lubricate the clutch bore and install the piston. The piston must be installed in the bore square and pushed all the way in until it bottoms. It is often necessary to work around the diameter of the piston, pressing in a little at a time, until the seal and piston are started in the bore square.

For installing lip-type seals, a tool can be made from copper tubing and 0.015 inch (0.38 mm) diameter music wire, figure 18-17. As shown, the tool is placed between the clutch

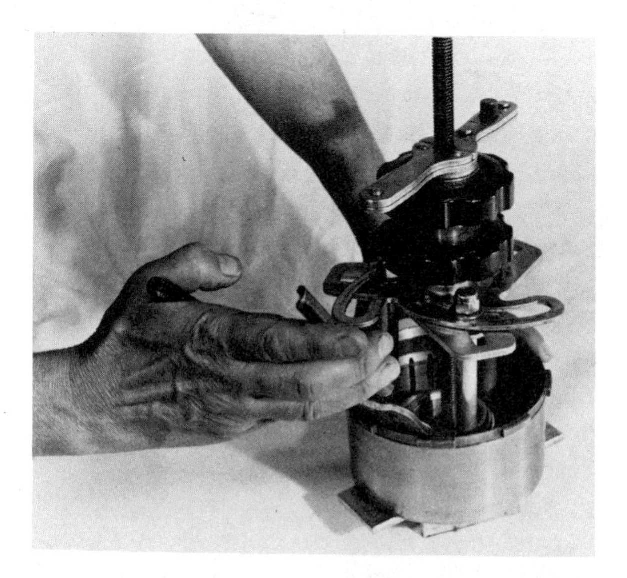

Fig. 18-13 Removing clutch piston return spring snap ring.

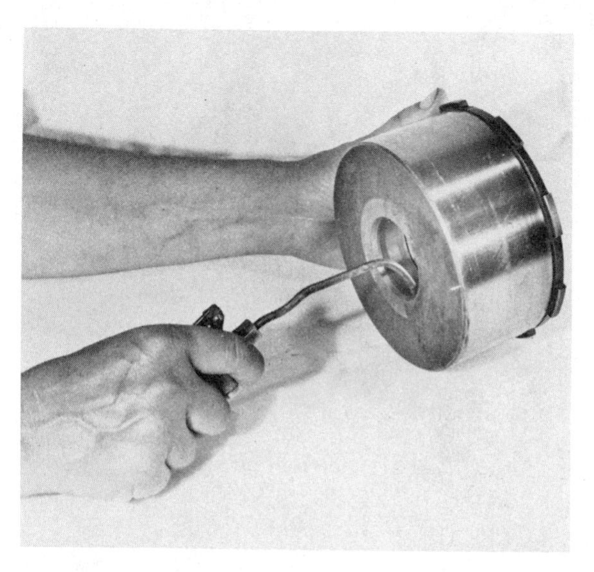

Fig. 18-14 Clutch pistons can be blown out with compressed air.

Fig. 18-15 Checking cast iron seal ring end-gap in clutch bore.

Fig. 18-16 Replacing a lathe-cut piston seal.

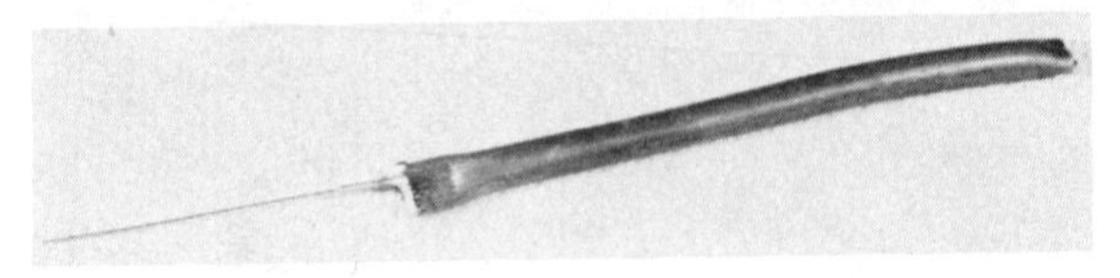

Fig. 18-17 A homemade tool being used to ease the seal into the clutch cylinder. A closeup of the tool, made from a guitar string, is also shown.

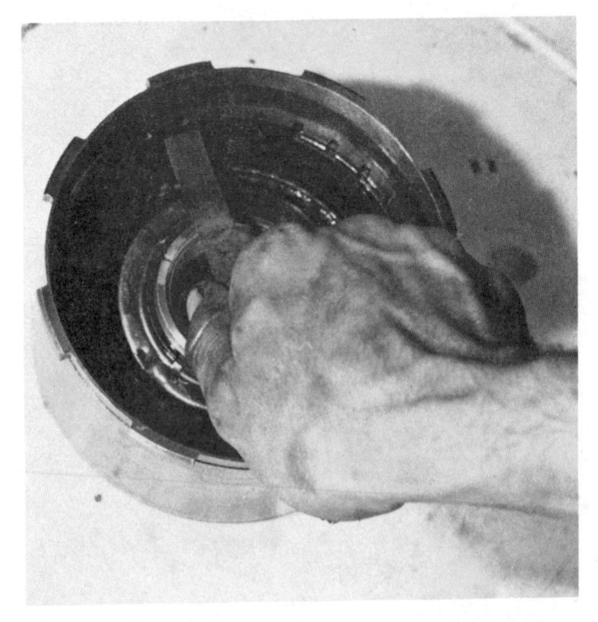

Fig. 18-18 Checking clutch pack clearance with a feeler gauge.

piston and bore. It is then worked around the diameter of the piston to help ease the lip of the seal into the bore.

Caution: Do not use a flat feeler gauge for this purpose. The sharp edge of the blade could cut the seal lip. With the piston pushed to the bottom of the bore, install the return spring, retainer, and snap ring. Be sure the snap ring is seated properly in the groove and retainer.

Install the clutch discs and pressure plates in the reverse order of removal, making sure a steel surface is in contact with each side of the lined discs. Also be sure that the same number of discs are installed as were removed. Install the snap ring, and check the clearance between the snap ring and pressure plate with a feeler gauge, figure 18-18. The correct clutch pack clearance can be found in the proper shop manual. Thicker or thinner snap rings and/or steel clutch plates are available to bring the clearance to proper specifications.

To complete clutch service, an air pressure test should be made to check for leaks. Lubricate and install new cast iron seal rings on the front pump clutch support. Locate the clutch apply holes in the pump housing by blowing compressed air through the clutch apply holes between the steel rings on the clutch support.

Install the high clutch on the clutch support and test for leaks and proper operation with compressed air and a rubber-tipped blowgun. Install the forward clutch and check in the same way, figure 18-19. The clutches should apply and release fully, and there should be no air leaks at the clutch piston seals. Small air leaks at the steel clutch support rings are normal since these are controlled leakage seals. However, a large air leak could mean that the clutch support grooves or clutch support bore are worn or out of round.

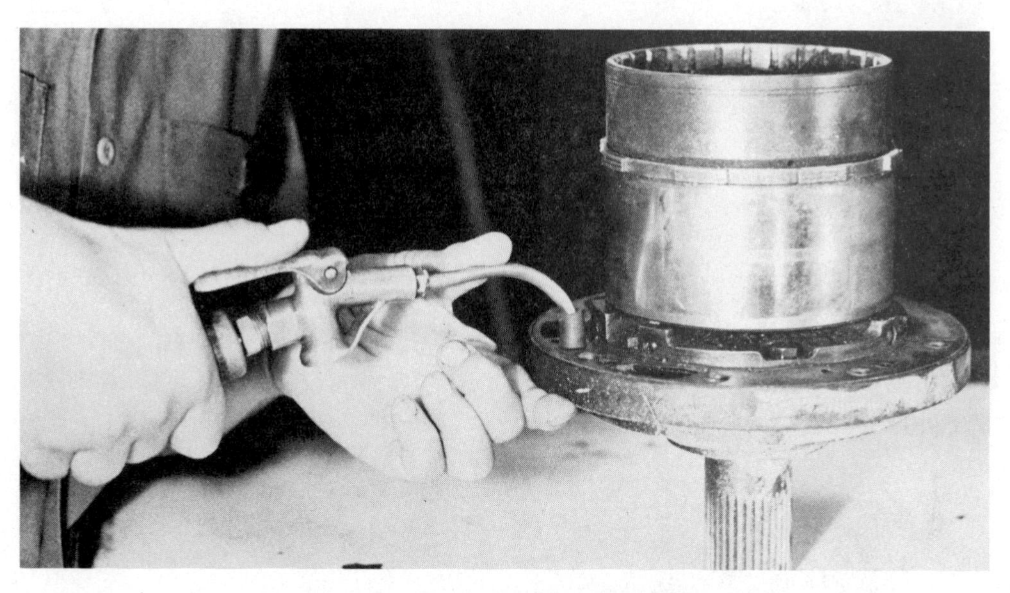

Fig. 18-19 Air pressure testing assembled clutches.

Servo and Band Service

The servo pistons and bores should be checked and serviced in the same way as the clutches. Always install new seals and soak new bands in transmission fluid before installing. After transmission assembly, the servos can be air-pressure tested as described in unit 16. (*Note:* Accumulators should be checked and serviced in the same way as servos.)

Planetary Gear Set Service

Because planetary gear sets are always in mesh, gear tooth failure is rare. However, bushing or thrust washer wear can take place, and these parts should be inspected and replaced if necessary. Planetary pinions are usually mounted on needle bearings. These should be checked for free operation and excess play. Kits are available to recondition the planetary carrier assembly if necessary.

One-way Clutch Service

The one-way clutch should be disassembled and inspected for wear as if it were a roller bearing. That is, there must be no galling of the races, and the rollers must be smooth with no flat spots. The energizer springs should be inspected for tension and cracks. If the transmission has very many miles on it, it is safest to use new energizer springs for reassembly. Be sure that the energizer springs are placed on the proper side of the rollers, and check for proper operation. The planet carrier race should turn freely in a clockwise direction. It should lock up, with no backlash, when turned counterclockwise. If the rollers and races are in good condition, but backlash is felt or there is no lockup in a counterclockwise direction, then the energizer springs are weak or have been installed backwards.

Check the parking gears and linkage for wear and proper operation. Install new linkage shaft seals. The transmission is then ready for assembly. The valve body, governor, and modulator assembly will have been serviced as described in unit 17.

TRANSMISSION ASSEMBLY

Transmission rebuilding is precision work. Again, keeping the reconditioned parts clean

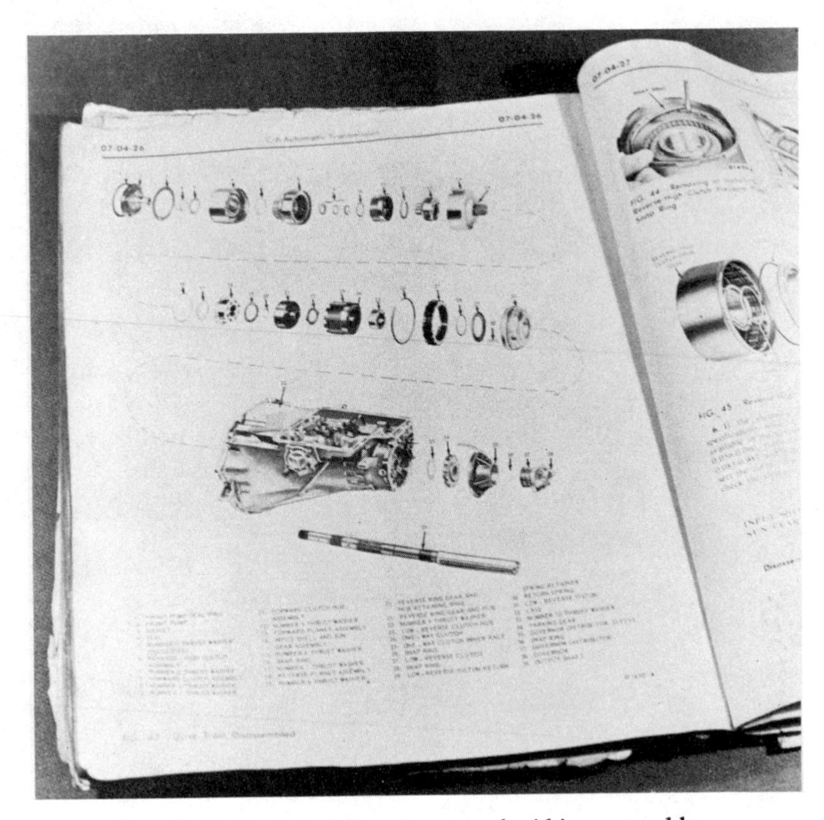

Fig. 18-20 Exploded views in manuals aid in reassembly.

is all important. Automatic transmissions and parts are costly and the fit and tolerance on some parts are very close. Even small amounts of dirt or grit can ruin these parts in a short time. A first-rate job of rebuilding cannot be done without strict attention to cleanliness. The work area and bench should be cleaned of every scrap of dirt, grit, oil or grease. Transmission rebuilding or assembly should not take place in an area where grinding operations, paint spraying, body work, or other activities could contaminate the parts.

Correct Bolt Torque

All bolts and nuts must be tightened evenly and to the correct torque. Failure to heed this advice could lead to warpage and leaks in such places as servo covers, valve body-to-case, extension housing, or front pump-to-case. To arrive at the correct torque, the bolts must be lubricated with transmission fluid and thread freely into their holes.

Diagrams and Exploded Views

Most shop manuals have detailed diagrams or exploded views showing the relation of each part to the others and the order of assembly, figure 18-20. They help the mechanic find the proper position of thrust washers, snap rings, gear sets, clutches, and so forth. Although the illustrations are important, the step-by-step assembly directions should also be read. Important steps, such as clearance and tolerance checks, are found in the text.

Lubrication

For lubrication purposes during reassembly, use only automatic transmission fluid of the correct type, or petroleum jelly.

Petroleum jelly can also be used to lubricate seals and to keep thrust washers from slipping out of place during assembly, figure 18-21. Do not use chassis lube or any other grease for these purposes. Other substances may have additives that could cause sticky valves or clog small passages.

As a general rule, transmission assembly should take place in the reverse order of disassembly. If a transmission stand is being used, the case can be turned straight up so that parts will stay in place. The gear train, for example, can be put together on the bench and installed as an assembly in the transmission case, figure 18-22. This prevents thrust washers from falling out of place and keeps clutch discs in proper mesh. If a transmission stand is not available, a **clean** gear or chassis lubrication drum can be used. After installing the output shaft and extension housing, place the transmission, with the converter end up, in the clean drum.

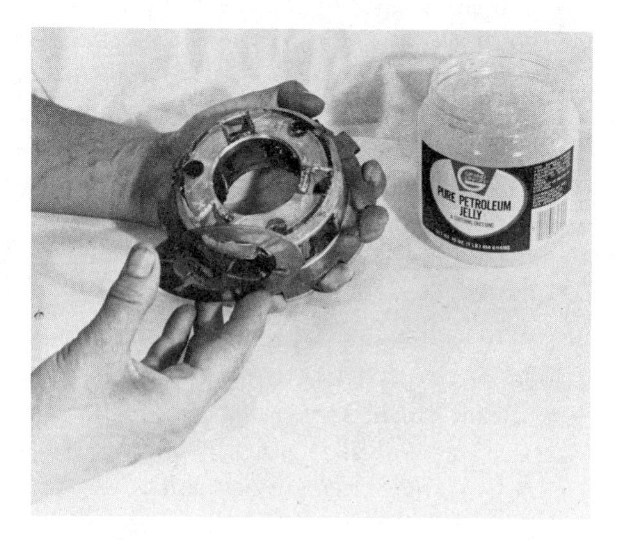

Fig. 18-21 Only petroleum jelly should be used to hold thrust washers in place.

Front Pump Installation

Care must be taken when installing the front pump. If a thrust washer has slipped out of place or clutch discs are not fully meshed, damage to the clutches or sun gear shell could take place as the pump is torqued to the case. Damage can be avoided by using the following steps:

1. Select a new gasket (if one is used) and place it in the case so that the bolt and oil holes in the gasket line up with those in the case. (*Note:* Some pumps use an O-ring on the outer diameter of the pump.

Fig. 18-22 Gear train being installed as a unit.

Fig. 18-23 Front pumps must be torqued carefully while checking for binding.

In this case, a gasket may or may not be needed.)

2. Install two headless pilot bolts on opposite sides of the gasket to hold it in place. These bolts can be bought or they can be made by hacksawing the heads off of two 3-inch bolts of the proper diameter and thread pitch.
3. Place the pump over the pilot bolts and tap it gently and evenly into place with a plastic mallet or wooden hammer handle.
4. Install the pump attaching bolts in the open bolt holes. Remove the pilot bolts and install the two remaining bolts. Tighten all bolts finger tight.
5. Tighten all bolts to the correct torque evenly and gradually. Use at lease three stages to bring the bolts to the correct torque.

Caution: At each stage of torquing, the input shaft must be free to turn, figure 18-23. If at any point the input shaft binds up, stop the torquing process and be sure that the pump is square with the case. If the pump is not cocked and the input shaft binds or cannot be turned, something is wrong. Remove the pump and check for an out-of-place thrust washer or clutch disc.

End-play Check

After the pump has been torqued, a final end-play check should be made to be sure that the correct selective thrust washer or washers have been installed. If the end-play is within the proper limits, the bands should be adjusted, and a final air pressure check of servos and clutches made. Complete the assembly by installing the valve body and oil pan. After converter and cooler service, the transmission is ready to be replaced in the vehicle.

CONVERTER AND COOLER SERVICE

If the rebuilt transmission is to provide good service, the converter and cooler must be clean and in serviceable condition. This service can be carried out by the mechanic, although some tasks are best left to specialty shops. Services done by the mechanic include cleaning and wear checks.

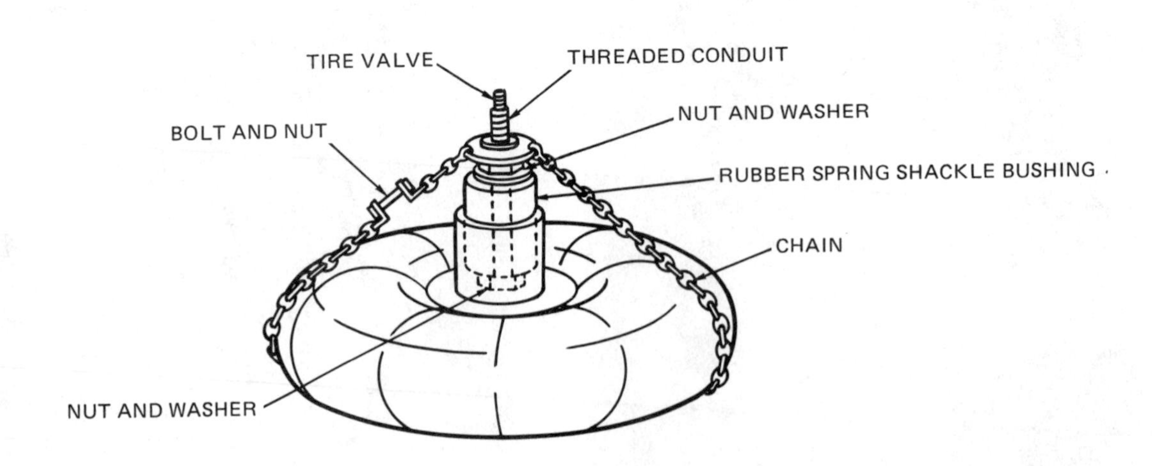

Fig. 18-24 Setup for pressure testing converter.

Converter Service

The converter drive bolts should be checked for thread damage and wear. Worn drive bolts could cause the converter to be mounted off center. This in turn could cause vibration and further damage to the converter and drive plate. The converter pump drive hub must be free of deep scoring in the seal area; the fingernail test (see unit 17) can be used to check this. Slight score marks in the seal area should be polished out with #600 crocus cloth. Any deep scoring means that the converter must be replaced.

Converter Leak Test

If the converter shows signs of leakage, an air pressure test should be made. Special tools can be bought for this purpose or the setup shown in figure 18-24 can be made in the field. The threaded conduit is of the type used in lamp construction. A nut and washer is soldered or brazed to one end, and a metal tire valve is threaded or brazed to the other end. With the rubber spring shackle bushing and top nut and washer in place, the tool is pushed into the converter hub. Tightening the top nut expands the rubber bushing in the converter hub and forms a tight seal.

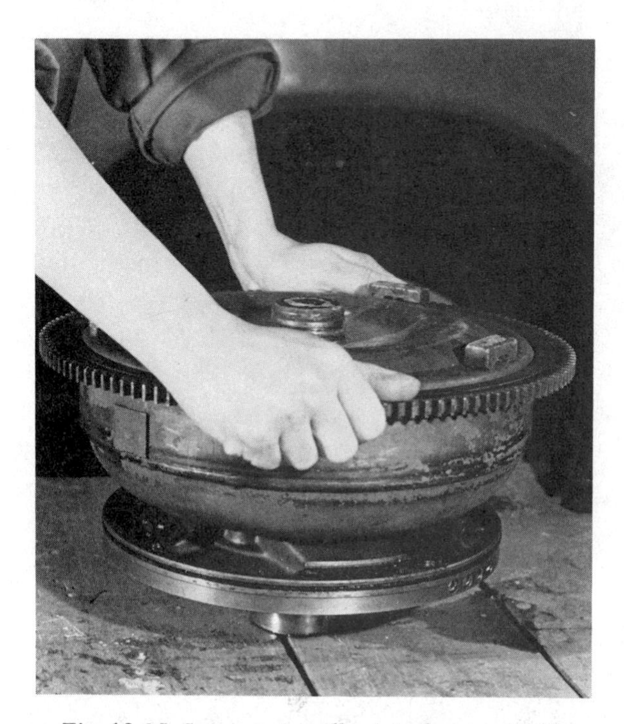

Fig. 18-25 Stator-to-impeller interference check.

For safety, the center link of a chain should be slipped over the conduit and bolted around the converter. The converter is then pressurized with air to about 25 psi (172 kPa), placed in a tub of water, and the weld areas watched for air bubbles. If no air bubbles appear, it can be assumed that the welds are not leaking.

Caution: Do not use full line pressure from the shop compressor to pressurize the converter, or damage may result; 25 psi (172 kPa) should be enough pressure to reveal any leaks.

Interference Checks

Worn thrust washers inside the converter could cause interference or metal-to-metal contact between the turbine, stator, and impeller hubs or blades. To check for stator-to-impeller interference, place the transmission pump on the work bench with the stator support facing up. Place the converter on the pump in the normal operating position, figure 18-25. Be sure that the stator splines are fully meshed. Turn the converter clockwise. If scraping noises are heard or felt, replace the converter.

To check for stator-to-turbine interference, turn the pump and converter assembly over and insert an input shaft through the pump to mesh with the splines in the turbine. Turning the input shaft will reveal any interference between the stator and turbine.

Turbine and Stator End-play

A quick check for end-play between the turbine, stator, and impeller can be made using the following procedure. Place the converter on a workbench with the drive end down. Use a pair of long, thin, external snap ring pliers to grip the turbine hub, figure 18-26(A). An upward pull on the pliers will take up the slack between the turbine, stator, and impeller. This is the total end-play between these parts. End-play specifications vary between manufacturers, but, in general should be between 0.015 and 0.060 inches (0.38 and 1.52 mm).

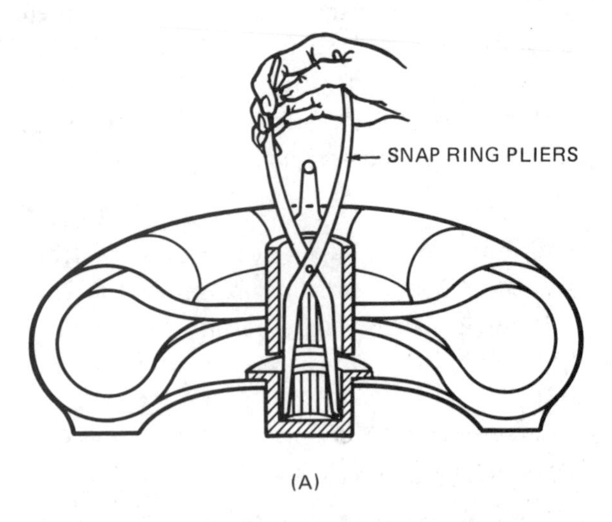

(A)

(B)

Fig. 18-26 Two methods of checking converter end-play.

If, on a quick check, the converter seems to have too much end-play, a more accurate check can be made by using a standard transmission clutch aligning tool and a dial indicator, figure 18-26(B). The split bushing of the clutch aligning tool is pushed in to the splined turbine hub and tightened in place. End-play can then be measured on the dial indicator as shown.

Converter and Cooler Cleaning

When a transmission has suffered a burnout, the converter and cooler are contaminated with burned friction material and metal filings. Special converter and cooler flushing machines are available to remove this kind of contamination. If the shop is not equipped for this work, the converter and cooler should be sent out to a job shop that does this type of cleaning. If a burnout has not occurred, the cleaning can be done satisfactorily by a mechanic.

First, drain old fluid from the converter and replace the drain plug. Next, pour two quarts or liters of clean solvent through the opening in the converter hub, and place the converter on the floor. Spin the turbine using a turbine shaft, adapter, and 1/2-inch drill. Adapters can be bought or made in the field, or an old input shaft can be turned down to fit a 1/2-inch drill. Figure 18-27 shows how to make a drill adapter for the turbine shaft. Drain the solvent and repeat the steps until the drained solvent is clean and clear of any contaminants. Then repeat the steps once more using two quarts or liters of clean transmission fluid instead of solvent. Drain and discard this fluid. This removes any harmful traces of solvent.

The cooler should be flushed with a pressure or siphon gun and solvent until the solvent runs clean and clear. To remove

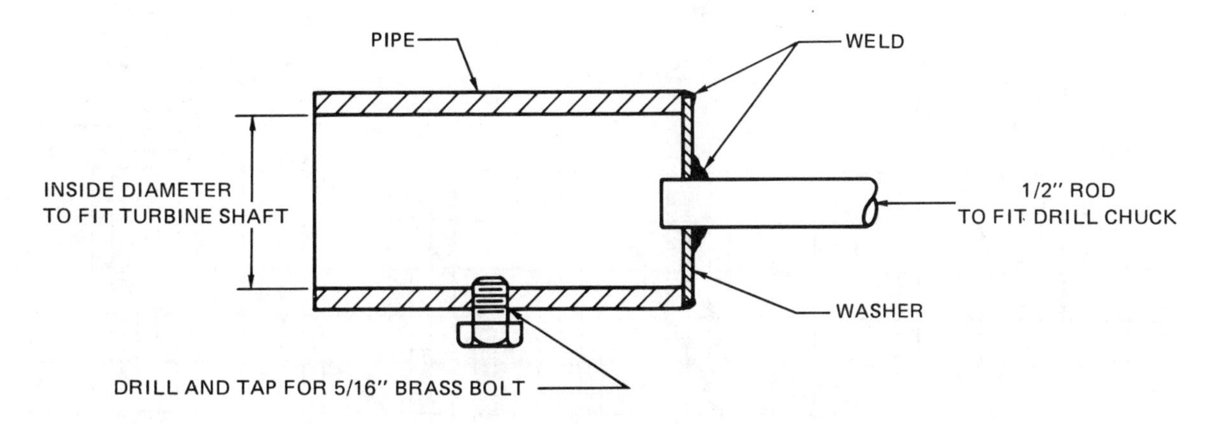

Fig. 18-27 Drill adapter for turbine shaft.

traces of the solvent, pump one quart or liter of clean transmission fluid through the cooler.

INSTALLING THE TRANSMISSION

To replace the transmission, simply reverse the steps used to remove it. However, certain steps can be taken to make the job easier and to avoid damage to any of the parts. First, be certain that the torque converter is fully installed on the transmission. Install the converter by using a back and forth twisting motion while, at the same time, pushing in on the converter. Three clear bumps should be felt or heard:

- the stator support splines meshing with the inner race of the stator one-way clutch.
- the input shaft splines meshing with the hub of the turbine.
- The converter pump drive hub meshing with the pump drive gear.

Proper installation of the converter can be checked by measuring the distance from the bell housing face to the face of the converter drive lugs. This measurement is given in most shop manuals.

Caution: Never try to install the transmission with the converter bolted to the drive plate on the engine. This will be almost certain to result in a ruined front pump.

Attach the transmission to a transmission jack and raise it close to the same height as the engine, so that the bell housing and engine mounting plate are about 3/4 of an inch (19 mm) apart. Using the adjustments on the jack, align the transmission to match the exact height and tilt of the engine. If the transmission has been carefully aligned with the engine, very little force will be needed to bring the bell housing flush with the engine block or mounting plate. If difficulty is encountered, recheck the alignment and be sure that the converter has not slipped forward. With the bell housing flush with the block, there should be some clearance between the converter drive lugs and the drive plate. After completing the installation, fill the transmission to the proper level and road test it for correct operation.

TUNING FOR SPECIAL USE

For certain uses, the operation of automatic transmissions can be changed or *tuned* to give better performance under special use conditions. Transmissions modified for racing are the most common example of special tuning. However, recreational vehicles and

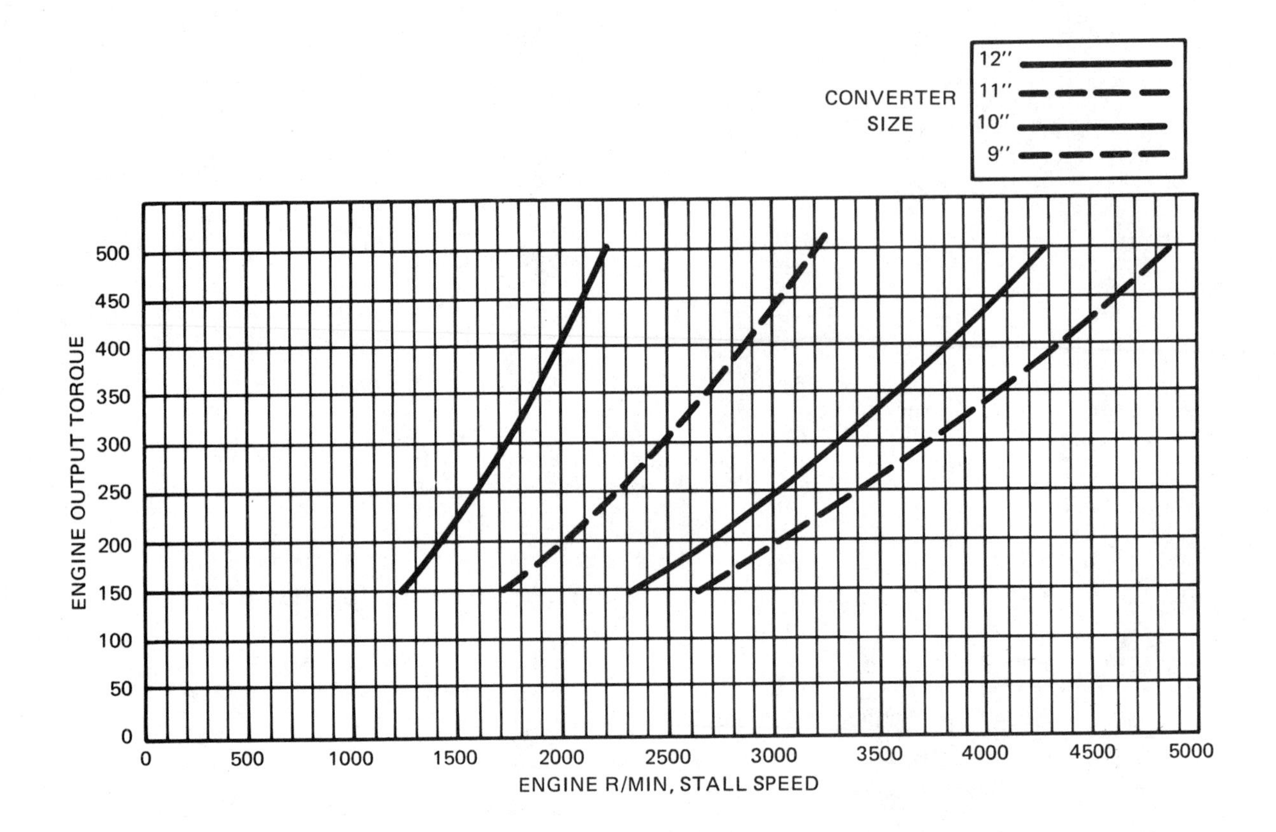

Fig. 18-28 Engine torque and stall speed curves.

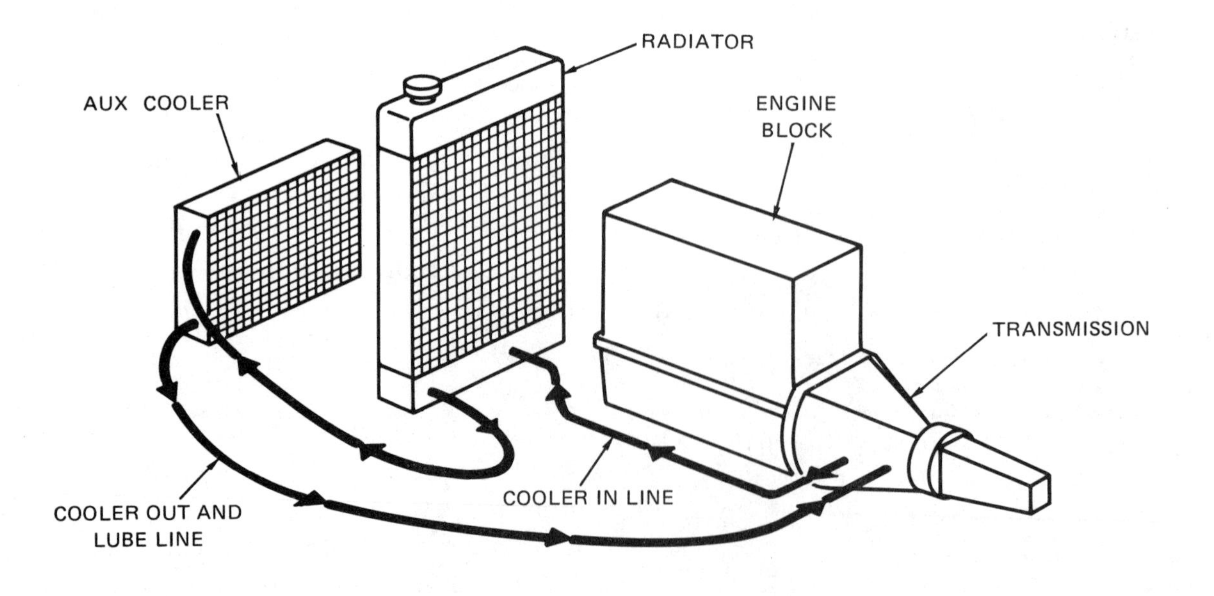

Fig. 18-29 Auxiliary cooler installation.

even cars used for everyday driving may benefit from certain modifications. For example, if a vehicle is driven mainly on backcountry, mountainous roads, slightly later shift points could help to provide better performance. This change is easily made by a simple throttle linkage or modulator adjustment.

Other modifications require replacement of certain parts and/or changes in the hydraulic system. The following paragraphs outline some of the basic modifications that are commonly made on transmissions.

SPECIAL CONVERTER

When an engine is modified for high performance, the torque curve of the engine changes. That is, after modification, the point of maximum torque occurs at a higher engine speed (r/min) than in a standard engine. To take full advantage of engine modifications, the torque converter stall speed should be matched to the engine's torque curve. Figure 18-28 shows that a decrease in converter size results in an increase in stall speed. By choosing the right converter, engine revolutions per minute can be kept close to the point of maximum torque, with a resulting increase in performance. This, of course, also results in more slippage and poor fuel economy. However, these factors usually are not considered important in the search for high performance.

At the other end of the scale, a converter that is larger than stock will give less slippage and better gas mileage, but the engine will be less responsive "off the line." Recreational vehicle owners or the economy-minded driver can benefit from this modification.

Increased Cooling

Heat buildup can be a problem in race cars or recreational vehicles. Higher stall speeds or heavy loading in slow, mountainous driving can cause heat buildup. The installation of an auxiliary cooler handles this extra heat buildup and helps to extend transmission life, figure 18-29.

Valve Body Modifications

Valve body modifications are usually thought of in terms of racing or "hot rodding" use. However, drivers of both passenger cars and recreational vehicles are finding that firm, crisp shifts are not annoying and will give increased performance and longer transmission life. The kits available for valve body modification make it possible to change transmission characteristics to suit individual needs. The kits include total performance packages that give neck-snapping, harsh shifts; those that provide full manual "stick shift performance"; or, for the best of two worlds, those that combine "stick shift" and full automatic shift, at the choice of the driver. These kits usually allow the driver to select manual first or second at any speed, and therefore, must be used with care.

SUMMARY

This unit is not intended as a replacement for the shop manual for a particular vehicle. Shop manuals are updated with each year or model change, and the specifications given in the shop manuals should be carefully followed. Instead of giving specific information, this unit presents basic skills and practices that have been developed by those in the field. Years of experience have shown that these methods are easy, safe, and efficient ways to approach transmission service.

Careful workmanship, attention to detail, and, above all, cleanliness, insure successful transmission service. In many cases, the shop manuals show the use of special factory tools for certain operations. Some of these tools are essential and must be used. In other cases, with a little imagination and effort, special tools can be made or adapted from materials at hand.

REVIEW

Select the best answer from the choices offered to complete the statement or answer the question.

1. Transmission front pumps are usually removed by using a

   a. hammer to tap on the output shaft.
   b. hammer to tap on the stator support.
   c. slide hammer puller threaded into the pump housing.
   d. punch and hammer to tap on the inside of the pump housing.

2. Gear train end-play is adjusted by

   a. selective-fit snap rings.
   b. selective-fit thrust washers.
   c. bearing shims.
   d. selective-fit pump gaskets.

3. Mechanic A says an air pressure test is used to check the front pump. Mechanic B says an air pressure test is used to check one-way clutch action. Which mechanic is right?

   a. A only
   b. B only
   c. both A and B
   d. neither A nor B

4. Clutch-pack clearance is corrected by changing the

   a. selective-fit snap ring.
   b. selective-fit thrust washer.
   c. return spring.
   d. spring pressure.

5. Mechanic A says that a few, shallow score marks in the pump gears will not hurt pump performance.
   Mechanic B says that slightly scored pump gears should be removed and installed upside down in the housing.
   Which mechanic is right?

   a. A only
   b. B only
   c. both A and B
   d. neither A nor B

6. In the interests of cost and customer satisfaction, a transmission that has had a burnout should

   a. have only the burned clutch or band replaced.
   b. be completely overhauled and cleaned.
   c. have the cooler replaced.
   d. not be used.

7. When removing an automatic transmission, the torque converter should be:

   (I) Left bolted to the drive plate.
   (II) Removed as a unit with the transmission.

   a. I only
   b. II only
   c. either I or II
   d. neither I nor II

8. A customer complains of late shifting. To eliminate this problem the mechanic will most likely

   a. repair the valve body.
   b. replace the torque converter.
   c. overhaul the entire transmission.
   d. adjust the bands.

9. In general, pump gear face-to-housing clearance should be no more than

   a. 0.001″.
   b. 0.0015″.
   c. 0.0025″.
   d. 0.003″.

10. In checking over a transmission, it is found that the fluid is black, contains foreign material, and has a burned odor.
    Mechanic A says that this is the sign of a burned band or clutch.
    Mechanic B says that the transmission should be completely overhauled.
    Which mechanic is right?

    a. A only
    b. B only
    c. both A and B
    d. neither A nor B

11. Modifying automatic transmissions for special use could result in:

    (I) harsh shifting.
    (II) increased performance and service life.

    a. I only
    b. II only
    c. either I or II
    d. neither I nor II

12. Mechanic A says an auxiliary cooler should be installed in series with the cooler return or lubrication line.
    Mechanic B says an auxiliary cooler should be mounted between the radiator and engine block.
    Which mechanic is right?

    a. A only
    b. B only
    c. both A and B
    d. neither A nor B

13. During transmission assembly, thrust washers should be held in place with

    a. gasket cement.
    b. lithium grease.
    c. chassis lube.
    d. petroleum jelly.

14. Mechanic A says that old clutch discs or bands should be washed in strong solvent before reusing.
Mechanic B says that new clutch discs or bands should be soaked in transmission fluid for one-half hour before being installed.
Which mechanic is right?

a. A only
b. B only
c. both A and B
d. neither A nor B

15. A mechanic has trouble removing part of the gear train. What should be done?

(I) Check for a snap ring, clip or set screw that may be holding the part in place.
(II) Use a slide hammer puller to force the part out of the case.

a. I only
b. II only
c. both I and II
d. neither I nor II

16. The **best** way to dry transmission parts after cleaning is with

a. compressed air.
b. paper towels.
c. clean shop towels.
d. old bath towels.

17. Mechanic A says that a properly assembled, one-way roller clutch should have 1/16 of an inch backlash before lockup.
Mechanic B says that a properly assembled planet carrier race of a one-way clutch should turn free in a counterclockwise direction.
Which mechanic is right?

a. A only
b. B only
c. both A and B
d. neither A nor B

18. To be sure that the torque converter is installed correctly, the mechanic should:

(I) Bolt the converter to the drive plate before installing the transmission.
(II) Measure the distance from the bell housing face to the converter drive lugs.

a. I only
b. II only
c. both I and II
d. neither I nor II

19. Mechanic A says an end-play check should be made during disassembly, before the front pump is removed.
Mechanic B says an end-play check should be made after the final assembly of the transmission.
Which mechanic is right?

a. A only
b. B only
c. both A and B
d. neither A nor B

20. Which of the following methods **cannot** be used to find cracks in an aluminum transmission case?

    a. dye penetrant test
    b. air pressure test
    c. visual inspection
    d. magnetic crack detection

EXTENDED STUDY PROJECTS

1. Visit several transmission repair shops and ask permission to watch rebuilding and repair operations.
2. Write a report comparing the techniques used by different mechanics.
3. By replying to advertisements in motoring magazines, obtain catalogs listing automatic transmission modification parts.
4. Using the catalogs, write a report on how you would modify a transmission for racing, recreational vehicle, or street use.

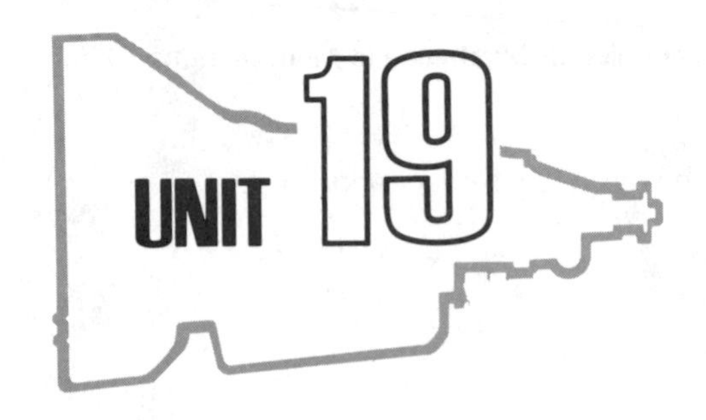

TURBO HYDRA-MATIC 125 AUTOMATIC TRANSAXLE

General Motors has designed the 125 Turbo Hydra-Matic automatic transaxle for use as a transverse-mounted, front-wheel drive unit in the smaller "X" body cars. The unit is lighter in weight than the small engine transmissions used in the past. The unit has three forward speeds and one reverse, plus neutral and park.

Like the other Turbo Hydra-Matics, the 125 makes use of a Simpson gear train. The 125, however, has the final drive and differential unit contained within the transmission case, figure 19-1. A forward clutch, intermediate band, direct clutch, low and reverse clutch, and a low roller clutch control the gear train. The control of the gear train and the power flow is the same as that described for the Ford C-6 and the THM 200 and 250 transmissions.

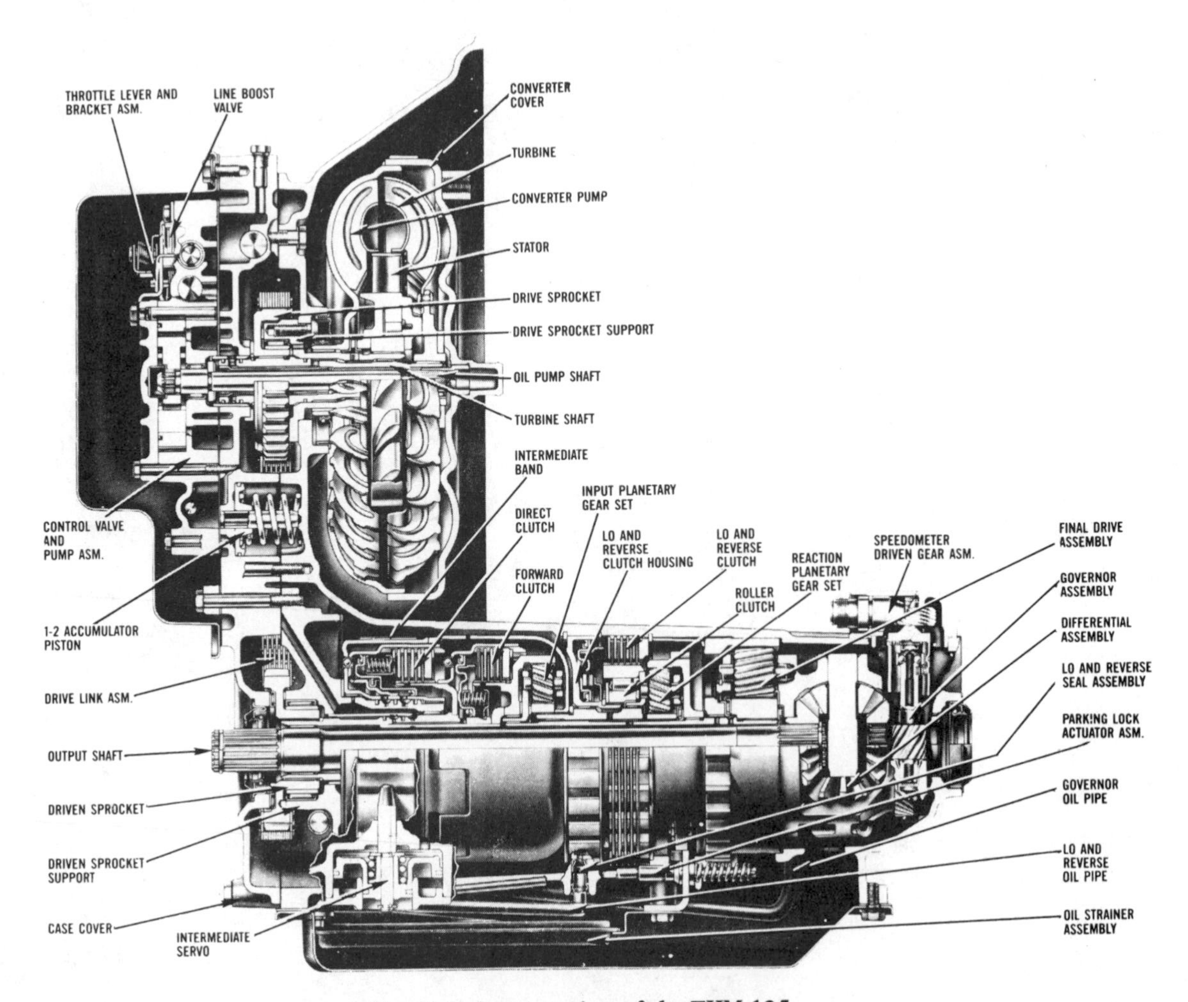

Fig. 19-1 Cutaway view of the THM 125.

TORQUE CONVERTER AND DRIVE LINK ASSEMBLY

The torque converter couples the engine to the transmission through a *drive link assembly*. The drive link assembly is made up of a turbine shaft and drive sprocket, a driven sprocket (splined to the transmission input shaft), and a link assembly, figure 19-2. The link assembly is constructed very much like the timing chain used on some engines to drive the camshaft. The drive and driven sprockets may have a different number of

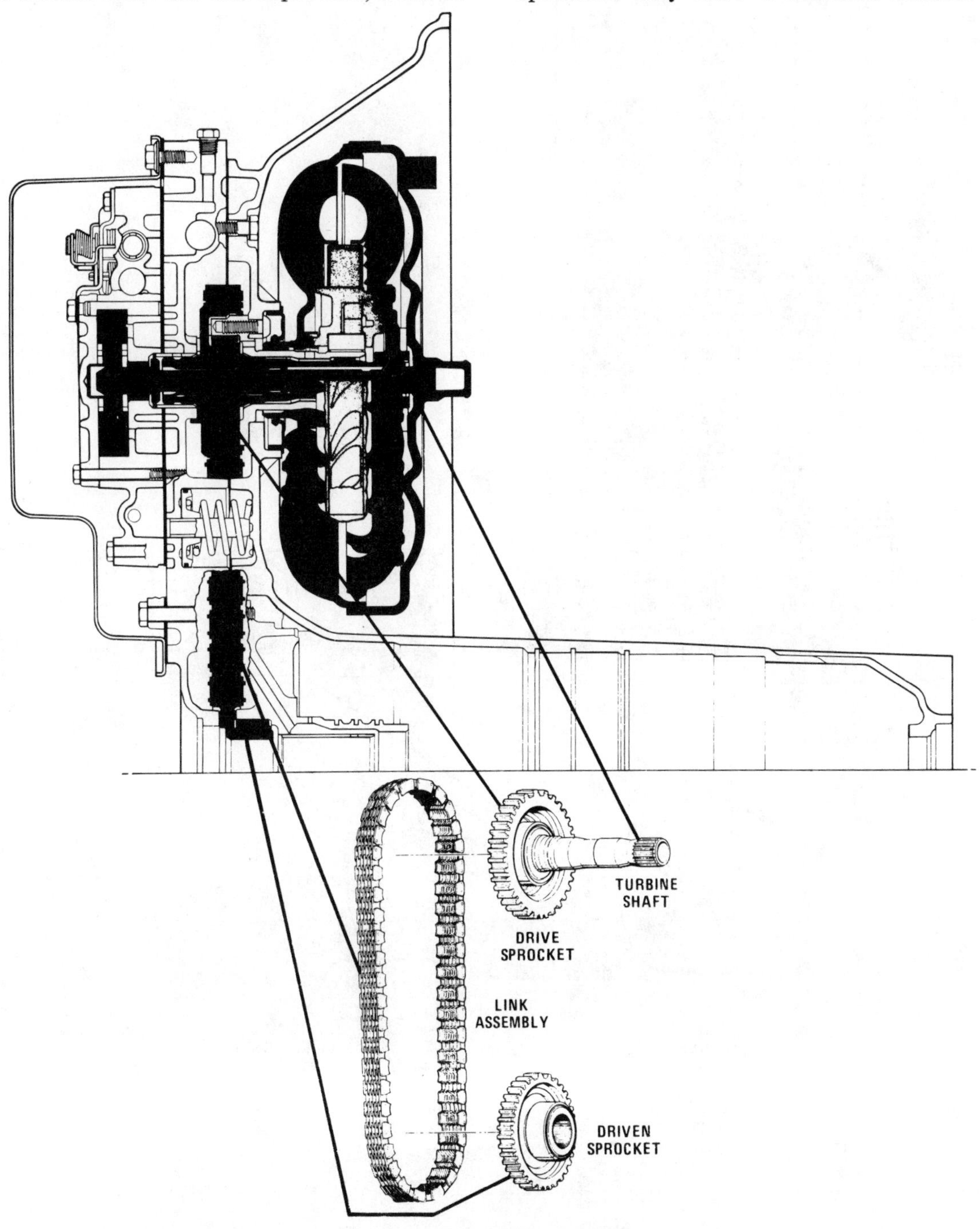

Fig. 19-2 Drive link assembly.

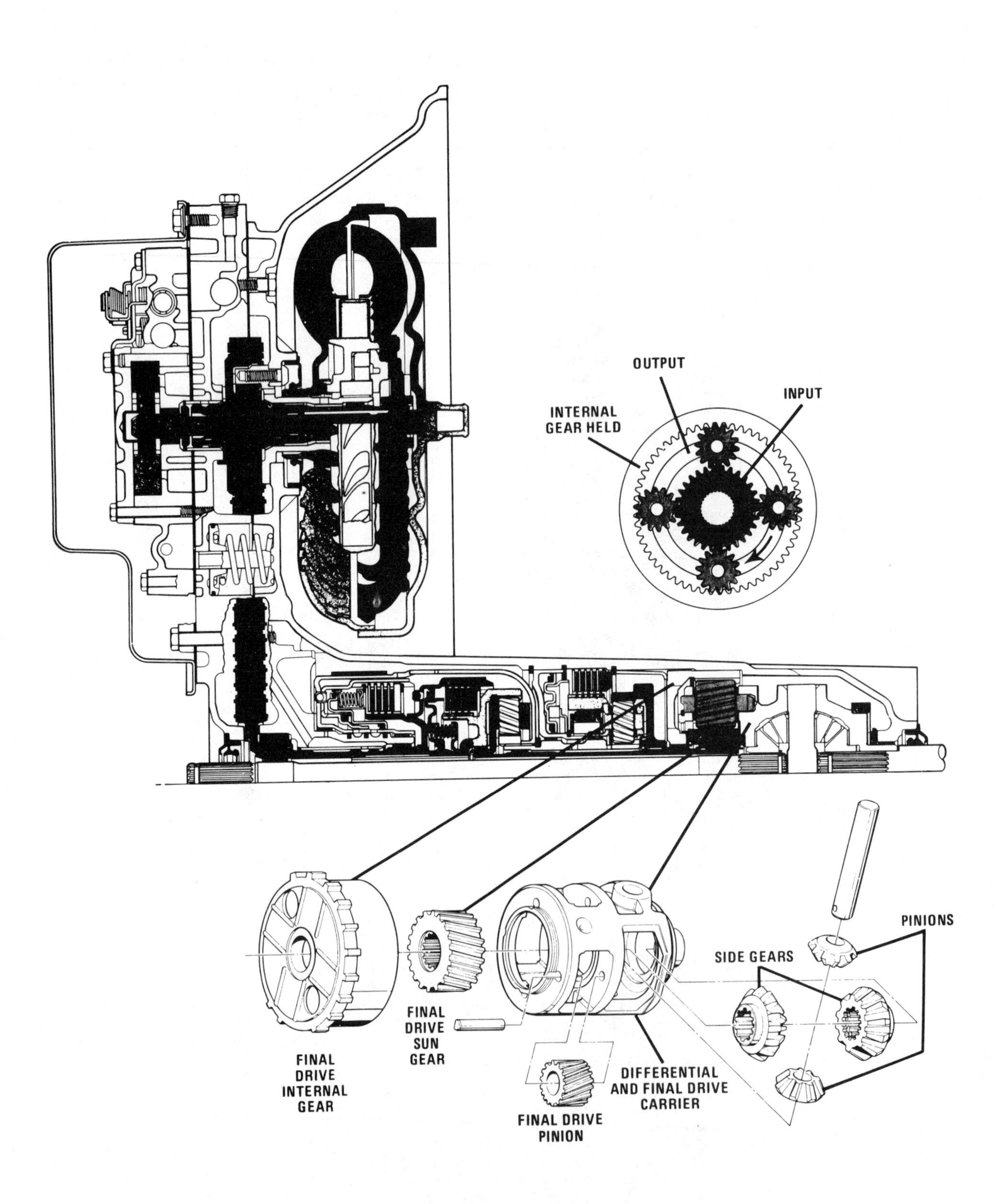

Fig. 19-3 Final drive and differential assemblies.

teeth. This allows the torque to the transmission input shaft to be varied according to engine size. The torque converter, of course, will also multiply engine torque at a ratio of about 1.95:1 at full stall.

FINAL DRIVE

In addition to the compound planetary gear set of the Simpson gear train, a simple planetary gear set is used as a final drive, figure 19-3. As shown, the output shaft from the rear ring gear is splined to the sun gear of the final drive gear set. The final drive internal (ring) gear is splined to the transmission case making the internal gear the held member of the final drive gear set. The planetary carrier is part of the differential assembly to which the two output (axle) shafts are splined.

This is the same setup as that found in a simple planetary gear set operating in its lowest, forward reduction (see unit 3). The final drive ratio of about 2.84:1 serves the same function as that given by the ring and pinion gears of a conventional drive axle unit.

DIFFERENTIAL

When going around corners, the outside wheel of a vehicle must move through a larger circle than the inner wheel. In order to do this the outer wheel must turn faster than the inner wheel. A *differential* allows this difference in speed, figure 19-4.

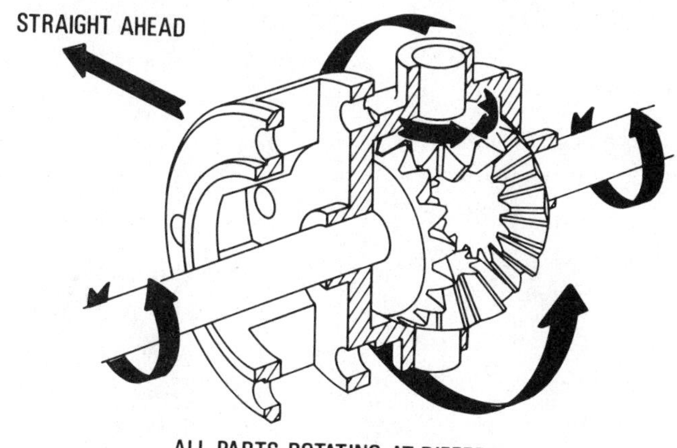

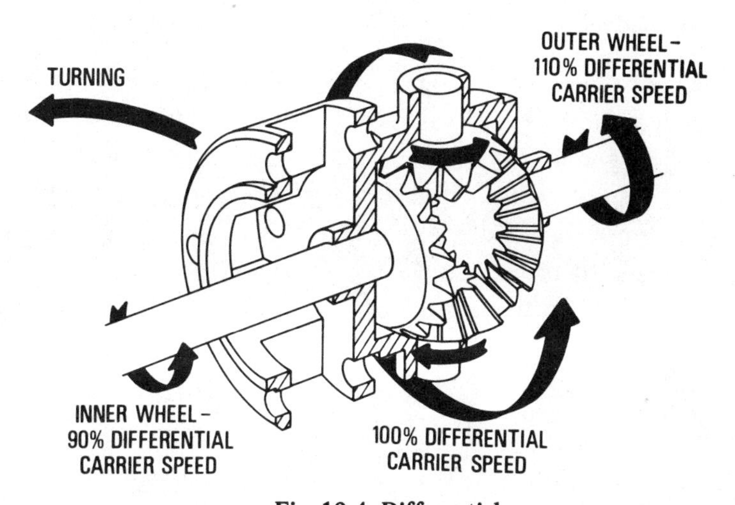

Fig. 19-4 Differential.

A differential assembly consists of a case, which in this instance is part of the final drive output carrier; differential side gears and pinions. As long as the vehicle is traveling in a straight line, the pinions serve merely as a link between the side gears, and torque is transmitted equally to each axle. Thus, the case, carrier and axle assemblies rotate as a unit.

When the vehicle is rounding a turn, the differential operates in the following way. While the carrier is still rotating, the pinions revolve enough to allow the outer wheel and axle to rotate at a faster speed than the inner wheel and axle. In the example shown in figure 19-4, page 279, the outer wheel is turning at 110 percent of the carrier speed while the inner wheel is turning at 90 percent of the carrier speed. In a sharp turn the speed difference between the inner and outer wheels would be even greater.

To further explain this, assume that one wheel of a vehicle is resting on hard, dry pavement while the other wheel is on a patch of ice. If the vehicle were prevented from moving, the wheel resting on the ice would turn at 200 percent of the carrier speed while the wheel on dry pavement would not turn at all. During normal, dry pavement use, the pinions will revolve the correct amount to balance the speed difference between the inner and outer wheels for the turn being made.

HYDRAULIC SYSTEM

The hydraulic system of the THM 125 has the same function as the hydraulic systems found in other automatic transmissions. That is, transmission fluid, under pressure, is directed by a valve body in order to provide full automatic and manual shifts. There are, however, some important technical differences that will be discussed in the following paragraphs.

Fluid Level

Due to design considerations in the THM 125, the oil pan is too shallow to contain all the fluid necessary, and another oil reservoir is needed, figure 19-5. The upper reservoir houses the pump and valve body assembly while the lower reservoir or sump contains the filter. Due to the design of this unit, the fluid level will vary according to operating temperature. When cold (65°-85°F or 18°-29°C), the thermostatic (temperature sensitive) element is open, and oil from the upper reservoir drains to the lower sump. At normal operating temperature (190°-200°F or 88°-93°C), the thermostatic element closes and the oil level increases in the upper reservoir as shown. Since the dipstick measures the oil level in the lower sump, the fluid level when cold will be **higher** than when hot.

This variation in oil level and the design of the dipstick tube make it necessary to employ the following procedure for checking the fluid level:

1. Shift the transmission through all ranges and let it stand in park, with the engine idling, for at least 5 minutes.
2. After the engine has idled at least 5 minutes, and with the transmission still in park, remove the dipstick and determine if the fluid is hot or cold. If the fluid is cold, the end of the dipstick will feel cool or warm to the touch; if the end of the dipstick is too hot to hold with the fingers, the fluid is hot.
3. Wipe the dipstick clean; push the dipstick completely into the tube and, once again, remove it.
4. If the fluid is hot, the level should be between the ADD and FULL marks on the dipstick.
5. If the fluid is cold, the level should be 7 mm (about 1/4 inch) **above** the FULL mark (between the two dimples above the full mark on the dipstick).

Due to the design of the dipstick tube, care should be taken to avoid a false fluid-level

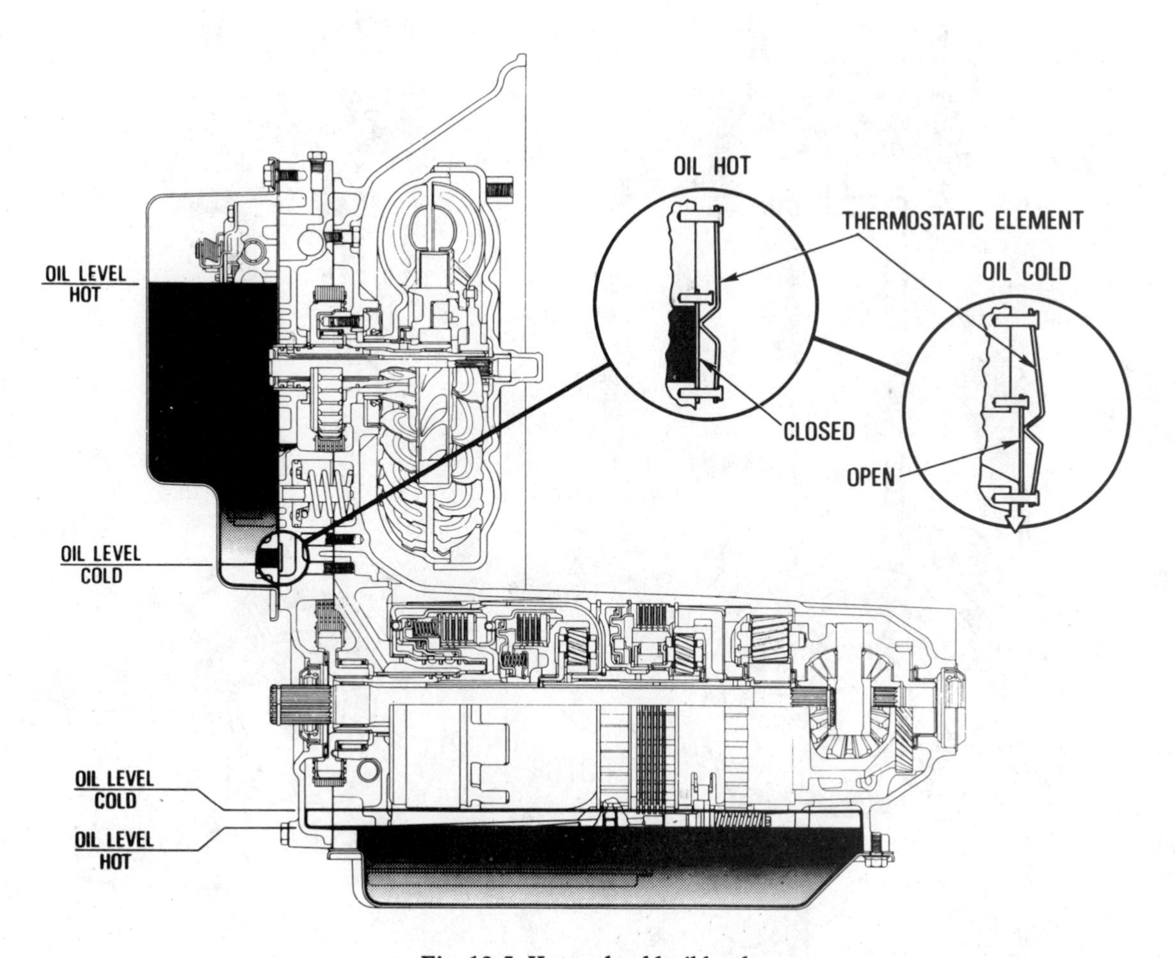

Fig. 19-5 Hot and cold oil levels.

reading. The following procedures help to avoid this problem:

- Be sure a full circle of fluid is shown on the dipstick (no dry spots).
- If the vehicle has been driven hard, let it stand, with the engine off, for 30 minutes before checking.
- If the vehicle has just come off the road or has been shifted through all the ranges, it must be idled in park for 5 minutes before checking.

If the fluid level is low, it should be brought to the correct level using Dexron II automatic transmission fluid. Finally, always keep in mind that the fluid level will be higher when cold.

Pump Assembly and Pressure Regulation

The THM 125 is unique among automatic transmissions in that it uses a variable capacity vane-type pump, figure 19-6, page 282. It is located within, and is a part of, the valve body, and is referred to as the *control valve and pump assembly* (see figure 19-1, page 276).

This pump is actually a part of the main pressure regulating system that automatically regulates pump output to the needs of the transmission. With the engine off, or just started, the priming spring holds the pump slide in the position of maximum output, (figure 19-6A). During operation, line pressure is sent to the pressure regulator valve.

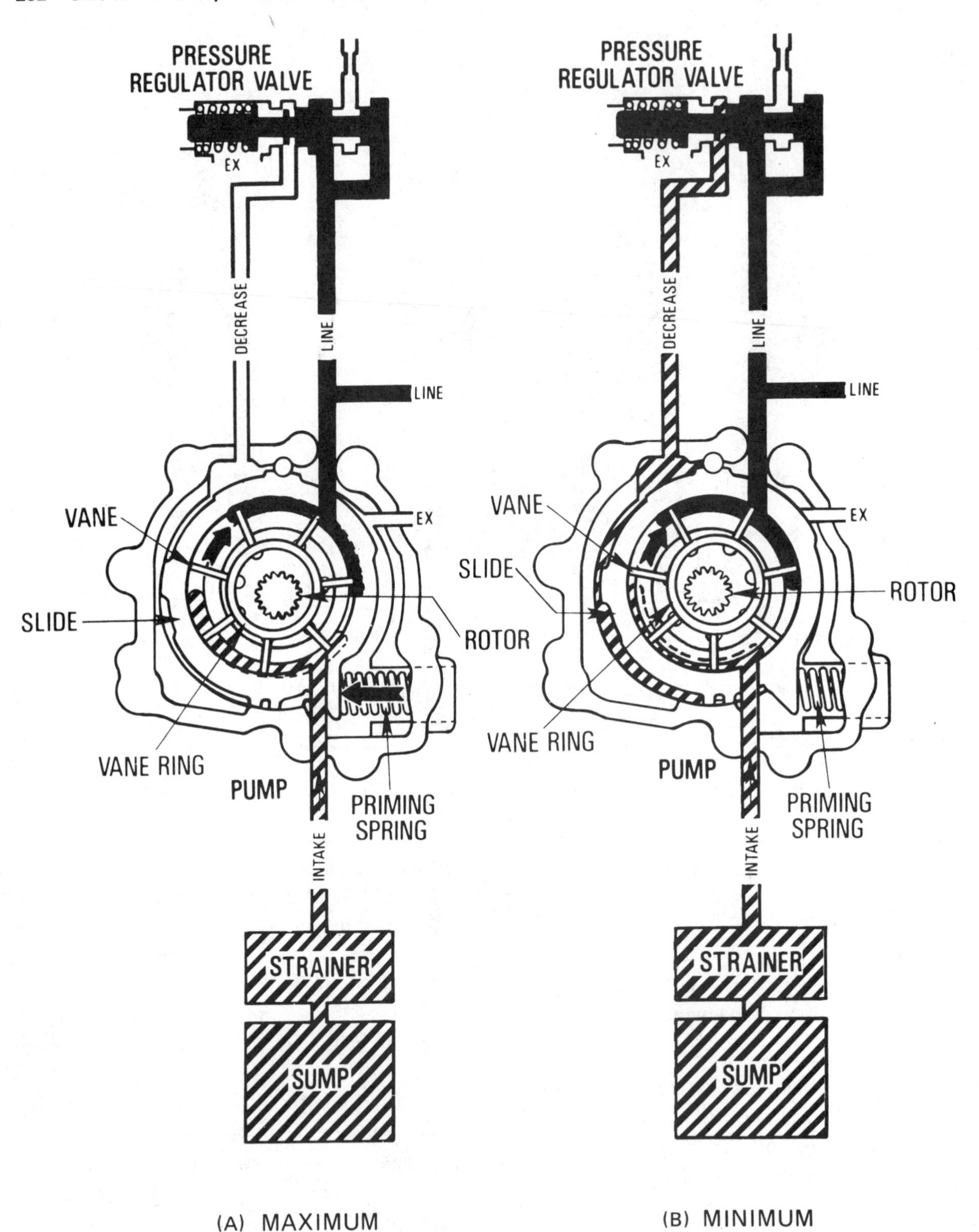

Fig. 19-6 Maximum and minimum pump output.
[For color diagram, see page 345]

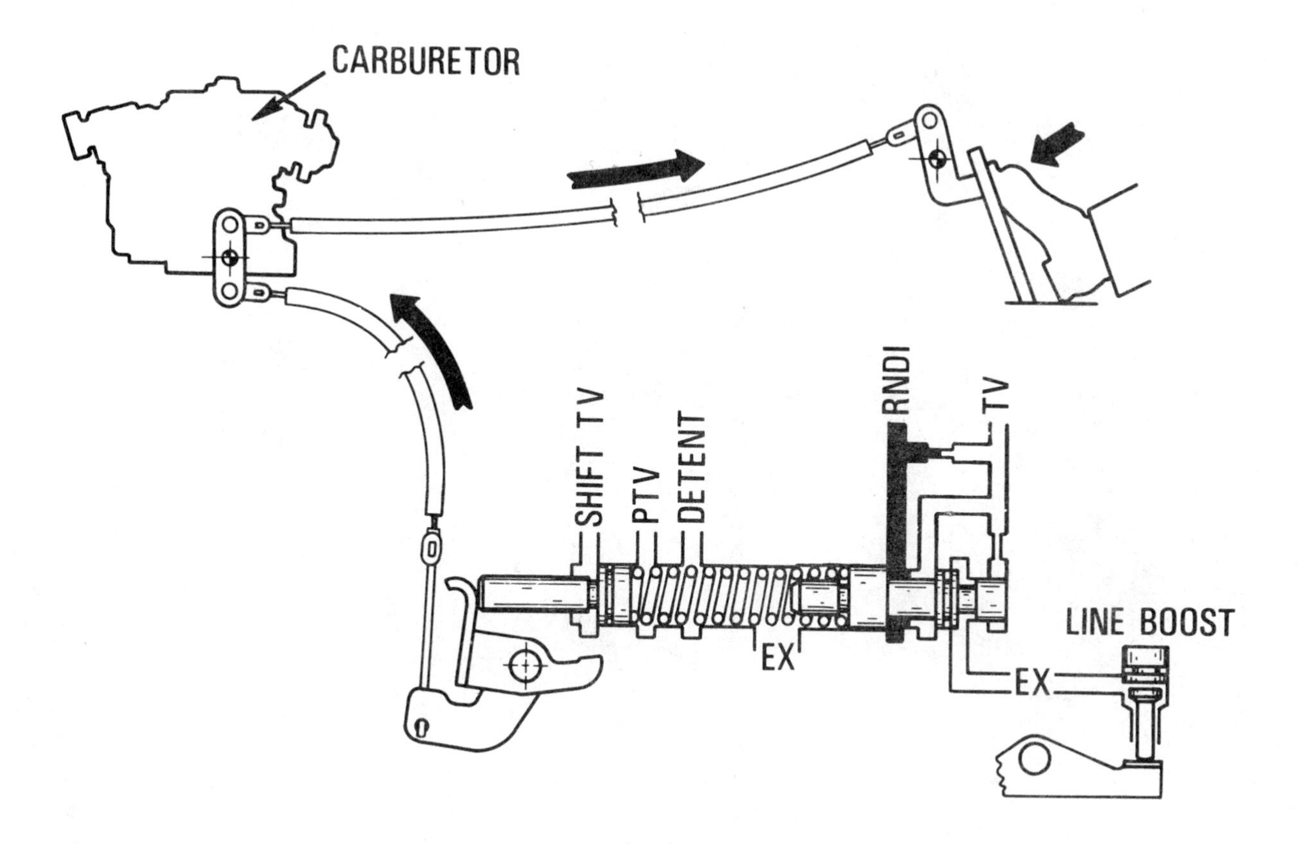

Fig. 19-7 Accelerator, carburetor and TV linkage.

The pressure regulator is a balanced bypass valve. In this case, however, the bypass is directed back to the pump slide through the decrease passage. As pressure builds, the bypass fluid, still under pressure, moves the slide to the right, compressing the priming spring (figure 19-6B). This increases the area of the pressure cavity in the pump, which results in a decrease of pump output.

The illustrations in figure 19-6 show the pump and pressure regulator in the maximum and minimum positions. With the transmission in use, however, both the pump and pressure regulator will seek a position to match the pressure needs of the transmission for various driving conditions. As in other transmissions, variable throttle pressure and intermediate and reverse boost passages are used to automatically adjust line pressure to driving conditions and the range selected.

Throttle Valve

The throttle valve is connected, through a linkage and cable arrangement, to the carburetor and accelerator pedal, figure 19-7. Throttle pressure is, therefore, regulated by throttle opening and will vary between 0 and 105 psi (723 kPa). The throttle valve is fed by line pressure in reverse, neutral, drive and intermediate (identified as RNDI in figure 19-7). Due to the absence of RNDI oil in park or low, TV pressure is cut off in these ranges.

This throttle valve works on the bypass principle: that is, bypassed RDNI oil is used as TV pressure. Three orificed exhaust ports in the TV system are used to control the flow of bypassed fluid: one at the shift TV passage,

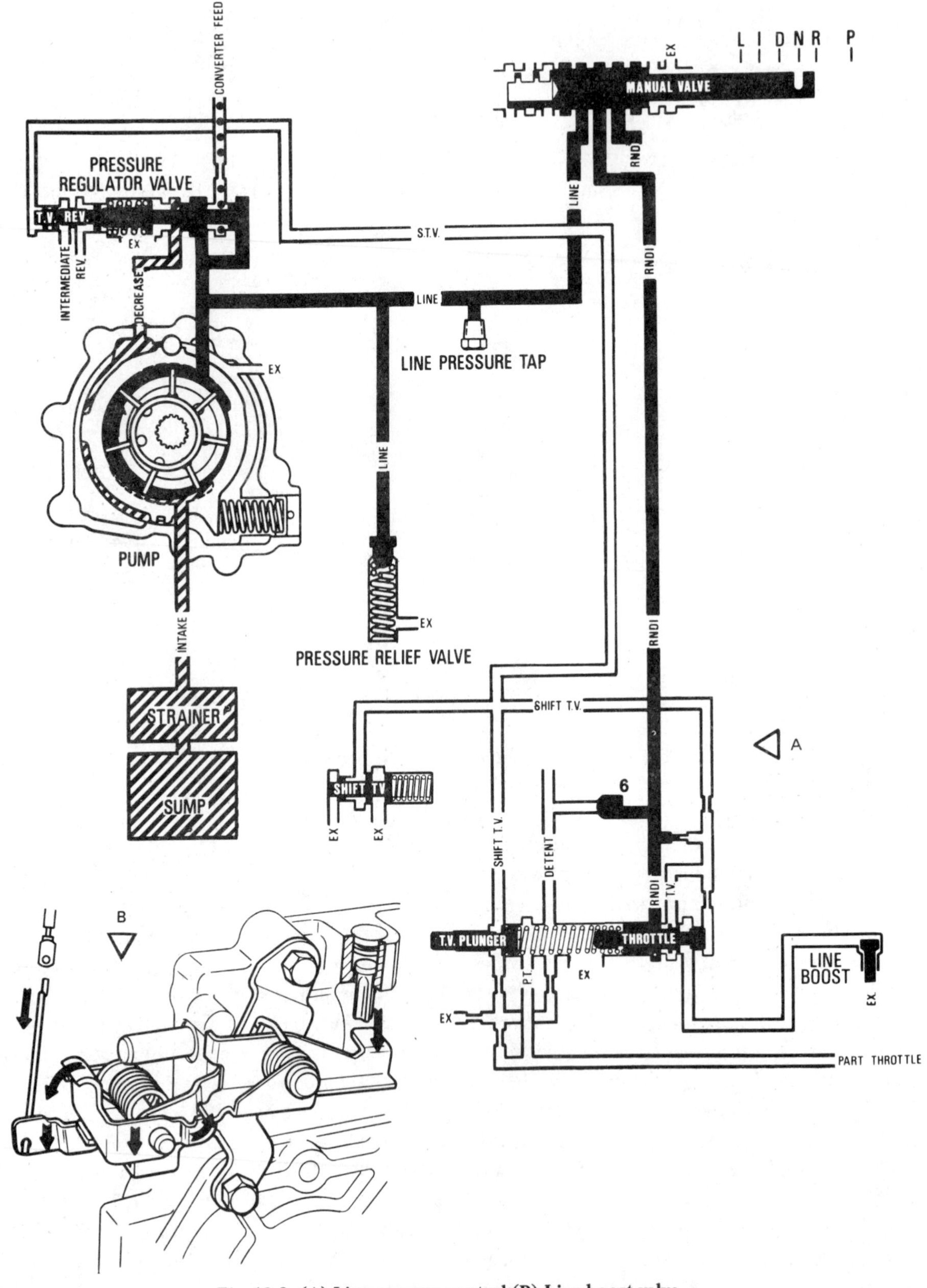

Fig. 19-8 (A) Line pressure control (B) Line boost valve.

one at the detent passage, and one at the line boost valve, figure 19-8A. The line boost valve, figure 19-8B, is normally held in open position. If the TV linkage, however, is not adjusted properly or becomes disconnected or broken, the line boost valve will close and seal off the exhaust of TV pressure at this point. This will boost TV and line pressure in order to prevent clutch or band slippage due to low pressure. Shift TV pressure at the TV plunger acts as a force to reduce the accelerator pedal effort necessary to move the TV plunger.

Shift TV Valve

The shift TV valve limits TV pressure to a maximum of 90 psi (620 kPa) by exhausting pressure over this amount. Part throttle (PT) and detent pressure will be higher, however. This will happen when the TV plunger is moved far enough to the right to block off the detent exhaust port. As in other automatic transmissions, TV pressure acts on the pressure regulator to boost line pressure according to throttle opening, and on the shift valves as an opposing force to governor pressure.

Governor

Instead of the usual restriction-type valve, the governor of the THM 125 works on the bypass principle, figure 19-9. Two check balls, a primary weight, a secondary

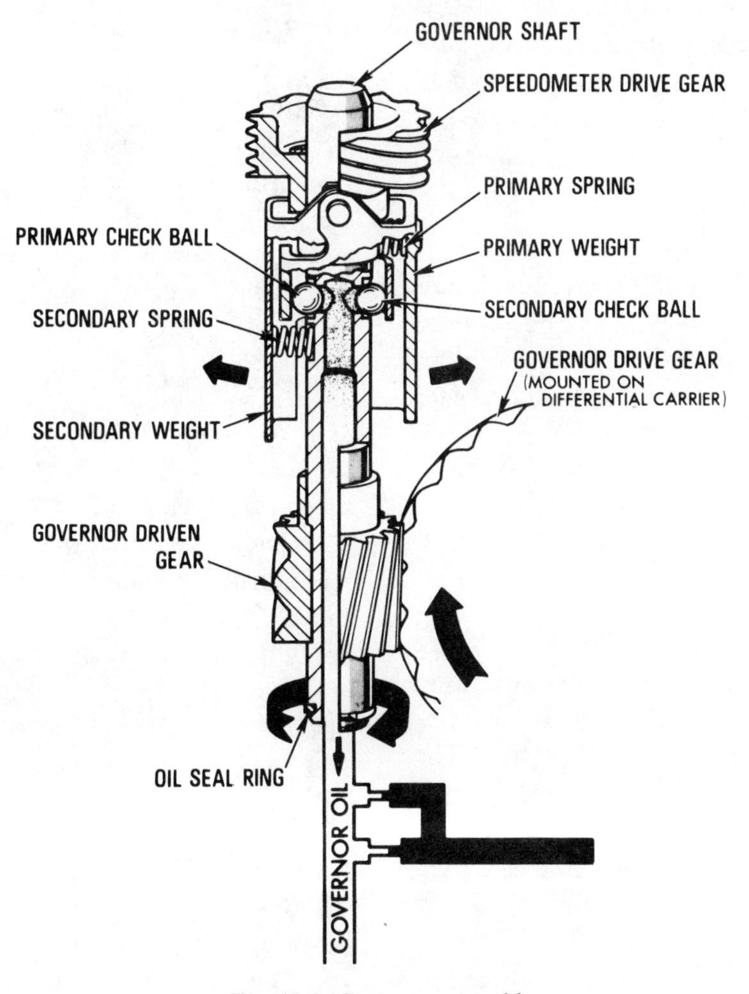

Fig. 19-9 Governor assembly.

weight and a primary and secondary spring are used to regulate governor pressure. Drive oil (line pressure) is metered through two orifices to the governor passage. With the vehicle at a standstill, governor oil is exhausted at the check balls and, due to the orificing, there is little or no governor pressure.

At low vehicle speeds and governor r/min, the heavier primary weight tends to seat the primary check ball and governor pressure increases. As the rotational speed of the governor increases with vehicle speed, the primary weight completely seats the primary check ball. From this point on, governor pressure is controlled by the lighter secondary weight and spring. The use of primary and secondary weights causes governor pressure to be proportional to vehicle speed.

HYDRAULIC CONTROL AND POWER FLOW

Except for the differences noted previously, the hydraulic system and power flow of the THM 125 is basically the same as in other transmissions using a Simpson gear train. Refer as needed to figures 19-10 and 19-11, pages 287 and 288, when studying the following paragraphs.

Park and Neutral

In park, fluid flow is to the pressure regulator valve, pump slide, converter and lubrication system, manual valve, line pressure tap and the pressure relief valve. The pressure regulator valve and pump slide limit line pressure to about 70 psi (483 kPa). A pressure relief valve is used in the system and limits line pressure to about 300 to 450 psi (2 069 to 3 103 kPa).

With the selector lever in neutral, line pressure enters the RDN passage to a valley of the 2-3 shift valve and the RDNI passage to the throttle valve. In this position, line pressure varies between 70 to 140 psi (483 to 965 kPa) depending on throttle opening. Shift TV pressure also acts on the 1-2 and 2-3 shift valves and the 1-2 accumulator valve.

Drive Range First

When the selector lever is moved to D, line pressure enters the drive passage. Drive oil applies the forward clutch and is sent to the governor, the 1-2 shift valve and the 1-2 accumulator valve. The 1-2 accumulator valve is controlled by a spring force and shift TV pressure, and regulates drive oil to a variable 1-2 accumulator pressure. This variable accumulator pressure acts on the 1-2 accumulator piston in preparation for a 1-2 shift.

With the forward clutch applied, the front ring gear is driving. The rear planetary carrier is being held by the low roller clutch, and power flow is the same as in other Simpson gear trains for drive range first.

Drive Range Second

When the 1-2 shift valve moves in response to governor pressure, drive oil (line pressure) enters the 2nd passage. Pressure in the 2nd passage strokes the intermediate servo piston to apply the band and strokes the 1-2 accumulator piston for smooth band apply. The intermediate band stops the sun gear, and the front planetary gear set is in intermediate or second speed.

Drive Range Third

When the 2-3 shift valve moves, RND oil (line pressure) enters the direct clutch passage. Direct clutch oil, at the inner piston area, applies the direct clutch, seats the accumulator check valve and releases the

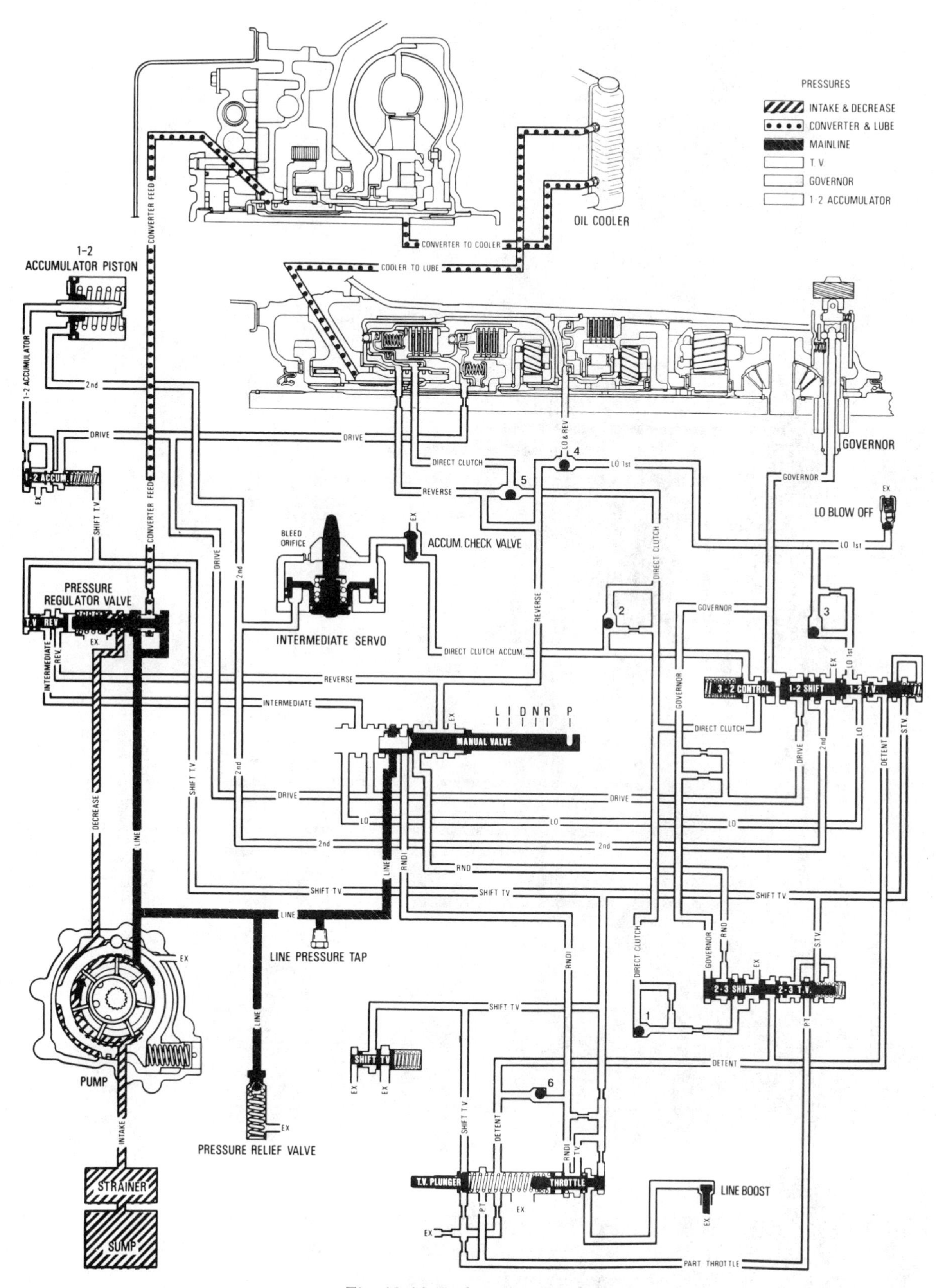

Fig. 19-10 Park-engine running.

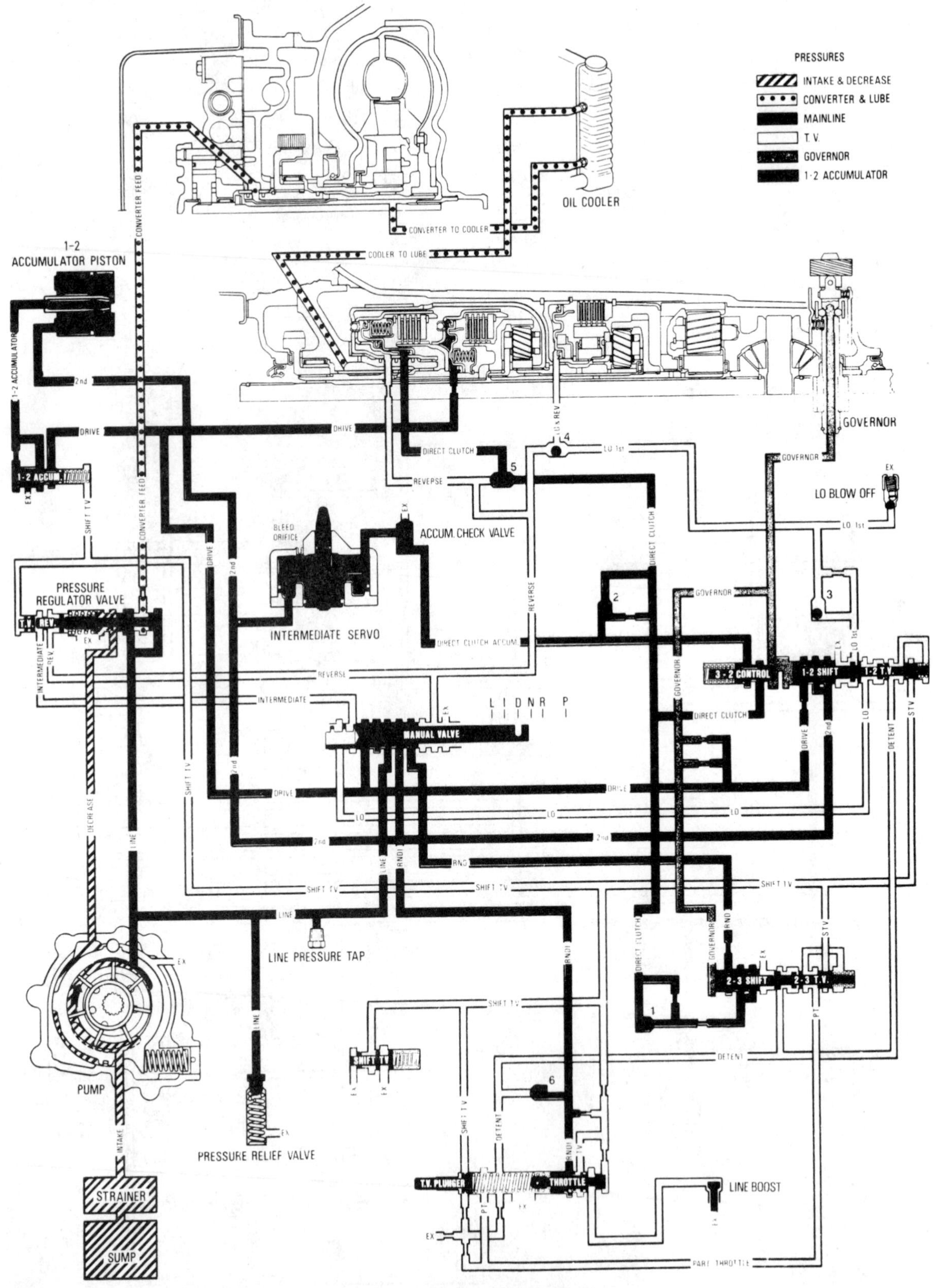

Fig. 19-11 Drive range-third gear.

[For color diagram, see page 346.]

intermediate servo. Intermediate servo release acts as an accumulator for direct clutch apply. With the direct clutch applied, the front ring gear and sun gear are driving.

Direct clutch oil also passes through the 3-2 control valve. The 3-2 control valve is regulated by governor pressure and controls the release rate of direct clutch oil for part-throttle and detent 3-2 downshifts.

Part-throttle 3-2 Downshift

Below speeds of about 50 mph (80 km/h), a part-throttle 3-2 downshift takes place when the driver depresses the accelerator pedal to nearly wide-open throttle. This opens the PT passage (at the TV plunger) to shift TV pressure. PT pressure moves the 2-3 valve to the left, and opens the direct clutch passage to exhaust.

Detent Downshift

At speeds below about 65 mph (105 km/h) a forced or detent 3-2 downshift will take place when the throttle is pushed wide open. This moves the TV plunger to the right and opens the PT and detent passages to maximum shift TV pressure. This pressure works on differential force areas of the 2-3 valve, and, combined with the spring force, will downshift the valve to the second speed position. With the valve in the second speed position, direct clutch oil is exhausted and intermediate servo oil (already present) applies the servo and band. At road speeds above 50 mph (80 km/h) governor pressure moves the 3-2 valve to the left. This routes the exhausting direct clutch accumulator oil through an orifice and slows up band apply for a smooth 3-2 shift.

At wide-open throttle, a detent 2-1 or 3-1 downshift will take place at speeds below about 30 mph (48 km/h). The combined forces of detent pressure and the spring will move the valve to the left. This exhausts 2nd oil, releasing the intermediate band, and the transmission will be in first speed.

Intermediate Range

In intermediate range, RDN oil is exhausted at the manual valve and the transmission will downshift to second regardless of road speed. The exhaust of RND oil releases the direct clutch and direct clutch accumulator oil. The band is then applied by 2nd oil acting on the intermediate servo.

Intermediate oil acts on the reverse boost valve at the pressure regulator and increases line pressure to 120 psi (327 kPa). This increased pressure gives added holding force to the forward clutch and intermediate band.

L Range

With the manual valve in L position, RND and RNDI oil is exhausted at the manual valve. Lo oil, from the manual valve, will downshift the 1-2 shift valve at speeds below about 40 mph (64 km/h). Since drive oil is present, the forward clutch remains applied. Lo 1st oil applies the low and reverse clutch for engine braking. Intermediate oil is still present at the pressure regulator valve which tends to keep line and Lo 1st pressure at 120 psi (828 kPa). Lo 1st oil, however, is also directed to the Lo blow-off valve. This unseats the Lo blow-off valve and exhausts Lo 1st oil. Since Lo 1st oil seats check ball 3 and must pass through an orifice, Lo 1st pressure will be reduced to about 35 psi (241 kPa).

Reverse

When reverse is selected, line pressure is sent to three passages: the reverse, the RNDI and the RND. Reverse oil applies the direct clutch at both the inner (direct clutch) and outer (reverse clutch) areas for greater holding power. The combination of reverse and shift TV oil at the pressure regulator valve, will boost line pressure to between 120 psi (827 kPa) to a maximum of 210 psi (1 447 kPa) depending on throttle opening. Due to the design of the manual valve, RND oil is present at the 2-3 shift valve, but serves no function in reverse.

SUMMARY

The THM 125 automatic transaxle is a fully automatic transmission, using a Simpson gear train, with three forward speeds. Manual selection of reverse, park, neutral, intermediate and low is also provided for. Designed for front-wheel drive "X" body cars, the unit also includes a final drive and differential assembly.

The hydraulic system, while using basic hydraulic principles, departs from the usual design in its use of a variable capacity vane-type pump. The pump, together with the pressure regulator valve, regulates line pressure. Due to the use of two oil reservoirs and a thermostatic valve, checking the fluid level requires special procedures.

REVIEW

Select the best answer from the choices offered to complete the statement or answer the question.

1. In drive range second gear, the sun gear in the THM 125 transmission is held by

   a. the intermediate band.
   b. the intermediate clutch.
   c. the intermediate roller clutch.
   d. the forward clutch.

2. The drive and driven sprockets of the THM 125 transmission may:

   (I) Have a different number of teeth.
   (II) Operate at different speeds.

   a. I only b. II only c both I and II d. neither I nor II

3. The drive sprocket of the THM 125 transmission is part of the

   a. input shaft.
   b. impeller.
   c. converter housing.
   d. turbine shaft.

4. The driven sprocket of the THM 125 transmission is splined to the

   a. front pump.
   b. input shaft.
   c. drive axle.
   d. converter hub.

5. The hydraulic pump of the THM 125 transmission has

   a. internal and external gears.
   b. a rotor and vanes.
   c. a sprocket and chain.
   d. a priming valve.

6. The fluid level on the THM 125 transmission

   a. is higher when cold.
   b. is higher when hot.
   c. cannot be checked cold.
   d. must be checked in D.

7. In order to help regulate pressure, the pump of the THM 125 transmission has:

   a. a variable capacity.
   b. a fluid coupling.
   c. a throttle connected cable.
   d. a low blow-off valve.

8. In the final drive unit of the THM 125 transmission, the sun gear is held by:

   (I) A roller clutch.
   (II) The differential carrier.

   a. I only b. II only c. both I and II d. neither I nor II

9. The purpose of a differential is to:

   (I) Provide a lower gear ratio to the drive wheels.
   (II) Allow one drive wheel to turn faster than the other.

   a. I only b. II only c. both I and II d. neither I nor II

10. In the final drive unit of the THM 125 transmission, the ring gear is

    a. driven by a clutch.
    b. held by a clutch.
    c. splined to the case.
    d. held by a band.

11. In the THM 125 hydraulic system, the valve used to prevent low line pressure should the throttle cable become broken is called the

    a. lo blow-off valve.
    b. TV plunger.
    c. line boost valve.
    d. TV boost valve.

12. A broken or disconnected TV cable in the THM 125 transmission would most likely cause

    a. low line pressure.
    b. low TV pressure.
    c. slipping shifts.
    d. harsh shifts.

EXTENDED STUDY PROJECTS

1. On blank oil circuit diagrams, complete each of the following exercises:

   a. Indicate the oil circuit for neutral.
   b. Indicate the oil circuit for D range low.
   c. Indicate the oil circuit for D range intermediate.
   d. Indicate the oil circuit for manual 2.
   e. Indicate the oil circuit for manual 1.
   f. Indicate the oil circuit for reverse.

2. Using the oil circuit diagrams and your knowledge of the Simpson gear train, answer the following questions:

   a. State three reasons for a no drive condition in D.
   b. State three reasons for late, harsh shifts.
   c. Assuming that the oil level is correct, give two reasons for no upshift.

CONVERSION FACTORS

Torque	lb-in × 0.11298 = newton-meters (N•m) lb-ft × 1.3558 = N•m
Pressure	in Hg × 3.368 = kilopascals (kPa) psi × 6.895 = kPa (newton/m^2 = pascal)
Energy or Work	ft-lb × 1.3558 = joules (J) calories × 4.187 = J Btu × 1055 = J Btu × 1.055 = kJ
Power	Horsepower × 0.746 = kilowatts (kW) ft-lb/min × 0.226 = watts (W)
Velocity	mph × 1.6093 = km/h ft/sec × 0.3048 = m/s
Linear	in × 25.4 = mm ft × 0.3048 = m
Area	in^2 × 645.16 = mm^2 in^2 × 6.452 = cm^2
Volume	in^3 × 16,387 = mm^3 in^3 × 16.387 = cm^3 in^3 × 0.01639 = L (Use only for fluids, salt, etc.)
Mass	ounces × 28.35 = grams (g) lb × 0.4536 = kg
Force	ounces (of force) × 0.278 = newtons (N) lb (of force) × 4.448 = N kilograms (of force) × 9.807 = N

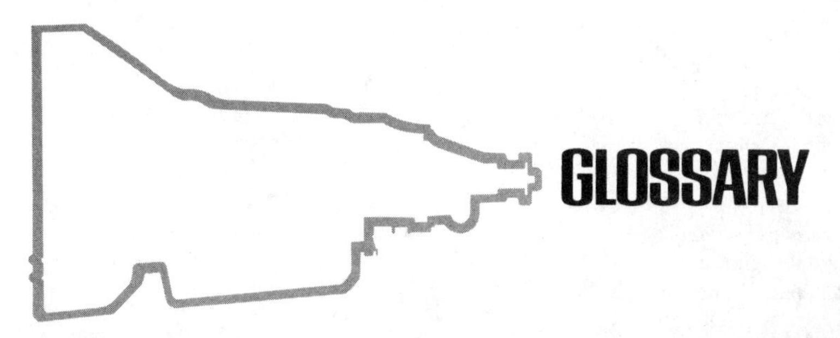

GLOSSARY

ATF – automatic transmission fluid

Accumulator – a piston and cylinder arrangement used to control pressure during the apply of a clutch or band

Accumulator valve – a valve used to control the flow and/or pressure of fluid to an accumulator, clutch or servo

Atmospheric pressure – pressure exerted by the surrounding air; at sea level, normal atmospheric pressure is said to be 14.7 psi, 29.92 in Hg, or, in SI units, 101 kPa.

Backlash – 1) the clearance between meshing teeth of two gears; or 2) the amount of free motion in a mechanical system such as a gear train. (*See end play*).

Balanced valve – a valve in which hydraulic pressure exactly balances an opposing force; usually a spring force varied by mechanical linkage, hydraulic pressure, vacuum (atmospheric pressure) or a combination of these forces

Band – a friction device used to hold a member of a planetary gear set by tightening around the members outer circumference

Bellows, aneroid – a pleated chamber sensitive to small changes in atmospheric pressure; used in some modulators to sense changes in altitude

Bore – the diameter of a hole; a machined or bored hole in a valve body or clutch cylinder

Burr – a rough edge on a piece of metal

Bypass valve – a valve that regulates hydraulic pressure by shunting or bypassing part of the incoming pressure to the sump or pump inlet

Check valve (ball check) – a valve that permits the flow of fluid in one direction only

Clutch – a friction device that can be used to either hold or drive a member of a planetary gear set

Coefficient of friction – the force required to overcome friction divided by the weight of the moving body

Detent – 1) a shallow depression, notch or hole in a shaft or plate into which a spring-loaded ball or plunger fits serving to lock the shaft or plate in one or more positions; or 2) in GM automatic transmissions, used to denote wide-open throttle shifts.

Diaphragm – a thin sheet of rubber or neoprene used to separate a pressure chamber, such as a modulator into a high (atmospheric) and low (vacuum) side.

Diaphragm spring – a disc-shaped spring used as a piston return spring or spacer in some clutch assemblies

End play – the amount of free motion in a mechanical system; in automatic transmissions, the working clearance between the components of a gear train

Friction – the resistance to motion caused by two surfaces in contact with one another.

Governor – in automatic transmissions, a device used to control shift points in relation to vehicle speed

Hydraulics – a branch of physics dealing with the use of liquids to transfer motion or force

Modulator – in an automatic transmission, a vacuum device used to control pressure and shift points

Multiple-disc clutch – a clutch using several friction discs; allows the frictional area to be increased while keeping a relatively small diameter

One-way clutch – a clutch using bearing races that allows rotation in one direction only; rollers or sprags allow free motion in one direction, but jam the races to prevent motion in the opposite direction

Planetary gear set (compound) – a planetary gear set made up of two, or parts of two or more simple planetary sets; more flexible and easier to control (for automatic transmission use) than a simple set

Planetary gear set (simple) – a gear set that consists of a sun gear, planet gears and carrier, and a ring gear; changes in ratio are made without taking the gears out of mesh

Regulator valve – a valve used to regulate hydraulic pressure

Restriction valve – a valve that regulates hydraulic pressure by restricting or blocking part of the incoming pressure

Servo – in an automatic transmission, a hydraulically operated cylinder/piston arrangement used to apply a band

Shift valve – a valve used to direct hydraulic fluid to a clutch or servo for automatic shifting

Shuttle valve – a valve used to direct hydraulic fluid to different parts of the valve body or hydraulic system

Spalling – bearing damage due to chipping or flaking of the case-hardened surfaces

Sprag – an hourglass-shaped bearing used in some one-way clutches

Vacuum – an absence or air; pressures below atmospheric

INDEX

11-83(3C294I)

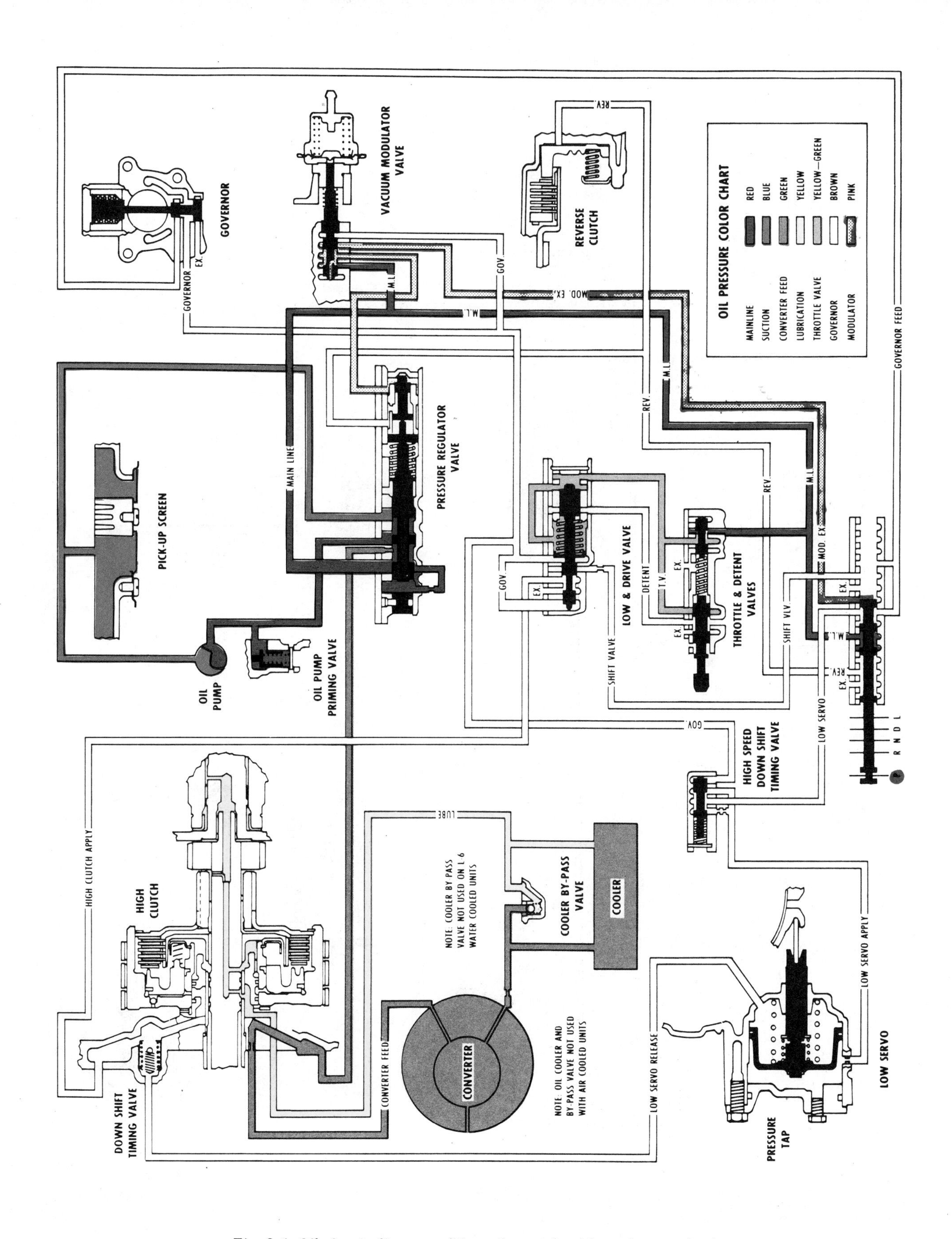

Fig. 8-6 Oil circuit diagram. (Neutral or park with engine running.)

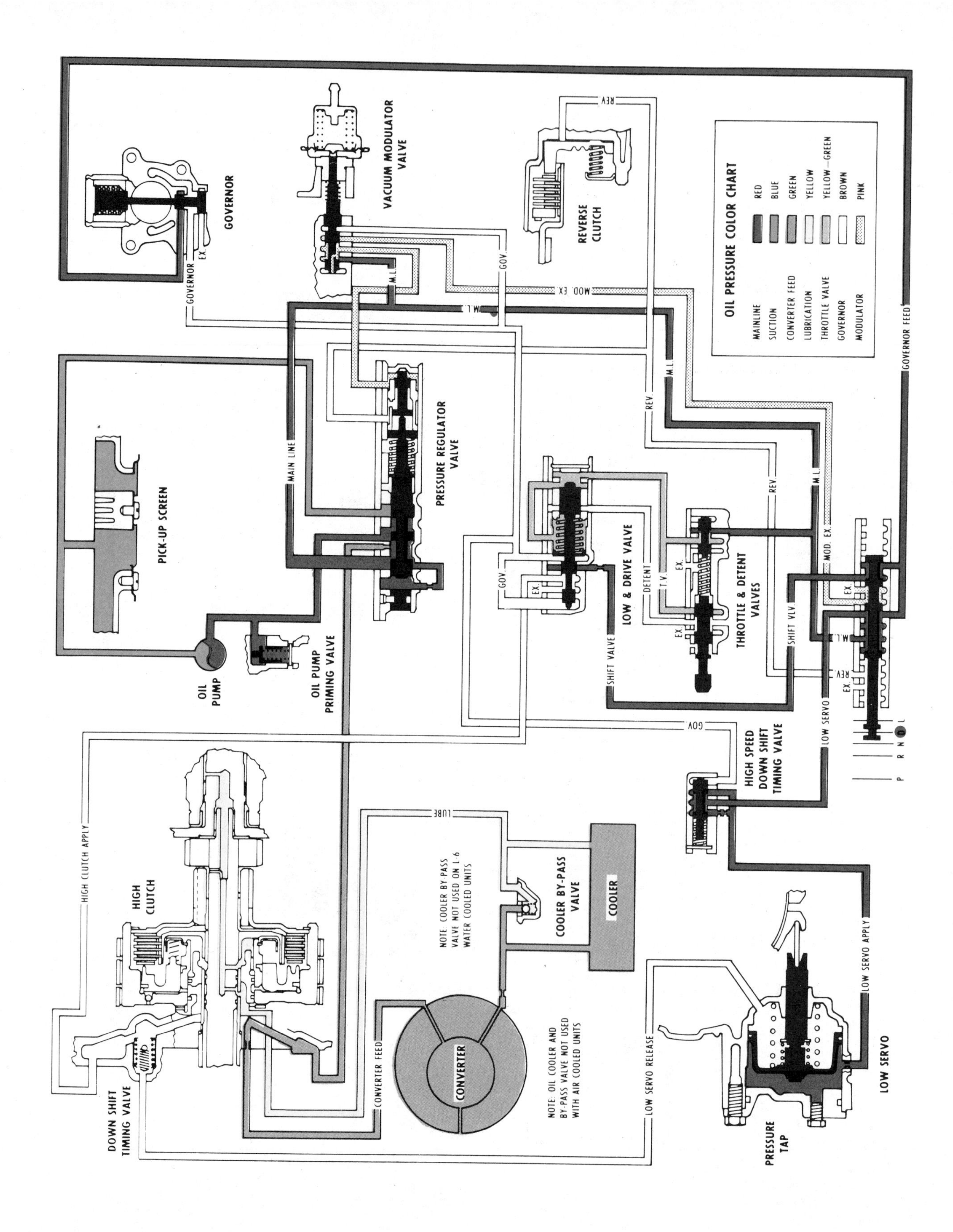

Fig. 8-8 Oil circuit diagram - drive range low.

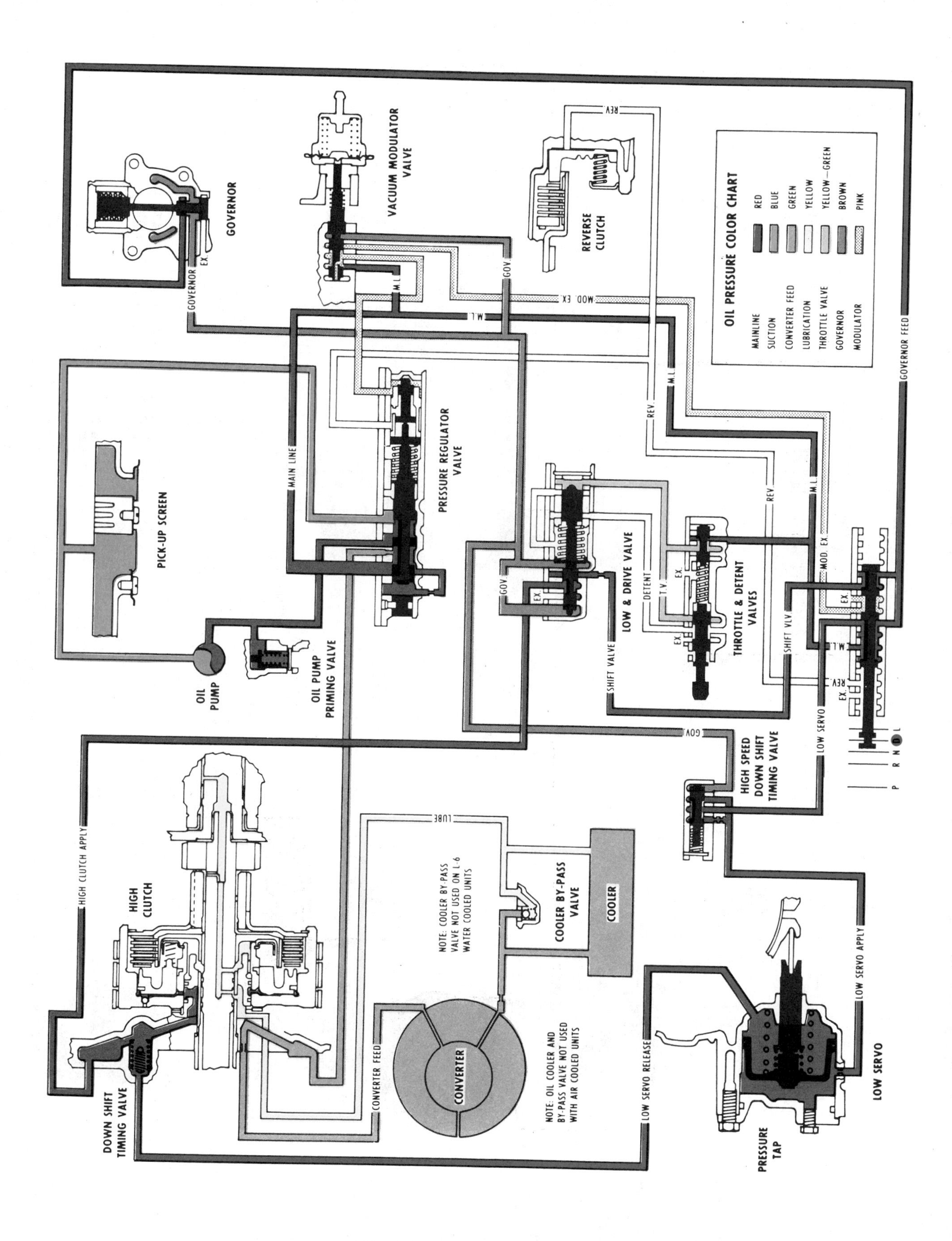

Fig. 8-12 Drive range high.

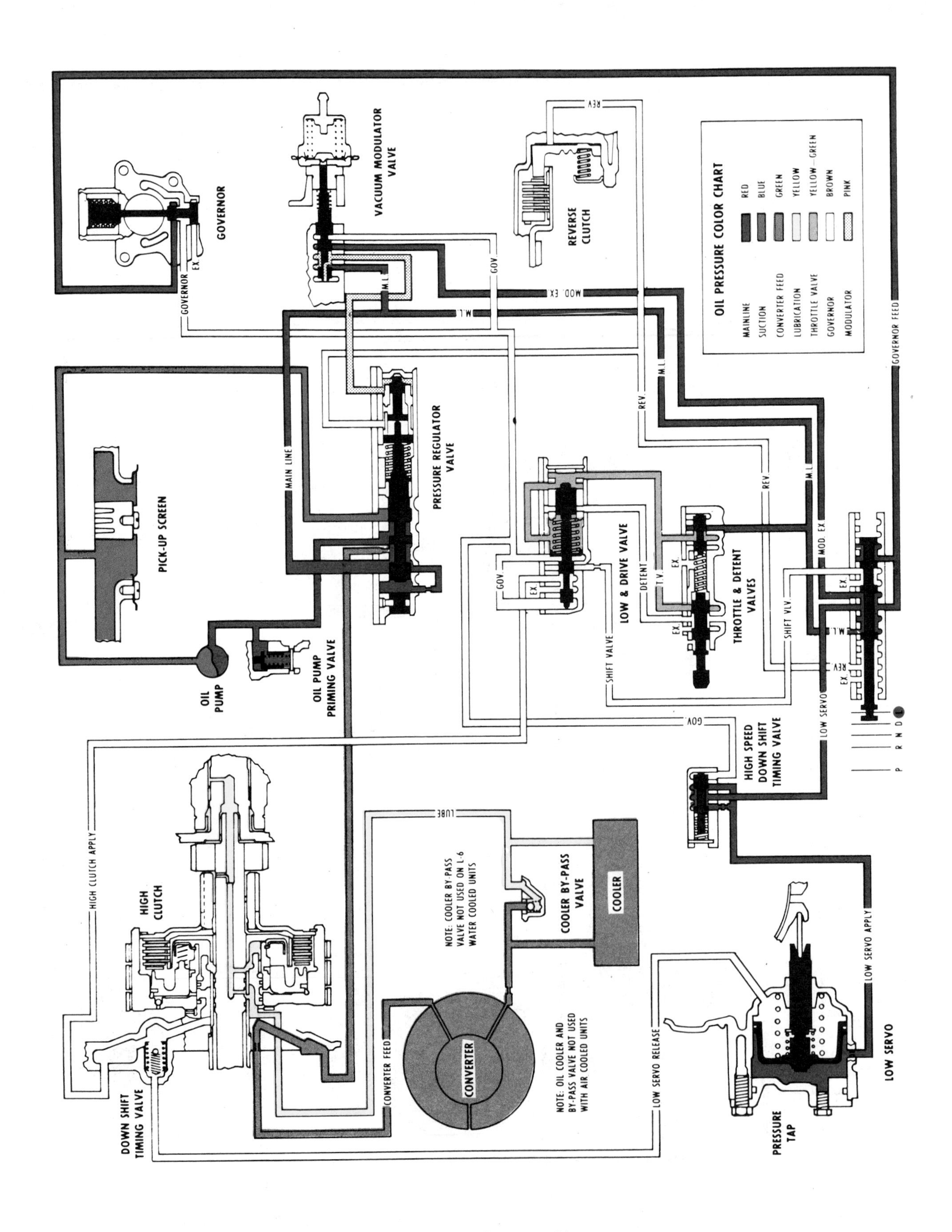

Fig. 8-16 Oil circuit for manual low.

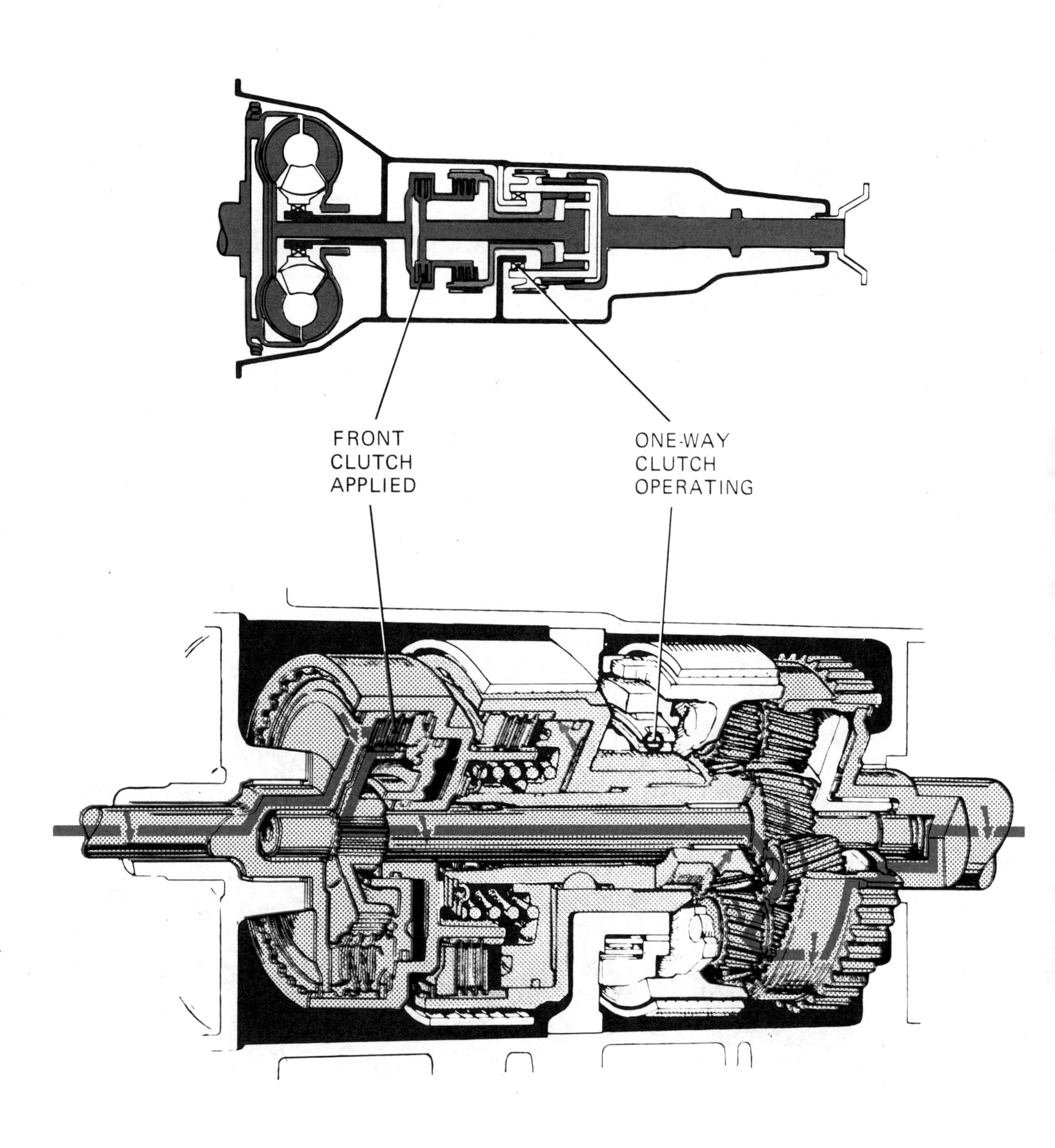

Fig. 9-4 Schematic for drive range low.

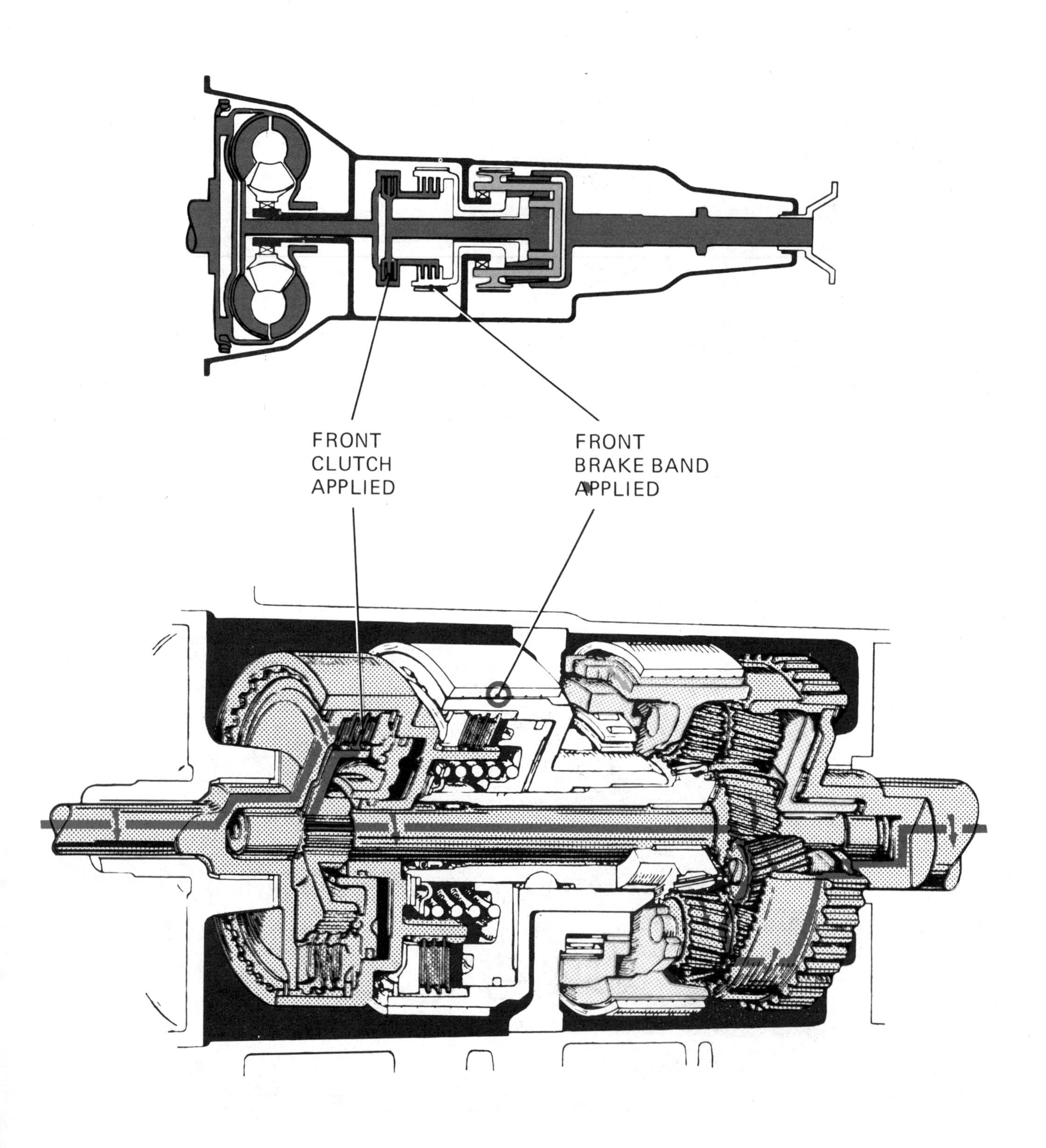

Fig. 9-7 Schematic for second gear.

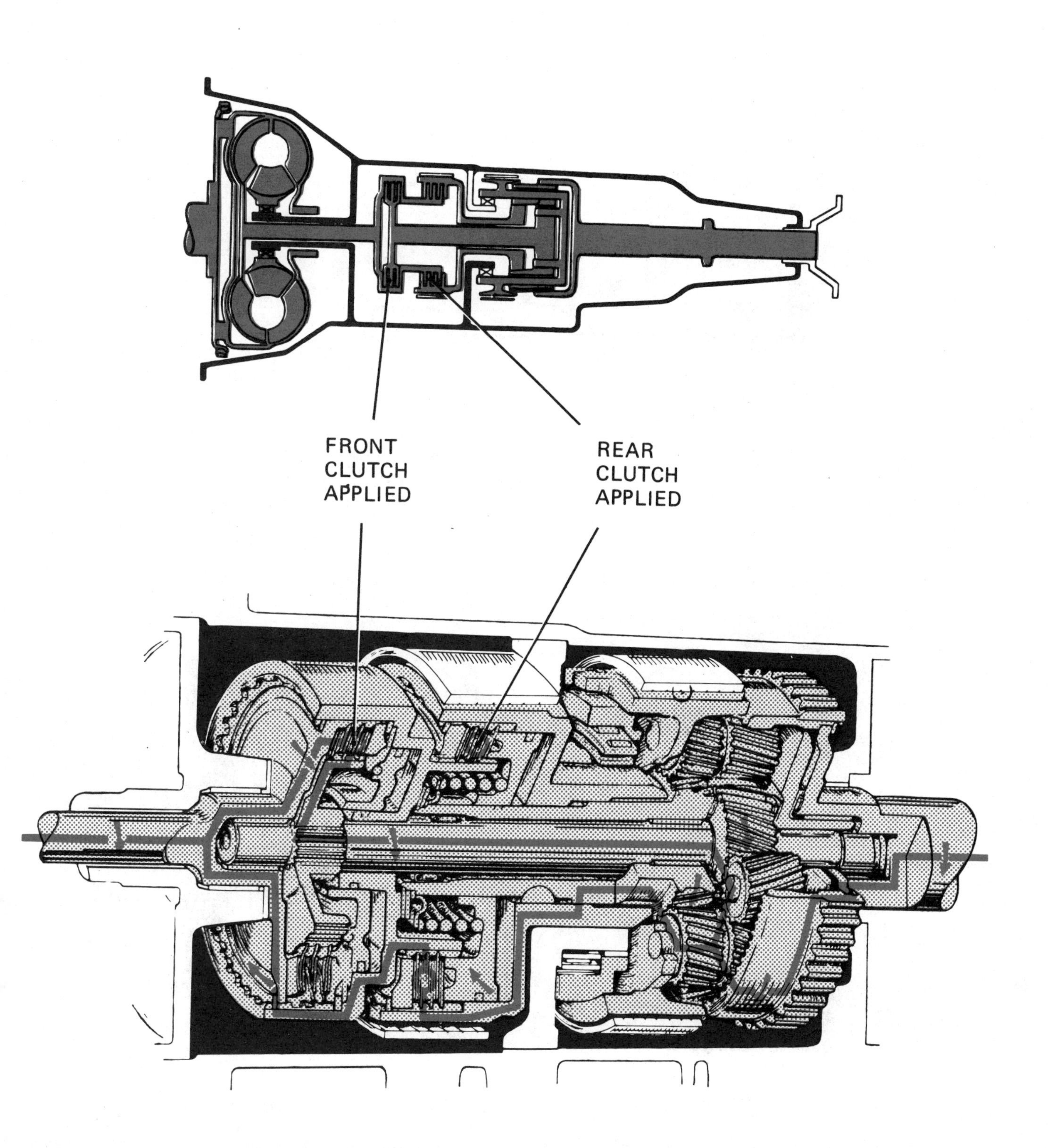

Fig. 9-9 Schematic for third gear or direct.

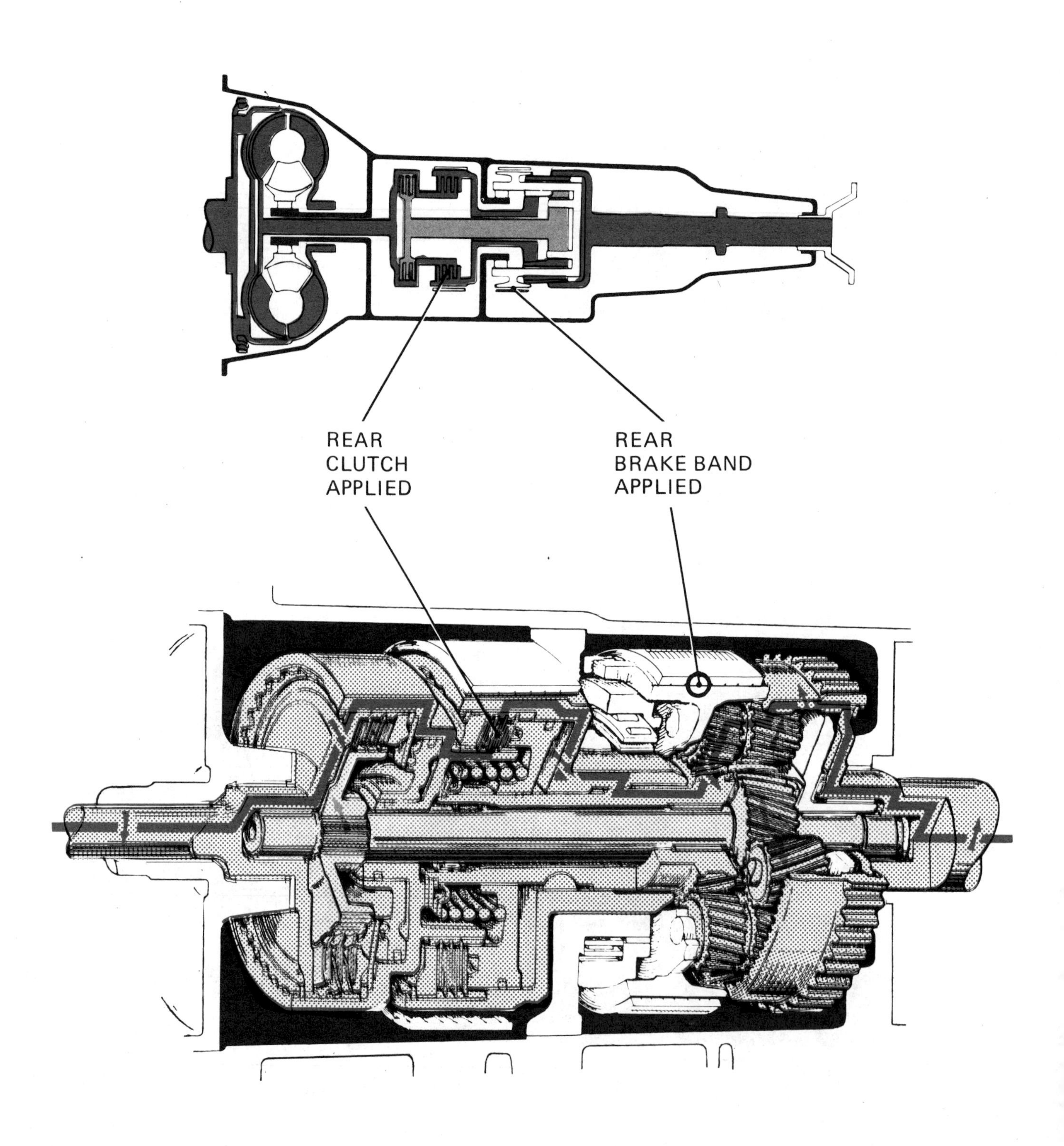

Fig. 9-11 Schematic for reverse.

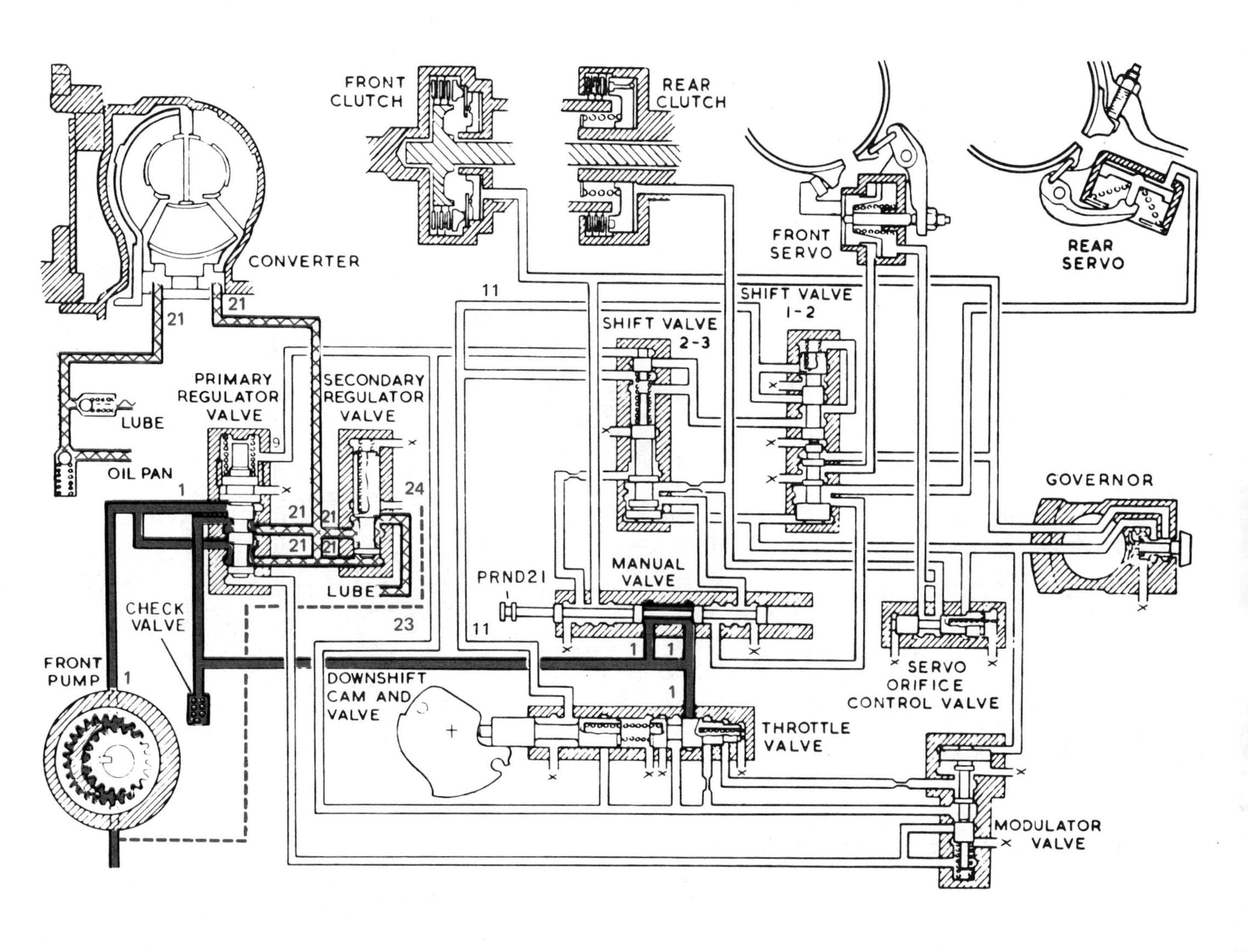

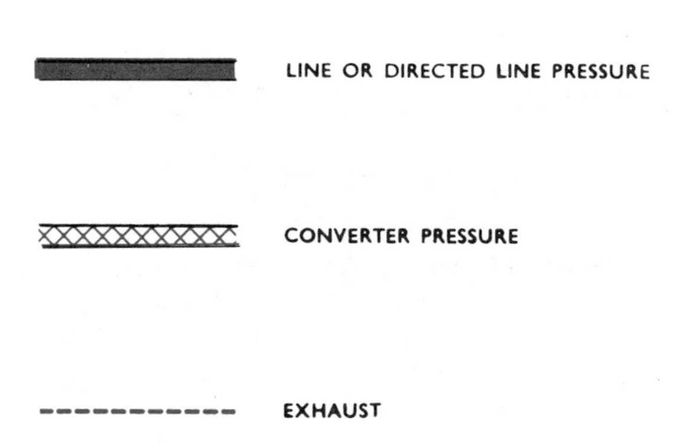

Fig. 9-13 Hydraulic circuit in neutral.

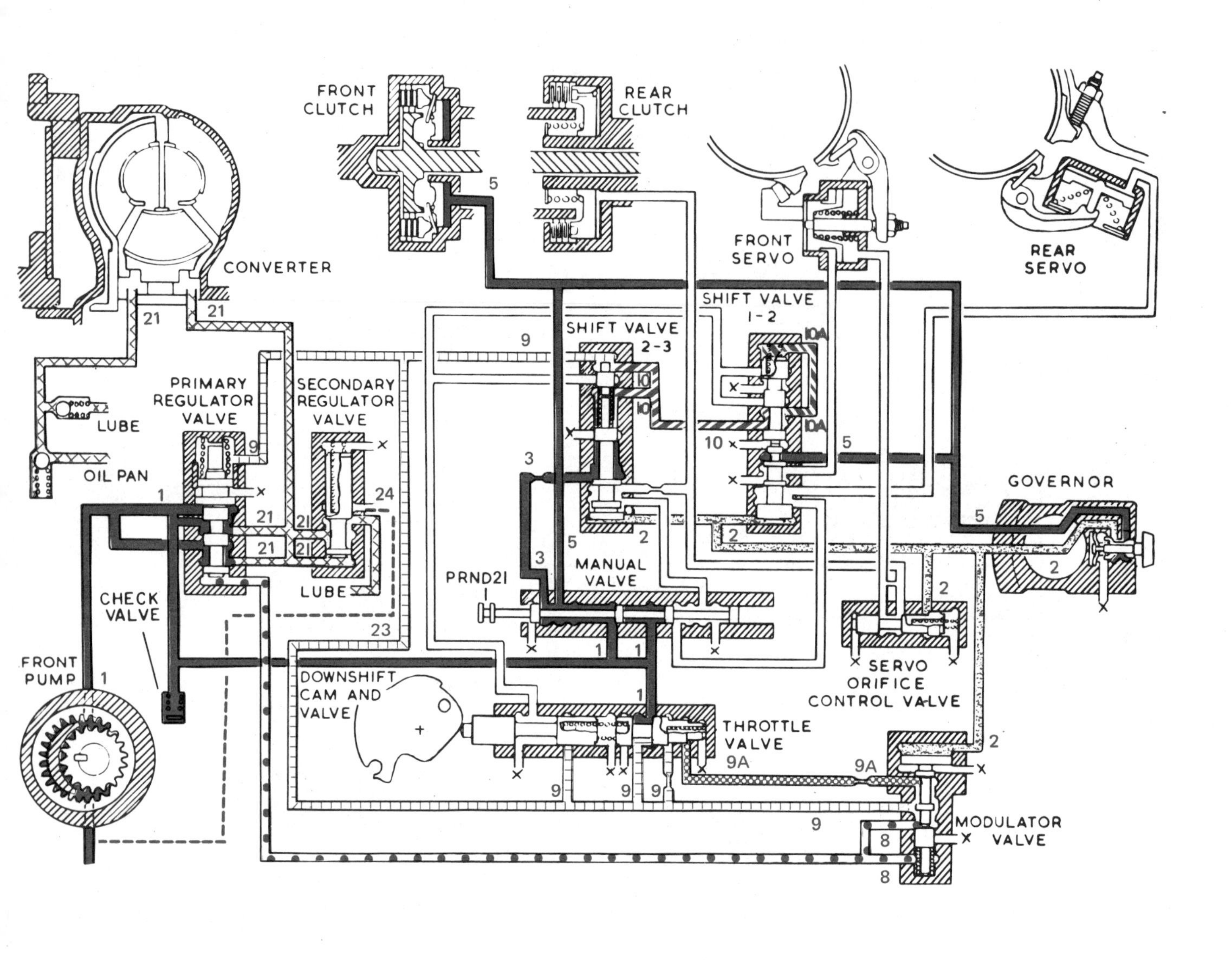

LINE OR DIRECTED LINE PRESSURE

CONVERTER PRESSURE

GOVERNOR PRESSURE

THROTTLE PRESSURE

FORCED THROTTLE PRESSURE

MODULATED THROTTLE PRESSURE

THROTTLE PRESSURE CONTROLLED BY MODULATOR VALVE

SHIFT VALVE PLUNGER PRESSURE

EXHAUST

Fig. 9-14 TV at half-open throttle, modulator operating.

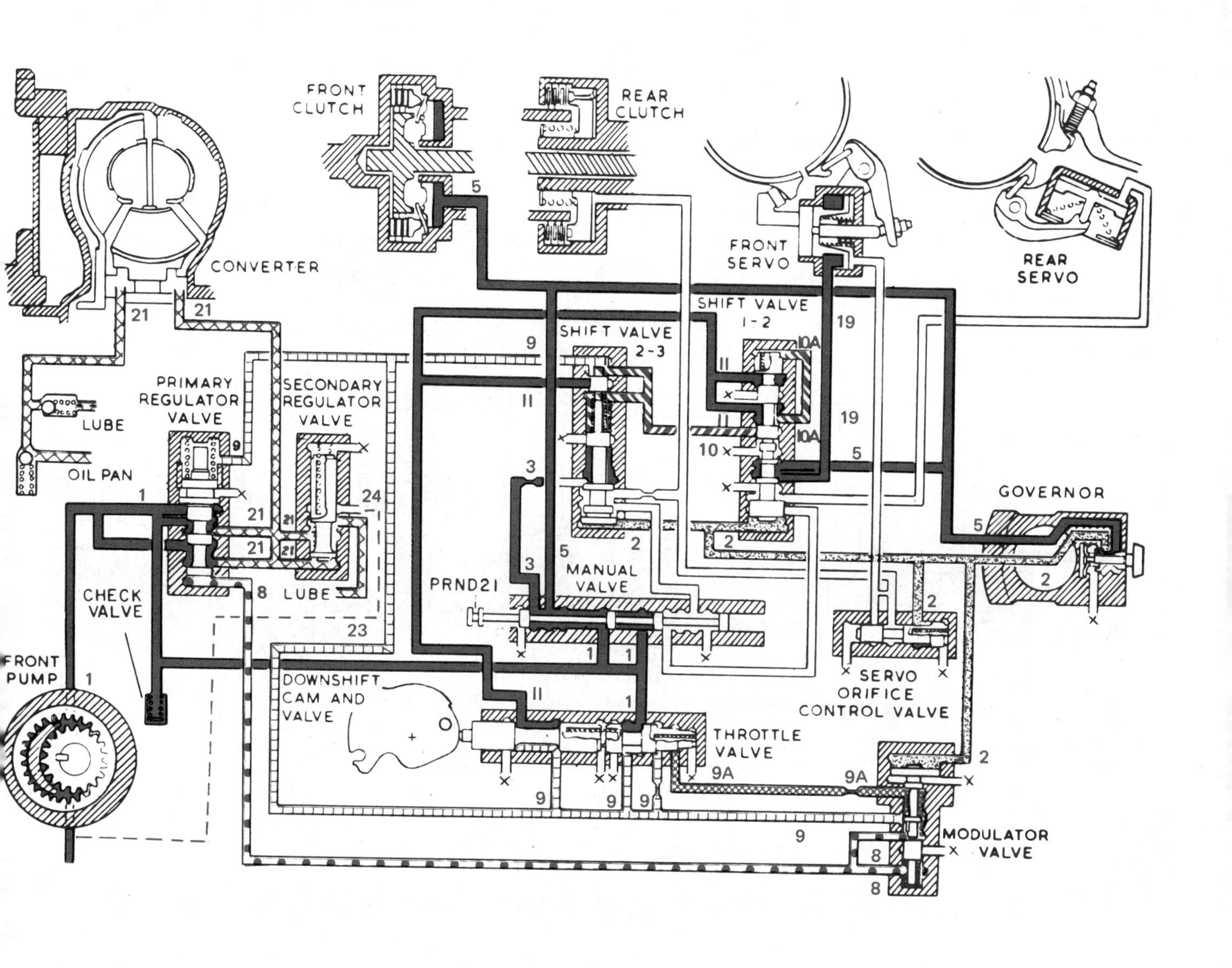

LINE OR DIRECTED LINE PRESSURE

CONVERTER PRESSURE

GOVERNOR PRESSURE

THROTTLE PRESSURE

FORCED THROTTLE PRESSURE

MODULATED THROTTLE PRESSURE

THROTTLE PRESSURE CONTROLLED BY MODULATOR VALVE

SHIFT VALVE PLUNGER PRESSURE

EXHAUST

Fig. 9-17 Hydraulic circuit for D intermediate or 2nd gear.

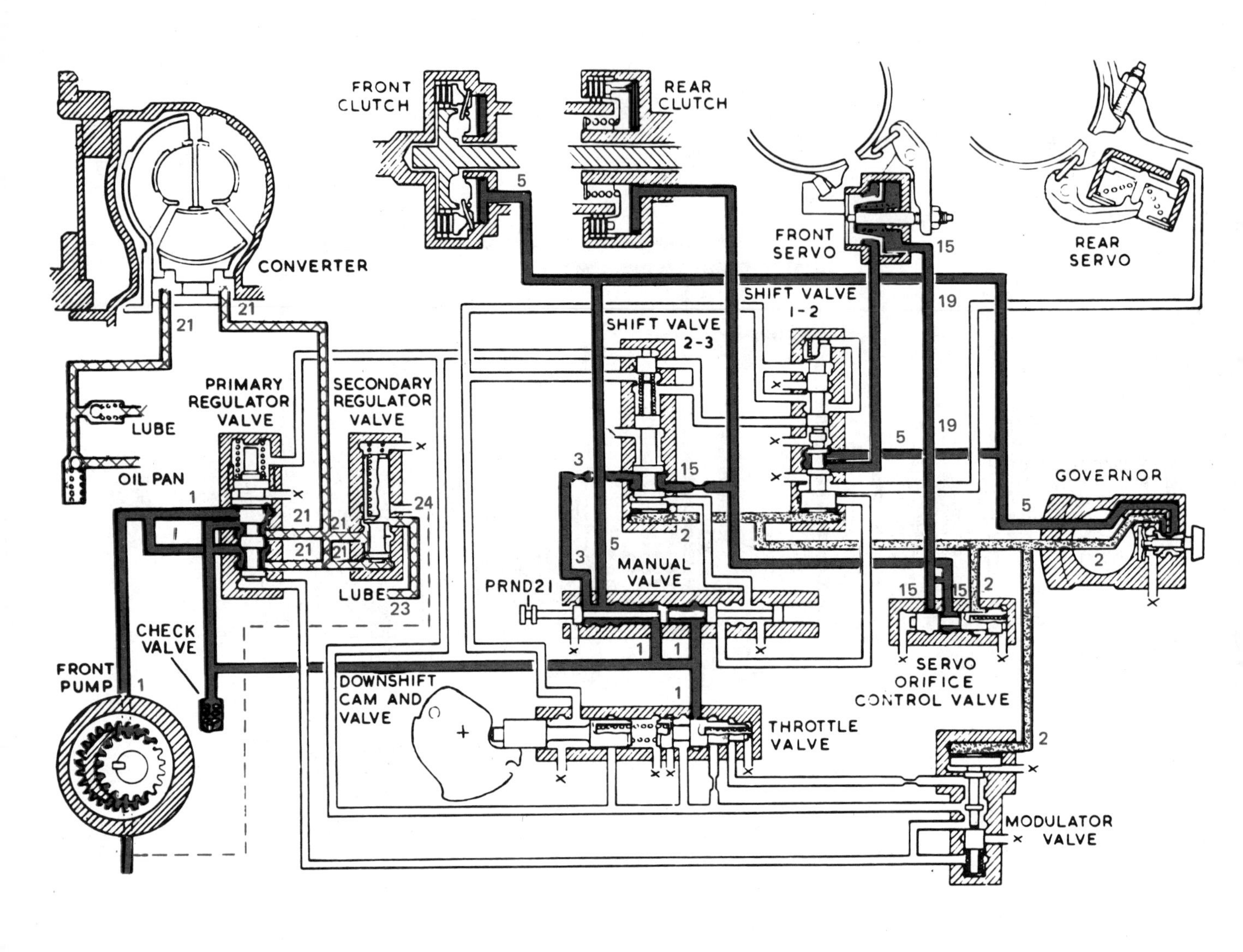

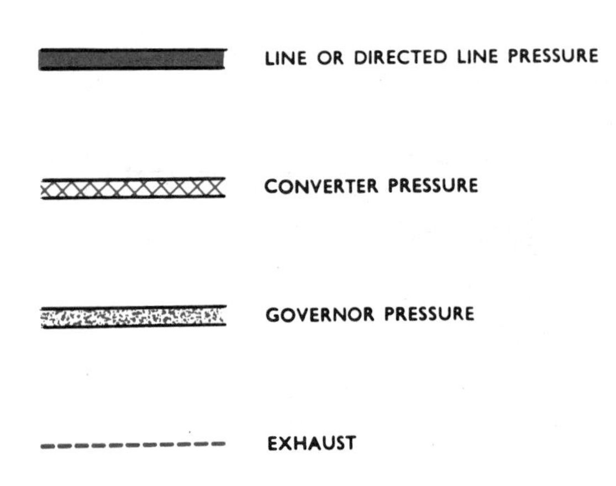

Fig. 9-18 Hydraulic circuit in D high.

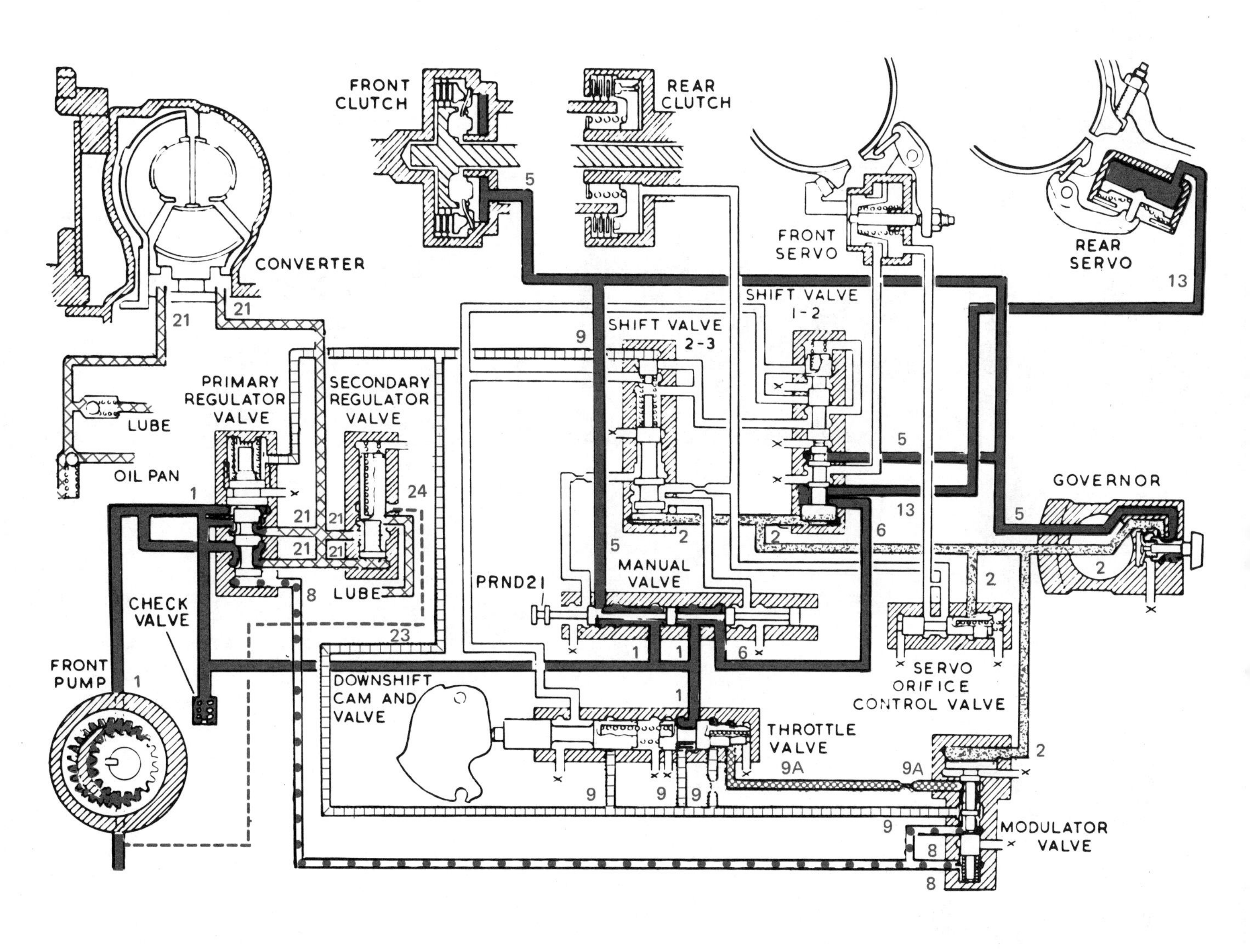

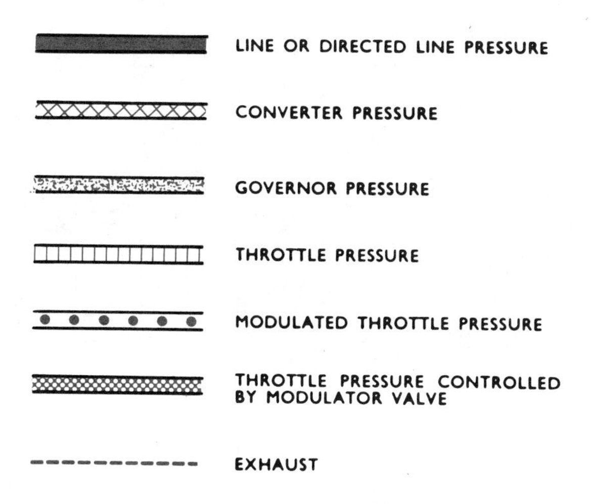

Fig. 9-19 Hydraulic circuit, Manual 1.

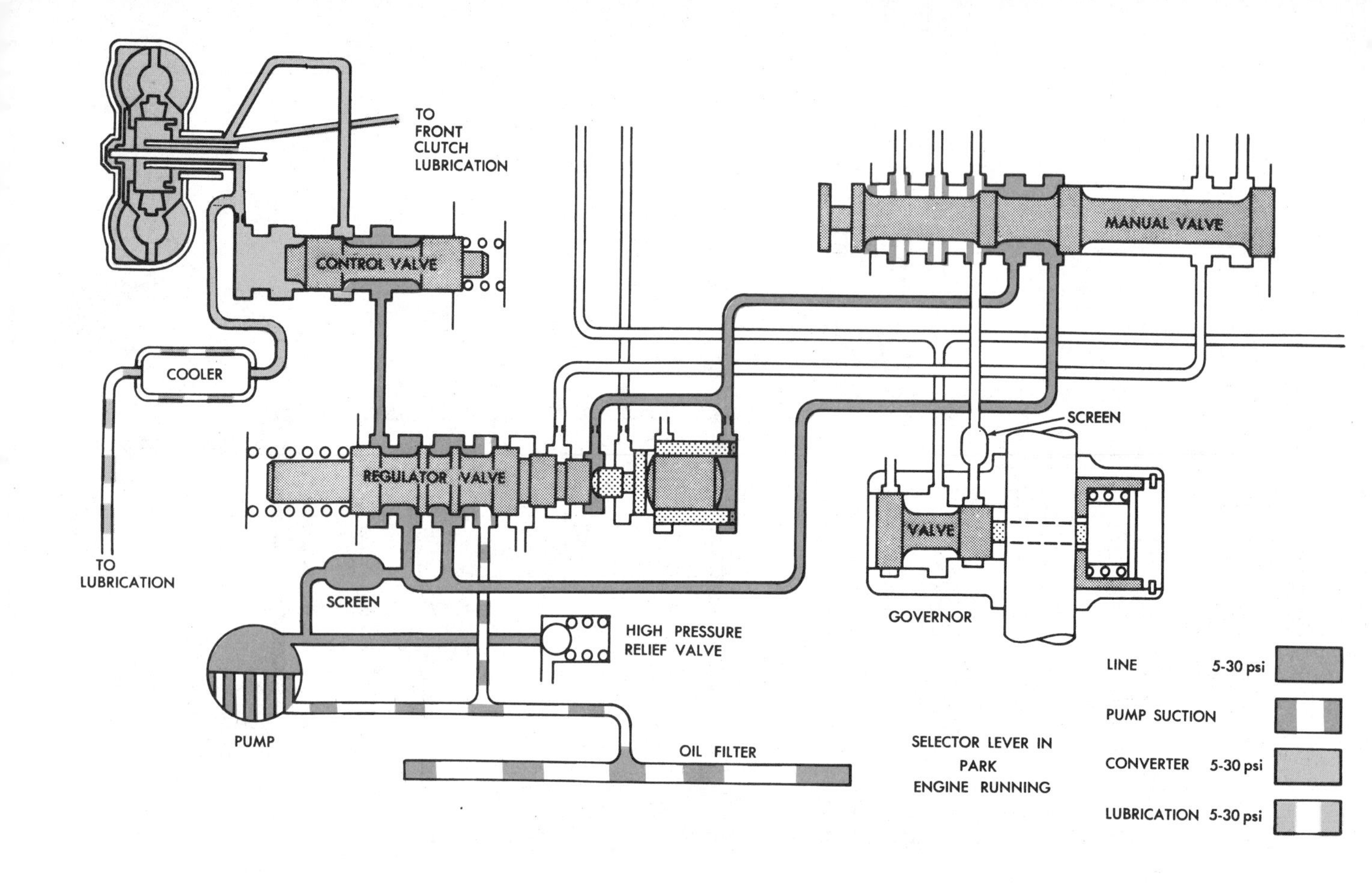

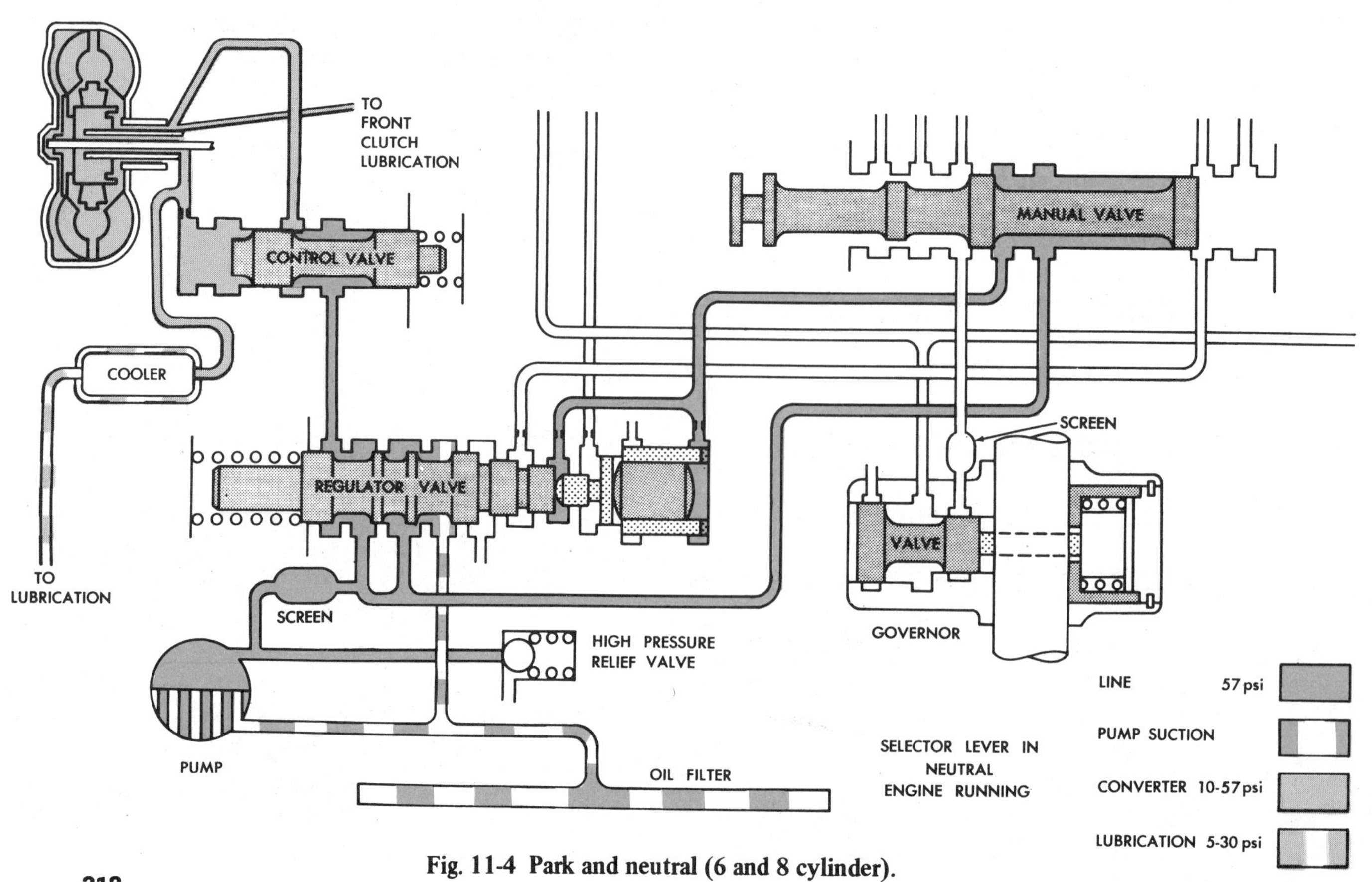

Fig. 11-4 Park and neutral (6 and 8 cylinder).

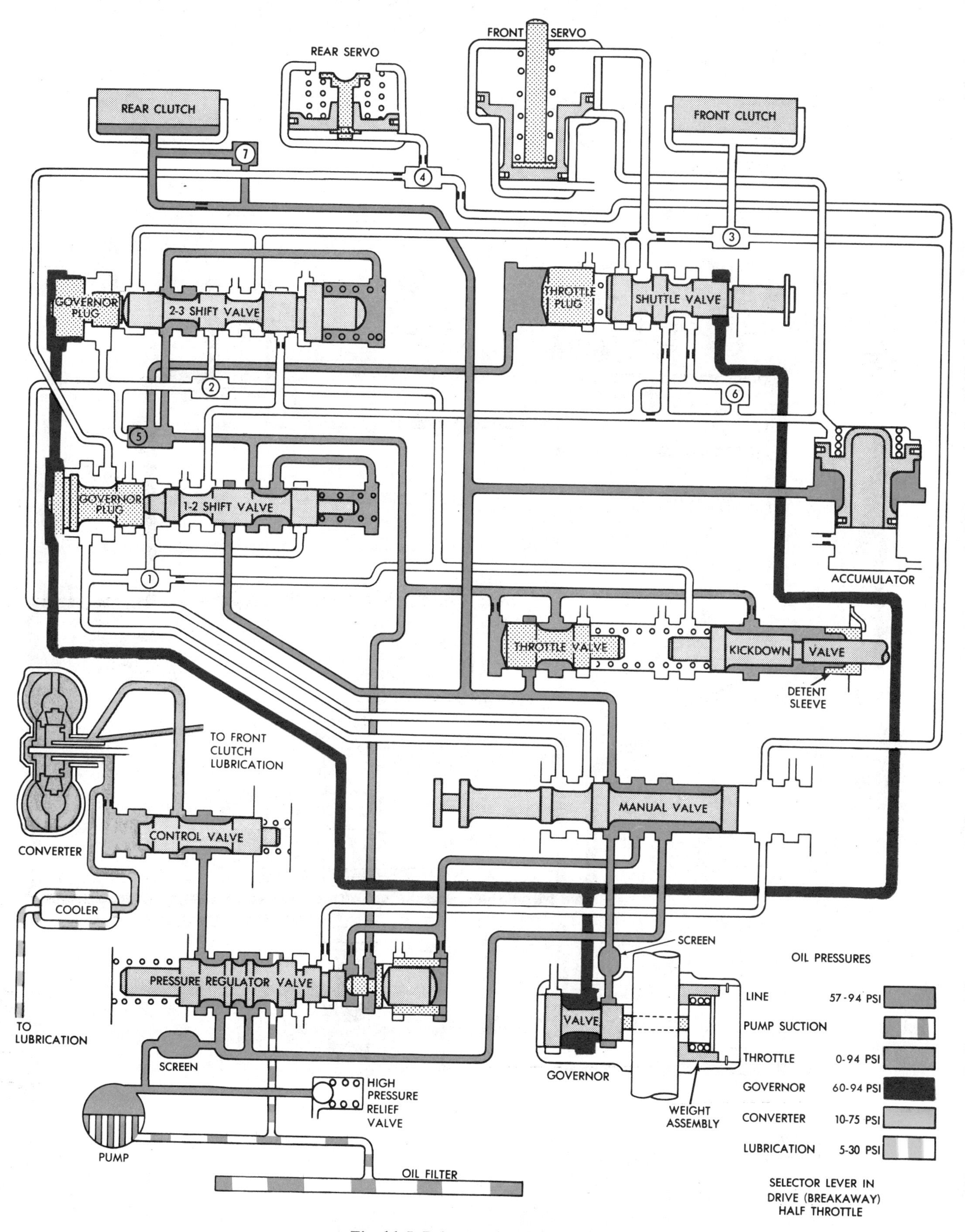

Fig. 11-5 Drive-Breakaway (6 cylinder).

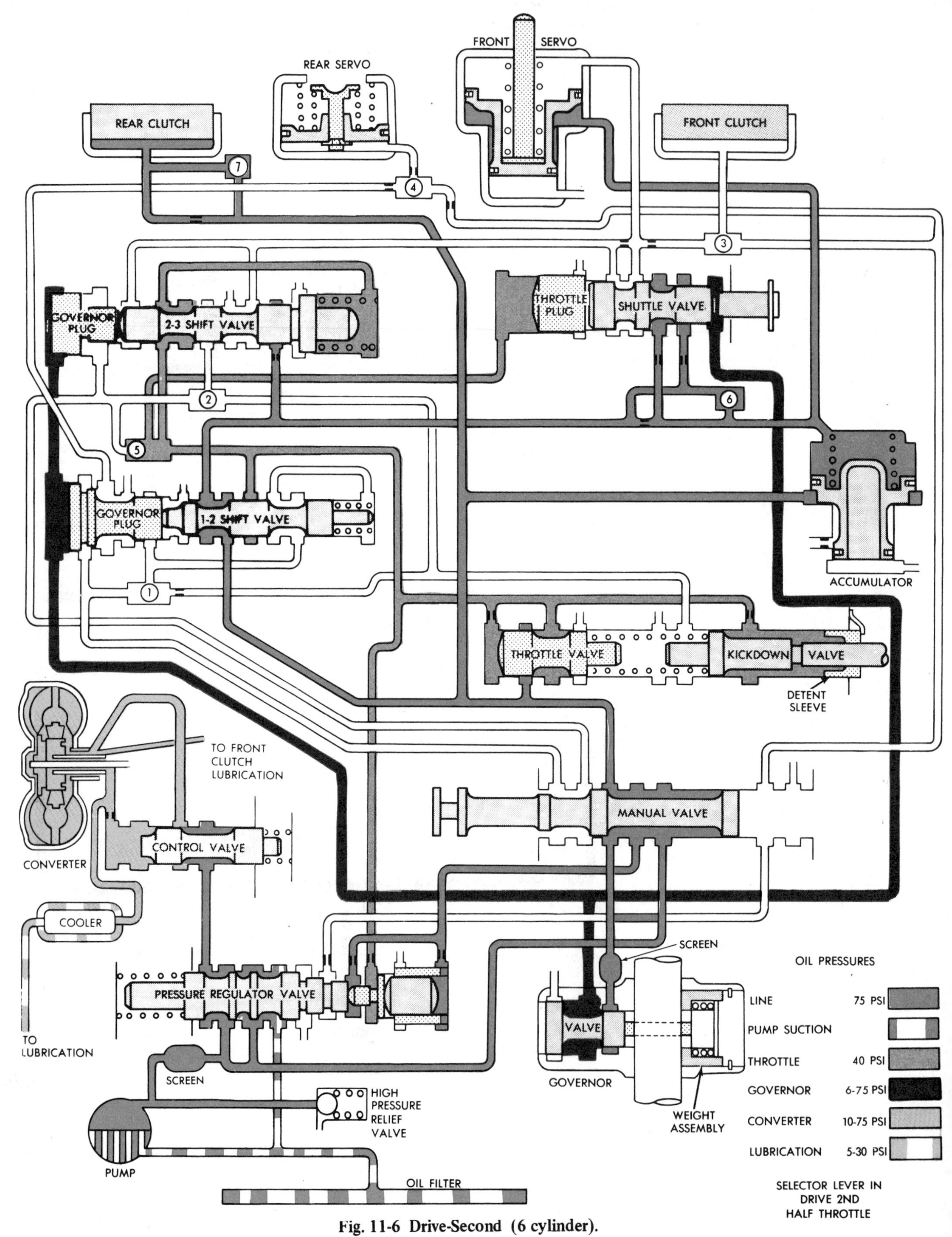

Fig. 11-6 Drive-Second (6 cylinder).

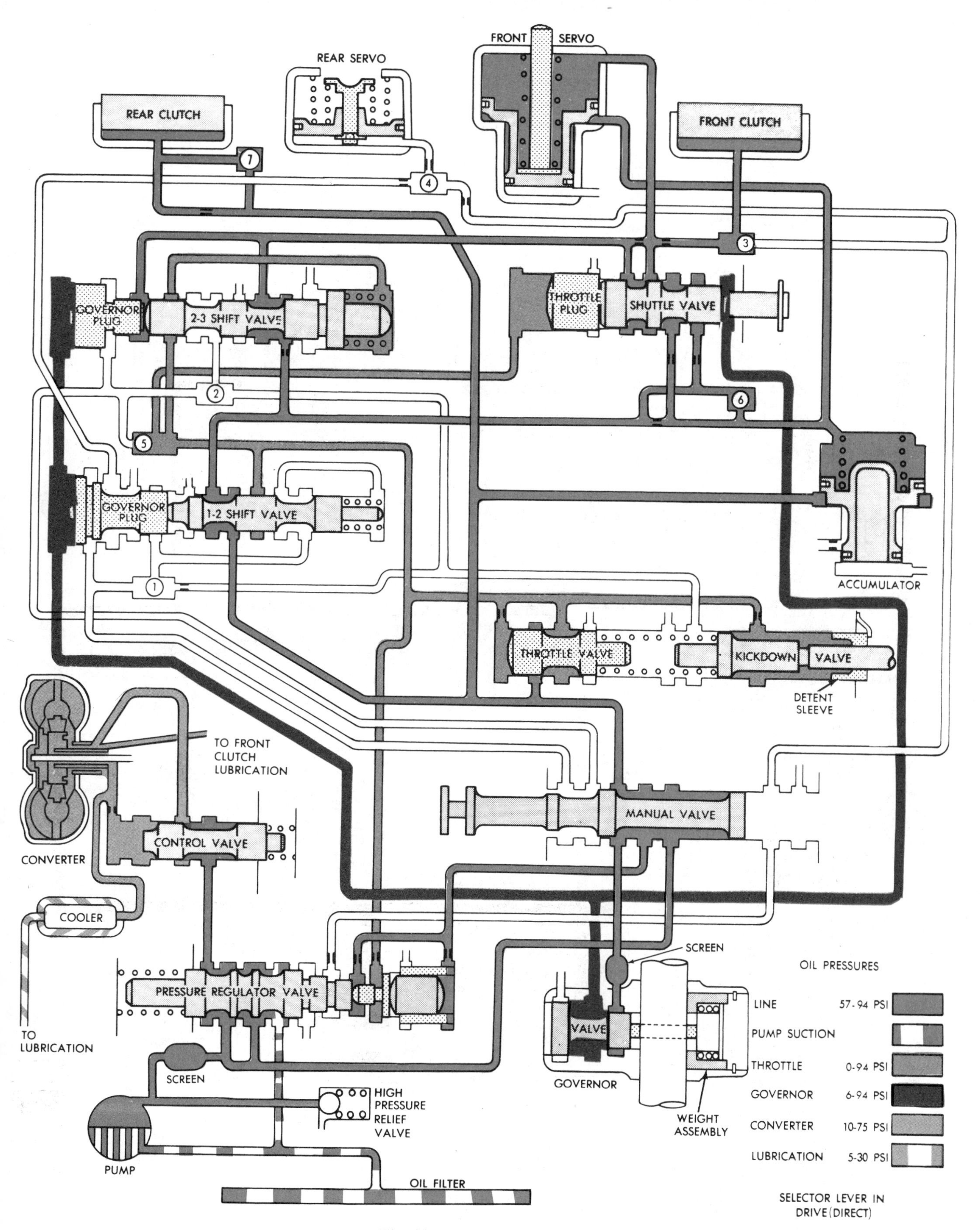

Fig. 11-7 Drive-Direct (6 cylinder).

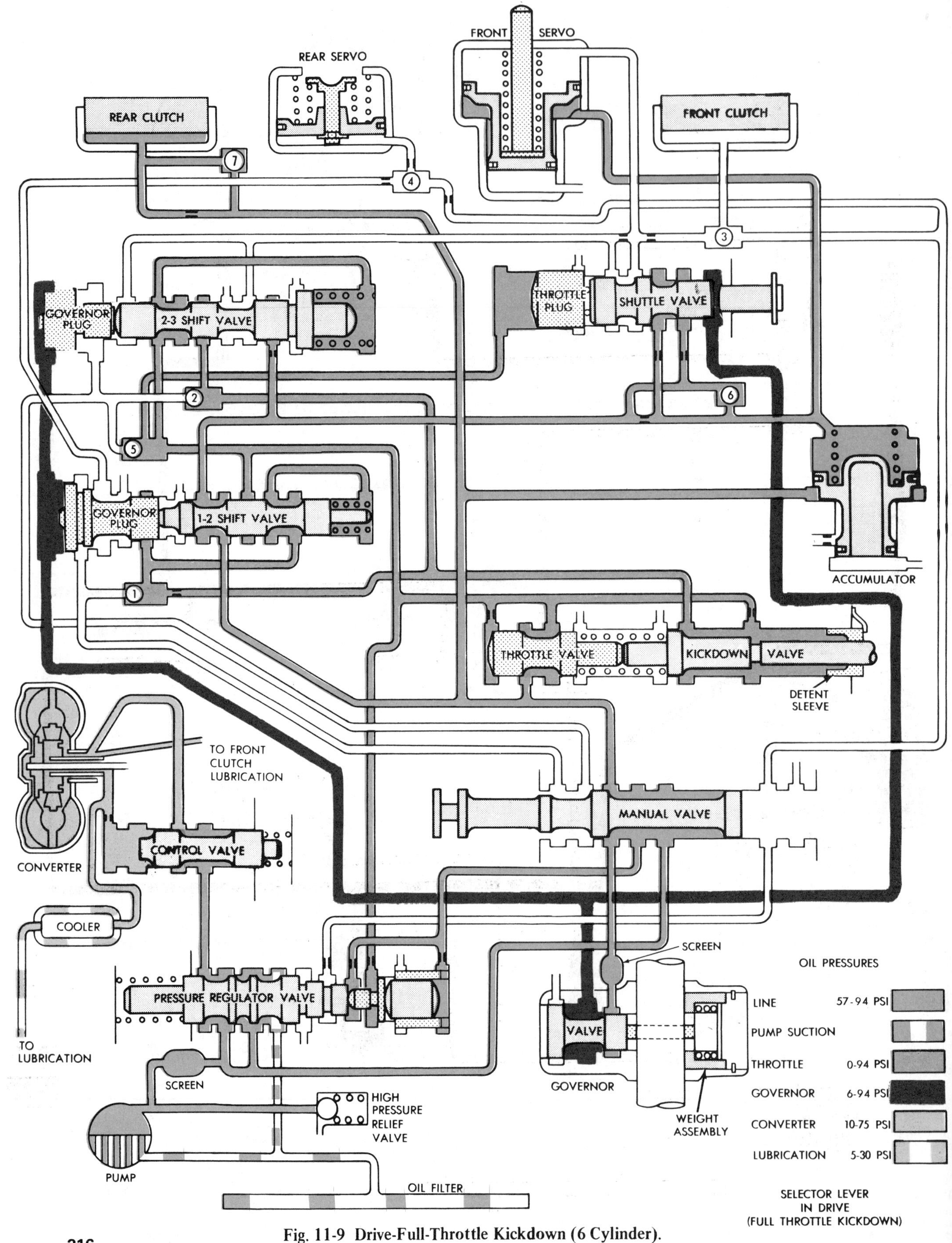

Fig. 11-9 Drive-Full-Throttle Kickdown (6 Cylinder).

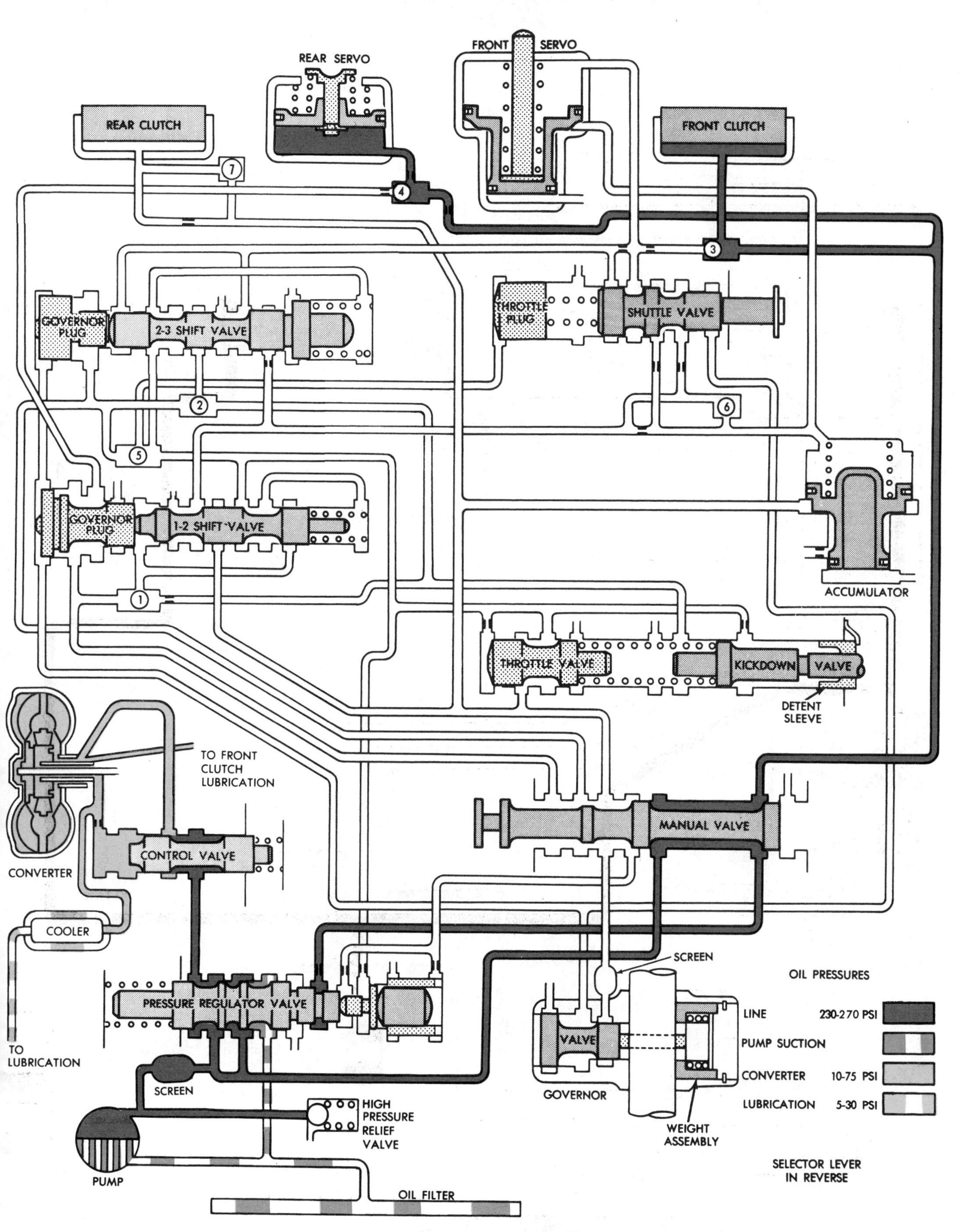

Fig. 11-11 Reverse (6 Cylinder).

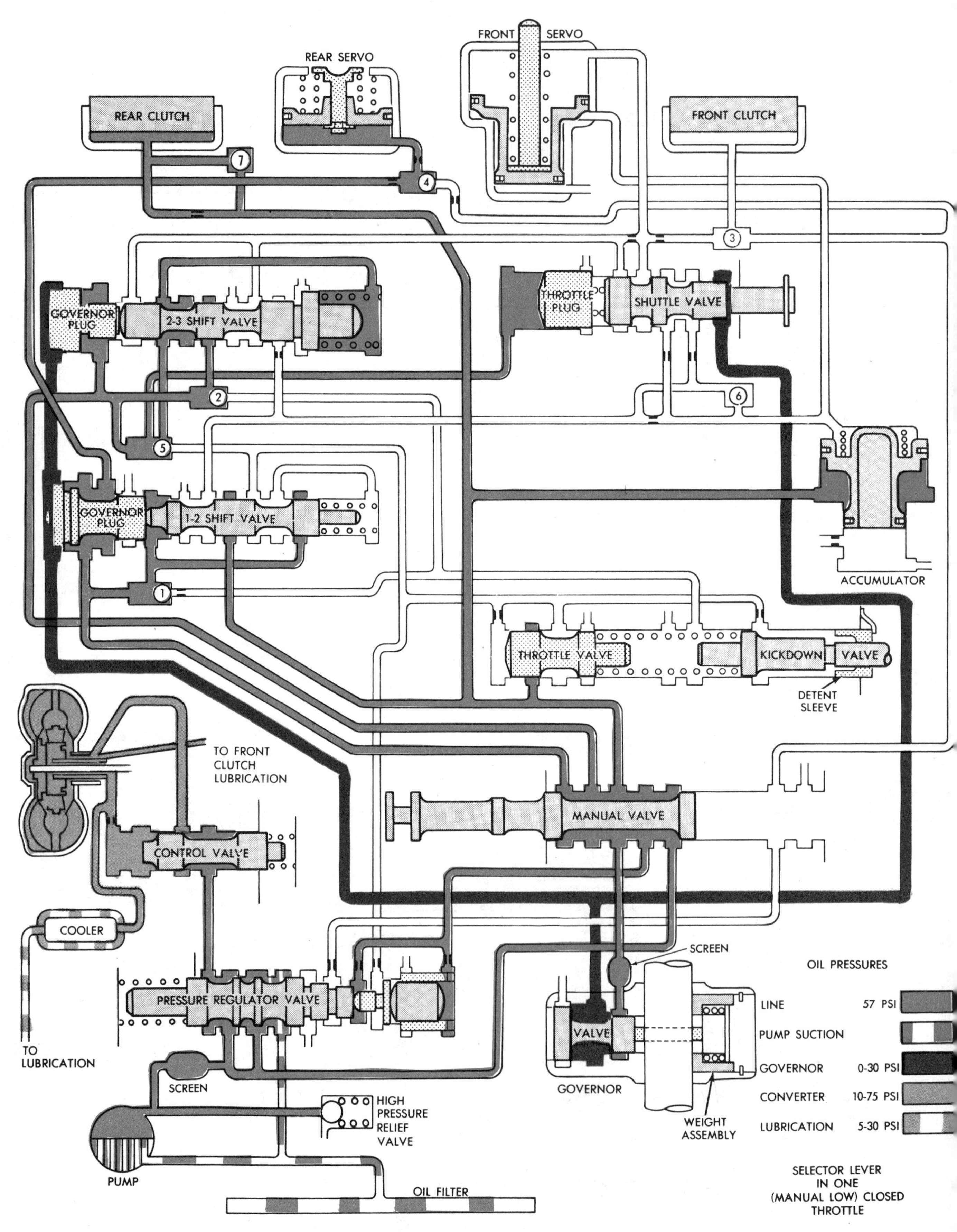

Fig. 11-12 Manual 1 (6 Cylinder).

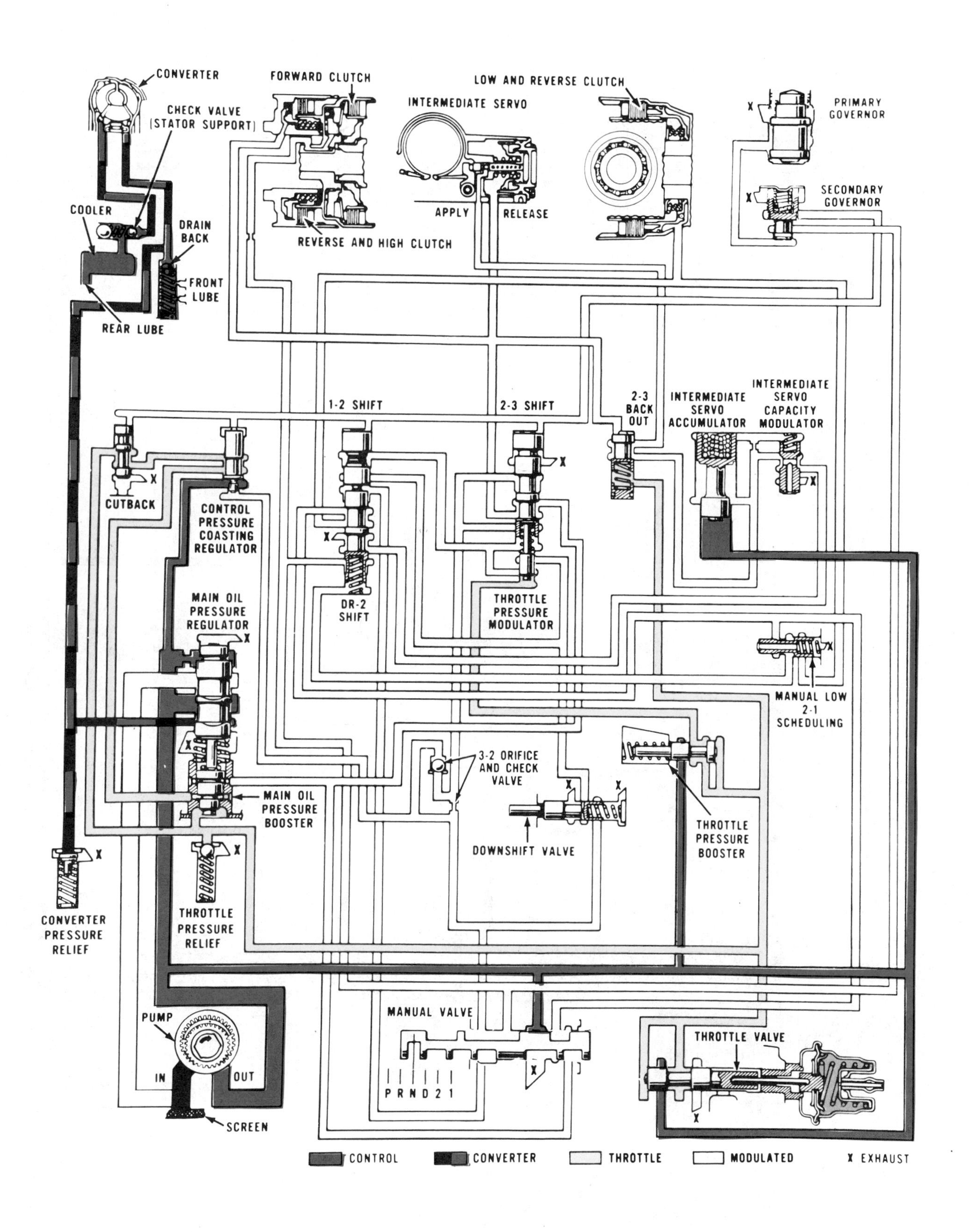

Fig. 12-2 Control pressure flow and regulation (engine idling and car stationary).

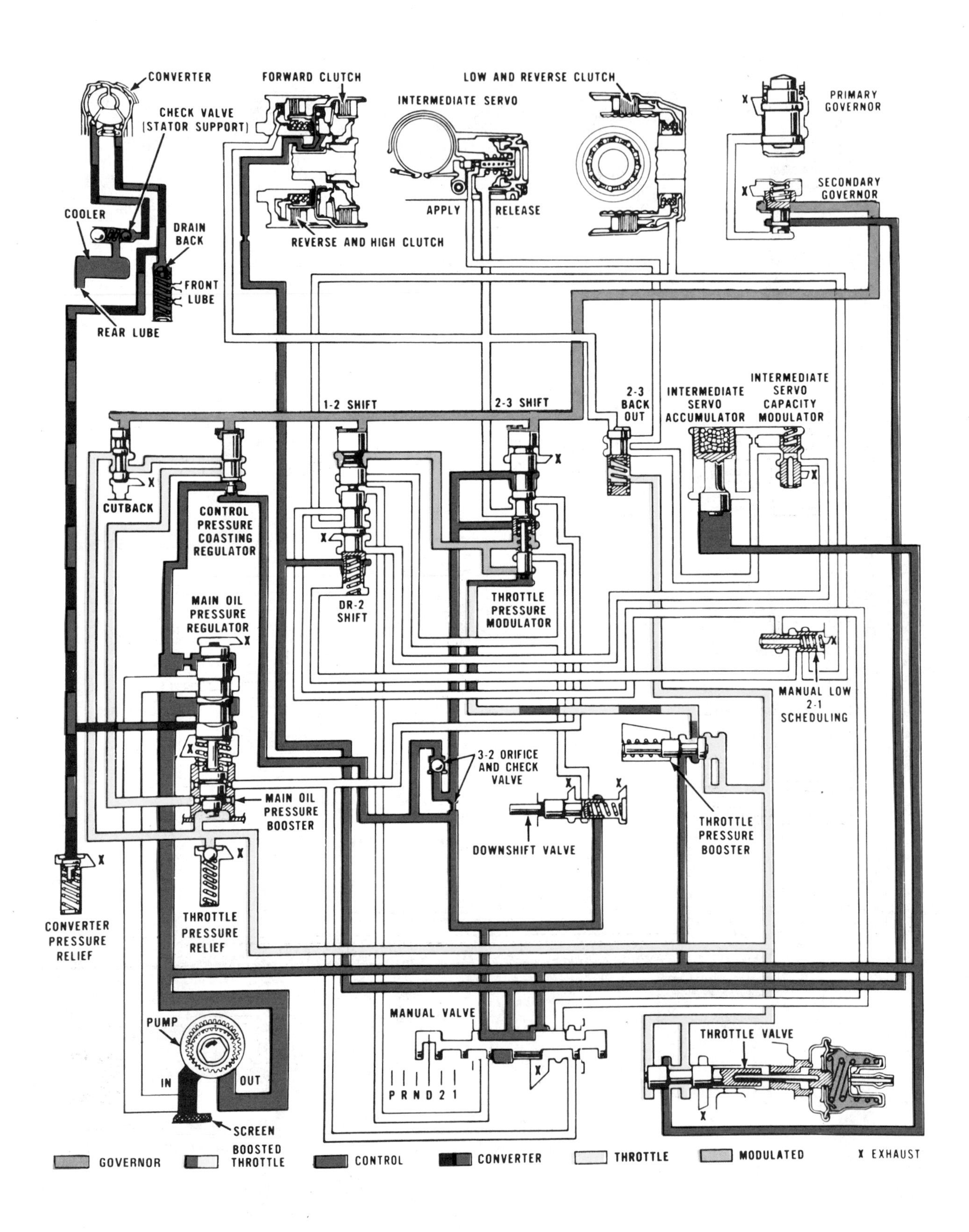

Fig. 12-3 First gear-D.

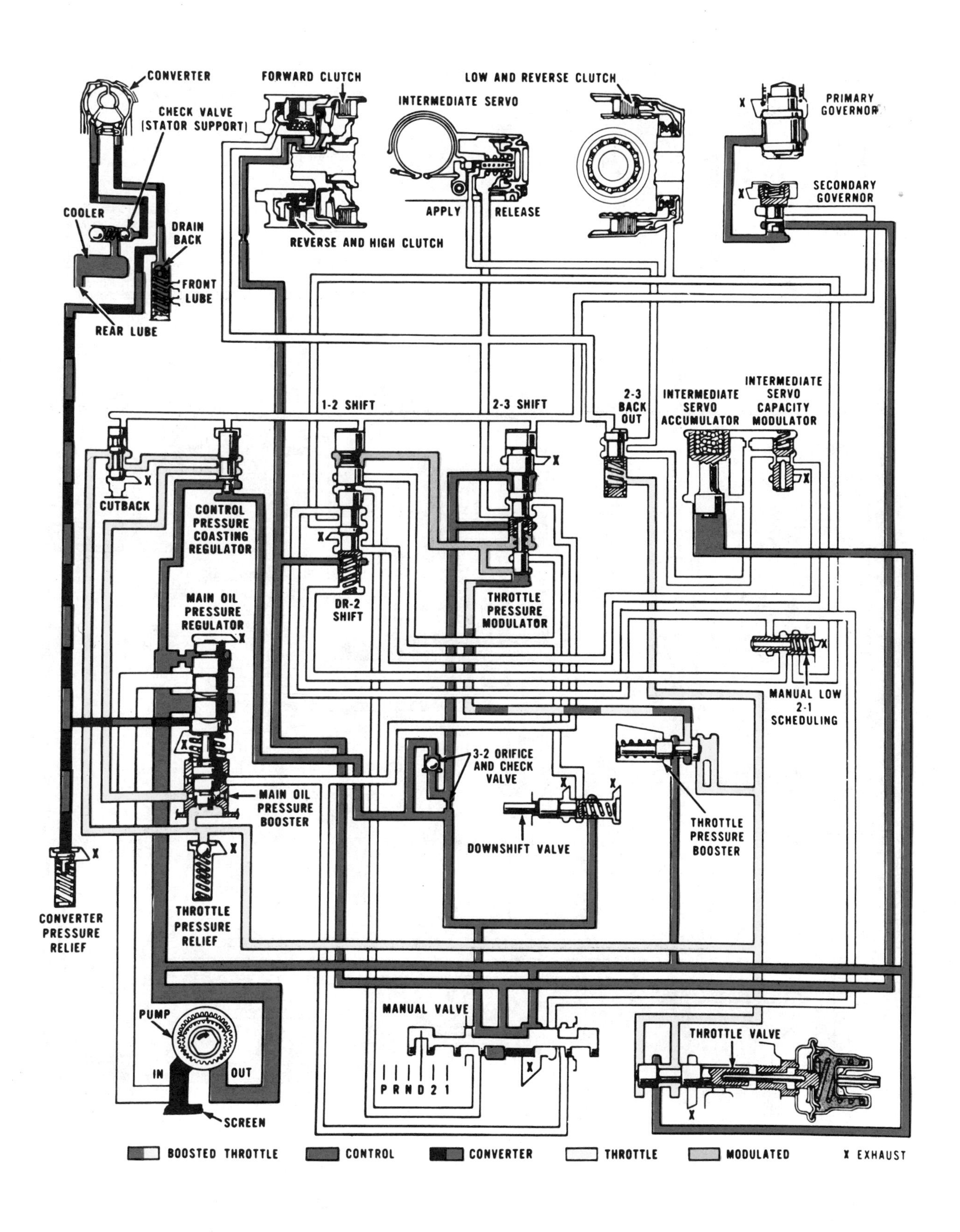

Fig. 12-5 Transmission pressures at less than one inch of vacuum, throttle pressure booster in operation.

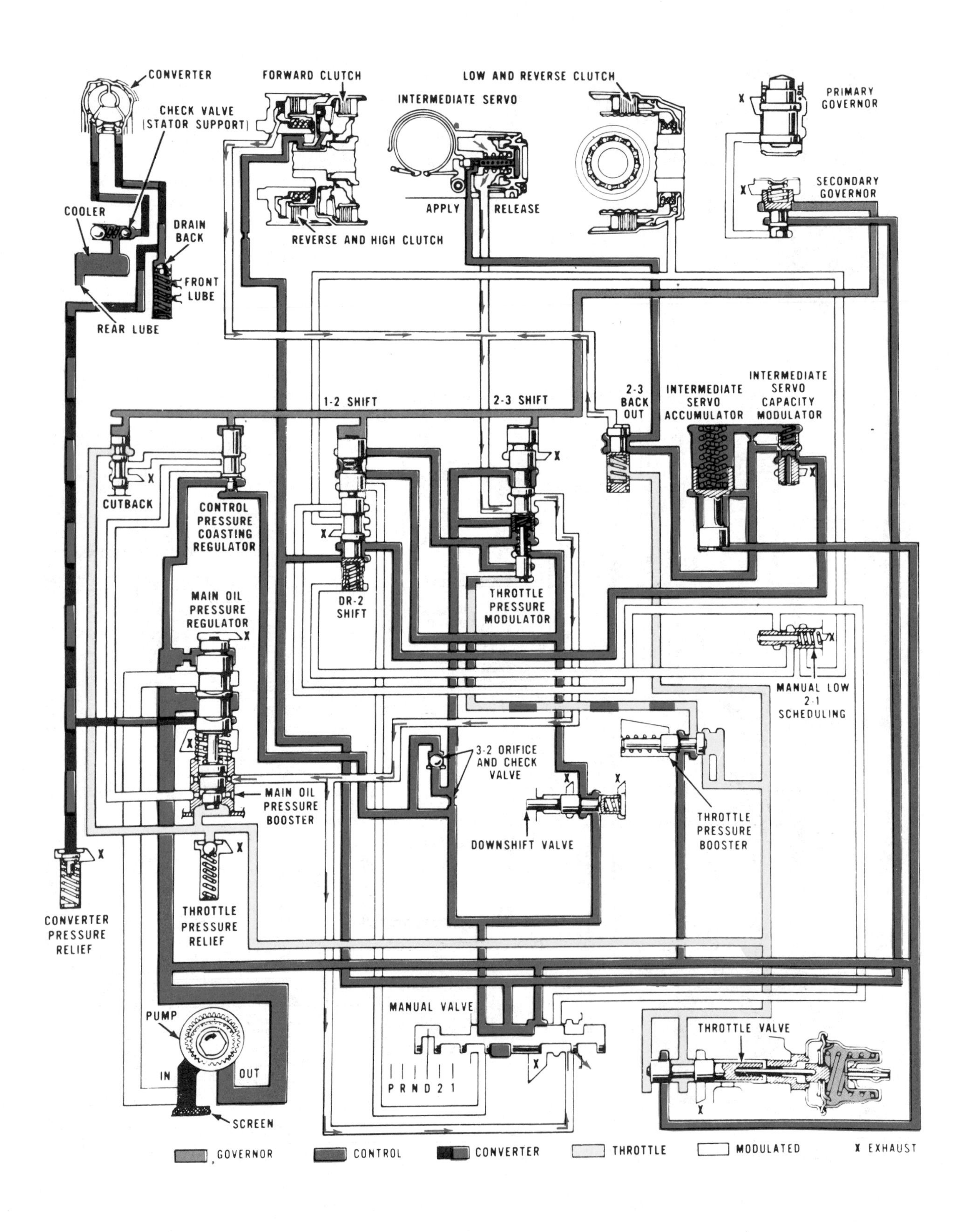

Fig. 12-6 3-2 downshift.

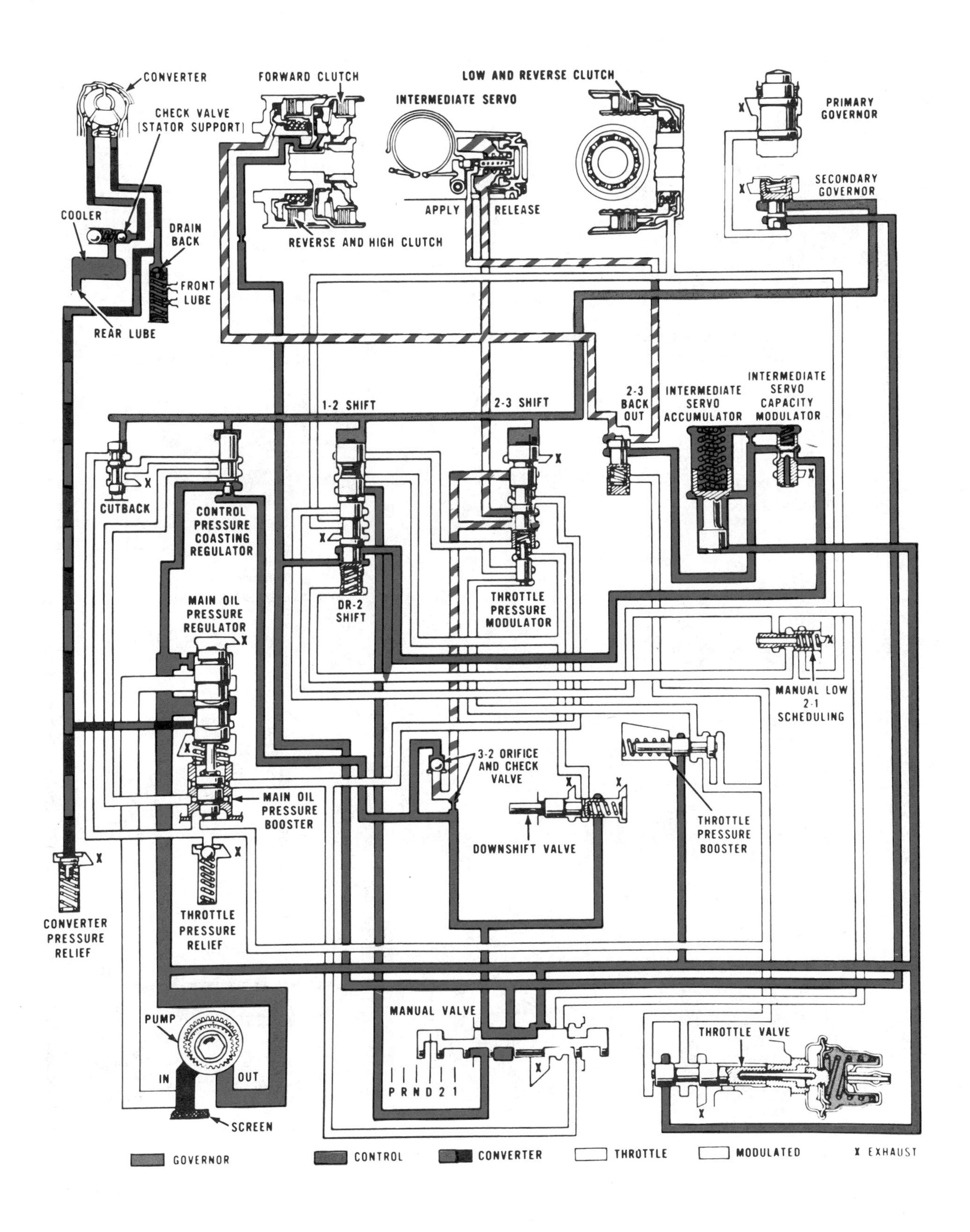

Fig. 12-7 2-3 backout shift.

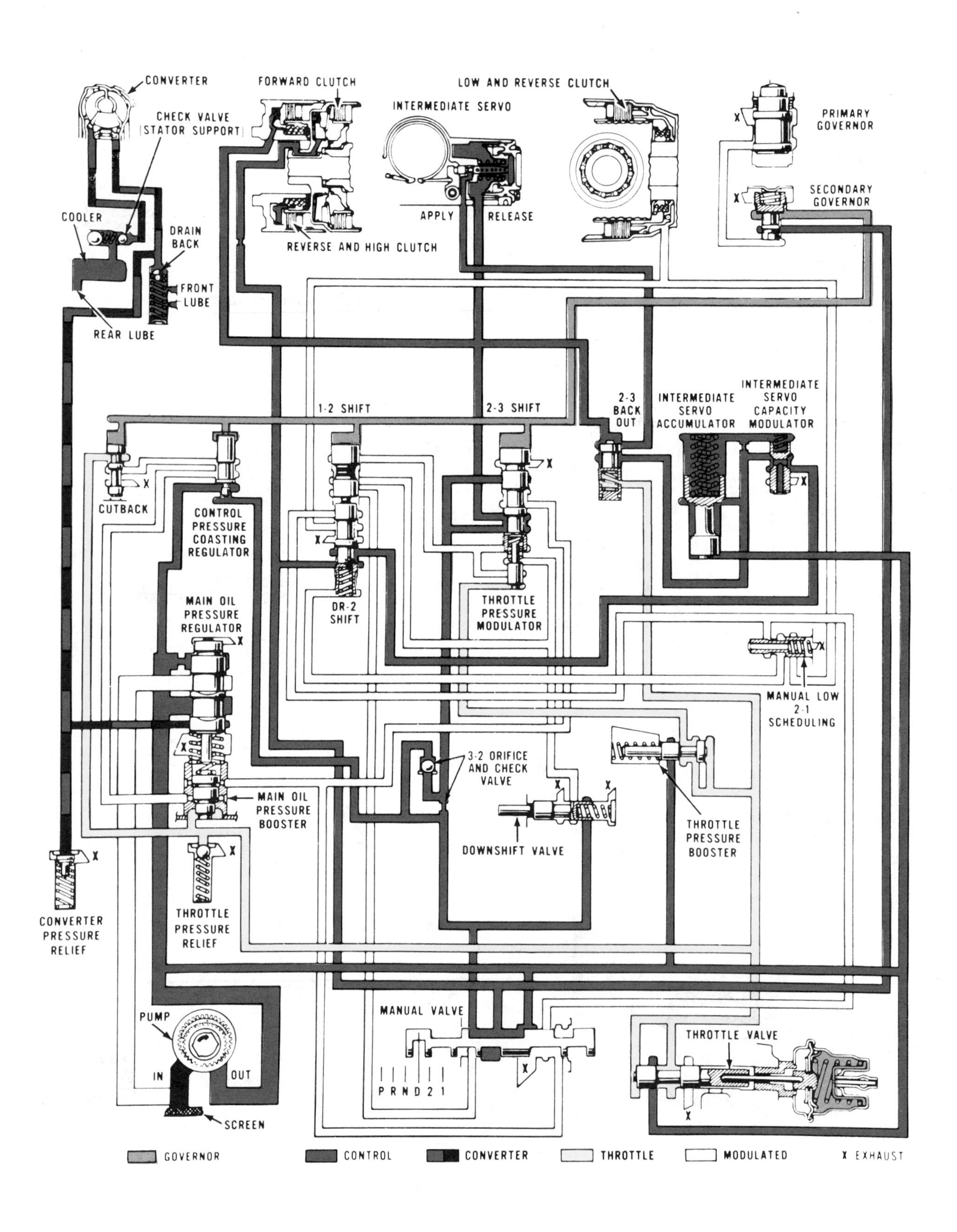

Fig. 12-8 High gear–D, cutback in use.

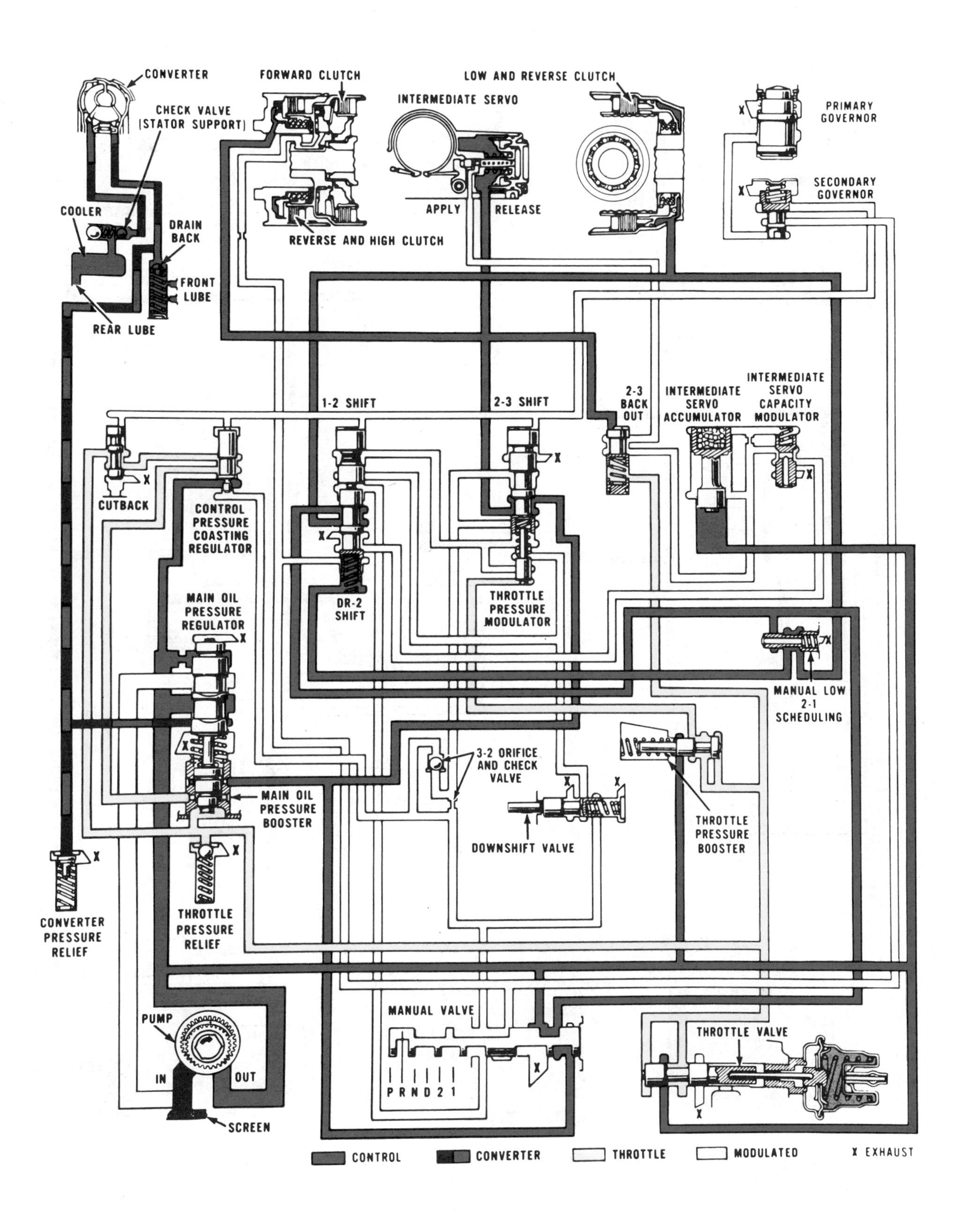

Fig. 12-9 Reverse.

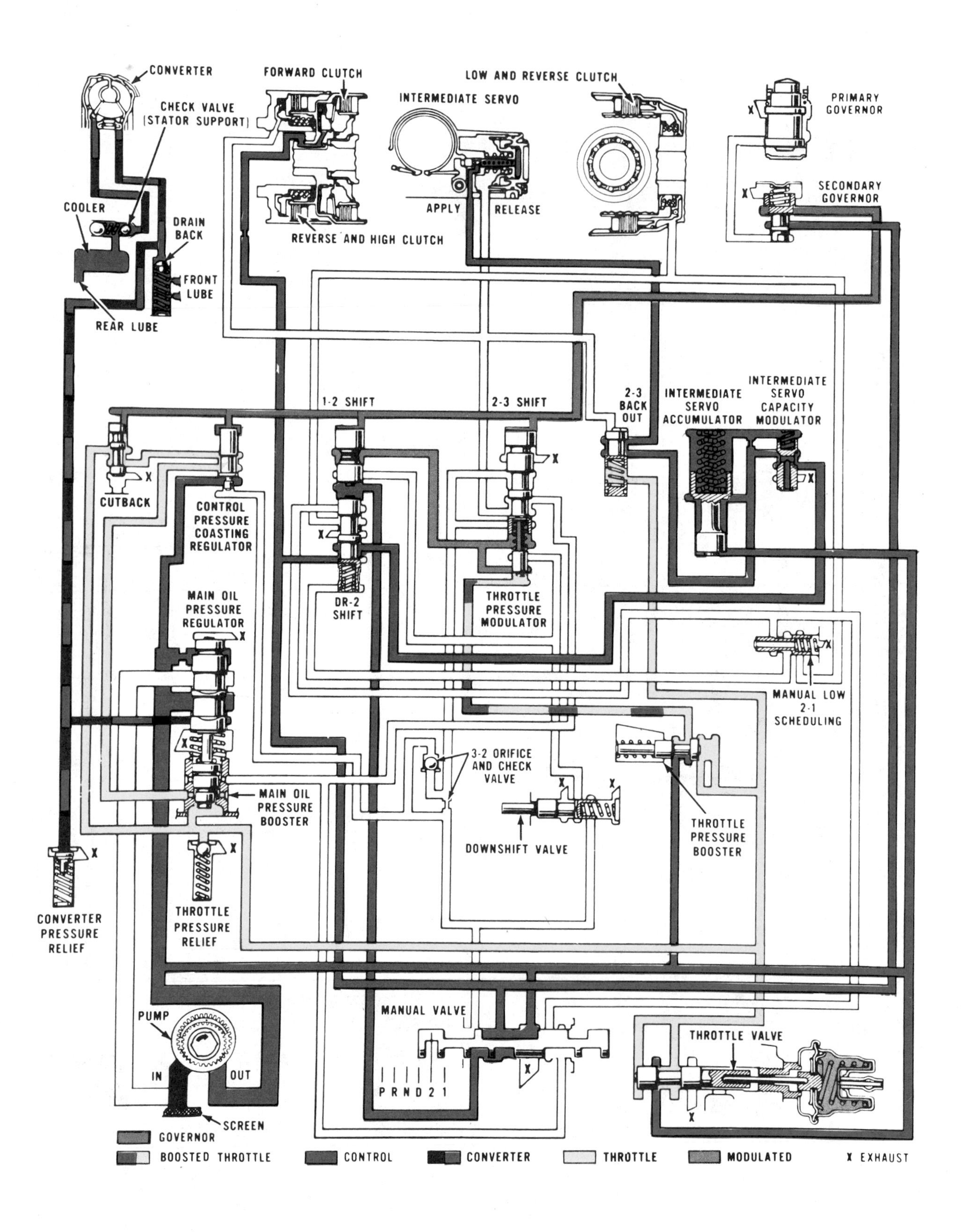

Fig. 12-10 Second gear – 2.

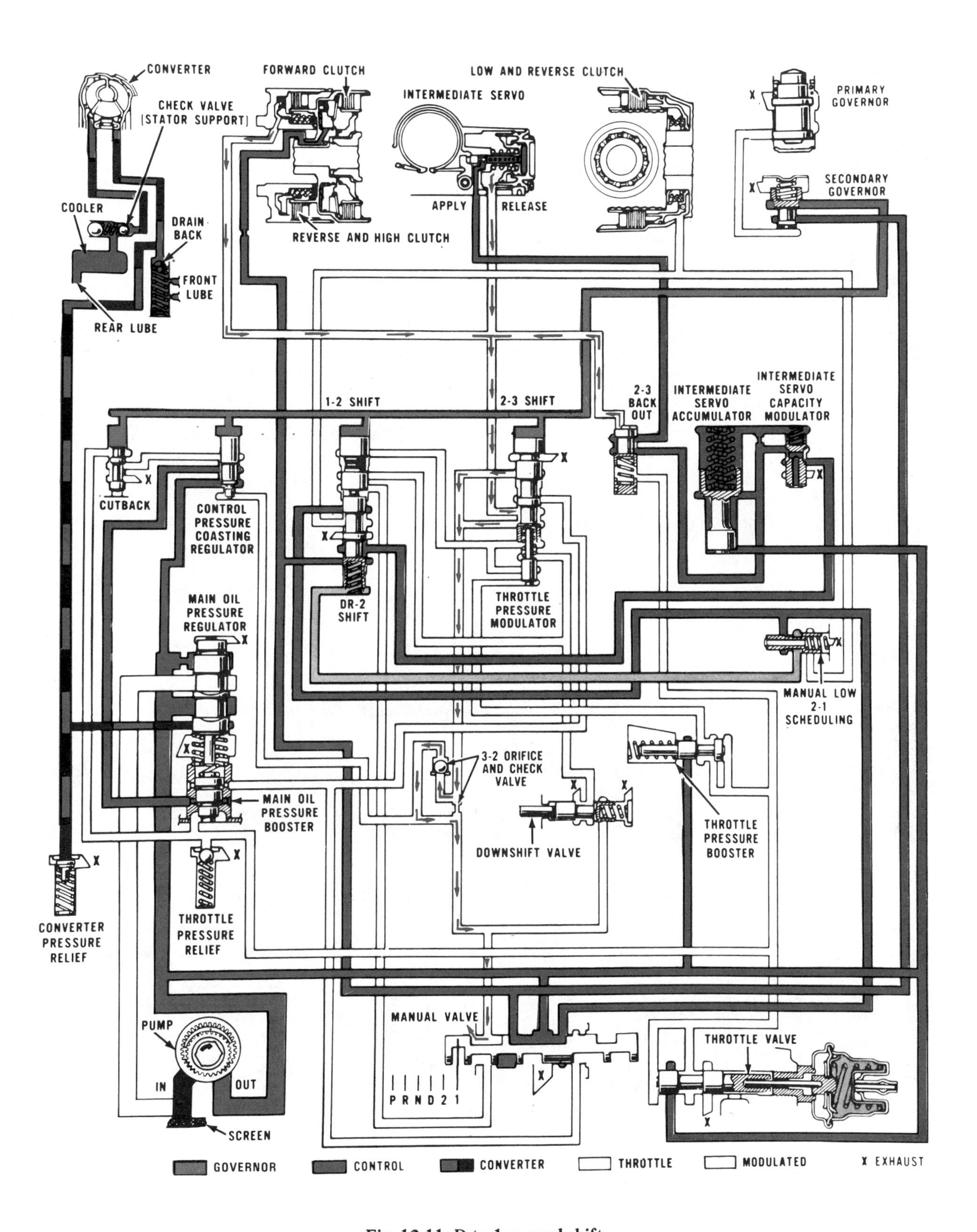

Fig. 12-11 D to 1 manual shift.

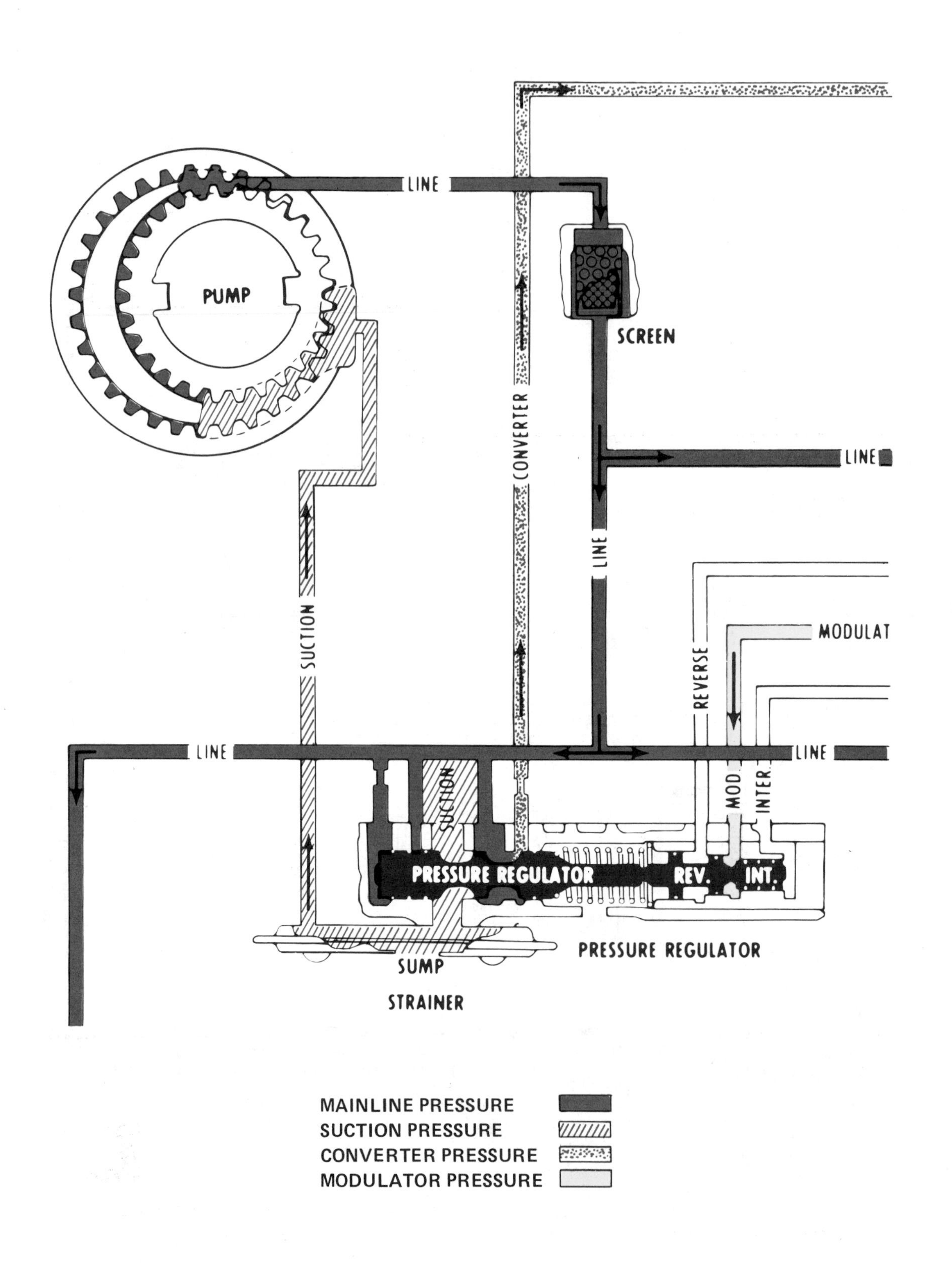

Fig. 13-4 Pump pressure regulation.

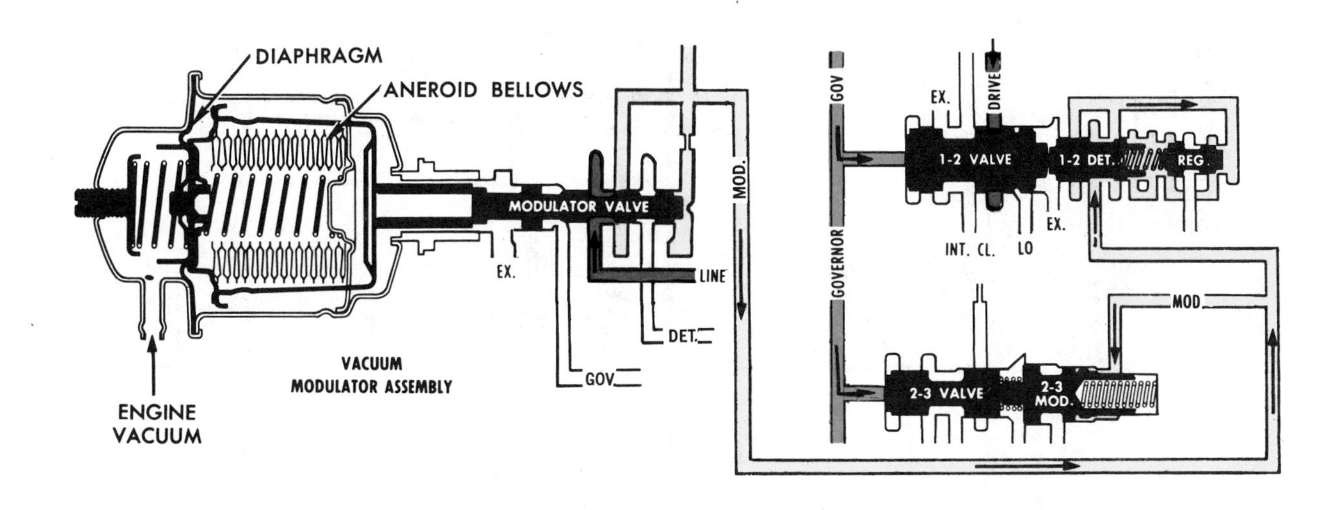

Fig. 13-6 Typical General Motors modulator circuit.

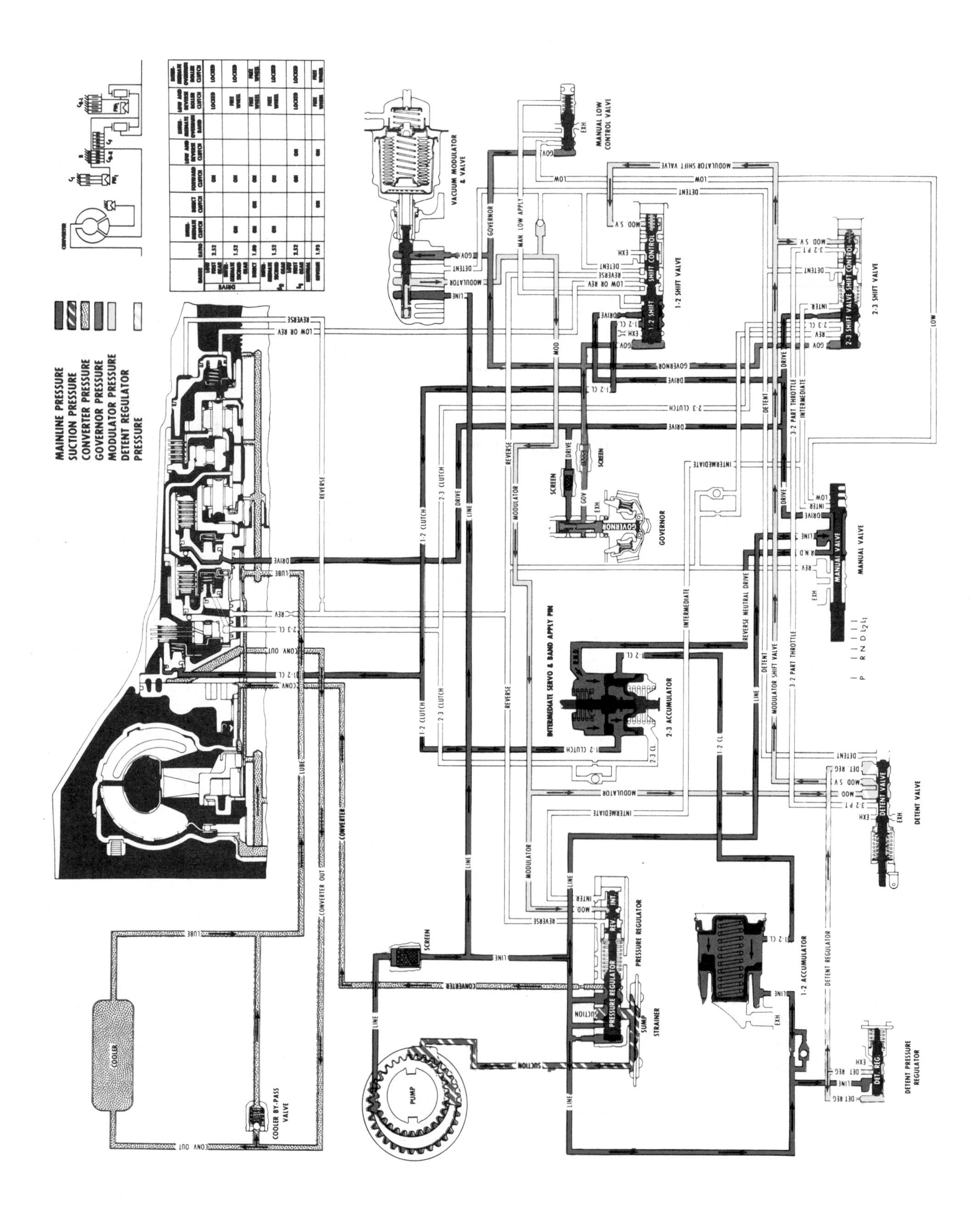

Fig. 13-9 1-2 shift valve in the upshift position and the 1-2 accumulator stroked against line pressure.

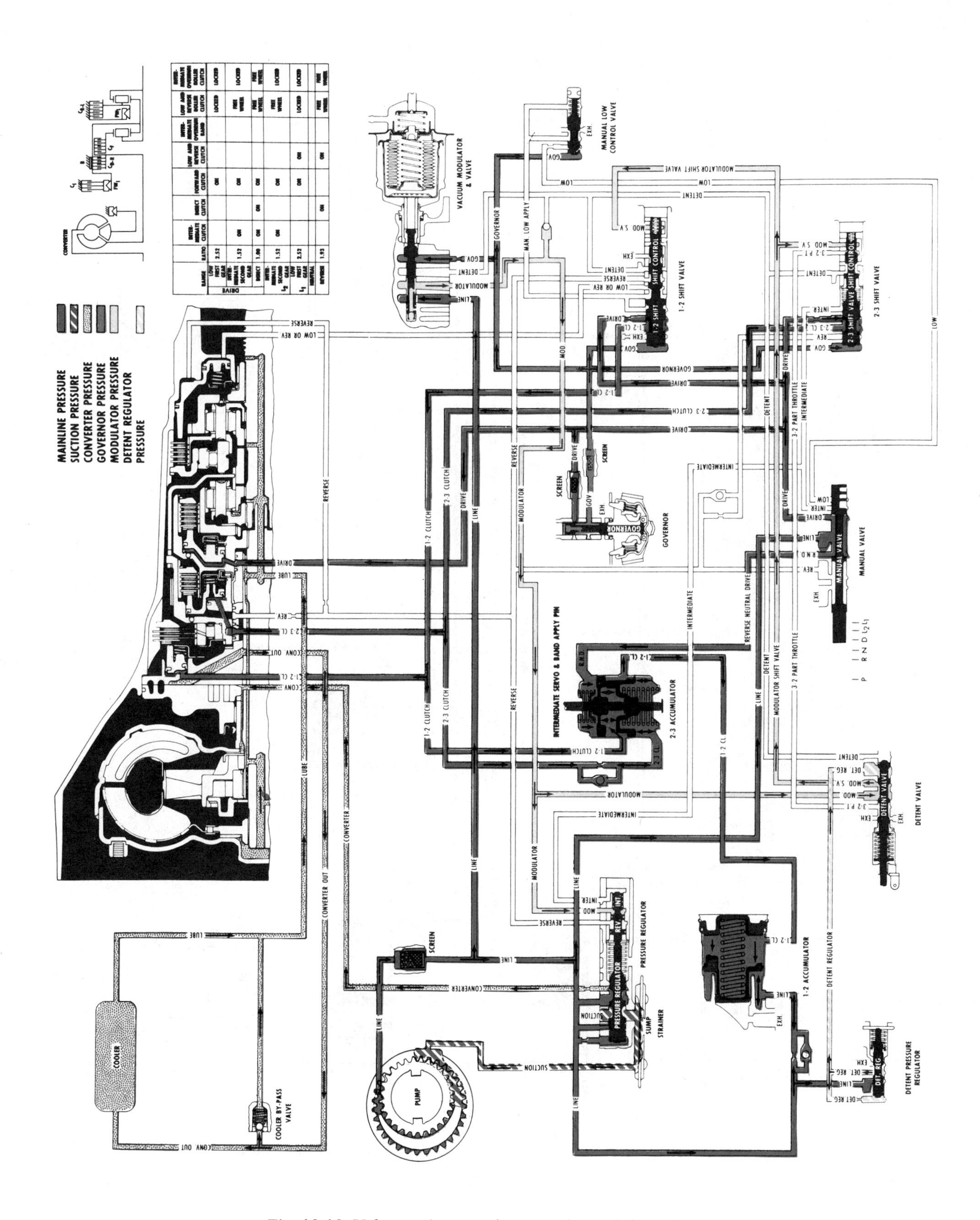

Fig. 13-10 Valves and accumulators in the upshift position.

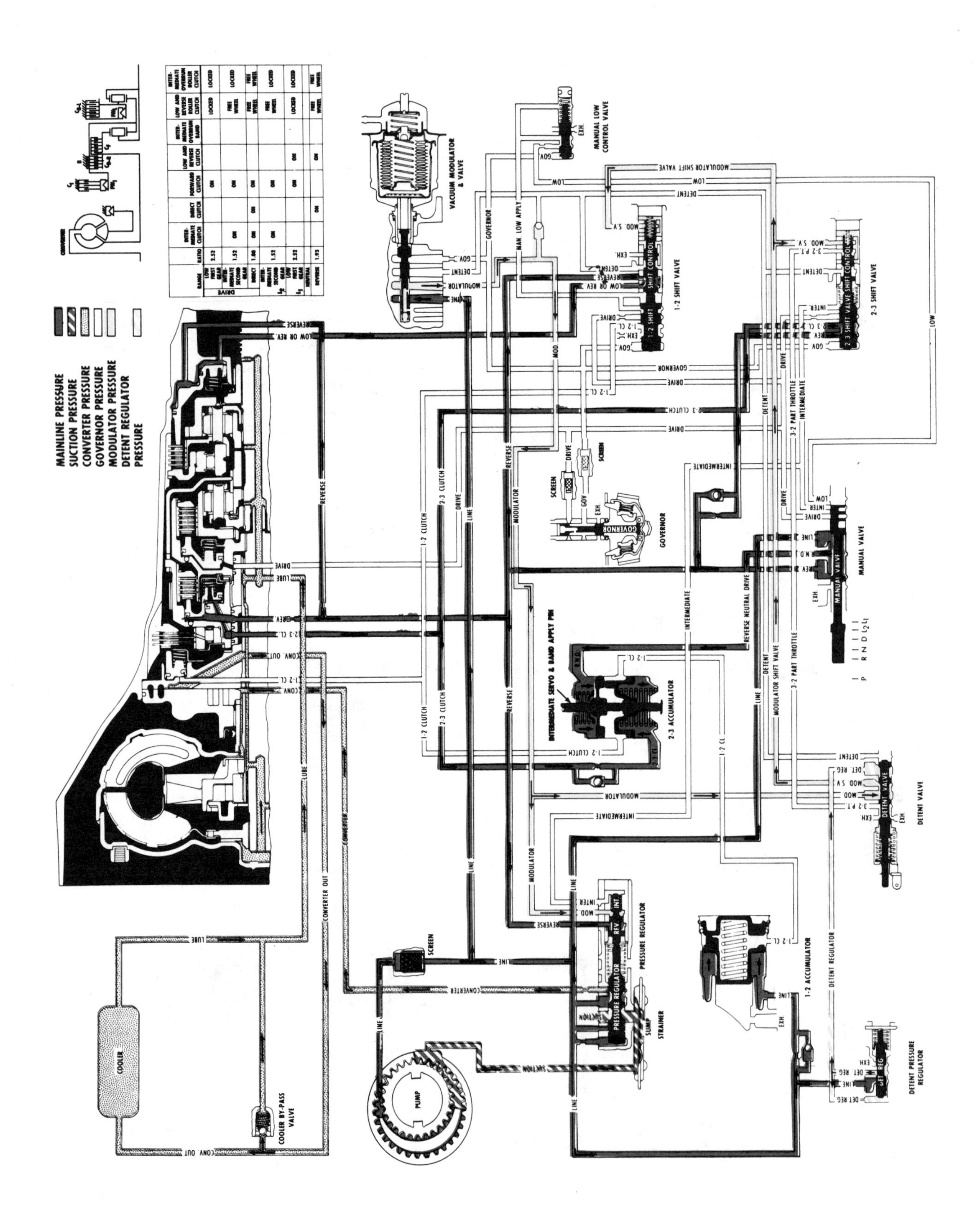

Fig. 13-11 Reverse range.

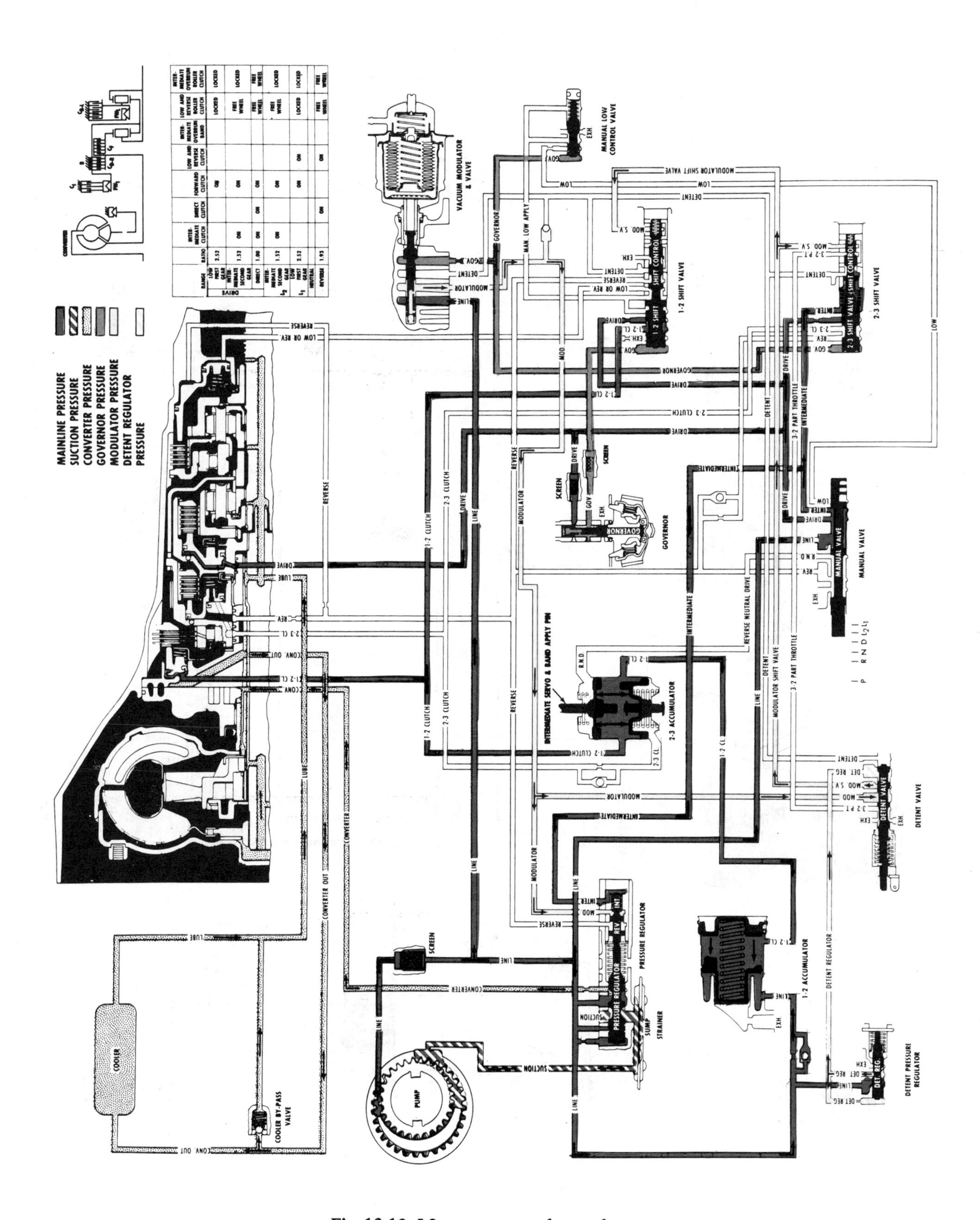

Fig. 13-12 L2 range: manual second gear.

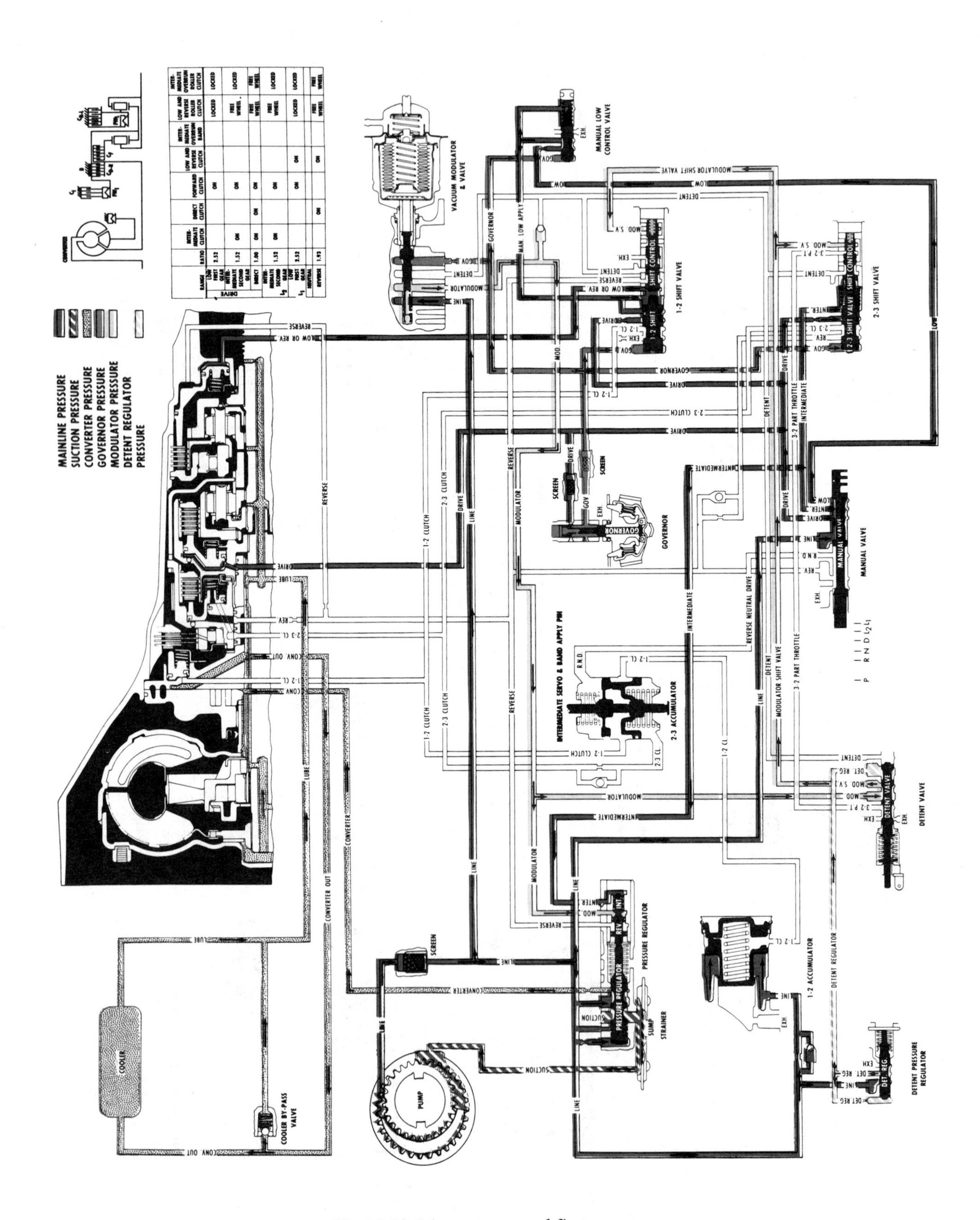

RANGE		RATIO	INTERMEDIATE CLUTCH	DIRECT CLUTCH	FORWARD CLUTCH	LOW AND REVERSE CLUTCH	INTERMEDIATE OVERRUN BAND	LOW AND REVERSE ROLLER CLUTCH	INTERMEDIATE OVERRUN ROLLER CLUTCH
DRIVE	LOW FIRST GEAR	2.52			ON			LOCKED	LOCKED
DRIVE	INTERMEDIATE SECOND GEAR	1.52	ON		ON			FREE WHEEL	LOCKED
DRIVE	DIRECT	1.00	ON	ON	ON			FREE WHEEL	FREE WHEEL
L2	INTERMEDIATE SECOND GEAR	1.52	ON		ON			FREE WHEEL	LOCKED
L1	LOW FIRST GEAR	2.52			ON	ON		LOCKED	LOCKED
NEUTRAL									
REVERSE		1.93		ON		ON		FREE WHEEL	FREE WHEEL

Fig. 13-13 L1 range: manual first gear.

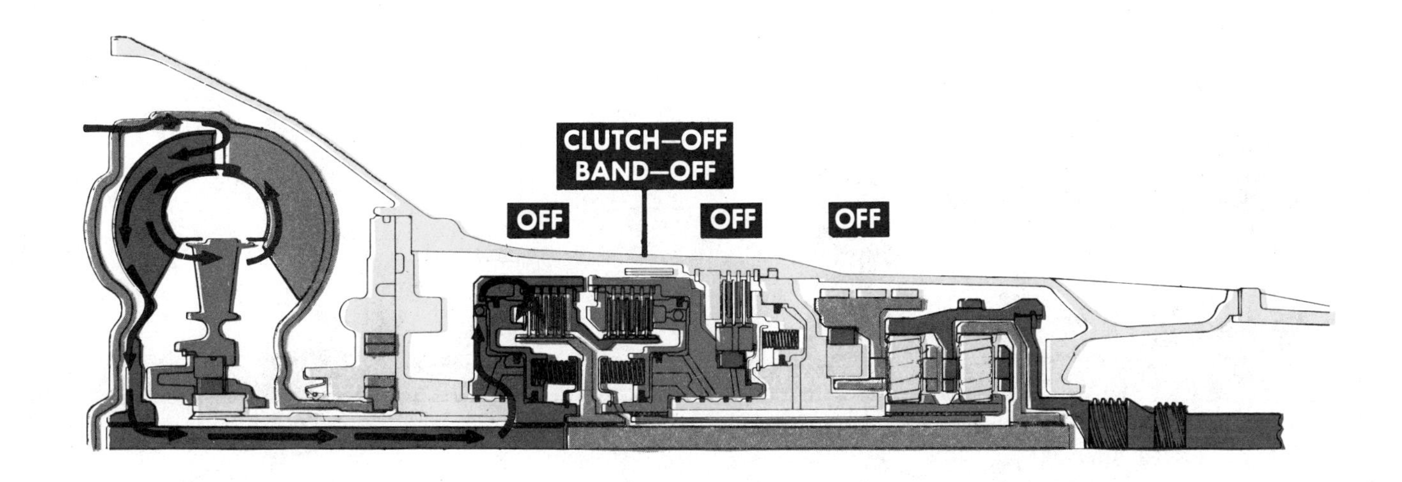

Neutral - Engine Running

NEUTRAL—ENGINE RUNNING

Forward Clutch - Released

Lo Roller Clutch - Ineffective

Direct Clutch - Released

Front Band - Released

Rear Band - Released

Intermediate Clutch - Released

Intermediate Sprag - Ineffective

Fig. 14-3 Members of the gear set in the Turbo Hydra-Matic 400 (neutral).

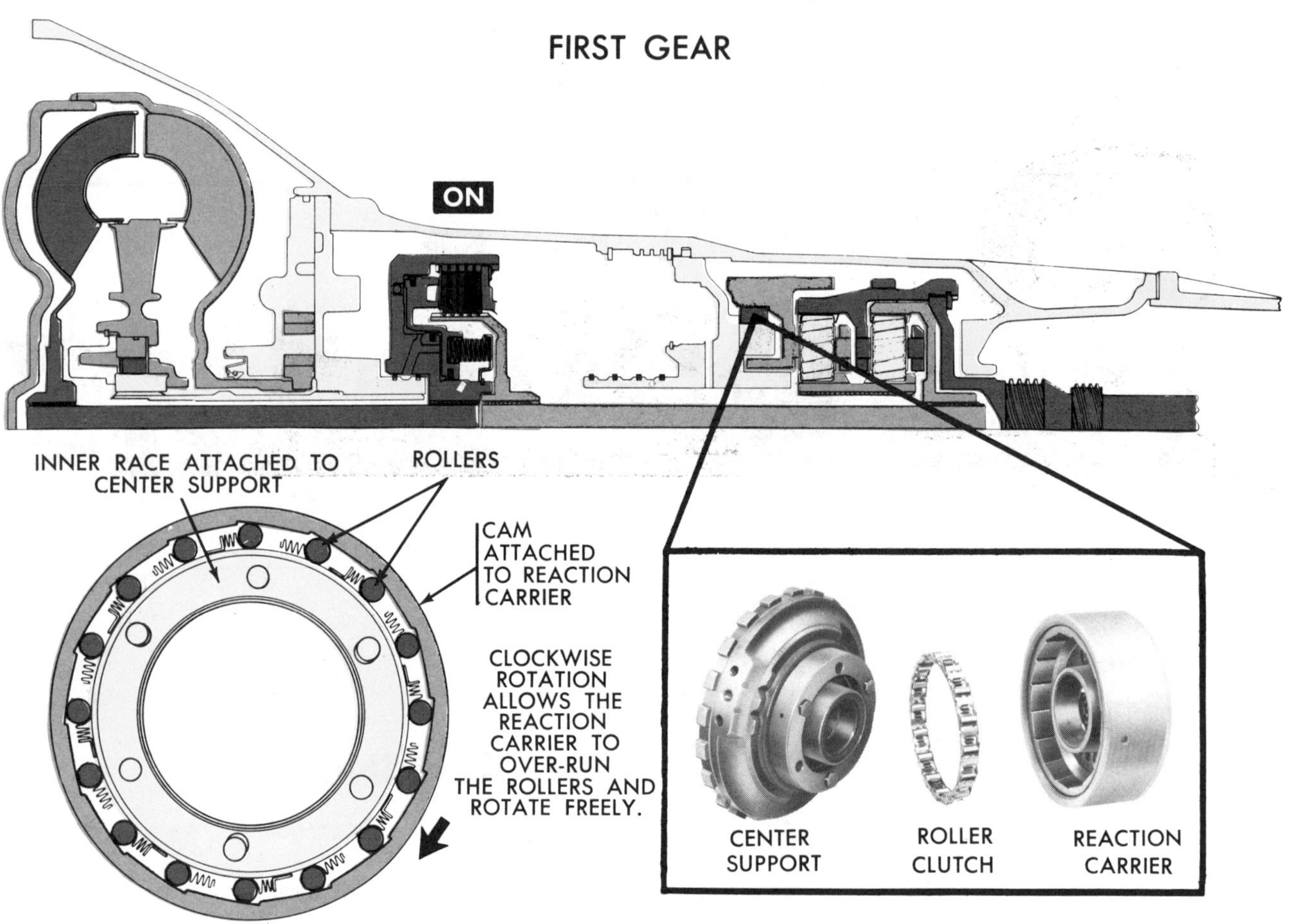

Fig. 14-4 Drive range first gear.

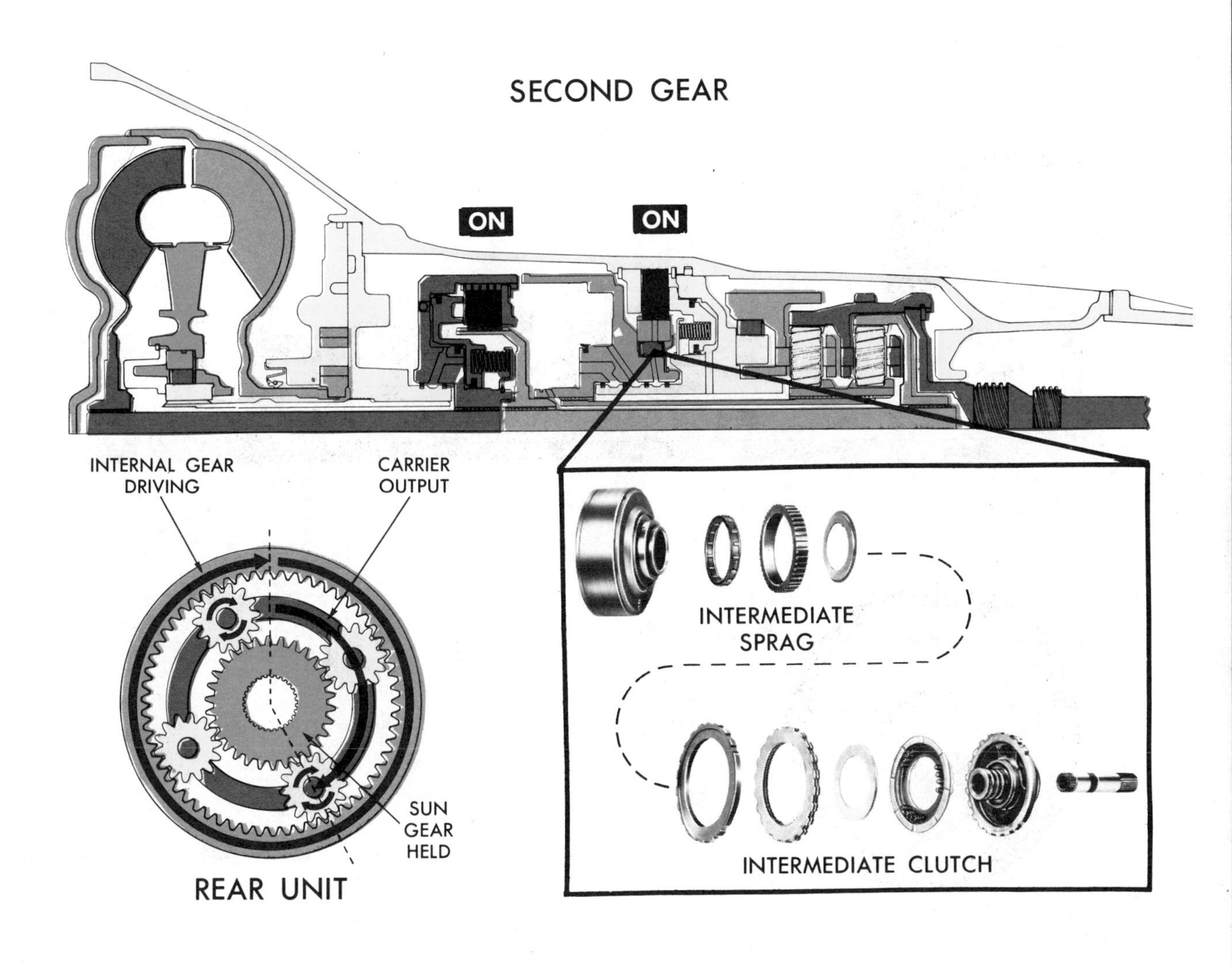

Fig. 14-5 Drive range second gear.

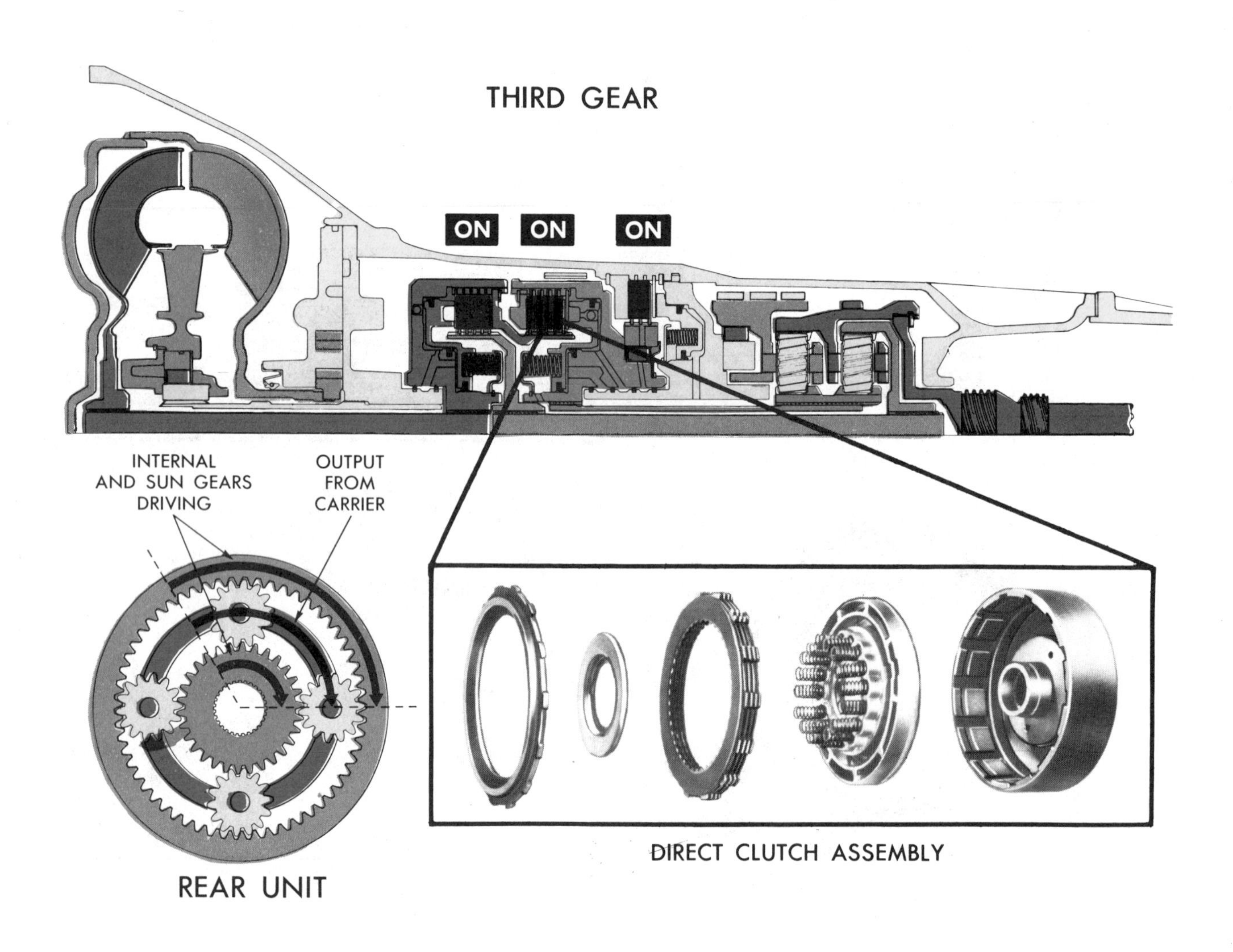

Fig. 14-6 Drive range third gear.

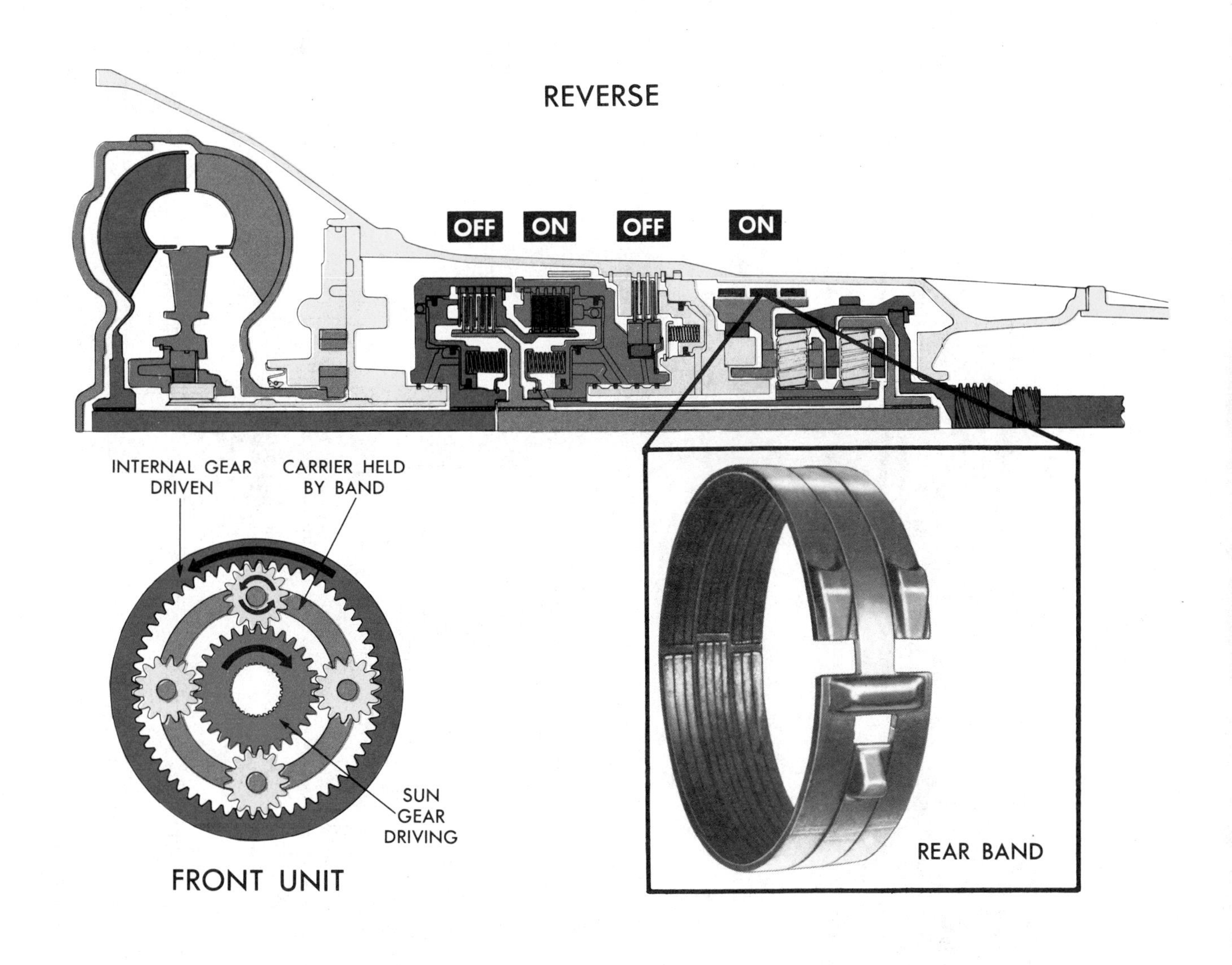

Fig. 14-7 Reverse.

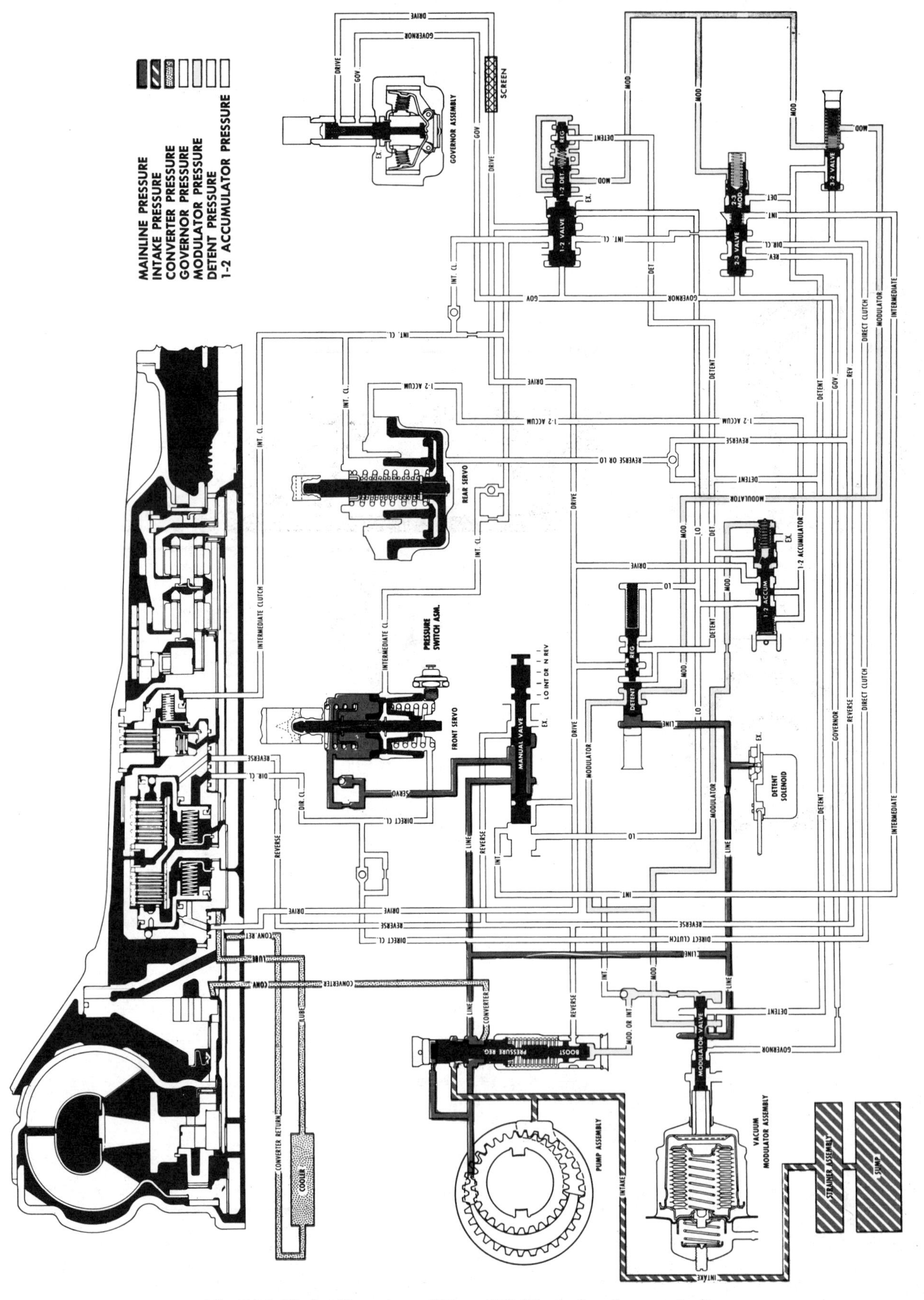

Fig. 14-8 Hydraulic system of Type 400 (Neutral engine running).

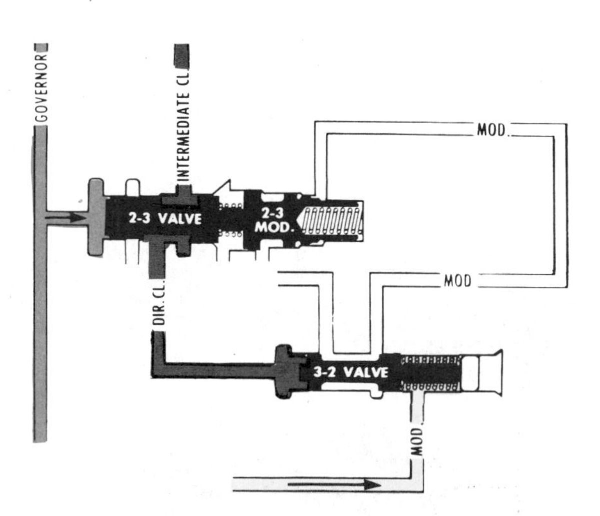

VALVES IN 3RD. GEAR POSITION MODULATOR PRESSURE UNDER APPROX. 90 P.S.I.

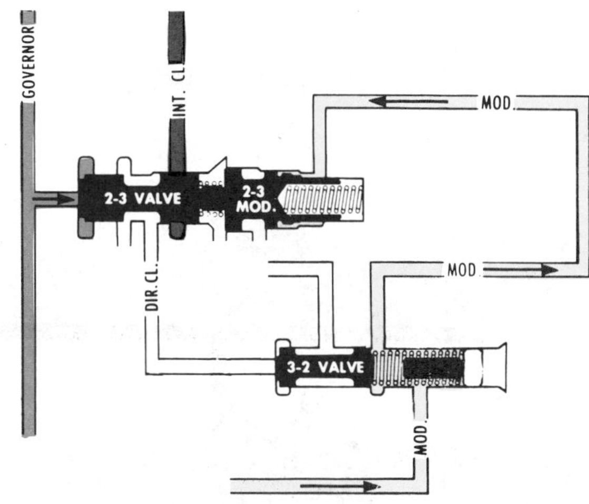

PART THROTTLE DOWNSHIFT VALVES IN 2ND. GEAR POSITION MODULATOR PRESSURE OVER 90 P.S.I.

Fig. 14-9 Part-throttle downshift 3-2.

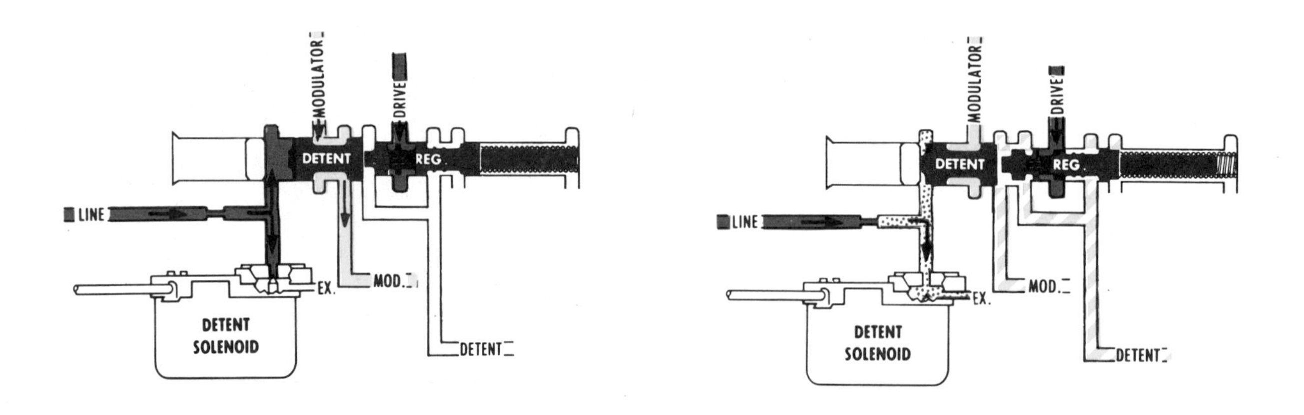

DETENT VALVE CLOSED

DETENT VALVE OPEN

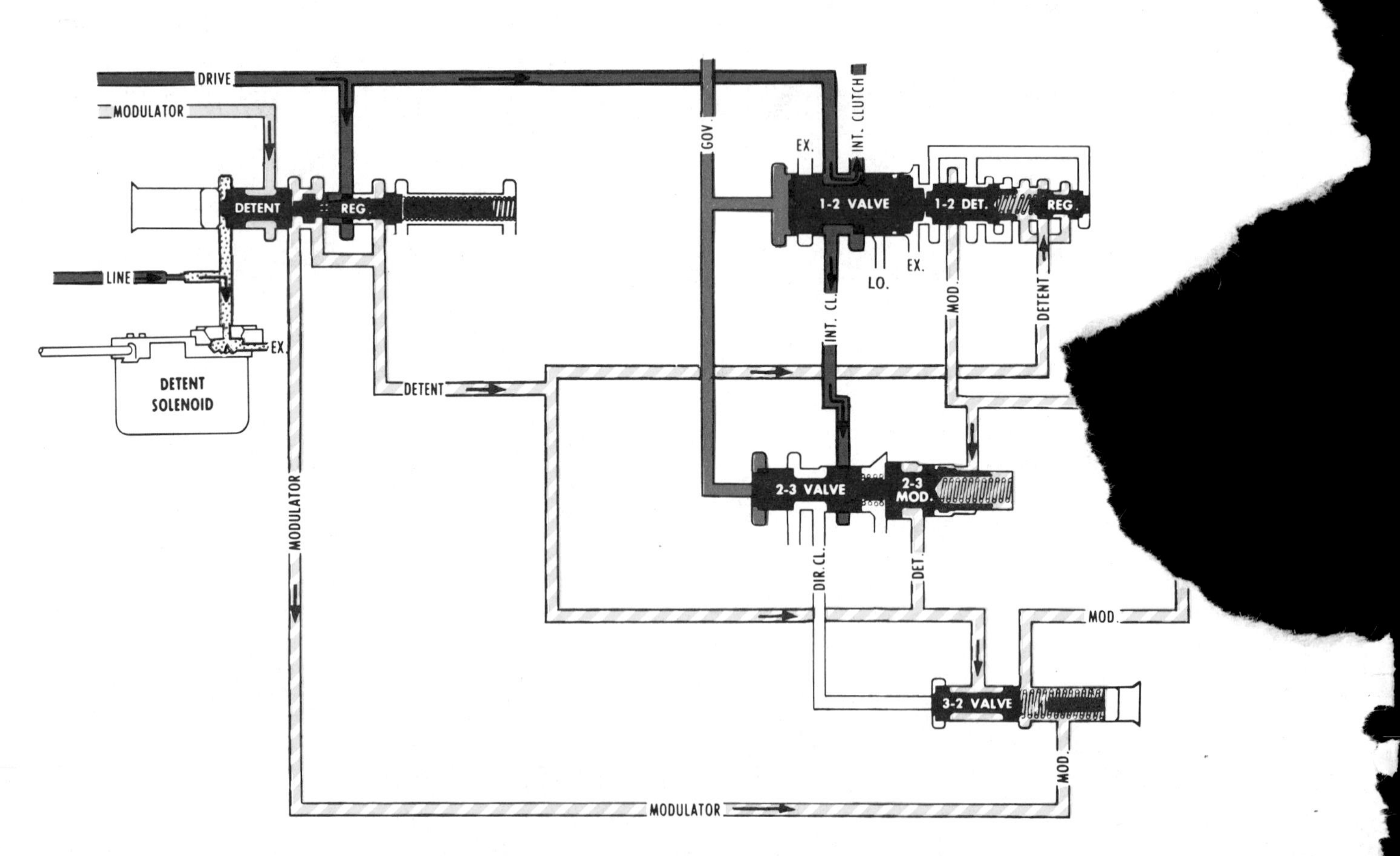

DETENT DOWNSHIFT – VALVES IN SECOND GEAR POSITION

Fig. 14-10 Detent downshift

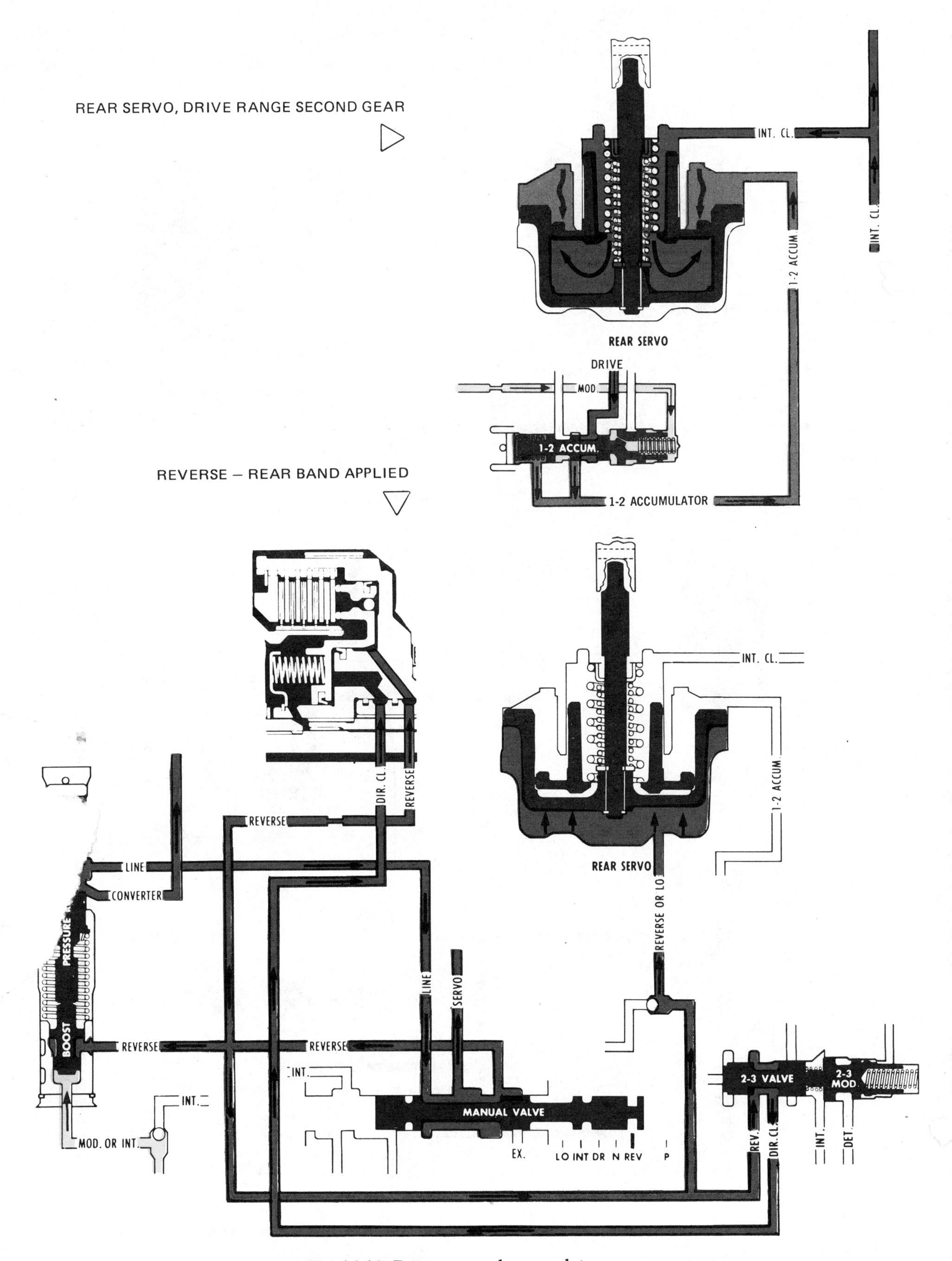

Fig. 14-12 Rear servo and accumulator.

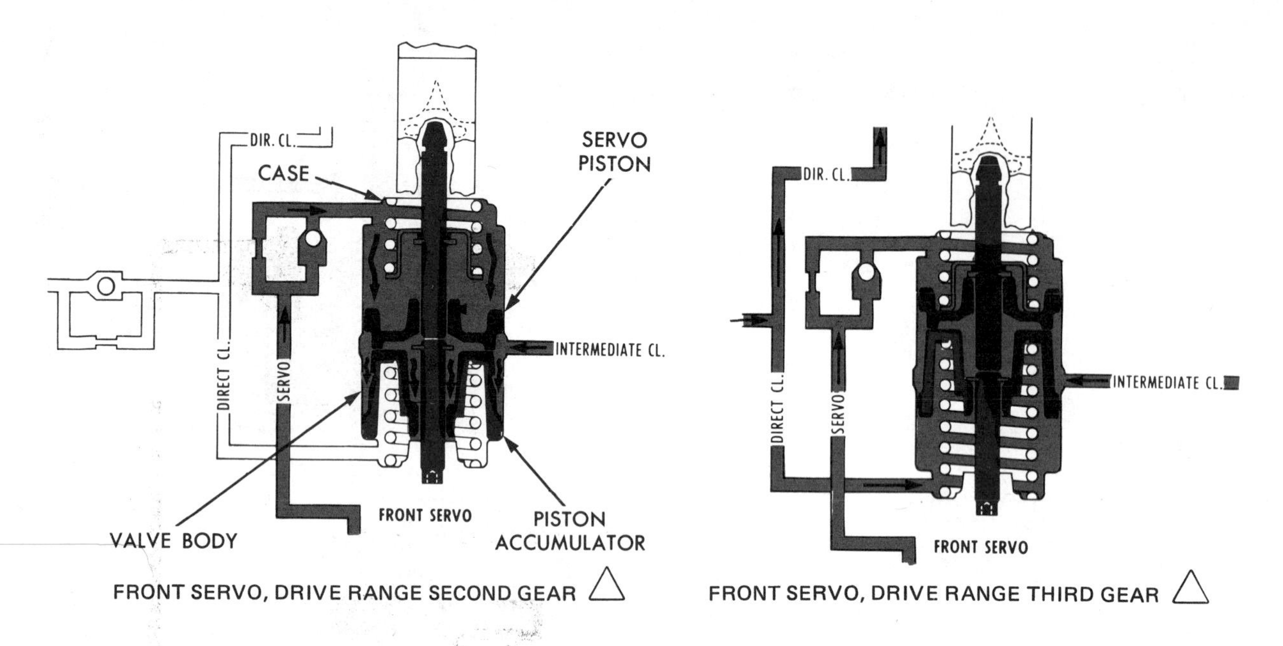

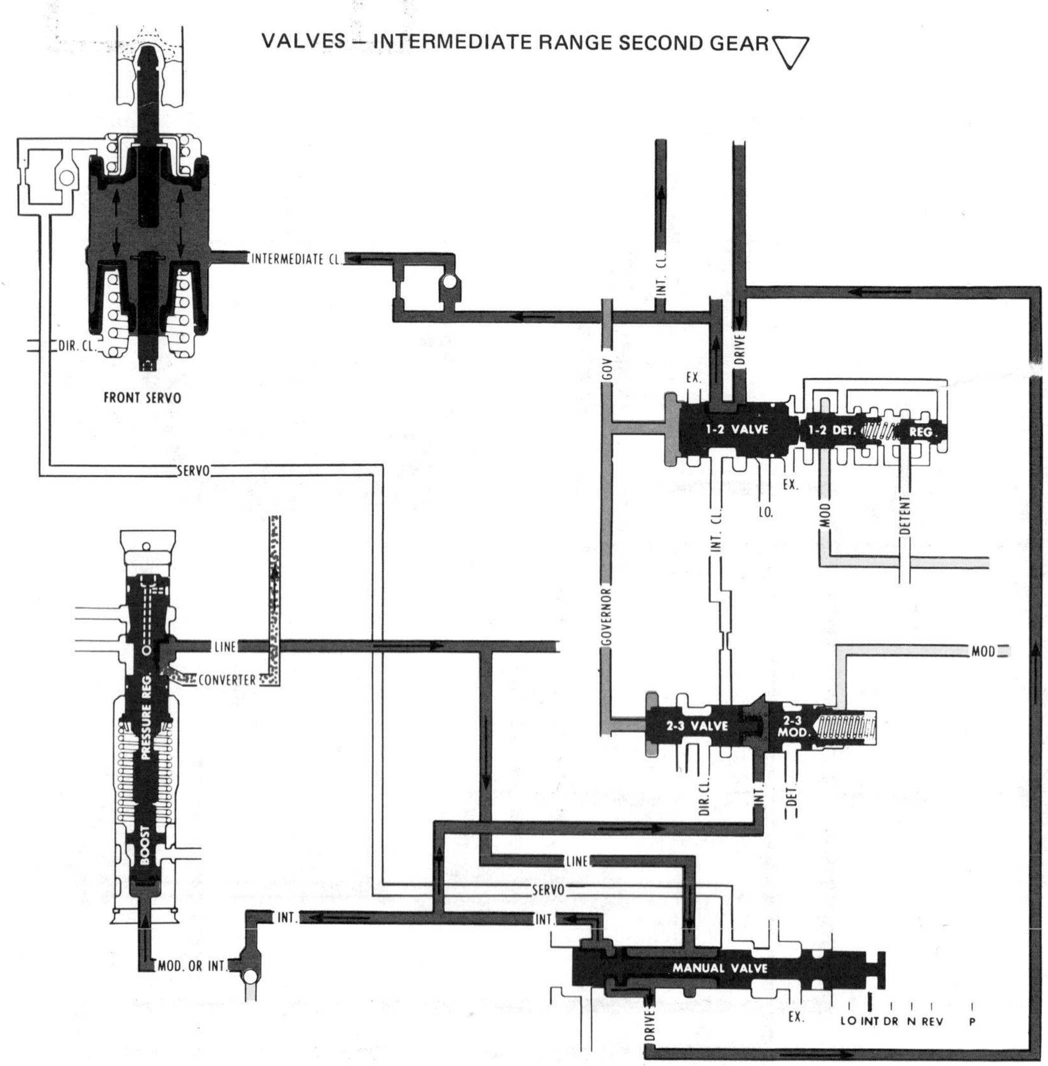

Fig. 14-13 Front servo and accumulator.

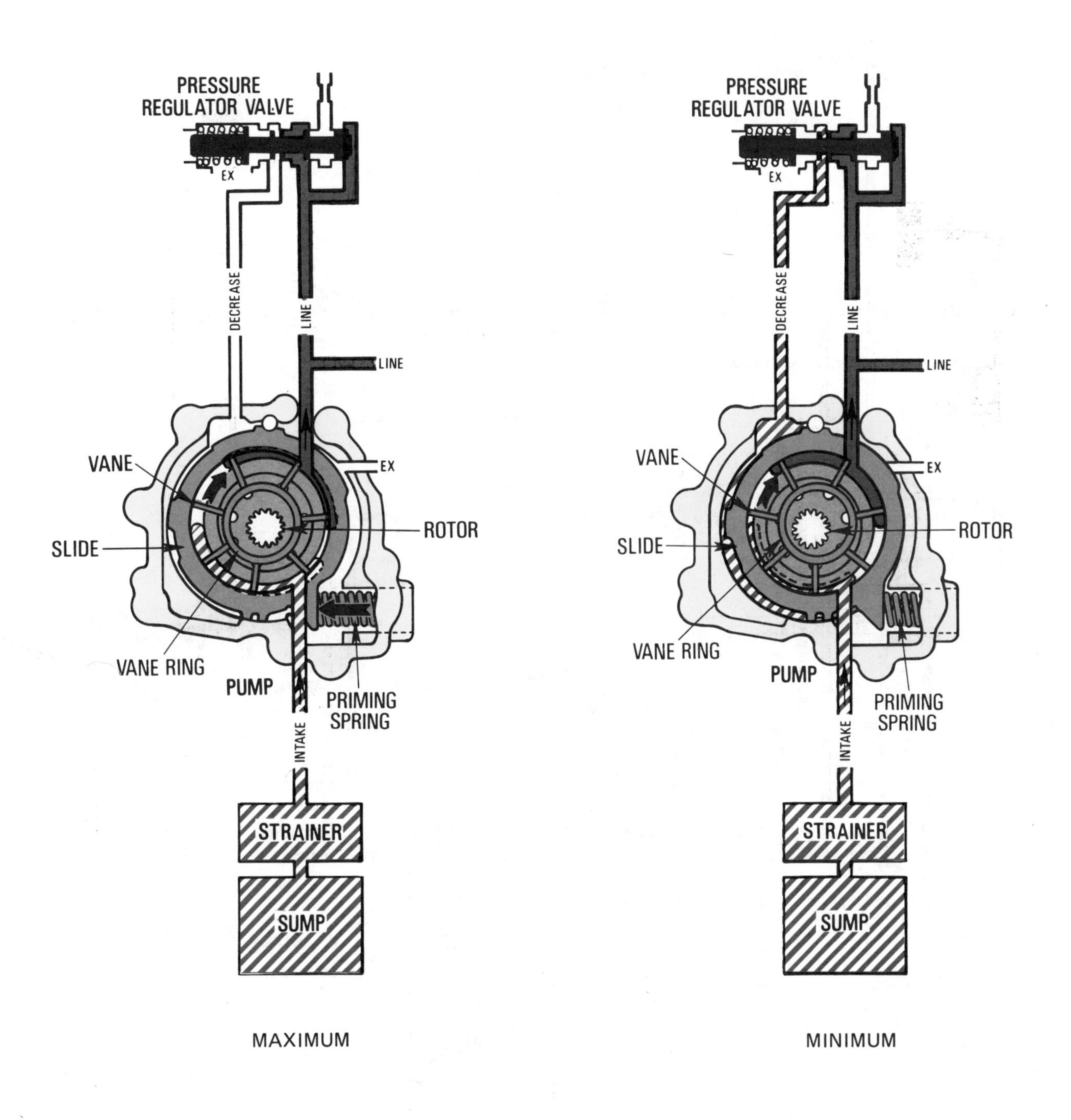

Fig. 19-6 Maximum and minimum pump output.

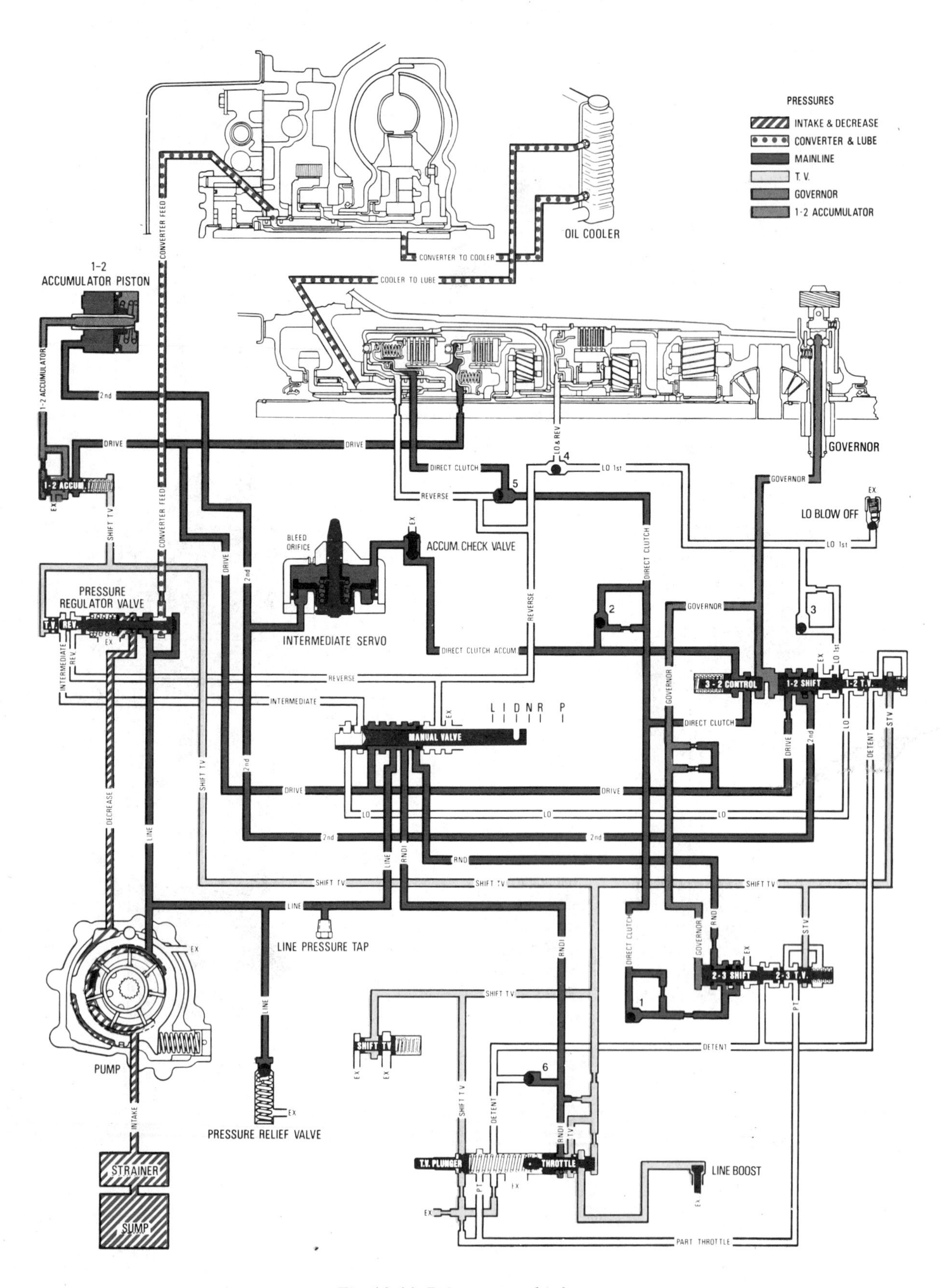

Fig. 19-11 Drive range–third gear.